Luis M. Vaschetto
Editor

miRNAs, Human Health and Diseases

Springer

Editor
Luis M. Vaschetto
The Ronin Institute for Independent Scholarship
Montclair, NJ, USA

ISSN 2191-2262 ISSN 2191-2270 (electronic)
Epigenetics and Human Health
ISBN 978-3-031-64787-1 ISBN 978-3-031-64788-8 (eBook)
https://doi.org/10.1007/978-3-031-64788-8

This Springer imprint is published by the registered company Springer Nature Switzerland AG
The registered company address is: Gewerbestrasse 11, 6330 Cham, Switzerland

Preface

The classical paradigm in gene expression refers to the idea that messenger RNAs (mRNAs) and their products (transcription factors) are the unique players responsible for controlling protein production through canonical pathways. The originally proposed signaling mechanisms involved mRNA carrying the genetic code from DNA to ribosomes, where the code was translated into proteins, with well-defined amounts of transcribed product being produced.

Fortunately, over the past decades, the emergence of non-coding RNA (ncRNA) sequences has greatly shed light on the amazing repertoire of molecules involved in gene expression. The discovery of ncRNAs held such significant importance that the Nobel Prize in Physiology or Medicine was awarded in 2006 to Dr. Andrew Fire and Dr. Craig C. Mello for their work on RNA interference pathways. In particular, the discovery of microRNAs (miRNAs), an evolutionary conserved class of small non-coding RNAs (ncRNAs), has revolutionized our vision of transcription. These sequences (miRNAs) act as master regulators of gene expression, influencing everything from development and growth to cell death and disease. While miRNAs primarily function by binding to specific mRNAs, these molecules possess a broader repertoire of action. They can also act as competing endogenous RNAs (ceRNAs), preventing other RNAs from targeting their intended mRNA targets. Additionally, miRNAs may interact with DNA or proteins involved in transcription, potentially influencing which genes are turned on or off.

The recent advancements in high-throughput next-generation sequencing have significantly enhanced our understanding on miRNA-mediated regulation pathways. Bioinformatics has become crucial in navigating huge genomic-transcriptomic data sets. In particular, in silico transcriptomics has emerged as a powerful tool across diverse miRNA research areas, including discovery and prediction, RNA structure analysis, target prediction, expression analysis, and functional annotation.

Both bioinformatic and experimental analyses have robustly demonstrated the regulatory potential of miRNAs, making them attractive therapeutic targets. The co-development of synthetic miRNAs loaded in appropriate delivery systems represents a particularly promising strategy. Introducing these molecules can functionally replace aberrantly downregulated or upregulated genes, actively restoring cellular homeostasis and impeding disease progression. Analysis of aberrant miRNA profiles can also shed light on the pathogenic mechanisms underlying diseases, including autoimmune disorders, neurodegenerative (Alzheimer's, Parkinson's) and cardiovascular illnesses, diabetes, and more. In diseased states, specific miRNA signatures emerge as potential diagnostic biomarkers. Furthermore, certain miRNAs are a hallmark of aging and act as modulators of longevity mechanisms. By analyzing miRNA profiles in tissues or biofluids, clinical geneticists may achieve early disease detection and contribute to patient's overall well-being.

miRNAs also represent crucial regulators in inter-kingdom communication, ultimately impacting human health and disease. For instance, host cell miRNAs can directly target viral RNA, hindering viral protein translation, and ultimately suppressing viral replication. As a counterpart, viruses can manipulate the host cell's miRNA processing machinery to promote their own replication and facilitate disease pathogenesis, thereby driving the coevolution of immune repertories and specific pathogens.

miRNAs, Human Health, and Diseases is the latest volume of the internationally acclaimed *Epigenetic and Human Health* book series. It represents a reference work, the culmination of collaborative efforts by leading experts in gene regulation from around the world. We are sincerely grateful for the input of the reviewers, Prof. León, Dr. Gholizadeh, Dr. Gallardo-Pérez, Prof. Purohit, Prof. Papaneophytou, Dr. Velandia-Huerto, Dr. Zhijin Fan, Dr. Martinez, and Dr. Sessa, whose insightful comments significantly improved the quality of this book.

Alta Gracia, Córdoba, Argentina — Luis M. Vaschetto

Contents

About the Editor and Contributors

About the Editor

Luis M. Vaschetto obtained his Ph.D. in Biological Sciences from the Universidad Nacional de Córdoba (Argentina). He was a postdoctoral researcher at the Consejo Nacional de Investigaciones Científicas y Técnicas (CONICET) and a Research Associate at the South Dakota State University (SDSU). He has been the first author of more than 10 academic articles and also published several book chapters. His research mainly focuses on understanding the epigenetic mechanisms involved in plant and animal development, as well as their effects on human health and diseases.

Contributors

Hoda Y. Abdallah Medical Genetics Unit, Histology and Cell Biology Department, Faculty of Medicine, Suez Canal University, Ismailia, Egypt

Center of Excellence in Molecular and Cellular Medicine, Faculty of Medicine, Suez Canal University, Ismailia, Egypt

Sarah Ahmed Aglan Chemical Pathology Department, Medical Research Institute, Alexandria University, Alexandria, Egypt

Jens Allmer Medical Informatics and Bioinformatics, Hochschule Ruhr West, Mülheim adR, Germany

Eduardo Andrés-León Bioinformatics Unit, Institute of Parasitology and Biomedicine López-Neyra (IPBLN), CSIC, Granada, Spain

Kazem Baesi Hepatitis and AIDS Department, Pasteur Institute of Iran, Tehran, Iran

Wei Bai Lab for Bone Metabolism, Xi'an Key Laboratory of Special Medicine and Health Engineering, Key Lab for Space Biosciences and Biotechnology, Research

Center for Special Medicine and Health Systems Engineering, NPU-UAB Joint Laboratory for Bone Metabolism, School of Life Sciences, Northwestern Polytechnical University, Xi'an, Shaanxi, China

Zhihao Chen Lab for Bone Metabolism, Xi'an Key Laboratory of Special Medicine and Health Engineering, Key Lab for Space Biosciences and Biotechnology, Research Center for Special Medicine and Health Systems Engineering, NPU-UAB Joint Laboratory for Bone Metabolism, School of Life Sciences, Northwestern Polytechnical University, Xi'an, Shaanxi, China

Müşerref Duygu Saçar Demirci Department of Bioengineering, Faculty of Life and Natural Sciences, Abdullah Gül University, Kayseri, Turkey

Yılmaz Mehmet Demirci Department of Engineering Science, Faculty of Engineering, Abdullah Gül University, Kayseri, Turkey

Ajay Elangovan Human Cytogenetics and Stem Cell Laboratory, Department of Zoology, School of Basic Sciences, Central University of Punjab, Bathinda, Punjab, India

Rasha Mohamed Helmy Elkaffas Clinical and Chemical Pathology Department, Faculty of Medicine, Cairo University, Cairo, Egypt

Floriana M. Farina Institute for Cardiovascular Prevention (IPEK), Ludwig-Maximilians-Universität München, Munich, Germany

German Centre for Cardiovascular Research (DZHK), Partner Site Munich Heart Alliance, Munich, Germany

Iván Fernández-Rengel Bioinformatics Unit, Institute of Parasitology and Biomedicine López-Neyra (IPBLN), CSIC, Granada, Spain

GENYO, Centre for Genomics and Oncological Research: Pfizer—University of Granada—Andalusian Regional Government, Granada, Spain

Marcel Friedrichs Faculty of Technology, Bioinformatics Department, Bielefeld University, Bielefeld, Germany

Moisés Manuel Gallardo-Pérez Universidad Popular Autónoma del Estado de Puebla, Puebla, México

Centro de Hematología y Medicina Interna, Clínica Ruiz, Puebla, México

Solón Javier Garcés-Eisele Universidad Popular Autónoma del Estado de Puebla, Puebla, México

Laboratorios Ruiz, SYNLAB, Puebla, México

Anto George Human Cytogenetics and Stem Cell Laboratory, Department of Zoology, School of Basic Sciences, Central University of Punjab, Bathinda, Punjab, India

Maryam Ghanbari Department of Microbial Biotechnology, Faculty of Biological Science, Tehran North Branch, Islamic Azad University, Tehran, Iran

Aruana F. F. Hansel Fröse Laboratory of Basic Stem-Cell Biology, Carlos Chagas Institute, Oswaldo Cruz Foundation (ICC/FIOCRUZ-PR), Curitiba, Brazil

Faculty of Technology, Bioinformatics Department, Bielefeld University, Bielefeld, Germany

James M. Henderson Institute for Cardiovascular Prevention (IPEK), Ludwig-Maximilians-Universität München, Munich, Germany

German Centre for Cardiovascular Research (DZHK), Partner Site Munich Heart Alliance, Munich, Germany

Jonas Holdmann St. Marien-Hospital, Department of Internal Medicine, Cologne, Germany

Division of Functional Genomics, Chair of Biochemistry and Molecular Medicine, Witten/Herdecke University, Witten, Germany

Mahalaxmi Iyer Department of Biotechnology, Karpagam Academy of Higher Education (Deemed to be University), Coimbatore, Tamil Nadu, India

Department of Microbiology, Central University of Punjab, Bathinda, Punjab, India

Harsimrat Kaur Human Cytogenetics and Stem Cell Laboratory, Department of Zoology, School of Basic Sciences, Central University of Punjab, Bathinda, Punjab, India

Min-Sun Kim Department of Pharmacy, College of Pharmacy and Research Institute of Life and Pharmaceutical Sciences, Sunchon National University, Suncheon, Republic of Korea

Sakshi Kushwaha Human Cytogenetics and Stem Cell Laboratory, Department of Zoology, School of Basic Sciences, Central University of Punjab, Bathinda, Punjab, India

Lukas Markert Division of Functional Genomics, Chair of Biochemistry and Molecular Medicine, Center for Biomedical Education and Research (ZBAF), Witten/Herdecke University, Witten, Germany

Helios Park Hospital Leipzig, Department of Anesthesiology and Intensive Care, Leipzig, Germany

Sindduja Muthukumar Human Cytogenetics and Stem Cell Laboratory, Department of Zoology, School of Basic Sciences, Central University of Punjab, Bathinda, Punjab, India

Hai Duc Nguyen Division of Microbiology, Tulane National Primate Research Center, Covington, LA, USA

Department of Pharmacy, College of Pharmacy and Research Institute of Life and Pharmaceutical Sciences, Sunchon National University, Suncheon, Republic of Korea

Mehmet Emin Orhan Department of Bioengineering, Graduate School of Engineering and Science, Abdullah Gül University, Kayseri, Turkey

Jyoti Parkash Neurochemistry and Neuroendocrinology Lab, Department of Zoology, Central University of Punjab, Bathinda, Punjab, India

Suryaji Patil Lab for Bone Metabolism, Xi'an Key Laboratory of Special Medicine and Health Engineering, Key Lab for Space Biosciences and Biotechnology, Research Center for Special Medicine and Health Systems Engineering, NPU-UAB Joint Laboratory for Bone Metabolism, School of Life Sciences, Northwestern Polytechnical University, Xi'an, Shaanxi, China

Shima Poorghobadi Hepatitis and AIDS Department, Pasteur Institute of Iran, Tehran, Iran

Airong Qian Lab for Bone Metabolism, Xi'an Key Laboratory of Special Medicine and Health Engineering, Key Lab for Space Biosciences and Biotechnology, Research Center for Special Medicine and Health Systems Engineering, NPU-UAB Joint Laboratory for Bone Metabolism, School of Life Sciences, Northwestern Polytechnical University, Xi'an, Shaanxi, China

Donato Santovito Institute for Cardiovascular Prevention (IPEK), Ludwig-Maximilians-Universität München, Munich, Germany

German Centre for Cardiovascular Research (DZHK), Partner Site Munich Heart Alliance, Munich, Germany

Institute of Genetic and Biomedical Research (IRGB), Unit of Milan, National Research Council (CNR), Milan, Italy

Andreas Savelsbergh Division of Functional Genomics, Chair of Biochemistry and Molecular Medicine, Witten/Herdecke University, Witten, Germany

Laura C. Terrón-Camero Bioinformatics Unit, Institute of Parasitology and Biomedicine López-Neyra (IPBLN), CSIC, Granada, Spain

Ye Tian Lab for Bone Metabolism, Xi'an Key Laboratory of Special Medicine and Health Engineering, Key Lab for Space Biosciences and Biotechnology, Research Center for Special Medicine and Health Systems Engineering, NPU-UAB Joint Laboratory for Bone Metabolism, School of Life Sciences, Northwestern Polytechnical University, Xi'an, Shaanxi, China

Balachandar Vellingiri Human Cytogenetics and Stem Cell Laboratory, Department of Zoology, School of Basic Sciences, Central University of Punjab, Bathinda, Punjab, India

Dhivya Venkatesan Department of Biotechnology, Karpagam Academy of Higher Education (Deemed to be University), Coimbatore, Tamil Nadu, India

Mukesh Kumar Yadav Department of Microbiology, Central University of Punjab, Bathinda, Punjab, India

Xin Zhao Lab for Bone Metabolism, Xi'an Key Laboratory of Special Medicine and Health Engineering, Key Lab for Space Biosciences and Biotechnology, Research Center for Special Medicine and Health Systems Engineering, NPU-UAB Joint Laboratory for Bone Metabolism, School of Life Sciences, Northwestern Polytechnical University, Xi'an, Shaanxi, China

Chapter 1
MiRNomics: An Introduction to MicroRNA Biology

Lukas Markert, Andreas Savelsbergh, and Jonas Holdmann

Abstract As information increases, so does ignorance. This is, of course, nothing new. Nevertheless, it is particularly striking in the case of genetics and epigenetics. Decoding the genome once was thought to be the key to complete understanding of human beings, their traits, and their diseases. Today, it is generally understood that we are much more than our genetic makeup. Epigenetics is a key factor in enhancing our individuality and enabling us to respond more rapidly to the environment. Noncoding RNAs, such as microRNAs, are part of this. MiRNomics is a scientific branch that has earned the right to be considered a discipline in its own. As an omic-technology, it combines powerful measurement methods with computerized data acquisition systems and analysis software. In the following pages, we would like to highlight some of its fascinating features. After portraying a brief history of the field, we will describe the origin and function of microRNA. We will then give a brief overview of its implementation in research and diagnostics before the detailed discussions in the chapters that follow.

Keywords MicroRNA · MiRNomics · MiRNA biogenesis · Epigenetics · SmallRNA · Computational analysis · Database

---

L. Markert (✉)
Helios Park Hospital Leipzig, Department of Anesthesiology and Intensive Care, Leipzig, Germany

Division of Functional Genomics, Chair of Biochemistry and Molecular Medicine, Witten/Herdecke University, Witten, Germany
e-mail: lukas.markert@uni-wh.de

A. Savelsbergh · J. Holdmann
Division of Functional Genomics, Chair of Biochemistry and Molecular Medicine, Witten/Herdecke University, Witten, Germany

L. M. Vaschetto (ed.), *miRNAs, Human Health and Diseases*, Epigenetics and Human Health 13, https://doi.org/10.1007/978-3-031-64788-8_1

List of Abbreviations

A, C, G, T	Adenine (A), guanine (G), cytosine (C), and thymine (T)
ADP	Adenosine diphosphate
cDNA	Complementary DNA
DNA	Deoxyribonucleic acid
dsRNA	Double-stranded DNA
miRNA	MicroRNA
mRNA	messengerRNA
ncRNA	Noncoding RNA
NGS	Next-generation sequencing
PCR	Polymerase chain reaction
piRNA	Piwi-interacting RNA
PIWI	p-element-induced wimpy testis
pre-miRNA	Precursor miRNA
Pri-miRNA	Primary miRNA
RISC	RNA-induced silencing complex
RNA	Ribonucleic acid
RNAi	RNA interference
rRNA	Ribosomal RNA
RT	Reverse transcriptase
rt-PCR	Real-time PCR
scaRNA	Small Cajal body-specific RNA
siRNA	Small interfering RNA
snoRNA	Small nucleolar RNA
snRNA	Small nuclear RNA
ssRNA	Single-stranded RNA
TRBP	Transactivation response element RNA-binding protein
tRNA	Transfer–/transport-RNA
UK	United Kingdom
UMI	Unique molecular identifier
UTR	Untranslated region

1.1 Historical Setting

The history of heredity is as old as life itself, and the teachings about it have developed over a very long period. Even in the Old Testament, in texts almost 5000 years old, there are references to the appearance of animals (colorful, spotted) if they were descended from certain animals. At that time, there was still a mixture of the mystical and the magical. For example, the place and circumstances of one's birth were seen as determining external factors. However, it is already described here that strong animals give rise to strong animals and weak animals to frail ones. The transmission of certain external characteristics is already mentioned (Genesis 30,35: "That same day he removed all the male goats that were streaked or spotted, and all the speckled or spotted female goats (all that had white on them) and all the dark-colored lambs, and he placed them in the care of his sons." (...) Genesis 30,41: "Whenever the stronger females were in heat, Jacob would place the branches in the troughs in front of the animals so they would mate near the branches, but if the animals were weak, he would not place them there. So the weak animals went to Laban and the strong ones to Jacob"). Perhaps, appearance was initially more important than an understanding of yield. But the art of breeding was more and more developed and helped to create new and high-yielding variants of animals and plants. Today, it is difficult to make reliable statements about whether this breeding was more coincidental or planned. In 1865, it was the Moravian-Austrian priest Gregor Mendel who derived his well-known laws of heredity from observations of the shape and appearance of peas. Basically, these laws are still valid today (law of dominance and uniformity, law of segregation, and law of independent assortment) (Mendel 1866). In 1869, nucleic acid was discovered by Friedrich Miescher in fish sperm and human pus—initially without the intellectual connection to heredity though (Miescher 1897). Together with the realization that paired *chromosomes* (Waldeyer 1888) are carriers of hereditary information (Sutton 1902), the American biologist Thomas Morgan was finally able to detect *genes* and gene mapping on chromosomes in the fruit fly *Drosophila melanogaster*, which has developed to an important model organism in the field of genetics (Morgan 1910). In addition to molecular and experimental discoveries, there have been epidemiological considerations in the field. It was Garrod, for example, who in 1909 formulated what others may have already known: that there were diseases with family histories (Garrod 1909). J. B. S. Haldane, for instance, was able to prove this impressively for the coincidence of hemophilia and color blindness (Bell and Haldane 1936). Watson and Crick—building on the diligent and underappreciated work of Rosalin Franklin—were able to decipher the structure of DNA in 1953. This was followed by a major effort, the Human Genome Project, to sequence the entire genome, which was declared completed in 2003. The philosopher's stone for the knowledge of genetics seemed to be within reach. In fact, it was not until March 2022 that the human genome was completely decoded (Nurk et al. 2022)—with only modern sequencing methods providing the necessary ability to read the "repetitive sequences." But then, as now, too many questions remained unanswered. A list of about three billion pairs of letters A, C, G, and T had been deciphered, from which about 25,000 genes could be deduced—sequences that code for a particular protein. But how do these sequences give rise to the more than 200 different cell types that exist in the human body? Almost all cells have the same genetic material, and yet they develop so differently. Having the (almost) complete genome in front of us was not enough to clearly predict the actual inheritance of traits and the underlying regulatory network. There were also pathophysiological questions. For example, two people with the same genetic mutation might develop cancer in one person but not in the other. And this is where the parallel emerging story of epigenetics comes in. Its roots date back to the twentieth century: In 1939, the term "epigenetics" was first used by Conrad Hal Waddington, a British embryologist and geneticist, to describe the interaction between genes and the environment in the development of an organism (Waddington 1939). In neologizing the word, Waddington addressed a centuries-old difference in thinking between preformationism and epigenesis. Preformationism, which originated in the Greek ancient (Anaxagoras) and was brought up again in seventeenth century, held that the "end product," or embryo, in the fertilized egg was already preformed on a tiny scale and was only in the process of growth. Thereby, either the sperm or the oocyte was seen as the origin of this preformation. Meanwhile, epigenesis, which goes back to Aristotle, held that in the egg all new biological complexity arises from paternal and maternal germinal material. This was the dominant position for a long time, based on the idea that materiality is inherently divine. Since the Age of Enlightenment, however, epigenesis has been replaced by preformationism. And emerging microscopic studies also seemed to confirm rather small, preformed creatures. Waddington attempted to combine these seemingly opposing views into a unifying theory, but he also introduced genetics as a basis for his thinking. He coined the term epigenetics as a model of growth and development. Another source of this creation was Haecker, who in 1918 had already called phenogenetics the phenotypic differences of genetic traits (Haecker 1918). In 1942, before the structure of DNA was known, Waddington described epigenetics as "the branch of biology that studies the causal interactions between genes and their products that give rise to the phenotype" (Waddington 1942).

Building on this, Nanney introduced his concept of "epigenetics" in 1958 (Nanney 1958). He distinguished two cellular control systems: first, a "library of specificities" achieved by template replication mechanisms based on DNA sequences, the "genetic system," and second, "auxiliary mechanisms" involved in determining which specificities should be expressed in a given cell, that is, the control of gene expression.

In the 1980s, the importance of epigenetics in human cancers was discovered (Feinberg and Vogelstein 1983). However, the scientific focus has long been on the genetics of these diseases (Feinberg and Tycko 2004). In the 1990s and 2000s, the term "epigenetics" became more common in current usage. At a meeting at Cold Spring Harbor in 2008, a definition of an epigenetic trait was formulated as "a stably heritable phenotype resulting from changes in a chromosome without changes in the DNA sequence" (Berger et al. 2009), although alternative definitions that include nonheritable traits are still widely used. For example, the definition was based on apparently simple but brilliant experiments in which mice were conditioned to the smell of acetone by electric shocks, and eventually, the children and grandchildren of these mice became sensitive to the smell without ever having been conditioned themselves (Deichmann 2016). In the 2000s, this was transferred to humans through health data. Data from Sweden, for example, showed that heart disease and diabetes were more likely to be passed on transgenerationally from a starving generation than from a well-fed one (Pembrey et al. 2006). In the United Kingdom, epidemiological studies have shown that the children of men who start smoking very early weigh more on average, after adjusting for other social effects (Feinberg and Tycko 2004; Zeilinger et al. 2013). Today, epigenetics is a rapidly growing field of research with important implications for human health and disease. The study of epigenetics has led to the discovery of new mechanisms of gene regulation and has provided insights into the development of diseases such as cancer, diabetes, and neurological disorders.

1.2 Epigenetic Mechanisms

What are the genetic conditions and epigenetic mechanisms that allow us to have an epigenetic code? To answer this question, we need to look again at the structure and packaging of the genetic material in the cell nucleus: Ribonucleic acid (RNA) is a macromolecule made up of a sequence of nucleotides (adenine, uracil, guanine, and cytosine) attached to a backbone consisting of a sugar (ribose) and phosphate. The most prominent function of RNA is to translate genetic information into proteins. In this process, different types of RNA (tRNA, rRNA, and mRNA) carry out specific tasks such as transport, catalysis, or transfer. Most RNA is single-stranded (ssRNA) but due to its flexibility folds on itself to form partially double-stranded sections. Depending on the specific RNA structure, this can lead to a stable and defined three-dimensional structure (Wang et al. 2023). In RNA viruses, ribonucleic acid itself serves as the genome. In some cases, it is present as a double-stranded helix (dsRNA). In the eukaryotic cell nucleus, genetic information is "stored" in the form of DNA. It is arranged in the well-known double-stranded helix. This strand in humans would reach a length of about two meters per cell nucleus, so that a complex compaction by (hundreds of) thousands of histone complexes is required. Histone proteins can be modified by the attachment of a variety of small molecules, thus influencing their compaction and by this the accessibility of the DNA (Bannister and Kouzarides 2011). This is where epigenetics comes in. The most important modifications are methylation, phosphorylation, acetylation, ubiquitination, ADP-ribosylation, sumoylation, carbonylation, glycosylation, and biotinolation.

Probably, the best-known epigenetic mechanism is DNA-methylation. Methyl groups, consisting of one carbon and three hydrogen atoms (CH3), are attached to a DNA strand (mostly to cytosine) and prevent following sequences from being read and decoded by influencing histone methylation and thus histone-mediated DNA compaction. This is the principle of nucleosome compaction. Methylation occurs especially at so-called CpG islands. These are DNA segments in which the base sequence cytosine-guanine, or more precisely cytosine-phosphatidyl-guanosine, or CpG for short, is particularly common (Liu et al. 2008). Environmental exposures (e.g., smoking during pregnancy) have been shown to affect these methylation levels (Zeilinger et al. 2013). However, over time, the degree of methylation can vary by more than 20%. Therefore, epigenetic mechanisms may be modifiable or reversible in some cases.

A further epigenetic principle is RNA interference (RNAi). This is another form of gene silencing. In this process, short noncoding RNAs interfere with mRNA, whereupon the bound segments are cleaved, and information transfer is prevented. Andrew Z. Fire and Craig C. Mello were awarded the Nobel Prize in Physiology or Medicine in 2006 for their discovery of the mechanisms of RNA interference. In 2018, the first RNAi-based therapeutic was approved (Titze-de-Almeida et al. 2020). The number of current studies in this field is increasing rapidly. The mechanism is lately also being used to treat common diseases such as hypertension (Desai et al. 2023). So, we will potentially be dealing with this topic a lot more. Among other reasons, this is why the focus of this book is on these small noncoding RNAs, in particular microRNAs.

1.3 Small RNAs

Small RNAs are polymeric noncoding RNAs that are, in the most common definitions, less than 200 nucleotides long (Storz 2002). Their main representatives are miRNAs (micro-RNAs), piRNAs (piwi-interacting RNAs), siRNAs (small interfering RNAs), snRNAs (small nuclear RNAs), snoRNAs (small nucleolar RNAs), and tnRNAs (tiny noncoding RNAs). PiRNAs were described in 2006 by Hannon and Grivna et al. in mouse spermatogenic cells and later in humans. It is 26–31 nucleotides long, only slightly longer than miRNAs. In humans, piRNAs are mainly found in germ cells, where they bind to PIWI (P-element induces wimpy testis) proteins, which are essential for the suppression of transposons (Lau et al. 2006; Grivna et al. 2006; Teixeira et al. 2017; Andersen et al. 2017; Girard et al. 2006). It is therefore thought to prevent gene expression in spermatozoa (Joshi et al. 2023). Their presence in somatic cells, both nuclear and cytoplasmic, has also been described. These and other functions, including in women, are currently being discussed (Ghildiyal and Zamore 2009).

SnRNAs are 100–300 nucleotides long, which are localized in the nucleus. They are catalytically involved in the splicing of pre-mRNA introns. SnRNAs are also involved in the regulation of some transcription factors and the maintenance of telomeres (Lützelberger 2000; Knippers 2006). One of its subgroups is considered to be snoRNAs, which are found in the nucleolus and modifies mainly ribosomal RNAs (Scott and Ono 2011). Due to their similar properties, scaRNAs (small cajal body localized RNAs), which are found in the Cajal bodies of the nucleus, are sometimes assigned to them (Verheggen and Bertrand 2012).

SiRNAs are about 19–25 nucleotides long and, through its participation in the RNA-induced silencing complex (RISC), perform posttranscriptional gene silencing tasks (Siomi and Siomi 2009).

1.3.1 MicroRNA

MicroRNAs (miRNA) are short RNA molecules, typically consisting of 20–25 molecules only. They are noncoding. Their function is rather to attach themselves to other RNA molecules and thereby prevent their translation into proteins or even cause the degradation of the RNA molecules (Bartel 2009, 2018). With that MiRNAs play an important role in the regulation of gene expression (Lagos-Quintana et al. 2001; Lau et al. 2001; Lee and Ambros 2001). MiRNAs are at least partially complementary to mRNAs, which can be influenced in their function by binding. But there are differences here. Animal miRNAs typically complement a site in the 3′

untranslated region UTR (3'UTR) with a rather imperfect match, whereas plant miRNAs are usually near-perfect complementary to coding regions of mRNAs (Moxon et al. 2008; Wang et al. 2004). Different miRNAs can act on one mRNA, but one miRNA can also act on several mRNAs at the same time. It is estimated that two-thirds of all human genes are co-regulated by miRNAs, and nearly every cellular metabolic process can be affected (Bartel 2004; Filipowicz et al. 2008; Coordes et al. 2014). As mentioned before, this co-regulation of RNA in gene regulation is called RNA interference (i.e., RNAi or RNA silencing). In 1990, Napoli et al. first used this to explain the phenomenon of co-suppression of plant genes when an extra gene copy was inserted into a petunia. In this case, the color of the flower was not increased, but actually decreased (Napoli et al. 1990). The first miRNA was discovered in 1993 in the nematode *C. elegans* as a product of the lin-4 gene (Lee et al. 1993). Inhibition of protein expression by binding to lin-14 mRNA was also described in the same year (Wightman et al. 1993). Since then, it has become clear that miRNAs also play a role in gene regulation in viruses, in all studied animals and in humans (Hwang and Mendell 2007).

1.3.1.1 Biosynthesis of Canonical MicroRNA

In the following, the biogenesis of miRNAs is described, using mammals as example. The genes for these molecules originate from the genome. MircroRNAs are transcribed by the enzyme RNA polymerase II (or III) (Yang and Lai 2011). The resulting primary transcript has a length of 500–3000 nucleotides. The enzyme complex of Drosha (RNase III) and the dsRNA-binding protein DGCR8 then cuts out the precursor miRNA (pre-miRNA) from this primary miRNA (pri-miRNA) in the cell nucleus. Those have an approximate length of 70–80 nucleotides and form thereby a characteristic hairpin structure (Michlewski and Cáceres 2019). The pre-miRNA is actively transported into the cytoplasm by the transport protein exportin-5 with Ran-GTP as a cofactor (Han et al. 2006; Wu et al. 2018). Using the cytosolic RNase-III-enzyme Dicer, the still immature (double-stranded) miRNA is formed. Dicer then loads the ds-miRNA onto a protein called Argonaut 2 (AGO2), forming the RNA-induced silencing complex-loading complex (RISC), which consists of Dicer, the transactivation response element RNA-binding protein (TRBP), and AGO2 (Meister 2013). Thereupon RISC can bind to the 3′UTR of complementary mRNA segments (Jing et al. 2005) by searching for matching base sequences (Fig. 1.1). In general, the interaction of RISC or target mRNA requires a complementarity of about seven base pairs (Mohr and Mott 2015). The mechanism by which AGO2 selects the miRNA guide strand still remains unclear. However, it is known that it is targeted and influenced by two key factors. The nucleobase at the 5′ end is important—with uracil being preferred. Additionally, the RISC can more easily take up the strand with a thermodynamically unstable 5′ end. The mature miRNA, also known as guide RNA, is formed by the miRNA strand with the 5′ terminus at its end. The annotated passenger strand is marked as miR* (Ui-Tei et al. 2012; Suzuki et al. 2015). Finally, the (guide) miRNA in its complex can be involved

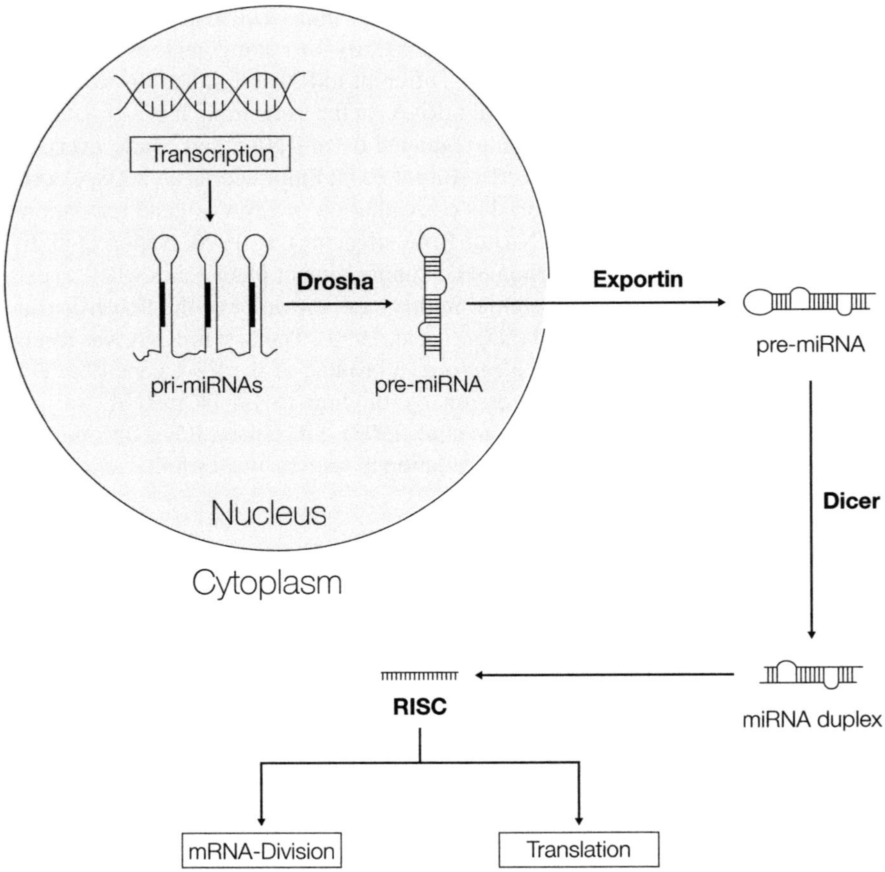

Fig. 1.1 MiRNA-biosynthesis

in the regulation of gene expression of different genes and thus in cell growth and apoptosis (Eddy 2001). In this context, the effect occurs posttranscriptionally via the binding of mRNAs in the cytoplasm, preventing ribosomal translation of the bound segments. Additionally, a translational effect has been described (Schubert et al. 2013).

MicroRNAs are mainly located intracellularly, but some are released from the cell, which facilitates diagnostic accessibility. Stabilization is thought to occur by binding to protein complexes and circulating RNAs (Gibbings et al. 2009). Release can occur by cell lysis or actively via exosomes. Exosomes are nanovesicles that are formed by endocytosis of the plasma membrane and then released into the extracellular space. They are thought to be used for cell-to-cell communication (Camussi et al. 2010). This may occur differently in healthy cells or cancer cells, in physiological and pathological ways (Johnstone et al. 1987; Mathivanan and Simpson 2009; Holdmann et al. 2022).

1.4 Modern Sequencing and Data Processing

Deciphering the diversity of small noncoding RNAs would hardly have been possible without the constant development of new and powerful sequencing methods.

A fundamental base for all nucleotide-handling techniques is the well-known polymerase chain reaction (PCR). It is based on the method developed by the biochemist Kary Mullis in 1983. PCR is an enzymatic technique used to rapidly produce DNA copies of a desired DNA segment. The mechanism of PCR is similar to natural DNA replication in organisms, with (a thermostable) DNA polymerase being the decisive enzyme responsible for building the DNA strands by attaching and joining nucleotides. The term "chain reaction" refers to the use of products from one reaction cycle in subsequent cycles. Each cycle always involves three reaction steps: denaturation, primer hybridization/annealing, and amplification/extension. In the laboratory, this reaction is then thermally controlled (in thermocyclers) (Fig. 1.2). If the template to be amplified is RNA, a reverse-transcription step is required,

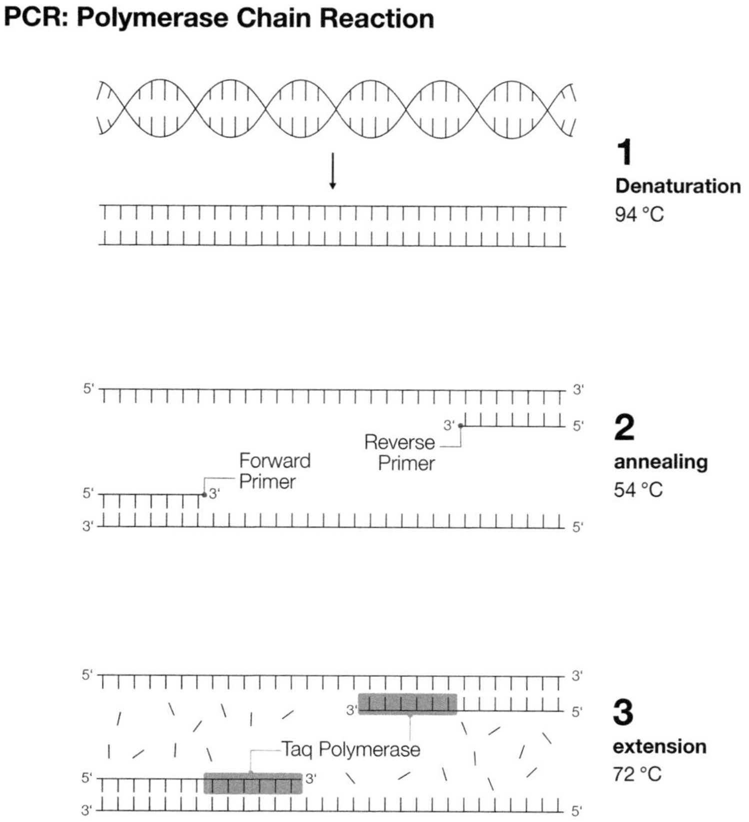

Fig. 1.2 Schematic cycle of a thermally controlled polymerase chain reaction

generating a DNA-copy of an RNA sequence. All modern sequencing techniques are fundamentally based on this principle.

Further developments in the following decades included agglutination PCR, digital PCR, immunoquantitative rt-PCR, inverse PCR, mass tag PCR, multiplex PCR, nested PCR, or quantitative rt-PCR. The sequencing method developed by Sanger was long considered the gold standard. In this process, it is not necessarily the entire DNA that is replicated but specific segments. As a result, it could also be performed more efficiently. But this method has several limitations in read-length and due to the use of a specific primer requires at least partial knowledge of the sequence environment to be analyzed. This requirement is overcome by methods called second or next generation sequencing (NGS). This term covers several newer sequencing methods that have been increasingly offered commercially since 2005 and are now far superior in terms of (time) efficiency. The methods of the different suppliers differ in function and evaluation. What they have in common, however, is that they can be used to sequence small gene segments as well as the entire genome almost simultaneously. Nucleotide strand formation is documented in real time. A commonly used system is provided by Illumina (San Diego, USA). Its procedure is a suitable example for the general approach. In Illumina's systems, the immobilized DNA from the library is first made single-stranded by denaturation and fixed to the sequencing support (FlowCell) by a sequence of primers. The four DNA nucleotides are then added in solution. The binding of a nucleotide to the strand is then catalyzed by DNA polymerase. The bases are labelled using fluorescent dyes. The binding of one of these to the complementary site along the DNA strand is registered by a camera. A protective group attached to the nucleotides also prevents the immediate binding of another base. This is known as cyclic reversible termination. After chemical removal of the terminating group and the reporter molecule, more nucleotides are added in solution, which in turn bind to the next site. NGS techniques rely on the amplification of the molecules to be analyzed by PCR-methods, thus producing multiple copies of the same sequence. Thereby one can distinguish between short and long reads. Short reads can capture the majority of genetic variation, while long-read sequencing allows the detection of complex structural variants that may be difficult to detect with short reads. Some special aspects are relevant for the extraction of (small) RNA. Library construction here begins with the reverse transcription (RT) of small RNAs, producing complementary DNA (cDNA) for amplification and sequencing. MiRNA molecules are extended through ligation or polyadenylation, which introduces primer-binding sites for RT and subsequent amplification (Benesova et al. 2021). Third-generation sequencing techniques such as nanopore focus in contrast on sequencing single molecules without the need of an amplification step. Instead, this method is based on transporting DNA through a nanometer-sized pore. The nanopore is typically made of a recombinant protein (now also made of silicon or graphene) embedded in a polymer membrane. An electrical voltage is applied to the membrane, and as the DNA passes through, the voltage changes in a specific manner for each of the four nucleobases, allowing for the sequence to be identified. A clear advantage here is the compactness of the measuring device, as well as readings with a length of up to two megapairs. However, a slightly higher susceptibility to errors is a factor (Deschamps et al. 2016; Pomerantz et al. 2018).

Whereas in the past reverse transcription was always an obstacle to the readout of RNA using these methods, RNA can now also be analyzed using nanopores (Stephenson et al. 2022). Alternatively to NGS, DNA chip technology has become highly relevant for miRNA research. DNA microarrays can be used to detect the amount of mRNA and noncoding RNA (ncRNA) of certain genes or rRNA of certain organisms. There are two main types of DNA microarrays: spotted microarrays and oligonucleotide microarrays. Spotted microarrays use cDNA, oligonucleotides, or fragments of PCR products corresponding to mRNA and ncRNA printed on the substrate, while oligonucleotide microarrays use synthetically produced oligonucleotides as probes applied to defined positions on a raster, such as glass slides (Yazaki et al. 2016). Both second- and third-generation sequencing techniques count as high-throughput technology because they allow the simultaneous formation and detection of many DNA fragments, that is, the entirety of the different segments of genetic material in solution. This generates large amounts of data, which then need to be bioinformatically processed (Metzker 2010; McCombie et al. 2018; van Dijk et al. 2014; Ng et al. 2009). That can be done using various informatic and bioinformatic tools and techniques. The main steps in miRNA analysis include the following:

(A) Raw sequencing data pre-processing: The raw sequencing data must be pre-processed to remove low-quality reads, adaptors, and contaminants and to trim adaptor sequences.
(B) Read mapping: The processed reads can be aligned to a reference genome or transcriptome using mapping tools such as Bowtie, BWA, and STAR (Li and Durbin 2009; Dobin et al. 2012; Langmead et al. 2009). The mapped reads can then be used to identify known miRNAs or to discover novel miRNAs. Tagging cDNA reads that span multiple exons presents a challenge for various algorithms used in RNA analysis. To predict or construct mRNA references, splicing occurs or junction information, or paired-end data is required (Otto et al. 2014).
(C) miRNA quantification: The abundance of miRNAs can be estimated using various tools such as miRDeep2, DESeq2, and edgeR (Ihaka and Gentleman 1996; Love et al. 2014; Kuhn 2019).
(D) miRNA target prediction: Several bioinformatic tools such as TargetScan, miRanda, and PicTar (McGeary et al. 2019; Betel et al. 2008; Krek et al. 2005) can be used to predict miRNA targets based on sequence complementarity and other features.
(E) Functional analysis: The predicted miRNA targets can be subjected to functional analysis to identify enriched biological pathways, gene ontology (GO) terms, and protein-protein interactions. Tools such as DAVID, Enrichr, and STRING (Sherman et al. 2022; Chen et al. 2013; Szklarczyk et al. 2018) can be used for functional analysis.

In addition to processing, it is of course crucial to list and link all the findings. Databases, which have been laboriously built up over the last few years, play an important role in this, providing comprehensive information on miRNA sequences, expression, targets, and functions and can be used for data integration and validation. Newly discovered molecules are mapped with associated information and can then be electronically matched to the base sequences found. The mirBase database is particularly important for microRNAs. It was launched in September 2010 with around 15,000 entries. Today, it contains more than 50,000 miRNAs from different species (Kozomara and Griffiths-Jones 2010; Kozomara et al. 2018). According to the founders of the platform, Ana Kozomara and Sam Griffiths-Jones, miRBase has five aims:

1. To provide a consistent naming system for microRNAs.
2. To provide a central place collecting all known microRNA sequences.
3. To provide human and computer readable information for each microRNA.
4. To provide primary evidence for each microRNA.
5. To aggregate and link to microRNA target information.

As useful as this database is, it also has some pitfalls. For example, if two or more miRNA entries have different names, accession numbers, and genome positions but identical sequences, they are indistinguishable from both a functional (e.g., targeting) and discovery perspective (Zhao et al. 2017). An additional potential challenge is that longer miRNA entries can overlap shorter miRNAs, leading to false-positive findings (Zhong et al. 2019). To address this concern, other databases, such as MirGenDB, use miRNA families (Fromm et al. 2019). There are more than 20 other miRNA databases, some of which are even better at listing pathways or gene relationships, and all of which have their advantages and disadvantages. Other useful databases include miRDB, mirPathDB, STarBase, mirTaBase, and TargetScan (Hsu et al. 2011; Yang et al. 2011; Chen and Wang 2020; Kehl et al. 2020; Fromm et al. 2019; Li et al. 2014; Huang et al. 2021).

1.5 Omics

Omics are high-performance technologies for the holistic analysis of groups of molecules. They combine interdisciplinary biochemical measurement methods with computerized data acquisition systems and analysis software. *Omics* is a term used as a suffix to describe various disciplines in biology that focus on the holistic analysis of a field. The term is derived from the Greek *ομική*, meaning female. Each particular subdiscipline is to be studied in depth. For example, proteomics is the study of the entire protein composition of an individual or sample, the proteome. Other examples are genomics, metabolomics, or lipidomics. Each of these disciplines studies the structures, function, expression, and clinical aspects of their targets (Conesa and Beck 2019). MiRNomics is accordingly the study of microRNA biology and their computational analysis. It involves the identification, characterization, and functional analysis of microRNAs and the development of computational tools for the analysis of microRNA data (Yousef and Allmer 2014).

The integration of different omics data, known as multi-omics, provides an opportunity to understand the flow of information underlying disease (Hasin et al. 2017). This is also known as systems biology. It could help to personalize the way diseases are diagnosed, treated, monitored, and prevented at an individual level.

1.6 Conclusion

MicroRNAs have become an important field of research in biology and beyond. Working with miRNAs presents several challenges, particularly due to the varying conditions involved. Many methods lack standardization and require expertise for optimal execution. Additionally, pre-analytical factors such as sample collection, transportation, and storage must be considered. It is important to address these challenges to ensure accurate and reliable results. The amplification techniques used can lead to several biases, for example, the quantitative bias if certain fragments of genetic material with amplification advantages are included. To overcome this, some techniques use unique molecular identifiers (UMIs), while others do not. Additionally, the choice of software can significantly impact the results. It is important to use statistical methods that can produce consistent results. The software powering these methods is often a black box for users. Therefore, comparing methods with other researchers is crucial in miRNomics. The advantage of *omics* studies lies in the combination of powerful measurement methods with computerized data acquisition systems and analysis software. This enables the exploration of research fields that were previously inaccessible. For instance, a single genomics analysis of a prostate biopsy might provide information on the expression values of approximately 25,000 genes. When analyzing, for example, 80 patient samples, a data matrix with two million data points is generated. To analyze this vast amount of data, integrative databases and software solutions are necessary. Pathway analysis can then be used to retrieve the biological functions of each stored molecule. If clinical data from patients are integrated into this analysis, a new molecular classification of seemingly homogeneous diseases may be possible. The information provided could go far beyond traditional risk factors and disease markers. So, based on a long history of research, there is currently an immense increase in knowledge about heredity and disease development. This has led to a professionalization of the field. In addition to high-throughput technologies for reading genetic samples, advances in bioinformatics have been particularly groundbreaking. This article discussed several of these methods. The fact that most of them are accessible to the public is mainly due to the ambition and honour of the researchers. The applications of artificial intelligence are likely to significantly impact omic technologies. Rapid transformation is expected in the field.

Acknowledgments We thank the editor of this work and the publisher for the good cooperation. It was a pleasure to make a small contribution to this project. Furthermore, we thank Mr. Hannes Wilke for the creation of attached graphics.

Compliance with Ethical Standards

Conflict of Interest LM, AS, and JH do hereby declare that they held a patent in the field of microRNA exploration. No monetary benefits have raised from this. The fact had no influence on the content of the present text. Besides that, there are no further conflicts of interest.

Ethical Approval This chapter contains no studies with human/animal participants performed by authors.

References

Andersen PR, Tirian L, Vunjak M, Brennecke J (2017) A heterochromatin-dependent transcription machinery drives piRNA expression. Nature 549(7670):54–59. https://doi.org/10.1038/nature23482

Bannister AJ, Kouzarides T (2011) Regulation of chromatin by histone modifications. Cell Res 21(3):381–395. https://doi.org/10.1038/cr.2011.22

Bartel DP (2004) MicroRNAs: genomics, biogenesis, mechanism, and function. Cell 116(2):281–297

Bartel DP (2009) MicroRNAs: target recognition and regulatory functions. Cell 136(2):215–233. https://doi.org/10.1016/j.cell.2009.01.002

Bartel DP (2018) Metazoan MicroRNAs. Cell 173(1):20–51. https://doi.org/10.1016/j.cell.2018.03.006

Bell J, Haldane JBS (1936) The linkage between the genes for colour-blindness and Haemophilia in man. Ann Hum Genet 50(1):3–34. https://doi.org/10.1111/j.1469-1809.1986.tb01935.x

Benesova S, Kubista M, Valihrach L (2021) Small RNA-sequencing: approaches and considerations for miRNA analysis. Diagnostics (Basel) 11(6). https://doi.org/10.3390/diagnostics11060964

Berger SL, Kouzarides T, Shiekhattar R, Shilatifard A (2009) An operational definition of epigenetics. Genes Dev 23(7):781–783. https://doi.org/10.1101/gad.1787609

Betel D, Wilson M, Gabow A, Marks DS, Sander C (2008) The microRNA.org resource: targets and expression. Nucleic Acids Res 36(Database issue):D149–D153. https://doi.org/10.1093/nar/gkm995

Camussi G, Deregibus MC, Bruno S, Cantaluppi V, Biancone L (2010) Exosomes/microvesicles as a mechanism of cell-to-cell communication. Kidney Int 78(9):838–848. https://doi.org/10.1038/ki.2010.278

Chen Y, Wang X (2020) miRDB: an online database for prediction of functional microRNA targets. Nucleic Acids Res 48(D1):D127–d131. https://doi.org/10.1093/nar/gkz757

Chen EY, Tan CM, Kou Y, Duan Q, Wang Z, Meirelles GV, Clark NR, Ma'ayan A (2013) Enrichr: interactive and collaborative HTML5 gene list enrichment analysis tool. BMC Bioinformatics 14:128. https://doi.org/10.1186/1471-2105-14-128

Conesa A, Beck S (2019) Making multi-omics data accessible to researchers. Scientific Data 6(1):251. https://doi.org/10.1038/s41597-019-0258-4

Coordes A, Lenarz M, Kaufmann AM, Albers AE (2014) Role of miRNA in malignoma of the head and neck. Laryngorhinootologie 93(3):201–209.; quiz 210. https://doi.org/10.1055/s-0033-1363960

Deichmann U (2016) Epigenetics: The origins and evolution of a fashionable topic. Dev Biol 416(1):249–254. https://doi.org/10.1016/j.ydbio.2016.06.005

Desai AS, Webb DJ, Taubel J, Casey S, Cheng Y, Robbie GJ, Foster D, Huang SA, Rhyee S, Sweetser MT, Bakris GL (2023) Zilebesiran, an RNA interference therapeutic agent for hypertension. N Engl J Med 389(3):228–238. https://doi.org/10.1056/NEJMoa2208391

Deschamps S, Mudge J, Cameron C, Ramaraj T, Anand A, Fengler K, Hayes K, Llaca V, Jones TJ, May G (2016) Characterization, correction and de novo assembly of an Oxford Nanopore genomic dataset from agrobacterium tumefaciens. Sci Rep 6(1):28625. https://doi.org/10.1038/srep28625

Dobin A, Davis CA, Schlesinger F, Drenkow J, Zaleski C, Jha S, Batut P, Chaisson M, Gingeras TR (2012) STAR: ultrafast universal RNA-seq aligner. Bioinformatics 29(1):15–21. https://doi.org/10.1093/bioinformatics/bts635

Eddy SR (2001) Non-coding RNA genes and the modern RNA world. Nat Rev Genet 2(12):919–929. https://doi.org/10.1038/35103511

Feinberg AP, Tycko B (2004) The history of cancer epigenetics. Nat Rev Cancer 4(2):143–153. https://doi.org/10.1038/nrc1279

Feinberg A, Vogelstein B (1983) Hypomethylation distinguishes genes of some human cancers from their normal counterparts. Nature 301:89–92. https://doi.org/10.1038/301089a0

Filipowicz W, Bhattacharyya SN, Sonenberg N (2008) Mechanisms of post-transcriptional regulation by microRNAs: are the answers in sight? Nat Rev Genet 9(2):102–114. https://doi.org/10.1038/nrg2290

Fromm B, Domanska D, Høye E, Ovchinnikov V, Kang W, Aparicio-Puerta E, Johansen M, Flatmark K, Mathelier A, Hovig E, Hackenberg M, Friedländer MR, Peterson KJ (2019) MirGeneDB 2.0: the metazoan microRNA complement. Nucleic Acids Res 48(D1):D132–D141. https://doi.org/10.1093/nar/gkz885

Garrod (1909) Inborn errors of metabolism. Nature 81(2073):96–97. https://doi.org/10.1038/081096c0

Ghildiyal M, Zamore PD (2009) Small silencing RNAs: an expanding universe. Nat Rev Genet 10(2):94–108. https://doi.org/10.1038/nrg2504

Gibbings DJ, Ciaudo C, Erhardt M, Voinnet O (2009) Multivesicular bodies associate with components of miRNA effector complexes and modulate miRNA activity. Nat Cell Biol 11(9):1143–1149. https://doi.org/10.1038/ncb1929

Girard A, Sachidanandam R, Hannon GJ, Carmell MA (2006) A germline-specific class of small RNAs binds mammalian Piwi proteins. Nature 442(7099):199–202. https://doi.org/10.1038/nature04917

Grivna ST, Beyret E, Wang Z, Lin H (2006) A novel class of small RNAs in mouse spermatogenic cells. Genes Dev 20(13):1709–1714. https://doi.org/10.1101/gad.1434406

Haecker V (1918) Entwicklungsgeschichtliche Eigenschaftsanalyse (Phänogenetik): gemeinsame Aufgaben der Entwicklungsgeschichte, Vererbungs-und Rassenlehre. Gustav Fischer

Han J, Lee Y, Yeom KH, Nam JW, Heo I, Rhee JK, Sohn SY, Cho Y, Zhang BT, Kim VN (2006) Molecular basis for the recognition of primary microRNAs by the Drosha-DGCR8 complex. Cell 125(5):887–901. https://doi.org/10.1016/j.cell.2006.03.043

Hasin Y, Seldin M, Lusis A (2017) Multi-omics approaches to disease. Genome Biol 18(1):83. https://doi.org/10.1186/s13059-017-1215-1

Holdmann J, Markert L, Klinger C, Kaufmann M, Schork K, Turewicz M, Eisenacher M, Degener S, Dreger NM, Roth S, Savelsbergh A (2022) MicroRNAs from urinary exosomes as alternative biomarkers in the differentiation of benign and malignant prostate diseases. J Circ Biomark 11:5–13. https://doi.org/10.33393/jcb.2022.2317

Hsu SD, Lin FM, Wu WY, Liang C, Huang WC, Chan WL, Tsai WT, Chen GZ, Lee CJ, Chiu CM, Chien CH, Wu MC, Huang CY, Tsou AP, Huang HD (2011) miRTarBase: a database curates experimentally validated microRNA-target interactions. Nucleic Acids Res 39(Database issue):D163–D169. https://doi.org/10.1093/nar/gkq1107

Huang H-Y, Lin Y-C-D, Cui S, Huang Y, Tang Y, Xu J, Bao J, Li Y, Wen J, Zuo H, Wang W, Li J, Ni J, Ruan Y, Li L, Chen Y, Xie Y, Zhu Z, Cai X, Chen X, Yao L, Chen Y, Luo Y, LuXu S, Luo M, Chiu C-M, Ma K, Zhu L, Cheng G-J, Bai C, Chiang Y-C, Wang L, Wei F, Lee T-Y, Huang H-D (2021) miRTarBase update 2022: an informative resource for experimentally validated miRNA–target interactions. Nucleic Acids Res 50(D1):D222–D230. https://doi.org/10.1093/nar/gkab1079

Hwang HW, Mendell JT (2007) MicroRNAs in cell proliferation, cell death, and tumorigenesis. Br J Cancer 96(Suppl):R40–R44

Ihaka R, Gentleman R (1996) R: a language for data analysis and graphics. J Comput Graph Stat 5(3):299–314. https://doi.org/10.2307/1390807

Jing Q, Huang S, Guth S, Zarubin T, Motoyama A, Chen J, Di Padova F, Lin SC, Gram H, Han J (2005) Involvement of microRNA in AU-rich element-mediated mRNA instability. Cell 120(5):623–634. https://doi.org/10.1016/j.cell.2004.12.038

Johnstone RM, Adam M, Hammond JR, Orr L, Turbide C (1987) Vesicle formation during reticulocyte maturation. Association of plasma membrane activities with released vesicles (exosomes). J Biol Chem 262(19):9412–9420

Joshi M, Sethi S, Mehta P, Kumari A, Rajender S (2023) Small RNAs, spermatogenesis, and male infertility: a decade of retrospect. Reprod Biol Endocrinol 21(1):106. https://doi.org/10.1186/s12958-023-01155-w

Kehl T, Kern F, Backes C, Fehlmann T, Stöckel D, Meese E, Lenhof HP, Keller A (2020) miRPathDB 2.0: a novel release of the miRNA pathway dictionary database. Nucleic Acids Res 48(D1):D142–d147. https://doi.org/10.1093/nar/gkz1022

Knippers R (2006) Molekulare Genetik. Thieme

Kozomara A, Griffiths-Jones S (2010) miRBase: integrating microRNA annotation and deep-sequencing data. Nucleic Acids Res 39(suppl_1):D152–D157. https://doi.org/10.1093/nar/gkq1027

Kozomara A, Birgaoanu M, Griffiths-Jones S (2018) miRBase: from microRNA sequences to function. Nucleic Acids Res 47(D1):D155–D162. https://doi.org/10.1093/nar/gky1141

Krek A, Grün D, Poy MN, Wolf R, Rosenberg L, Epstein EJ, MacMenamin P, da Piedade I, Gunsalus KC, Stoffel M, Rajewsky N (2005) Combinatorial microRNA target predictions. Nat Genet 37(5):495–500. https://doi.org/10.1038/ng1536

Kuhn MCJW, Weston S, Williams A, Keefer C, Engelhardt A, Cooper T, Mayer Z, Kenkel B, Benesty M, Lescarbeau R, Ziem A, Scrucca L, Tang Y, Candan C, Hunt T, the R Core Team (2019) Caret: classification and regression training. R package version 6:84. https://CRAN.R--project.org/package=caret

Lagos-Quintana M, Rauhut R, Lendeckel W, Tuschl T (2001) Identification of novel genes coding for small expressed RNAs. Science (New York, NY) 294(5543):853–858. https://doi.org/10.1126/science.1064921

Langmead B, Trapnell C, Pop M, Salzberg SL (2009) Ultrafast and memory-efficient alignment of short DNA sequences to the human genome. Genome Biol 10(3):R25. https://doi.org/10.1186/gb-2009-10-3-r25

Lau NC, Lim LP, Weinstein EG, Bartel DP (2001) An abundant class of tiny RNAs with probable regulatory roles in Caenorhabditis elegans. Science (New York, NY) 294(5543):858–862. https://doi.org/10.1126/science.1065062

Lau NC, Seto AG, Kim J, Kuramochi-Miyagawa S, Nakano T, Bartel DP, Kingston RE (2006) Characterization of the piRNA complex from rat testes. Science (New York, NY) 313(5785):363–367. https://doi.org/10.1126/science.1130164

Lee RC, Ambros V (2001) An extensive class of small RNAs in Caenorhabditis elegans. Science (New York, NY) 294(5543):862–864. https://doi.org/10.1126/science.1065329

Lee RC, Feinbaum RL, Ambros V (1993) The C. Elegans heterochronic gene lin-4 encodes small RNAs with antisense complementarity to lin-14. Cell 75(5):843–854

Li H, Durbin R (2009) Fast and accurate short read alignment with burrows-wheeler transform. Bioinformatics 25(14):1754–1760. https://doi.org/10.1093/bioinformatics/btp324

Li JH, Liu S, Zhou H, Qu LH, Yang JH (2014) starBase v2.0: decoding miRNA-ceRNA, miRNA-ncRNA and protein-RNA interaction networks from large-scale CLIP-Seq data. Nucleic Acids Res 42(Database issue):D92–D97. https://doi.org/10.1093/nar/gkt1248

Liu L, Li Y, Tollefsbol TO (2008) Gene-environment interactions and epigenetic basis of human diseases. Curr Issues Mol Biol 10(1-2):25–36

Love MI, Huber W, Anders S (2014) Moderated estimation of fold change and dispersion for RNA-seq data with DESeq2. Genome Biol 15(12):550. https://doi.org/10.1186/s13059-014-0550-8

Lützelberger M (2000) Prä-mRNA Splicing in der Spalthefe Schizosaccharomyces pombe : In vivo Charakterisierung der Funktion des srp2 Gens. TU Braunschweig - Institut für Genetik

Mathivanan S, Simpson RJ (2009) ExoCarta: a compendium of exosomal proteins and RNA. Proteomics 9(21):4997–5000. https://doi.org/10.1002/pmic.200900351

McCombie WR, McPherson JD, Mardis ER (2018) Next-generation sequencing technologies. Cold Spring Harb Perspect Med. https://doi.org/10.1101/cshperspect.a036798

McGeary SE, Lin KS, Shi CY, Pham TM, Bisaria N, Kelley GM, Bartel DP (2019) The biochemical basis of microRNA targeting efficacy. Science (New York, NY) 366(6472). https://doi.org/10.1126/science.aav1741

Meister G (2013) Argonaute proteins: functional insights and emerging roles. Nat Rev Genet 14(7):447–459. https://doi.org/10.1038/nrg3462

Mendel G (1866) Versuche über Pflanzen-Hybriden. Verhandlungen des Naturforschenden Vereines in Brünn 4 Deutsches Textarchiv:3-47

Metzker ML (2010) Sequencing technologies - the next generation. Nat Rev Genet 11(1):31–46. https://doi.org/10.1038/nrg2626

Michlewski G, Cáceres JF (2019) Post-transcriptional control of miRNA biogenesis. RNA 25(1):1–16

Miescher F (1897) Die Histochemischen und Physiologischen Arbeiten von Friedrich Miescher. Erster Band Verlag von F. C. W. Vogel

Mohr AM, Mott JL (2015) Overview of microRNA biology. Semin Liver Dis 35(1):3–11. https://doi.org/10.1055/s-0034-1397344

Morgan TH (1910) Chromosomes and heredity. Am Nat 44(524):449–496

Moxon S, Jing R, Szittya G, Schwach F, Pilcher RLR, Moulton V, Dalmay T (2008) Deep sequencing of tomato short RNAs identifies microRNAs targeting genes involved in fruit ripening. Genome Res 18(10):1602–1609

Nanney DL (1958) Epigenetic control systems[*]. Proc Natl Acad Sci 44(7):712–717. https://doi.org/10.1073/pnas.44.7.712

Napoli C, Lemieux C, Jorgensen R (1990) Introduction of a chimeric Chalcone synthase gene into petunia results in reversible co-suppression of homologous genes in trans. Plant Cell 2(4):279–289. https://doi.org/10.1105/tpc.2.4.279

Ng SB, Turner EH, Robertson PD, Flygare SD, Bigham AW, Lee C, Shaffer T, Wong M, Bhattacharjee A, Eichler EE, Bamshad M, Nickerson DA, Shendure J (2009) Targeted capture and massively parallel sequencing of 12 human exomes. Nature 461(7261):272–276. https://doi.org/10.1038/nature08250

Nurk S, Koren S, Rhie A, Rautiainen M, Bzikadze AV, Mikheenko A, Vollger MR, Altemose N, Uralsky L, Gershman A, Aganezov S, Hoyt SJ, Diekhans M, Logsdon GA, Alonge M, Antonarakis SE, Borchers M, Bouffard GG, Brooks SY, Caldas GV, Chen N-C, Cheng H, Chin C-S, Chow W, de Lima LG, Dishuck PC, Durbin R, Dvorkina T, Fiddes IT, Formenti G, Fulton RS, Fungtammasan A, Garrison E, Grady PGS, Graves-Lindsay TA, Hall IM, Hansen NF, Hartley GA, Haukness M, Howe K, Hunkapiller MW, Jain C, Jain M, Jarvis ED, Kerpedjiev P, Kirsche M, Kolmogorov M, Korlach J, Kremitzki M, Li H, Maduro VV, Marschall T, McCartney AM, McDaniel J, Miller DE, Mullikin JC, Myers EW, Olson ND, Paten B, Peluso P, Pevzner PA, Porubsky D, Potapova T, Rogaev EI, Rosenfeld JA, Salzberg SL, Schneider VA, Sedlazeck FJ, Shafin K, Shew CJ, Shumate A, Sims Y, Smit AFA, Soto DC, Sović I, Storer JM, Streets A, Sullivan BA, Thibaud-Nissen F, Torrance J, Wagner J, Walenz BP, Wenger A, Wood JMD, Xiao C, Yan SM, Young AC, Zarate S, Surti U, McCoy RC, Dennis MY, Alexandrov IA, Gerton JL, O'Neill RJ, Timp W, Zook JM, Schatz MC, Eichler EE, Miga KH, Phillippy AM (2022) The complete sequence of a human genome. Science (New York, NY) 376(6588):44–53. https://doi.org/10.1126/science.abj6987

Otto C, Stadler PF, Hoffmann S (2014) Lacking alignments? The next-generation sequencing mapper segemehl revisited. Bioinformatics 30(13):1837–1843. https://doi.org/10.1093/bioinformatics/btu146

Pembrey ME, Bygren LO, Kaati G, Edvinsson S, Northstone K, Sjöström M, Golding J, The AST (2006) Sex-specific, male-line transgenerational responses in humans. Eur J Hum Genet 14(2):159–166. https://doi.org/10.1038/sj.ejhg.5201538

Pomerantz A, Peñafiel N, Arteaga A, Bustamante L, Pichardo F, Coloma LA, Barrio-Amorós CL, Salazar-Valenzuela D, Prost S (2018) Real-time DNA barcoding in a rainforest using nanopore sequencing: opportunities for rapid biodiversity assessments and local capacity building. GigaScience 7(4). https://doi.org/10.1093/gigascience/giy033

Schubert M, Spahn M, Kneitz S, Scholz CJ, Joniau S, Stroebel P, Riedmiller H, Kneitz B (2013) Distinct microRNA expression profile in prostate cancer patients with early clinical failure and the impact of let-7 as prognostic marker in high-risk prostate cancer. PLoS One 8(6):e65064. https://doi.org/10.1371/journal.pone.0065064

Scott MS, Ono M (2011) From snoRNA to miRNA: dual function regulatory non-coding RNAs. Biochimie 93(11):1987–1992. https://doi.org/10.1016/j.biochi.2011.05.026

Sherman BT, Hao M, Qiu J, Jiao X, Baseler MW, Lane HC, Imamichi T, Chang W (2022) DAVID: a web server for functional enrichment analysis and functional annotation of gene lists (2021 update). Nucleic Acids Res 50(W1):W216–w221. https://doi.org/10.1093/nar/gkac194

Siomi H, Siomi MC (2009) On the road to reading the RNA-interference code. Nature 457(7228):396–404. https://doi.org/10.1038/nature07754

Stephenson W, Razaghi R, Busan S, Weeks KM, Timp W, Smibert P (2022) Direct detection of RNA modifications and structure using single-molecule nanopore sequencing. Cell Genomics 2(2):100097. https://doi.org/10.1016/j.xgen.2022.100097

Storz G (2002) An expanding universe of noncoding RNAs. Science (New York, NY) 296(5571):1260–1263. https://doi.org/10.1126/science.1072249

Sutton WS (1902) On the morphology of the chromoso group in Brachystola magna. Biol Bull 4(1):24–39

Suzuki HI, Katsura A, Yasuda T, Ueno T, Mano H, Sugimoto K, Miyazono K (2015) Small-RNA asymmetry is directly driven by mammalian Argonautes. Nat Struct Mol Biol 22(7):512–521. https://doi.org/10.1038/nsmb.3050

Szklarczyk D, Gable AL, Lyon D, Junge A, Wyder S, Huerta-Cepas J, Simonovic M, Doncheva NT, Morris JH, Bork P, Jensen LJ, Christian M (2018) STRING v11: protein–protein association networks with increased coverage, supporting functional discovery in genome-wide experimental datasets. Nucleic Acids Res 47(D1):D607–D613. https://doi.org/10.1093/nar/gky1131

Teixeira FK, Okuniewska M, Malone CD, Coux RX, Rio DC, Lehmann R (2017) piRNA-mediated regulation of transposon alternative splicing in the soma and germ line. Nature 552(7684):268–272. https://doi.org/10.1038/nature25018

Titze-de-Almeida SS, Brandão PRP, Faber I, Titze-de-Almeida R (2020) Leading RNA interference therapeutics part 1: silencing hereditary Transthyretin amyloidosis, with a focus on Patisiran. Mol Diagn Ther 24(1):49–59. https://doi.org/10.1007/s40291-019-00434-w

Ui-Tei K, Nishi K, Takahashi T, Nagasawa T (2012) Thermodynamic control of small RNA-mediated gene silencing. Front Genet 3. https://doi.org/10.3389/fgene.2012.00101

van Dijk EL, Auger H, Jaszczyszyn Y, Thermes C (2014) Ten years of next-generation sequencing technology. Trends in Genetics 30(9):418–426. https://doi.org/10.1016/j.tig.2014.07.001

Verheggen C, Bertrand E (2012) CRM1 plays a nuclear role in transporting snoRNPs to nucleoli in higher eukaryotes. Nucleus 3(2):132–137. https://doi.org/10.4161/nucl.19266

Waddington CH (1939) An introduction to modern genetics. An introduction to modern genetics

Waddington CH (1942) The Epigenotype. Int J Epidemiol 41(1):10–13. https://doi.org/10.1093/ije/dyr184

Waldeyer W (1888) Ueber Karyokinese und ihre Beziehungen zu den Befruchtungsvorgängen. Arch Mikrosk Anat 32(1):1–122. https://doi.org/10.1007/BF02956988

Wang X-J, Reyes JL, Chua N-H, Gaasterland T (2004) Prediction and identification of Arabidopsis thaliana microRNAs and their mRNA targets. Genome Biol 5(9):R65. https://doi.org/10.1186/gb-2004-5-9-r65

Wang X, Yu S, Lou E, Tan YL, Tan ZJ (2023) RNA 3D structure prediction: Progress and perspective. Molecules 28(14). https://doi.org/10.3390/molecules28145532

Wightman B, Ha I, Ruvkun G (1993) Posttranscriptional regulation of the heterochronic gene lin-14 by lin-4 mediates temporal pattern formation in C. elegans. Cell 75(5):855–862

Wu K, He J, Pu W, Peng Y (2018) The role of Exportin-5 in MicroRNA biogenesis and cancer. Genomics Proteomics Bioinformatics 16(2):120–126. https://doi.org/10.1016/j.gpb.2017.09.004

Yang J-S, Lai EC (2011) Alternative miRNA biogenesis pathways and the interpretation of core miRNA pathway mutants. Mol Cell 43(6):892–903

Yang JH, Li JH, Shao P, Zhou H, Chen YQ, Qu LH (2011) starBase: a database for exploring microRNA-mRNA interaction maps from Argonaute CLIP-Seq and Degradome-Seq data. Nucleic Acids Res 39(Database issue):D202–D209. https://doi.org/10.1093/nar/gkq1056

Yazaki J, Galli M, Kim AY, Nito K, Aleman F, Chang KN, Carvunis A-R, Quan R, Nguyen H, Song L, Alvarez JM, Huang S-sC, Chen H, Ramachandran N, Altmann S, Gutiérrez RA, Hill DE, Schroeder JI, Chory J, LaBaer J, Vidal M, Braun P, Ecker JR (2016) Mapping transcription factor interactome networks using HaloTag protein arrays. Proc Natl Acad Sci 113(29):E4238–E4247. https://doi.org/10.1073/pnas.1603229113

Yousef M, Allmer J (2014) miRNomics: MicroRNA biology and computational analysis. https://doi.org/10.1007/978-1-62703-748-8

Zeilinger S, Kühnel B, Klopp N, Baurecht H, Kleinschmidt A, Gieger C, Weidinger S, Lattka E, Adamski J, Peters A, Strauch K, Waldenberger M, Illig T (2013) Tobacco smoking leads to extensive genome-wide changes in DNA methylation. PLoS One 8(5):e63812. https://doi.org/10.1371/journal.pone.0063812

Zhao S, Gordon W, Du S, Zhang C, He W, Xi L, Mathur S, Agostino M, Paradis T, von Schack D, Vincent M, Zhang B (2017) QuickMIRSeq: a pipeline for quick and accurate quantification of both known miRNAs and isomiRs by jointly processing multiple samples from microRNA sequencing. BMC bioinformatics 18(1):180. https://doi.org/10.1186/s12859-017-1601-4

Zhong X, Heinicke F, Rayner S (2019) miRBaseMiner, a tool for investigating miRBase content. RNA Biol 16(11):1534–1546. https://doi.org/10.1080/15476286.2019.1637680

Chapter 2
MicroRNA Databases

Aruana F. F. Hansel Fröse, Marcel Friedrichs, and Jens Allmer

Abstract MicroRNAs (miRNAs) play a pivotal role in posttranscriptional gene regulation, making computational methods crucial for studying miRNAs. However, with the exponential growth of available data, accessing and effectively utilizing the vast amount of information has become a significant challenge for precise miRNA study. In this chapter, we address this issue through two approaches. Firstly, we present an extensive compilation of over 190 miRNA databases and repositories of various species, phenotypes, and groupings of interest. Secondly, we highlight a practical and comprehensive use case that underscores the application of the databases and their utility in addressing different questions and objectives related to miRNA research. We discuss the challenges of how the availability and maintenance of miRNA repositories are indispensable for repeatability and reproducible research in the bioinformatics field. This chapter provides not only an updated overview of present miRNA databases but also practical application of said repositories. Additionally, we made the miRNA databases list available as a supplement to this work and online at miRNAverse.de. By bridging the gap between theoretical knowledge and practical application, this chapter aims to equip researchers with the necessary tools to navigate the ever-evolving field of miRNA research.

Supplementary Information The online version contains supplementary material available at https://doi.org/10.1007/978-3-031-64788-8_2.

A. F. F. Hansel Fröse
Laboratory of Basic Stem-Cell Biology, Carlos Chagas Institute, Oswaldo Cruz Foundation (ICC/FIOCRUZ-PR), Curitiba, Brazil

Faculty of Technology, Bioinformatics Department, Bielefeld University, Bielefeld, Germany

M. Friedrichs
Faculty of Technology, Bioinformatics Department, Bielefeld University, Bielefeld, Germany

J. Allmer (✉)
Medical Informatics and Bioinformatics, Hochschule Ruhr West, Mülheim adR, Germany
e-mail: jens@allmer.de

L. M. Vaschetto (ed.), *miRNAs, Human Health and Diseases*, Epigenetics and Human Health 13, https://doi.org/10.1007/978-3-031-64788-8_2

Keywords MicroRNA · MiRNA database · Repository · Knowledgebase · Family database · Use case · Target gene · Target prediction · Data analysis · Bioinformatics

Abbreviations

UTR	Untranslated region
RNA	Ribonucleic acid
mRNA	Messenger RNA
tRNA	Transfer RNA
miRNA	microRNA
pri-miRNA	Primary miRNA
pre-miRNA	Precursor miRNA
RISC	RNA-induced silencing complex
miRISC	miRNA-induced silencing complex
DB	Database
FAIR	Findable, accessible, interoperable, and reusable
DNA	Desoxyribonucleic acid

2.1 Introduction

Diseases are often manifested through the dysregulation of gene expression. Gene expression regulation can occur on multiple levels, starting with DNA methylation and ending with the ubiquitylation of proteins (a degradation signal). Posttranscriptional regulation, the modulation of translation efficiency, is performed by microRNAs (miRNAs). However, other molecules, such as small interfering RNAs, short hairpin RNAs, piwi-interacting RNAs, and long noncoding RNAs, also influence gene expression on the transcriptomic level. While miRNAs may not be the only short-regulating RNAs on this level, they appear essential, and therefore, we focus on these molecules in this work.

MicroRNAs are small noncoding RNAs whose mature form is typically between 18 and 24 nucleotides in length. They perform an essential posttranscriptional regulatory function of gene expression. The miRNA biogenesis starts with transcribing the primary miRNA (pri-miRNA) via RNA polymerase II. RNAs, including pri-miRNAs, typically fold upon themselves and may contain many hairpin structures in their folded form. Drosha can subsequently cleave such hairpins, leading to three products: (1) the RNA preceding the hairpin, the hairpin itself, and the RNA following the hairpin structure. After this first cleavage, the cutout hairpin structure, the precursor (pre-mRNA), contains about 70 nucleotides and is exported to the cytosol, where Dicer can process it. Dicer produces a short double-stranded RNA, and one of the strands is then incorporated into the RNA-induced silencing complex (RISC). The mature miRNA, incorporated in miRISC (miRNA-induced silencing complex), conveys specificity to the regulation through base-pairing with its target mRNAs (Krol et al. 2010).

Targeting by the miRNA occurs via base-pairing between the mature miRNA in the miRISC and the target mRNAs. Binding usually occurs in the 3′ untranslated region (UTR) of its target mRNA but can also occur in other areas of the mRNA, such as the coding region or the 5'UTR (Tucker et al. 2021; Ryczek et al. 2023). In the latter case, it may lead to an activation of translation. Binding in the 3′UTR region is associated with translational repression, deadenylation, or mRNA decay (Krol et al. 2010). The majority (approximately 98%) of the human genome is composed of noncoding DNA, and some parts of it are transcribed to noncoding RNAs, including functional RNAs such as translational RNAs (tRNAs) and miRNAs, post-transcriptional regulators of protein expression (Krol et al. 2010). Through interaction with dozens or even hundreds of targets, miRNAs can influence many cellular processes ranging from basic housekeeping functions to tissue differentiation and development. It needs to be kept in mind, though, that a miRNA can only regulate those of its expressed targets that are co-expressed in the same space and time (Lai et al. 1912).

When miRNA regulation is dysfunctional, miRNAs can be associated with diseases and syndromes. Therefore, they can be biomarkers for various diagnoses or prognoses (Chen et al. 2019).

While not highly relevant to this work, it is of note that regulation through miRNA in plants and animals can differ by modes of action, accessory proteins, and how they are generated (Krol et al. 2010). Bioinformatic tools aid the study of miRNAs in many ways by assessing coherent annotation, target prediction and identification, and discovering functional and relevant associations. Bioinformatics tools for miRNA analysis are discussed in Chap. 3. Both chapters are synergistic in that tools depend on data, and databases can be generated with the help of tools. This chapter targets non-bioinformaticians but can also help bioinformaticians with knowledge about databases for data integration. Knowing where and how to search for miRNAs is leverage for improving robust research, specifically considering large volumes of data stored and differently organized in numerous databases. We identified almost 200 databases related to miRNAs. Ranging from miRBase (Griffiths-Jones et al. 2006), whose early aim was to provide a central repository that provides names and IDs to miRNAs to miRNA-disease associations. Not all these databases are continuously maintained, and many have vanished or have been updated since their inception. Therefore, we decided to make a short list of the most cited, accessible, and recently updated databases presented in this manuscript (Table 2.1) and provide the complete list as a supplement. A database is defined as a managed collection of data. Management typically entails a database management system and a formal language, such as the structured query language, to interact with the data. In this work, we collected all resources pertaining to miRNAs, whether they can be formally considered databases or not, since often, the authors of a resource refer to their spreadsheet as a database. While the latter, formally, is not to be considered a database, its content may add value to some search strategies.

Table 2.1 A non-comprehensive list of databases that include miRNA-related information

Database name	Note	Database URL	Last update	N° of scholar citations	Year of publication	Reference
TargetScanS	Database of miRNA-gene targets	https://www.targetscan.org/	2021	13,995	2005	Lewis et al. (2005) and McGeary et al. (2019)
starBase	Experimentally validated database from large-scale CLIP-Seq comprising miRNA-ceRNA, miRNA-ncRNA, and other protein-RNA interactions	http://starbase.sysu.edu.cn/	2023	3678	2014	Li et al. (2014a)
miRDB	miRNA target prediction and functional annotation	http://mirdb.org/miRDB/	2020	1662	2015	Wong and Wang (2015)
Rfam	RNA family database, includes miRNAs.	https://rfam.org/	2022	1645	2003	Griffiths-Jones et al. (2003)
miRTarBase	Database of experimentally validated miRNA-target interactions	http://mirtarbase.cuhk.edu.cn/	2022	1307	2011	Hsu et al. (2011)
mirWalk	Predicted and experimentally validated interactions between miRNAs and mRNAs	http://mirwalk.umm.uni-heidelberg.de/	2022	888	2014	Dweep et al. (2014) and Sticht et al. (2018)
mirDIP	Integrated database of miRNA target predictions in human	http://ophid.utoronto.ca/mirDIP/	2023	393	2018	Tokar et al. (2018) and Hauschild et al. (2023)
ChIPBase	Experimentally validated transcription factor targeted miRNA and ncRNA regulation database	https://rna.sysu.edu.cn/chipbase/	2022	339	2013	Yang et al. (2013) and Huang et al. (2023)
multiMiR	miRNA targets database interaction from other 14 human and mouse databases	http://multimir.ucdenver.edu/	2020	339	2014	Ru et al. (2014)
DIANA-mir-path	Integrates miRNA in pathways by modeling their impact considering targets and gene multiplicity	https://diana-lab.e-ce.uth.gr/app/miRPathv4	2022	322	2009	Papadopoulos et al. (2009)

The comprehensive table can be found in the supplement to this work and online at http://mirnaverse.de/. Here, the top ten databases are presented. All databases were updated after 2019, and all of them are available as of summer 2023. They are ranked by their number of citations in Google Scholar

We also think it is important to address how the resources can add value to research, as we did previously for RNA databases in cancer (Allmer 2019). Therefore, we provide selected use cases and show how they can be addressed using the available databases. We hope the selected list and the comprehensive list of miRNA-related databases will help researchers find the data they need for their research. The supplement with the complete list of miRNA databases and the online list available at miRNAverse.de should facilitate the search and access to the listed miRNA repositories. In the future, we aim to integrate some of the resources that haven't been updated in a long time and produce a fallback repository for such data.

2.2 Data Sources

There is an abundance of online repositories, websites, tools, and databases related to RNAs in general and miRNAs in particular. We will focus on miRNA databases but will also briefly explore databases that contain, for example, miRNA-related tools (for more details, refer to Chap. 3 in this book). Tools for miRNA research afford various analyses, including miRNA prediction, miRNA target prediction, phenotype association, functional investigation, and many more. On the other hand, miRNA databases contain data concerning miRNAs or information regarding miRNAs with their supporting data. Types of miRNA databases include functional, nomenclature, annotation, knowledgebases, interactions, regulation, and more. One can also find differences between predicted results and experimentally validated data for both tools and databases. Both categories can be helpful depending on the research objectives (Allmer 2010; Saçar Demirci et al. 2019). The diversity among miRNA-related databases can be seen in Fig. 2.1, where we used the titles of the publications associated with the resources we found to create a word cloud.

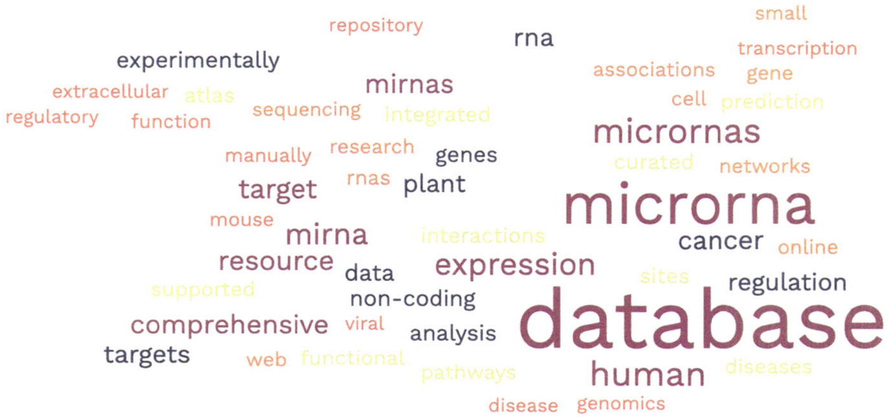

Fig. 2.1 Word cloud of the titles of the 195 databases collected in Supplementary Table 1. Only the 50 most frequently used words are displayed, and only lowercase characters are used. We used https://www.freewordcloudgenerator.com/generatewordcloud to generate the word cloud

Here, we focused on manually compiling an up-to-date list of miRNA databases, which amounted to more than 190 different repositories. This list is available in the supplement, while Table 2.1 lists our selection of filtered databases and repositories. To compile this list, we used various approaches. First, we started with our previous works, such as our chapters and books (Yousef and Allmer 2014; Allmer and Yousef 2022). Then, we augmented the list by mining the literature by querying for microRNA and/or databases since 2019, both in PubMed and Google Scholar. Additionally, we searched "miRNA" within the catalogs of Database Commons (Ma et al. 2022), Tools4Mirs (Lukasik et al. 2016), Noncoding RNA Databases Resource (NRDR) (Paschoal et al. 2012), NAR DB status (Friedrichs and Königs 2023), and recent miRNA tools review (Chen et al. 2019). We manually selected the appropriate miRNA databases from the union of all these results. We eliminated redundancy and checked if the repository was still available. We also noted the number of references in Google Scholar to the resources to rank the list. Additional categorization of our miRNA database list was made considering the species for which miRNAs are available in the identified resources. The latter is important since, for example, researchers working on humans may not be interested in databases containing mostly plant miRNAs and vice versa.

Many species are represented in the miRNA-related databases listed here, and miRBase contains 277 species, including virus strains, for example. However, miRBase aims to list and name all miRNAs. Other databases have different objectives and may only contain a select number of species. Most databases with additional information contain information on *Homo sapiens* (Fig. 2.2). The next big

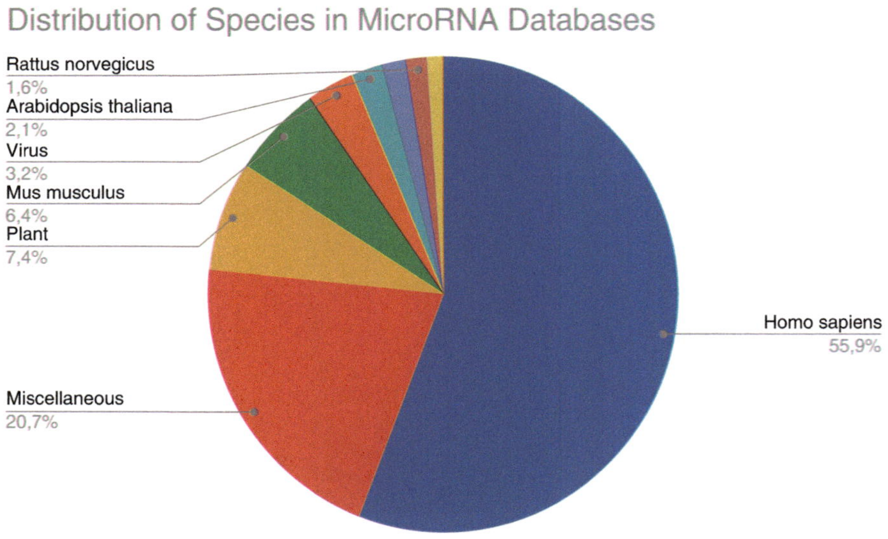

Fig. 2.2 Pie chart of the distribution of species in the 195 miRNA-related databases we selected for this work. *H. sapiens* is the most common and may also be represented in miscellaneous. Plants, mice, and viruses are also frequently available in databases

group is miscellaneous, which may contain any number of species and may also likely include *H. sapiens*.

Apart from the most common research subjects, *H. sapiens* and *Mus musculus*, there were many repositories for plants, and among plants, the most commonly included one is *Arabidopsis thaliana*. Other plants, such as rice and wheat, are attractive for agricultural research and are included in a few databases. Interestingly, many resources include information relating to viruses.

To aid the further filtering and use of the miRNA database, we also assessed whether the specific databases were still available during the preparation of this chapter. Unfortunately, many previously published databases were inaccessible during our survey (Fig. 2.3). As seen in our compilation, two of the ten most cited miRNA repositories in Google Scholar, smiRNAdb and microRNA.org, were unavailable at the time of publication (for up-to-date information on availability, visit http://mirnaverse.de/). There were also many databases published in 2020 that could not be accessed anymore. Despite miRNA databases becoming inaccessible over time, the number of published miRNA databases since 2004 increased, and consequently, the number of available repositories also increased. Out of the total of 195 published databases, 128 (65%) are still available (Fig. 2.3).

Another important metric is whether the database has recently been updated. Unfortunately, determining this is highly involved; we could only do that for a few databases (Table 2.1). Some of the largest and long-standing repositories, such as miRbase, cannot be found in Table 2.1 because there has not been an update after 2019. An excellent resource to check whether a database has seen a recent update is Database Commons (Ma et al. 2022). Database Commons confirms that miRBase

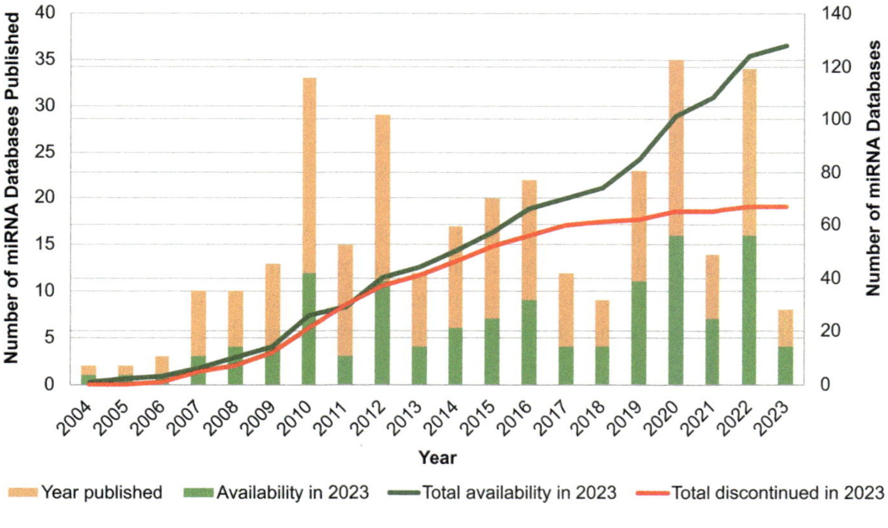

Fig. 2.3 Quantity of published miRNA databases per year since 2004 (orange bars, left axis) and the corresponding proportion of databases that are still available until 2023 (green bars, left axis). The green line represents the cumulative available database (right axis), and the orange line represents the accumulation of databases that are no longer available (right axis)

has not been updated since late 2018. While Database Commons is a great resource, it does not always represent the current availability and last updated status for some databases. In that case, the database websites, if available, or the underlying manuscripts need to be consulted. We urge the development of an open repository maintained by a larger organization where registered users can update all data and not just the authors of the related manuscript.

2.3 Use Cases

A multitude of questions can be answered using the currently available miRNA-related databases. In the following, we discuss a few opportunities for how such databases can be employed to answer miRNA-related questions.

Depending on the biological question, answers can be found in various databases. For non-bioinformaticians, the miRNA databases are more intuitive to use than miRNA tools since the data repositories are primarily online and provide user-friendly access via a website. Therefore, the researchers do not need to be bioinformaticians to benefit from using the miRNA databases.

Many questions can be answered using miRNA repositories, and we have described some common objectives in Table 2.2. Every miRNA database is unique; consequently, the steps to reach information for a specific miRNA vary. Some databases are online, some involve downloading the resource, some require an input file, and others support online queries to the database. Therefore, Table 2.2 and Fig. 2.4 can only provide limited guidance with a basic workflow that needs to be adapted to real-world scenarios. To visually orient miRNA research with databases, we drew an example workflow to show how to address a specific research question in Fig. 2.4.

Using multiple miRNA databases to answer a question is always beneficial, especially considering that different requirements cannot be met with only one repository sometimes. In such cases, filtering outcomes in every step to achieve the appropriate results is important. To exemplify, as detailed in Fig. 2.4, if the researcher wants to understand how the regulation between disease-associated miRNAs and target genes with regulatory relationships with lncRNAs is, first, they could start by selecting the disease-related miRNA databases and use it to select only the miRNA(s) of interest or the disease of interest. From these results, they can filter using the validated target genes and finally filter the results using the lncRNA-miRNA pair database as a reference.

A crucial but often forgotten detail about miRNA research is ensuring that the target genes are also expressed in the sample. Logically, the expression of miRNAs must be accompanied by the expression of their respective targets. Otherwise, the expected regulation cannot occur. The necessary step is to filter your target genes to maintain only the ones expressed in your samples. Results can be improved by checking other next-generation sequencing repositories to verify whether other studies' expression profiles are similar. The gene expression filtering steps can be done at the beginning of the analysis before the target prediction or at the end of the

Table 2.2 Use cases for various miRNA-related databases

Objective	Basic process	Example databases	Expected result
Find miRNA targets	Select the miRNA target databases of the species being studied. Select predicted targets or experimentally validated targets. Select databases specific to the phenotype of interest (healthy tissues, disease-related). Search the miRNA of interest in the databases	**TargetScanS** (Lewis et al. 2005), **MiRBase** (Griffiths-Jones et al. 2006), **MiRDB** (Wong and Wang 2015), miRTarBase (Hsu et al. 2011), TarBase (Sethupathy et al. 2006), multiMiR (Ru et al. 2014)	List of target genes predicted or experimentally validated, of the specific miRNA of interest
Find the miRNAs targeting an mRNA	Select the miRNA target databases of the species being studied. Select predicted targets or experimentally validated targets. Select databases specific to the phenotype of interest (healthy tissues, disease-related). Search the gene of interest in the databases	**TargetScanS** (Lewis et al. 2005)**, miRDB** (Wong and Wang 2015)**, miRTarBase** (Hsu et al. 2011), TarBase (Sethupathy et al. 2006), multiMiR (Ru et al. 2014)	List of miRNAs, predicted or experimentally validated, that target the gene of interest
Establish a regulatory network from a miRNA with other molecules **(lncRNA, RBPs, circRNA, etc)**	Select the miRNA regulatory databases of the regulation of interest, for example miRNA-lncRNA, miRNA-RBPs, miRNA-circRNA etc. select if predicted targets of experimentally validated relationships. Select the species being studied. Select the phenotype of interest (disease related, specific tissue, etc.). Search the miRNA of interest in the regulatory database	**StarBase** (Li et al. 2014a), **ComiRNet** (Pio et al. 2015), PlanTE-MIR DB (Alan et al. 2016), miRTarBase (Hsu et al. 2011), TransmiR (Wang et al. 2009), doRiNA (Anders et al. 2012; Blin et al. 2015), DIANA-LncBase (Paraskevopoulou et al. 2016; Karagkouni et al. 2020), ceRDB (Sarver and Subramanian 2012), PceRBase (Yuan et al. 2017), AtmiRNET (Chien et al. 2015)	List or molecule network connected to the miRNA of interest
Discover the **function** of a miRNA-target interaction	Select the miRNA functional annotation databases. Select for species of interest. Select for the relevant phenotype (disease-related, healthy tissues, etc.). Search for the miRNA of interest in the annotation database	**MiRDB** (Wong and Wang 2015), **miRSystem** (Lu et al. 2012), **PmiREN** (Guo et al. 2020; Guo et al. 2022), **LncTarD** (Zhao et al. 2020; Zhao et al. 2023), **miRFANs** (Liu et al. 2012), **dbMTS** (Li et al. 2020)	List of functions of the miRNA of interest

(continued)

Table 2.2 (continued)

Objective	Basic process	Example databases	Expected result
Investigate a miRNA's **expression profile**	Select the miRNA database related to the species and phenotype of interest (specific tissue, specific organ, disease, or health-associated phenotype) and filter for the miRNA of interest	**HOCTAR** (Gennarino et al. 2011), mirEX (Bielewicz et al. 2012), dbDEMC (Yang et al. 2010; Xu et al. 2022), miRNeye (Karali et al. 2010), ExprTargetDB (Gamazon et al. 2010), TissueAtlas (Koenig et al. 2016), WheatCENet (Li et al. 2022)	Expression profile related to the miRNA of interest in a particular tissue, organ, phenotype, and species
Find diseases associated with a miRNA (establish **biomarkers)**	Select disease-associated miRNA databases. Select the species of interest. Search for the miRNA of interest or get the complete list of disease-associated miRNAs	**DbPKD** (Wang et al. 2019), OncomiR (Wong et al. 2018), miR2Disease (Jiang et al. 2009), miRTarB (Krol et al. 2010), miRCancer (Xie et al. 2013), HMDD (Li et al. 2014b; Huang et al. 2019), dbDEMC (Yang et al. 2010; Xu et al. 2022)	List of miRNAs associated with that disease
Find miRNAs associated with a **phenotype** or other characteristic (location, SNP, methylation, e-QTL, etc.)	Select the appropriate database regarding the phenotype of interest. Search for the miRNA of interest. Alternatively, results can be improved with subsequent filters with other databases results regarding complementary characteristics such as expression profile, disease-oriented miRNAs, pathways, etc.	**EVmiRNA** (Liu et al. 2019), miREV (Hildebrandt et al. 2021), PuTmiR (Bandyopadhyay and Bhattacharyya 2010), PlanTE-MIR DB (Alan et al. 2016), miR-EdiTar (Laganà et al. 2012), ExcellmiRDB (Barupal et al. 2015), EVAtlas (Liu et al. 2022), dbMTS (Li et al. 2020), JAMIR-eQTL (Akiyama et al. 2021), IsomiR-eQL (Moradi et al. 2022)	List of miRNAs associated with that phenotype or characteristics
Find which **pathways** are related to a miRNA	Select pathway-related miRNA database. Filter for the species of interest. Search for the miRNA of interest in the database	**DIANA-mir-Path** (Papadopoulos et al. 2009), miRSystem (Lu et al. 2012), miRPathDB (Backes et al. 2017; Kehl et al. 2020), miR + Pathway (Pian et al. 2020)	List of pathways and ontologies related to the miRNA of study
Find **evolutionary** similar miRNAs	Select the miRNA homologues database of the specie of interest. Filter for the species of interest. Search for the specific miRNA	**MiRviewer** (Kiezun et al. 2012), ZooMir (Li et al. 2010), CircuitsDB (Friard et al. 2010), miROrtho (Gerlach et al. 2009), miRviewer (Kiezun et al. 2012), CoGemiR (Maselli et al. 2008)	Relation of homologs in different species regarding the miRNA of interest

Find the miRNA **family, precursors** and/or **seed sequences**	Select the miRNA family database. Filter for the species of interest. Search for the specific miRNA of interest. Find the relation of precursors and RNA family related to the miRNA	**Rfam** (Griffiths-Jones et al. 2003), miRDB (Wong and Wang 2015)**, mirtronDB** (Da Fonseca et al. 2019), miRBase (Griffiths-Jones 2006; Griffiths-Jones et al. 2008; Griffiths-Jones 2010; Kozomara et al. 2019)	Relation of precursors, RNA family, and sequences derived from the miRNA of interest
Investigate a miRNA's **sequence** and **secondary structure**	Select the miRNA database related to RNA structure. Filter for the species of interest. Filter for experimentally validated or predicted structure. Filter for the miRNA of interest. If searching for the associated molecules, first find the related molecules in the other use case C, and then filter the results	**RNA STRAND** (Andronescu et al. 2008), miRBase (Griffiths-Jones 2006; Griffiths-Jones et al. 2008; Griffiths-Jones 2010; Kozomara et al. 2019), miRDB (Wong and Wang 2015), mESAdb (Kaya et al. 2011), mirtronDB (Da Fonseca et al. 2019)	Secondary structure of the miRNA of interest
Literature, text mining and other **databases** for miRNAs	Select the database of interest. If literature database, search for the miRNA of interest. If searching for other databases, search for characteristics of wanted miRNAs	**MirPub** (Kanellos et al. 2015; Vergoulis et al. 2015), **HumiR** (Solomon et al. 2020), **AgingBank,** (Gao et al. 2022) **Database Commons** (Ma et al. 2022), Tools4Mirs (Lukasik et al. 2016)	Selected publications related to the miRNA of interest, or alternatively list of databases with the desired characteristics of the miRNAs

Given an objective, which databases can help answer the question, and what are the expected results?

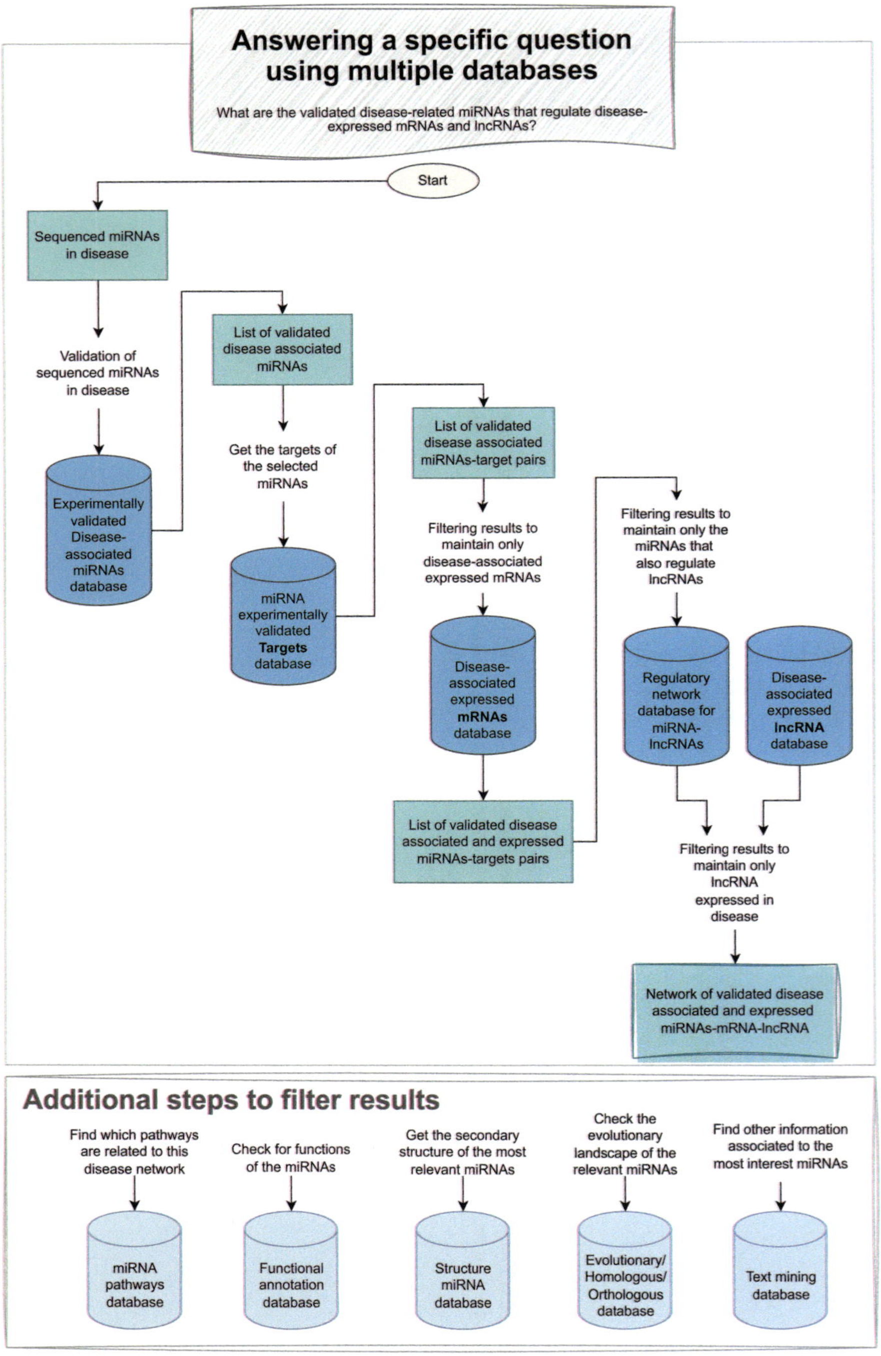

Fig. 2.4 Example data analysis workflow to answer a specific research question using multiple miRNA-related databases and extra steps to refine the results

workflow. While ensuring the co-expression of miRNA and its target in time and space, the target multiplicity is also essential to consider. For this, the number of co-expressed targets needs to be considered, and the more there are, the less effect can be expected. Additionally, lncRNAs can act as miRNA sponges, which diminishes the impact of a miRNA.

Exceptional circumstances may occur, such as the miRNA of interest is not present in the database, its name might be different, or the database that the researcher wants to use is no longer available. Considering that until today many miRNA-related databases have ceased to exist (67; 34%), problems with availability are quite common (Fig. 2.3). To avoid this, special attention is necessary to increase the chance of success. When delineating the analysis design, the researcher should plan to use more than one database for the same step. Redundancy at each analysis step also increases the results as data is not typically shared among databases. The use of multiple databases at each step can therefore lead to more robust results. It is also interesting to use one or more updated annotation database, to make sure that the name and classification of the miRNA is up to date, and most importantly that it is still annotated in the genome version of the species in question.

Many of the databases used in Fig. 2.4 have alternatives, and Table 2.2 lists several databases for each objective. It can be sufficient to use only one database in data analysis, but results can improve if multiple databases are queried. The number of available databases presents a formidable data integration challenge. This challenge would typically be tackled via integrative bioinformatics. We can only recommend the manual analysis if only a few miRNAs will be investigated, for example, the ones that were found to be differentially expressed between conditions.

2.4 Discussion

With this work, we aim to help researchers discover the variety of miRNA repositories and get the most out of the use of these databases, even if they are not familiar with bioinformatics. Through the comprehensive list of almost 200 miRNA-related databases and their detailed categorization in the supplement to this work and on the website miRNAverse.de, we hope to facilitate the discovery of the relevant database(s) for many research objectives. Using some of the most cited miRNA databases is a good start, but it is important to note that some of these resources have not been updated since 2019. Therefore, we suggest using Table 2.1 as the starting point before turning to the complete table in the supplement. Table 2.1 only displays resources that can be reached and that have been updated since 2020. Of these, we selected the top ten according to the references in Google Scholar. Interestingly, one of the most important databases in miRNA research, miRBase, has not been updated since 2019, although more than 400 manuscripts in PubMed contain "novel miRNA" or "novel microRNA" in their title or abstract since 2020. Hence, miRBase is not represented in Table 2.1, but the comprehensive list of miRNA databases can be found in the supplement and on our website miRNAverse.de.

2.5 Conclusions

MicroRNAs convey their function by binding mRNAs, which leads to the digestion of the mRNA or the modulation of the resulting protein abundance. Protein abundance influences regulatory networks and may alter cell homeostasis, leading to a specific phenotype. There is a lot of federated knowledge spread throughout databases. The more than 190 miRNA-related databases we compiled model different aspects of miRNA genesis and function.

Using some of the most cited miRNA databases is a good start in miRNA research, such as TargetScan and miRDB. Conveniently, sometimes, they provide more than one type of miRNA data information, such as miRBase. However, using specific repositories can help get a robust and precise result, such as a regulatory network with other molecules with miRTarBase or DIANA-lncBASE and spatial information of miRNA in vesicles in EVmiRNA. Not to forget specialized databases of miRNA information in different species and contexts, such as PlanTE-MIR DB and miRNeye. Additionally, we defined some common research questions that can be answered using miRNA databases. We used the miRNA database list we created to compile repositories that can be used to address that query. Most importantly, we discussed the importance of the availability of the published databases since even large resources and highly cited repositories may be discontinued. Finally, in this review, we provided a suggested general use case that outlines step-by-step how to use more than one miRNA database, integrating its results to answer a more specific question.

In an attempt to study and compile information about miRNAs, many bioinformatics tools, databases, and repositories were developed throughout the years. On one hand, it has enabled us to understand miRNA function in many different contexts better. On the other hand, we have reached the era of big data to the extent that knowing every tool and database is improbable before choosing the option that best fits our needs. In general, the organization and standardization of these bioinformatic tools are not consistent, which can impact research outcomes. As much as miRNA nomenclature standards are met, there is no database or tool maintenance model to follow, emphasizing a lack of reliability in this manner. A significant challenge lies in the fact that numerous databases lack regular updates and suffer from a loss of availability post-publication. This uncertainty greatly hampers repeatability, which consequently does not completely abide by reproducible research in the field of bioinformatics (de Amorim et al. 2022).

2.6 Future Perspectives

Efforts have been made to address the organizational and maintenance problem of deposited research data, such as the FAIR data principles: findability, accessibility, interoperability, and reusability (Wilkinson et al. 2016). Together, these basic principles aim to better manage repositories and databases so their information can be discovered, accessed, integrated, and reused. Adherence to these principles promotes efficient data management and, consequently, better use of the deposited information, which in turn benefits the whole scientific community with reliable and reproducible research.

With this limitation in mind, we first recommend that developers of miRNA databases could focus not only on the performance of the search engine but also in the integrity, update, and availability of their repository. Furthermore, developing integrative tools for multi-omics data would address integrated principles and aid in regulatory network research. Finally, funding agencies should support data integration, data sharing, and repository interoperability among miRNA databases. Overcoming these problems has great prospects of achieving better reproducible bioinformatic research and more robust research results.

In conclusion, this chapter highlights the strategic use of different miRNA databases that permit the advancement of the field of miRNA research, which in turn improves the understanding of gene regulation, disease mechanisms, and the overall function of miRNA and associated molecules in various species. Databases are a valuable information resource for bioinformatic research, but their impact can extend beyond species conservation to disease diagnosis or even precision medicine. Thus, continued improvement of these repositories and the knowledge of how to use them are essential for advancing miRNA research in many aspects.

Acknowledgments The authors would like to thank Ralf Hofestädt for his kind support. This study was financed in part by the Coordination for the Improvement of Higher Education Personnel (Coordenação de Aperfeiçoamento de Pessoal de Nível Superior - Brasil (Capes)), under the finance code 001. The funders had no role in the design, preparation, writing, or decision to publish this review article. The content and conclusions presented in this paper are solely those of the authors and have not been influenced by the funders.

Compliance with Ethical Standards This review article has been diligently prepared with complete adherence to ethical standards. Our analysis exclusively utilized information from publicly available sources, and no private or unpublished datasets were used. Funding sources for this review have been disclosed, and any potential conflicts of interest have been transparently addressed. All contributors to this work have been appropriately acknowledged and meet the authorship criteria as per established guidelines. We have respected the intellectual property rights of the database creators, ensuring that any material used from these databases adheres to fair use principles and necessary permissions. As this review did not involve direct experimental research on human or animal subjects, ethical approval and consent to participate were not applicable. However, it is noted that original studies from which data were extracted had obtained requisite ethical clearances. Throughout the preparation of this article, we have maintained strict compliance with ethical guidelines.

References

Akiyama S, Higaki S, Ochiya T, Ozaki K, Niida S, Shigemizu D (2021) JAMIR-eQTL: Japanese genome-wide identification of microRNA expression quantitative trait loci across dementia types. Database 2021. https://doi.org/10.1093/database/baab072

Alan AP, Gabriel GY, Paschoal AR, Domingues DS (2016) PlanTE-MIR DB: a database for transposable element-related microRNAs in plant genomes. Funct Integr Genomics 16. https://doi.org/10.1007/s10142-016-0480-5

Allmer J (2010) Computational approaches in microRNA analysis and storage. In: Genç Ş, Genç K (eds) MicroRNAs and the nerve system. Turkish Academy of Sciences (TÜBA) Publishing, pp 92–125. ISBN: 978-9944-252-54-6

Allmer J (2019) Noncoding RNA Databases. Curr Pharm Biotechnol (IF: 2.829) 24(7):825–831. https://doi.org/10.2174/1570180819666220520124511

Allmer J, Yousef M (2022) miRNomics: MicroRNA biology and computational analysis, vol 2. Series: methods in molecular biology: 2257. Humana Press. ISBN: 978-1-0716-1170-8. https://doi.org/10.1007/978-1-0716-1170-8

Anders G, Mackowiak SD, Jens M, Maaskola J, Kuntzagk A, Rajewsky N, Landthaler M, Dieterich C (2012) doRiNA: a database of RNA interactions in post-transcriptional regulation. Nucleic Acids Res 40. https://doi.org/10.1093/nar/gkr1007

Andronescu M, Bereg V, Hoos HH, Condon A (2008) RNA STRAND: the RNA secondary structure and statistical analysis database. BMC Bioinformatics 9. https://doi.org/10.1186/1471-2105-9-340

Backes C, Kehl T, Stöckel D, Fehlmann T, Schneider L, Meese E, Lenhof HP, Keller A (2017) MiRPathDB: a new dictionary on microRNAs and target pathways. Nucleic Acids Res 45. https://doi.org/10.1093/nar/gkw926

Bandyopadhyay S, Bhattacharyya M (2010) PuTmiR: a database for extracting neighboring transcription factors of human microRNAs. BMC Bioinformatics 11. https://doi.org/10.1186/1471-2105-11-190

Barupal JK, Saini AK, Chand T, Meena A, Beniwal S, Suthar JR, Meena N, Kachhwaha S, Kothari SL (2015) ExcellmiRDB for translational genomics: a curated online resource for extracellular MicroRNAs. OMICS 19. https://doi.org/10.1089/omi.2014.0106

Bielewicz D, Dolata J, Zielezinski A, Alaba S, Szarzynska B, Szczesniak MW, Jarmolowski A, Szweykowska-Kulinska Z, Karlowski WM (2012) mirEX: a platform for comparative exploration of plant pri-miRNA expression data. Nucleic Acids Res 40. https://doi.org/10.1093/nar/gkr878

Blin K, Dieterich C, Wurmus R, Rajewsky N, Landthaler M, Akalin A (2015) DoRiNA 2.0-upgrading the dorina database of RNA interactions in post-transcriptional regulation. Nucleic Acids Res 43. https://doi.org/10.1093/nar/gku1180

Chen L, Heikkinen L, Wang C, Yang Y, Sun H, Wong G (2019) Trends in the development of miRNA bioinformatics tools. Brief Bioinform 20:1836–1852. https://doi.org/10.1093/bib/bby054

Chien CH, Chiang-Hsieh YF, Chen YA, Chow CN, Wu NY, Hou PF, Chang WC (2015) AtmiRNET: a web-based resource for reconstructing regulatory networks of Arabidopsis microRNAs. Database 2015. https://doi.org/10.1093/database/bav042

Da Fonseca BHR, Domingues DS, Paschoal AR (2019) MirtronDB: a mirtron knowledge base. Bioinformatics 35. https://doi.org/10.1093/bioinformatics/btz153

de Amorim TS, Pedro DLF, Paschoal AR (2022) MicroRNA databases and tools. Methods Mol Biol 2257:131–166. https://doi.org/10.1007/978-1-0716-1170-8_7/FIGURES/9

Dweep H, Gretz N, Sticht C (2014) MiRWalk database for miRNA-target interactions. Methods Mol Biol 1182. https://doi.org/10.1007/978-1-4939-1062-5_25

Friard O, Re A, Taverna D, De Bortoli M, Corá D (2010) CircuitsDB: a database of mixed microRNA/transcription factor feed-forward regulatory circuits in human and mouse. BMC Bioinformatics 11. https://doi.org/10.1186/1471-2105-11-435

Friedrichs M, Königs C (2023) A web-based platform for the annotation and analysis of NAR-published databases. PLoS One 18. https://doi.org/10.1371/journal.pone.0293134

Gamazon ER, Im HK, Duan S, Lussier YA, Cox NJ, Dolan ME, Zhang W (2010) ExprTarget: an integrative approach to predicting human microRNA targets. PLoS One 5. https://doi.org/10.1371/journal.pone.0013534

Gao Y, Shang S, Guo S, Wang X, Zhou H, Sun Y, Gan J, Zhang Y, Li X, Ning S, Zhang Y (2022) AgingBank: a manually curated knowledgebase and high-throughput analysis platform that provides experimentally supported multi-omics data relevant to aging in multiple species. Brief Bioinform 23. https://doi.org/10.1093/bib/bbac438

Gennarino VA, Sardiello M, Mutarelli M, Dharmalingam G, Maselli V, Lago G, Banfi S (2011) HOCTAR database: a unique resource for microRNA target prediction. Gene 480. https://doi.org/10.1016/j.gene.2011.03.005

Gerlach D, Kriventseva EV, Rahman N, Vejnar CE, Zdobnov EM (2009) miROrtho: computational survey of microRNA genes. Nucleic Acids Res 37. https://doi.org/10.1093/nar/gkn707

Griffiths-Jones S (2006) miRBase: the microRNA sequence database. Methods Mol Biol 342. https://doi.org/10.1385/1-59745-123-1:129

Griffiths-Jones S (2010) MiRBase: MicroRNA sequences and annotation. Curr Protoc Bioinformatics. https://doi.org/10.1002/0471250953.bi1209s29

Griffiths-Jones S, Bateman A, Marshall M, Khanna A, Eddy SR (2003) Rfam: an RNA family database. Nucleic Acids Res 31. https://doi.org/10.1093/nar/gkg006

Griffiths-Jones S, Grocock RJ, van Dongen S, Bateman A, Enright AJ (2006) miRBase: microRNA sequences, targets and gene nomenclature. Nucleic Acids Res 34. https://doi.org/10.1093/nar/gkj112

Griffiths-Jones S, Saini HK, Van Dongen S, Enright AJ (2008) miRBase: tools for microRNA genomics. Nucleic Acids Res 36. https://doi.org/10.1093/nar/gkm952

Guo Z, Kuang Z, Wang Y, Zhao Y, Tao Y, Cheng C, Yang J, Lu X, Hao C, Wang T, Cao X, Wei J, Li L, Yang X (2020) PmiREN: a comprehensive encyclopedia of plant miRNAs. Nucleic Acids Res 48. https://doi.org/10.1093/nar/gkz894

Guo Z, Kuang Z, Zhao Y, Deng Y, He H, Wan M, Tao Y, Wang D, Wei J, Li L, Yang X (2022) PmiREN2.0: from data annotation to functional exploration of plant microRNAs. Nucleic Acids Res 50. https://doi.org/10.1093/nar/gkab811

Hauschild AC, Pastrello C, Ekaputeri GKA, Bethune-Waddell D, Abovsky M, Ahmed Z, Kotlyar M, Lu R, Jurisica I (2023) MirDIP 5.2: tissue context annotation and novel microRNA curation. Nucleic Acids Res 51. https://doi.org/10.1093/nar/gkac1070

Hildebrandt A, Kirchner B, Nolte-t Hoen ENM, Pfaffl MW (2021) miREV: an online database and tool to uncover potential reference RNAs and biomarkers in small-RNA sequencing data sets from extracellular vesicles enriched samples. J Mol Biol 433. https://doi.org/10.1016/j.jmb.2021.167070

Hsu SD, Lin FM, Wu WY, Liang C, Huang WC, Chan WL, Tsai WT, Chen GZ, Lee CJ, Chiu CM, Chien CH, Wu MC, Huang CY, Tsou AP, Da HH (2011) MiRTarBase: a database curates experimentally validated microRNA-target interactions. Nucleic Acids Res 39. https://doi.org/10.1093/nar/gkq1107

Huang Z, Shi J, Gao Y, Cui C, Zhang S, Li J, Zhou Y, Cui Q (2019) HMDD v3.0: a database for experimentally supported human microRNA-disease associations. Nucleic Acids Res 47. https://doi.org/10.1093/nar/gky1010

Huang J, Zheng W, Zhang P, Lin Q, Chen Z, Xuan J, Liu C, Wu D, Huang Q, Zheng L, Liu S, Zhou K, Qu L, Li B, Yang J (2023) ChIPBase v3.0: the encyclopedia of transcriptional regulations of non-coding RNAs and protein-coding genes. Nucleic Acids Res 51. https://doi.org/10.1093/nar/gkac1067

Jiang Q, Wang Y, Hao Y, Juan L, Teng M, Zhang X, Li M, Wang G, Liu Y (2009) miR2Disease: a manually curated database for microRNA deregulation in human disease. Nucleic Acids Res 37. https://doi.org/10.1093/nar/gkn714

Kanellos I, Vlachokyriakou V, Vergoulis T, Georgakilas G, Vassiliou Y, Hatzigeorgiou AK, Dalamagas T (2015) MirPub v2: Towards ranking and refining miRNA publication search results. In: Lecture Notes in Computer Science (including subseries Lecture Notes in Artificial Intelligence and Lecture Notes in Bioinformatics)

Karagkouni D, Paraskevopoulou MD, Tastsoglou S, Skoufos G, Karavangeli A, Pierros V, Zacharopoulou E, Hatzigeorgiou AG (2020) DIANA-LncBase v3: indexing experimen-

tally supported miRNA targets on non-coding transcripts. Nucleic Acids Res 48. https://doi.org/10.1093/nar/gkz1036
Karali M, Peluso I, Gennarino VA, Bilio M, Verde R, Lago G, Dollé P, Banfi S (2010) MiRNeye: a microRNA expression atlas of the mouse eye. BMC Genomics 11. https://doi.org/10.1186/1471-2164-11-715
Kaya KD, Karakülah G, Yakicier CM, Acar AC, Konu Ö (2011) MESAdb: MicroRNA expression and sequence analysis database. Nucleic Acids Res 39. https://doi.org/10.1093/nar/gkq1256
Kehl T, Kern F, Backes C, Fehlmann T, Stöckel D, Meese E, Lenhof HP, Keller A (2020) MiRPathDB 2.0: a novel release of the miRNA pathway dictionary database. Nucleic Acids Res 48. https://doi.org/10.1093/nar/gkz1022
Kiezun A, Artzi S, Modai S, Volk N, Isakov O, Shomron N (2012) MiRviewer: a multispecies microRNA homologous viewer. BMC Res Notes 5. https://doi.org/10.1186/1756-0500-5-92
Koenig EM, Fisher C, Bernard H, Wolenski FS, Gerrein J, Carsillo M, Gallacher M, Tse A, Peters R, Smith A, Meehan A, Tirrell S, Kirby P (2016) The beagle dog MicroRNA tissue atlas: identifying translatable biomarkers of organ toxicity. BMC Genomics 17. https://doi.org/10.1186/s12864-016-2958-x
Kozomara A, Birgaoanu M, Griffiths-Jones S (2019) MiRBase: From microRNA sequences to function. Nucleic Acids Res 47. https://doi.org/10.1093/nar/gky1141
Krol J, Loedige I, Filipowicz W (2010) The widespread regulation of microRNA biogenesis, function and decay. Nat Rev Genet 11:597–610. https://doi.org/10.1038/nrg2843
Laganà A, Paone A, Veneziano D, Cascione L, Gasparini P, Carasi S, Russo F, Nigita G, MacCa V, Giugno R, Pulvirenti A, Shasha D, Ferro A, Croce CM (2012) MiR-EdiTar: a database of predicted A-to-I edited miRNA target sites. Bioinformatics 28. https://doi.org/10.1093/bioinformatics/bts589
Lai X, Gupta SK, Vera J (1912) Computational biology of non-coding RNA methods and protocols, Methods in molecular biology. Humana Press. https://doi.org/10.1007/978-1-4939-8982-9
Lewis BP, Burge CB, Bartel DP (2005) Conserved seed pairing, often flanked by adenosines, indicates that thousands of human genes are microRNA targets. Cell 120. https://doi.org/10.1016/j.cell.2004.12.035
Li SC, Chan WC, Hu LY, Lai CH, Hsu CN, Lin W chang (2010) Identification of homologous microRNAs in 56 animal genomes. Genomics 96. https://doi.org/10.1016/j.ygeno.2010.03.009
Li JH, Liu S, Zhou H, Qu LH, Yang JH (2014a) StarBase v2.0: decoding miRNA-ceRNA, miRNA-ncRNA and protein-RNA interaction networks from large-scale CLIP-Seq data. Nucleic Acids Res 42:92–97. https://doi.org/10.1093/nar/gkt1248
Li Y, Qiu C, Tu J, Geng B, Yang J, Jiang T, Cui Q (2014b) HMDD v2.0: a database for experimentally supported human microRNA and disease associations. Nucleic Acids Res 42. https://doi.org/10.1093/nar/gkt1023
Li C, Mou C, Swartz MD, Yu B, Bai Y, Tu Y, Liu X (2020) dbMTS: a comprehensive database of putative human microRNA target site SNVs and their functional predictions. Hum Mutat 41. https://doi.org/10.1002/humu.24020
Li Z, Hu Y, Ma X, Da L, She J, Liu Y, Yi X, Cao Y, Xu W, Jiao Y, Su Z (2022) WheatCENet: a database for comparative co-expression networks analysis of allohexaploid wheat and its progenitors. Genomics Proteomics Bioinformatics. https://doi.org/10.1016/j.gpb.2022.04.007
Liu H, Jin T, Liao R, Wan L, Xu B, Zhou S, Guan J (2012) MiRFANs: an integrated database for Arabidopsis thaliana microRNA function annotations. BMC Plant Biol 12. https://doi.org/10.1186/1471-2229-12-68
Liu T, Zhang Q, Zhang J, Li C, Miao YR, Lei Q, Li Q, Guo AY (2019) EVmiRNA: a database of miRNA profiling in extracellular vesicles. Nucleic Acids Res 47. https://doi.org/10.1093/nar/gky985
Liu CJ, Xie GY, Miao YR, Xia M, Wang Y, Lei Q, Zhang Q, Guo AY (2022) EVAtlas: a comprehensive database for ncRNA expression in human extracellular vesicles. Nucleic Acids Res 50. https://doi.org/10.1093/nar/gkab668

Lu TP, Lee CY, Tsai MH, Chiu YC, Hsiao CK, Lai LC, Chuang EY (2012) MiRSystem: an integrated system for characterizing enriched functions and pathways of microRNA targets. PLoS One 7. https://doi.org/10.1371/journal.pone.0042390

Lukasik A, Wójcikowski M, Zielenkiewicz P (2016) Tools4miRs - one place to gather all the tools for miRNA analysis. In: Bioinformatics. Oxford University Press, Oxford, pp 2722–2724. https://doi.org/10.1093/bioinformatics/btw189

Ma L, Zou D, Liu L, Shireen H, Abbasi AA, Bateman A, Xiao J, Zhao W, Bao Y, Zhang Z (2022) Database commons: a catalog of worldwide biological databases. Genomics Proteomics Bioinformatics. https://doi.org/10.1016/J.GPB.2022.12.004

Maselli V, Di Bernardo D, Banfi S (2008) CoGemiR: A comparative genomics microRNA database. BMC Genomics 9. https://doi.org/10.1186/1471-2164-9-457

McGeary SE, Lin KS, Shi CY, Pham TM, Bisaria N, Kelley GM, Bartel DP (2019) The biochemical basis of microRNA targeting efficacy. Science 1979:366. https://doi.org/10.1126/science.aav1741

Moradi A, Whatmore P, Farashi S, Barrero RA, Batra J (2022) IsomiR-eQTL: a cancer-specific expression quantitative trait loci database of miRNAs and their isoforms. Int J Mol Sci 23. https://doi.org/10.3390/ijms232012493

Papadopoulos GL, Alexiou P, Maragkakis M, Reczko M, Hatzigeorgiou AG (2009) DIANA-mirPath: integrating human and mouse microRNAs in pathways. Bioinformatics 25. https://doi.org/10.1093/bioinformatics/btp299

Paraskevopoulou MD, Vlachos IS, Karagkouni D, Georgakilas G, Kanellos I, Vergoulis T, Zagganas K, Tsanakas P, Floros E, Dalamagas T, Hatzigeorgiou AG (2016) DIANA-LncBase v2: indexing microRNA targets on non-coding transcripts. Nucleic Acids Res 44. https://doi.org/10.1093/nar/gkv1270

Paschoal AR, Maracaja-Coutinho V, Setubal JC, Simões ZLP, Verjovski-Almeida S, Durham AM (2012) Non-coding transcription characterization and annotation: a guide and web resource for non-coding RNA databases. RNA Biol 9:274–282. https://doi.org/10.4161/RNA.19352/SUPPL_FILE/KRNB_A_10919352_SM0001.ZIP

Pian C, Zhang G, Gao L, Fan X, Li F (2020) MiR+pathway: the integration and visualization of miRNA and KEGG pathways. Brief Bioinform 21. https://doi.org/10.1093/bib/bby128

Pio G, Ceci M, Malerba D, D'Elia D (2015) ComiRNet: a web-based system for the analysis of miRNA-gene regulatory networks. BMC Bioinformatics 16. https://doi.org/10.1186/1471-2105-16-S9-S7

Ru Y, Kechris KJ, Tabakoff B, Hoffman P, Radcliffe RA, Bowler R, Mahaffey S, Rossi S, Calin GA, Bemis L, Theodorescu D (2014) The multiMiR R package and database: integration of microRNA-target interactions along with their disease and drug associations. Nucleic Acids Res 42. https://doi.org/10.1093/nar/gku631

Ryczek N, Łyś A, Makałowska I (2023) The functional meaning of 5′UTR in protein-coding genes. Int J Mol Sci 24. https://doi.org/10.3390/ijms24032976

Saçar Demirci MD, Yousef M, Allmer J (2019) Computational prediction of functional microRNA–mRNA interactions. Methods Mol Biol 1912:175–196. https://doi.org/10.1007/978-1-4939-8982-9_7

Sarver AL, Subramanian S (2012) Competing endogenous RNA database. Bioinformation 8:731–733. https://doi.org/10.6026/97320630008731

Sethupathy P, Corda B, Hatzigeorgiou AG (2006) TarBase: a comprehensive database of experimentally supported animal microRNA targets. RNA 12. https://doi.org/10.1261/rna.2239606

Solomon J, Kern F, Fehlmann T, Meese E, Keller A (2020) Humir: web services, tools and databases for exploring human microrna data. Biomol Ther 10. https://doi.org/10.3390/biom10111576

Sticht C, De La Torre C, Parveen A, Gretz N (2018) Mirwalk: an online resource for prediction of microrna binding sites. PLoS One 13. https://doi.org/10.1371/journal.pone.0206239

Tokar T, Pastrello C, Rossos AEM, Abovsky M, Hauschild AC, Tsay M, Lu R, Jurisica I (2018) MirDIP 4.1 - integrative database of human microRNA target predictions. Nucleic Acids Res 46. https://doi.org/10.1093/nar/gkx1144

Tucker AR, Salazar NA, Ayoola AO, Memili E, Thomas BN, Morenikeji OB (2021) Regulatory network of miRNA, lncRNA, transcription factor and target immune response genes in bovine mastitis. Sci Rep 11. https://doi.org/10.1038/s41598-021-01280-9

Vergoulis T, Kanellos I, Kostoulas N, Georgakilas G, Sellis T, Hatzigeorgiou A, Dalamagas T (2015) MirPub: a database for searching microRNA publications. Bioinformatics 31. https://doi.org/10.1093/bioinformatics/btu819

Wang J, Lu M, Qiu C, Cui Q (2009) TransmiR: a transcription factor microRNA regulation database. Nucleic Acids Res 38. https://doi.org/10.1093/nar/gkp803

Wang G, Ouyang J, Li S, Wang H, Lian B, Liu Z, Xie L (2019) The analysis of risk factors for diabetic nephropathy progression and the construction of a prognostic database for chronic kidney diseases. J Transl Med 17. https://doi.org/10.1186/s12967-019-2016-y

Wilkinson MD, Dumontier M, Aalbersberg IJ, Appleton G, Axton M, Baak A, Blomberg N, Boiten JW, da Silva Santos LB, Bourne PE, Bouwman J, Brookes AJ, Clark T, Crosas M, Dillo I, Dumon O, Edmunds S, Evelo CT, Finkers R, Gonzalez-Beltran A, Gray AJG, Groth P, Goble C, Grethe JS, Heringa J, Hoen PAC, Hooft R, Kuhn T, Kok R, Kok J, Lusher SJ, Martone ME, Mons A, Packer AL, Persson B, Rocca-Serra P, Roos M, van Schaik R, Sansone SA, Schultes E, Sengstag T, Slater T, Strawn G, Swertz MA, Thompson M, Van Der Lei J, Van Mulligen E, Velterop J, Waagmeester A, Wittenburg P, Wolstencroft K, Zhao J, Mons B (2016) Comment: the FAIR guiding principles for scientific data management and stewardship. Sci Data 3. https://doi.org/10.1038/sdata.2016.18

Wong N, Wang X (2015) miRDB: an online resource for microRNA target prediction and functional annotations. Nucleic Acids Res 43. https://doi.org/10.1093/nar/gku1104

Wong NW, Chen Y, Chen S, Wang X (2018) OncomiR: an online resource for exploring pan-cancer microRNA dysregulation. Bioinformatics 34. https://doi.org/10.1093/bioinformatics/btx627

Xie B, Ding Q, Han H, Wu D (2013) MiRCancer: a microRNA-cancer association database constructed by text mining on literature. Bioinformatics 29. https://doi.org/10.1093/bioinformatics/btt014

Xu F, Wang Y, Ling Y, Zhou C, Wang H, Teschendorff AE, Zhao Y, Zhao H, He Y, Zhang G, Yang Z (2022) dbDEMC 3.0: functional exploration of differentially expressed miRNAs in cancers of human and model organisms. Genomics Proteomics Bioinformatics 20. https://doi.org/10.1016/j.gpb.2022.04.006

Yang Z, Ren F, Liu C, He S, Sun G, Gao Q, Yao L, Zhang Y, Miao R, Cao Y, Zhao Y, Zhong Y, Zhao H (2010) DbDEMC: a database of differentially expressed miRNAs in human cancers. BMC Genomics 11. https://doi.org/10.1186/1471-2164-11-S4-S5

Yang JH, Li JH, Jiang S, Zhou H, Qu LH (2013) ChIPBase: a database for decoding the transcriptional regulation of long non-coding RNA and microRNA genes from ChIP-Seq data. Nucleic Acids Res 41. https://doi.org/10.1093/nar/gks1060

Yousef M, Allmer J (2014) miRNomics: MicroRNA biology and computational analysis, Methods in molecular biology. Humana Press. https://doi.org/10.1007/978-1-62703-748-8

Yuan C, Meng X, Li X, Illing N, Ingle RA, Wang J, Chen M (2017) PceRBase: a database of plant competing endogenous RNA. Nucleic Acids Res 45. https://doi.org/10.1093/nar/gkw916

Zhao H, Shi J, Zhang Y, Xie A, Yu L, Zhang C, Lei J, Xu H, Leng Z, Li T, Huang W, Lin S, Wang L, Xiao Y, Li X (2020) LncTarD: a manually-curated database of experimentally-supported functional lncRNA-target regulations in human diseases. Nucleic Acids Res 48. https://doi.org/10.1093/nar/gkz985

Zhao H, Yin X, Xu H, Liu K, Liu W, Wang L, Zhang C, Bo L, Lan X, Lin S, Feng K, Ning S, Zhang Y, Wang L (2023) LncTarD 2.0: an updated comprehensive database for experimentally-supported functional lncRNA-target regulations in human diseases. Nucleic Acids Res 51. https://doi.org/10.1093/nar/gkac984

Chapter 3
Bioinformatics Tools to Study the Role of miRNAs

Mehmet Emin Orhan, Yılmaz Mehmet Demirci, and Müşerref Duygu Saçar Demirci

Abstract The application of high-throughput technologies in molecular biology and health sciences has led to the generation of large amounts of experimental data that require the development of sophisticated computational technologies to extract information from these enormous data sets. To address this challenge, various tools and resources have been developed. However, finding the right tool to answer specific research questions accurately and efficiently can be a daunting task. MicroRNAs (miRNAs) are small, noncoding sequences that regulate gene expression posttranscriptionally. These sequences can be found in a wide range of organisms, including viruses and higher eukaryotes. Detecting miRNAs is a complex task that can be compared to finding needles in a haystack. In this chapter, we will review the available literature to guide the techniques that can be used for identifying miRNAs.

Keywords MiRNA · Bioinformatics · Databases · Tools · Prediction · Target gene

Abbreviations

miRNA	microRNA
RNA	Ribonucleic acid
RNA-Seq	RNA Sequencing
pre-miRNA	Precursor-miRNA

M. E. Orhan
Department of Bioengineering, Graduate School of Engineering and Science, Abdullah Gül University, Kayseri, Turkey

Y. M. Demirci
Department of Engineering Science, Faculty of Engineering, Abdullah Gül University, Kayseri, Turkey

M. D. S. Demirci (✉)
Department of Bioengineering, Faculty of Life and Natural Sciences, Abdullah Gül University, Kayseri, Turkey
e-mail: duygu.sacar@agu.edu.tr

L. M. Vaschetto (ed.), *miRNAs, Human Health and Diseases*, Epigenetics and Human Health 13, https://doi.org/10.1007/978-3-031-64788-8_3

R2DT	RNA 2D templates
UI	User interface
MTI	MicroRNA target interaction
UTR	Untranslated region
NLP	Natural language processing
NGS	Next-generation sequencing
Chip-Seq	Chromatin immunoprecipitation sequencing
mRNA	Messenger RNA
ceRNA	Competing endogenous RNA
lncRNA	Long noncoding RNA
circRNA	Circular RNA
CLIP-Seq	Cross-linking immunoprecipitation
ncRNA	Noncoding RNA
TF	Transcription factor
SNP	Single-nucleotide polymorphism
KEGG	Kyoto Encyclopedia of Genes and Genomes
HMDD	Human microRNA Disease Database

3.1 Introduction

MicroRNAs (miRNAs) are small (18–24 nucleotides) sequences involved in posttranscriptional regulation of gene expression (Bartel 2004). Since miRNA is one of the most studied noncoding RNAs, many published databases and tools are designed specifically for analyzing miRNA sequences. At first glance, this seems very helpful for researchers trying to discover new miRNAs and/or searching for a specific miRNA involved in a particular pathway. However, various aspects need to be considered before starting an experiment. In this chapter, we provide a list of available tools and compare them in terms of their methods, update date, platform, etc. to guide the users.

3.2 miRNA Sequence Repositories

Since the discovery of the first miRNAs, numerous miRNA sequences have been identified through wet-lab experiments and/or bioinformatics approaches, leading to the construction of miRNA sequence databases (Table 3.1; Kozomara et al. 2019).

3.2.1 miRBase

MirBase, established in 2004 by Griffiths-Jones, stands as the first database in the literature dedicated to managing and collecting miRNA sequences and annotations (Griffiths-Jones 2004). The pioneer repository focuses exclusively on miRNAs,

Table 3.1 Main miRNA sequence repositories. For further information about the miRNA databases, please refer to Chap. 2

Database	Summary	Sequence collection	Features	Latest update	Website
miRBase (Griffiths-Jones et al. 2006)	Central repository for the storage of miRNA sequence and annotations	38.589 hairpin precursor and 48.860 mature microRNA sequences from 271 species	• Literature review supported with word cloud • Secondary structure of sequence • Predicted target links • AI generated summary of related microRNA • Gene family and disease association	2019 (release 22)	https://www.mirbase.org/
RNAcentral (Bateman et al. 2011)	Comprehensive database for all noncoding RNAs	30 million sequences from 250 species including humans and mice	• Annotations of sequences • Open access publication list • Localization data of related sequence (protein or microRNA) • mRNA interactions related sequence • Precursor secondary structures	2023 (release 22.0)	https://rnacentral.org/
Rfam (Kalvari et al. 2021)	RNA families' collection including miRNA	Over 4000 families from 8302 species	• RNA families information • Secondary structure information • Summary of related RNA • Taxonomy information	2022 (release 14.9)	https://rfam.org/

assigns unique IDs to individual sequences, and establishes connections with other repositories. The current version of miRBase contains 48,860 entries for mature miRNAs and 38,589 miRNA hairpin precursors from 271 species (Kozomara et al. 2019).

MiRNA data sets within miRBase are curated from multiple databases and/or extracted through text-mining techniques. The repository provides gene family information from Rfam (Kalvari et al. 2018) and expression levels for each tissue from TissueAtlas (Keller et al. 2022). Additionally, miRBase incorporates secondary structure information predicted by RNAstructure (Mathews 2014) and compiles sequence verification methods from relevant publications. In addition to being a comprehensive repository for miRNA sequences, miRBase provides a downloadable gff format for genome annotations that can be used for small RNA-seq annotations (Kozomara and Griffiths-Jones 2014).

3.2.2 RNAcentral

RNAcentral, a comprehensive database developed by the European Bioinformatics Institute in 2013, serves as a resource not only for miRNAs but also for other non-coding RNA sequences. The 22nd version, which was released in early 2023, comprises a vast collection of over 30 million sequences from 250 different species. These sequences have been sourced from 51 databases, making it one of the most comprehensive collections of its kind (Sweeney et al. 2020).

RNAcentral is a comprehensive miRNA resource with over 400,000 pre-miRNA inputs and 130,000 miRNA sequences. In addition, it contains secondary structures, related publications, annotation details, genomic positions, and taxonomy information (Sweeney et al. 2021b).

In contrast to similar sequence databases, RNAcentral offers two different tools: (1) RNA 2D Templates (R2DT) (Sweeney et al. 2021a), which employ a specialized algorithm for the prediction of secondary structures of RNAs, and (2) nhmmer (Wheeler and Eddy 2013), a local alignment tool utilized for sequence-based searching. In addition, nhmmer is integrated into miRBase to search for miRNA sequences. These tools might be helpful, especially for researchers who do not have experience in bioinformatics analysis of miRNAs.

3.3 miRNA Detection Tools

While thousands of miRNAs have been identified in hundreds of organisms, a considerable number of potential miRNA sequences remain to be identified (Chen et al. 2019). Therefore, effective computational miRNA prediction and annotation tools (Table 3.2) are critical for the detection of novel miRNAs. In this section, we delve into two noteworthy tools: miRDeep2 (Friedländer et al. 2012), a well-known algorithm for miRNA identification, and miRNAFold (Tav et al. 2016), an easy-to-use server for the comprehensive characterization of miRNAs on a genome-scale.

Table 3.2 A selection of miRNA detection tools. SA stands for Stand-alone, WB: Web Based, and PK: Programming Language Package in Platform column

Name	Summary	Clade	Methodology	Published Year	Platform	Website	Reference
MiRscan	Homology-based miRNA prediction	Worms	Conservation Structure	2003	WB, PK	http://hollywood.mit.edu/mirscan/index.html	Lim et al. (2003)
Triplet-SVM	Machine learning approach for miRNA detection	Any	Sequence Machine learning	2005	SA	http://bioinfo.au.tsinghua.edu.cn/mirnasvm	Xue et al. (2005)
RNAmicro	Combined sequence and structure predictions for minimizing false positive	Any	Sequence Machine learning	2006	SA	Not available: http://www.tbi.univie.ac.at/~jana/software/RNAmicro.html	Hertel and Stadler (2006)
MiRFinder	Pairwise genome search for miRNA detection	Animal	Conservation Machine learning	2007	SA	https://www.bioinformatics.org/mirfinder/	Huang et al. (2007b)
miRPred	Machine learning approach combined with random forest and triplet-SVM	Any	Machine learning	2007	SA	Article supplementary file	Brameier and Wiuf (2007)
miRanalyzer	Improved machine learning models with training known miRNAs	31 species (including 6 plants)	Machine learning NGS application	2009	WB, SA	https://bioinfo5.ugr.es/miRanalyzer/miRanalyzer.php	Hackenberg et al. (2011)
miRDeep2	Structure and NGS based application	Animal	NGS application	2012	SA	https://github.com/rajewsky-lab/mirdeep2	Friedländer et al. (2012)
miRNAFold	Hairpin structure and sequence complementary based pre-miRNA detection algorithm	78 species (plant and animals)	Sequence Conservation	2012	WB	http://evryrna.ibisc.univ-evry.fr/miRNAFold	Tempel and Tahi (2012)
izMiR	Ensemble machine learning-based pre-miRNA detection	Any	Sequence Machine learning	2017	PK	https://data.mendeley.com/datasets/mgh5r9wny7/1	Saçar Demirci et al. (2017)

3.3.1 miRDeep2

miRDeep2 tool is a stand-alone software designed for miRNA detection in deep sequencing data (Friedländer et al. 2008). miRDeep2 provides a user-friendly interface, well-documented outcomes, efficient memory usage, and easy integration with other downstream analysis tools.

The core part of the algorithm is based on detecting known and novel miRNA predictions using the expression profiles of each miRNA. In contrast to some of the existing methods, such as MIReNA (Mathelier et al. 2011), miRanalyzer (Hackenberg et al. 2009), and miRTRAP (Hendrix et al. 2010), which demonstrated limited coverage across species, miRDeep2 can be applied to deep sequencing data from all animal species.

Furthermore, in terms of memory and time utilization, miRDeep2 seems efficient, according to the analysis of approximately 30 million RNAs that are completed in less than 5 h while requiring 3 GB of memory (Friedländer et al. 2012). The main focus of the mirDeep2 prediction function is to accurately detect miRNAs by considering the secondary structure and compatibility of the pre-miRNA positions. This accurate prediction is achieved by using a cutoff value to filter out unsuitable or unpromising precursors, thereby reducing the frequency of false positives and improving the overall accuracy of the software.

3.3.2 miRNAFold

miRNAFold is a computational method developed for the identification and characterization of miRNAs on a genome-wide scale (Tav et al. 2016). The primary goal of miRNAFold is to facilitate the identification of miRNA precursors on a large scale within genomes, enabling quick and highly accurate prediction (Tempel and Tahi 2012). The tool provides a user-friendly web server with an easy-to-use user interface (UI) as well as a standalone version.

Compared to similar applications, miRNAFold offers significant advantages, such as the ability to process sequences of unlimited size and the availability of a well-established web server (Tav et al. 2016). Furthermore, the stand-alone version of the application is capable of directly predicting miRNAs using next-generation sequencing data. It is also reported to be five times faster than mirinho, which is an algorithm designed for miRNA identification in sRNA-seq (short RNA sequences) (Higashi et al. 2015).

3.4 miRNA Structure Tools

MiRNAs participate in gene silencing by interacting with target molecules, and their ability depends on site accessibility, which is linked to the presence of secondary structures (O'Brien et al. 2018). The secondary structure of a miRNA contains hairpin stem loops, bulges, base-paired regions, and internal loops (Liu et al. 2016). These structural elements may enable or prevent miRNA-target interactions (Kertesz et al. 2007).

To visualize these secondary structures based on sequence information, various tools have been developed, including RNAfold (Lorenz et al. 2016), R2DT (Sweeney et al. 2021a), and RNAStructure (Mathews 2014).

3.4.1 The ViennaRNA Package

The ViennaRNA package contains a variety of tools designed for the prediction of RNA secondary structures (Gruber et al. 2008). Noteworthy tools within this package include RNAup, employed for predicting site accessibility in a target region, and RNAcofold, which is utilized for predicting the structure of single-stranded RNA and DNA molecules (Lorenz et al. 2011).

Among these tools, RNAfold is the primary secondary structure prediction tool. It calculates secondary structures based on the minimum free-energy principle. RNAfold software is available in both the standalone and online web server versions. Limitations exist in the web version of RNAfold, where each computation is restricted to 10,000 nucleotides, while the stand-alone version of the ViennaRNA package does not have such constraints.

3.5 miRNA Targeting Tools

Determining miRNA target interactions (MTIs) through in silico analysis enhances our understanding of miRNA-gene connections and facilitates the detection of post-transcriptionally regulated genes (Lee and Shin 2012). Moreover, target prediction tools can be used to identify the most suitable candidates for gene enrichment analysis, which significantly contributes to determining the essential miRNAs in various biological processes (Sayed and Abdellatif 2011; Table 3.3).

The development of new computational methods and increased knowledge of miRNAs have significantly improved the precision of miRNA-target prediction tools. When reviewing the existing algorithms in the literature, six important features have been identified for miRNA target interaction prediction, namely, seed match complementary of target sequences, evolutionary conservation of miRNA-target interaction, thermodynamic stability of interactions, site

Table 3.3 Selected miRNA target prediction tools

Name	Algorithm	Platform	Current version	Website	Reference
miRanda	SM, CH, ML, CM	WB, SA	v3.3a	http://cbio.mskcc.org/miRNA2003/miranda.html	Enright et al. (2003)
RNAhybrid	SM, CH	WB, SA	v2.1.2	https://bibiserv.cebitec.uni-bielefeld.de/rnahybrid	Rehmsmeier et al. (2004)
TargetScan	SM, EC, CM	WB, SA	v8.0	https://www.targetscan.org	Lewis et al. (2005)
psRNAtarget	SM, CM	WB	V2	https://www.zhaolab.org/psRNATarget/	Dai et al. (2018)
COMIR	SM, CH, ML, CM	SA, WB	–	www.benoslab.pitt.edu/comir/	Coronnello and Benos (2013)
GAMUT	SM, CH, ML, CM	SA	–	https://sourceforge.net/projects/cudamiranda/	Wang et al. (2014)
iDeep	ML, SM, CM	SA	–	http://www.csbio.sjtu.edu.cn/bioinf/iDeep/	Pan and Shen (2017)
MiRTDL	ML	WB, SA	–	http://nclab.hit.edu.cn/ccrm	Cheng et al. (2016)
deepTarget	ML	SA	–	http://data.snu.ac.kr/pub/deepTarget/	Lee et al. (2016)
Genmir++	EX, ML	SA	–	http://www.psi.toronto.edu/genmir/	Huang et al. (2007a)
RNA22	CM, CH	WB	v2.0	https://cm.jefferson.edu/rna22/	Miranda et al. (2006)
TargetFinder	CM	SA	–	https://github.com/carringtonlab/TargetFinder	Fahlgren and Carrington (2010)
miRWalk	IA, TM	WB	v3.0	http://mirwalk.umm.uni-heidelberg.de/	Sticht et al. (2018)
miRTarBase	MC	WB	v9.0	https://mirtarbase.cuhk.edu.cn/	Huang et al. (2022)
TarBase	MC, IA	WB	v8.0	http://www.microrna.gr/tarbase	Karagkouni et al. (2018)

Algorithm category: Seed matching (SM), complement matching (CM), compensatory hybridization (CH), evolutionary conservation (EC), machine learning (ML), integrated approach (IA), manually curated (MC), text mining (TM), expression correlation (EX). Platform category: Stand-alone (SA), Web-based (WB) accessibility of the target of miRNA, 3′UTR region, and local AU content (Peterson et al. 2014).

In addition to classical algorithms, machine learning and deep learning algorithms are frequently applied to investigate the interaction between miRNAs and their targets (Chen et al. 2019). These machine learning models can be trained with multiple parameters. It has been observed that models trained with multiple features such as iDeep (Pan and Shen 2017), MiRTDL (Cheng et al. 2016), deepTarget (Lee et al. 2016), Genmir++ (Huang et al. 2007a) showed higher precision than alternative algorithms. However, effective implementation of algorithms requires experience and/or knowledge in the concept of machine learning. Another issue is that most targeting tools are designed to work with a limited number of species (Table 3.3).

In this section, we present some algorithms that are both user-friendly and comprehensive. Among these, miRTarBase stands out as a source containing experimentally validated interactions (Huang et al. 2020). TargetScan is a widely recognized tool that forms the basis of many algorithms (Lewis et al. 2005). Lastly, we examined COMIR, developed as a combined prediction tool (Coronnello and Benos 2013) and the plant-focused psRNAtarget tool, which can predict targets based on two sequences (Dai et al. 2018).

3.5.1 *TargetScan*

TargetScan, the first miRNA targeting tool for vertebrates, was released in 2003 (Lewis et al. 2005). Developed as a target prediction method, it utilizes conserved regions and 3′UTR regions to identify interactions between miRNAs and their target molecules.

The web server of TargetScan, equipped with pre-calculated data, a search option, and functional analysis of miRNA target interactions (MTI), incorporates various features. The continuously updated database encompasses datasets from ten different species, including human, mice, and worm (Agarwal et al. 2015). These comprehensive features have positioned TargetScan as one of the most frequently used tools for miRNA-target prediction. Notably, the latest version of TargetScan (released in 2021) contains more than 700,00 predicted targets, underlining its significance in the dynamic field of miRNA research (McGeary et al. 2019).

3.5.2 *miRTarBase*

MiRTarBase is a miRNA target interaction (MTI) database containing experimentally validated miRNA target interactions (Hsu et al. 2011). The primary objective behind the establishment of miRTarBase is to facilitate the use of scientifically validated MTI data, thus enabling researchers to make significant advancements in their respective fields of study. The initial version of miRTarBase was created in 2011 by manually collecting interaction data from publications with experimental validations. The current curation of miRTarBase (Release 9.0 beta) has undergone continuous improvement, and a text mining and natural language processing (NLP) system have been developed to identify novel validated interactions. Furthermore, the system is classified based on MTIs, and it has developed a rating system using experimental verification methods such as next-generation sequencing (NGS) and chromatin immunoprecipitation sequencing (ChIP-seq).

3.5.3 psRNATarget

psRNATarget is a specialized prediction tool for identifying miRNA targets in plants (Dai et al. 2018), which utilizes a dynamic programming method with a modified Smith-Waterman algorithm for sequence alignment. Additionally, it includes the RNAup (Mückstein et al. 2006) algorithm to predict target site accessibility.

PsRNAtarget has been developed for the analysis of miRNA-target interactions in plants. However, its versatility extends beyond plants and can be used to analyze miRNA-target interactions in various organisms. The tool's salient feature is its ability to handle small RNAs and the upload of target candidate sequences for analysis. Consequently, this feature makes it a valuable tool for the scientific community at large (Srivastava et al. 2014).

The online server, designed to be user-friendly and regularly updated, incorporates a configurable scoring system aimed at enhancing prediction accuracy. This approach makes psRNATarget a valuable tool for researchers exploring miRNA-target interactions across various species.

3.5.4 ComiR

The combinatorial miRNA target prediction program (ComiR) is utilized to determine the mRNA being targeted by a set of miRNAs (Coronnello and Benos 2013). ComiR employs miRNA expression data and combines different miRNA target prediction tools, miRanda (Enright et al. 2003), PITA (Kertesz et al. 2007), TargetScan (Lewis et al. 2005), and mirSVR (Betel et al. 2010), to improve the accuracy of the target prediction. This powerful tool has been specifically designed to cater to the needs of researchers working with four different species: *Homo sapiens*, *Mus musculus*, *Drosophila melanogaster*, and *Caenorhabditis elegans*. Users can easily access this tool through the ComiR website. It is noteworthy that the integration of composite scores from the four algorithms is facilitated by support vector machine models (Coronnello and Benos 2013).

3.6 miRNA Network and Competing Endogenous RNA (ceRNA) Analysis

The initial investigations about miRNAs mostly focused on elucidating the relationships between miRNAs and their target mRNA molecules (Krol et al. 2010). miRNAs repress protein-coding genes in the posttranscriptional phase through cleavage or destabilization of the targeted mRNA molecules. Later, it was shown that miRNAs not only target mRNA but also bind to other RNA molecules, such as long noncoding RNAs and circular RNAs. All these RNA molecules compete with each other for miRNA binding (Pasquinelli 2012). This competition among RNA molecules is known as the competing endogenous RNA (ceRNA) hypothesis in the literature (Salmena et al. 2011).

Detailed studies conducted during developmental stages underscore the importance of identifying critical targets of miRNAs and analyzing ceRNA networks (Table 3.4). For instance, mutations in long noncoding RNAs may cause major developmental problems (Engreitz et al. 2016). Hence, using miRNA and analyzing potential networks hold promise in enhancing disease identification and expanding the range of therapeutic alternatives (Ala 2020).

3.6.1 StarBase

The Encyclopedia of RNA Interactomes (ENCORI/StarBase) is a database for miRNA-related biological interactions providing comprehensive analysis of miRNA-protein, miRNA-mRNA, miRNA-lncRNA, miRNA-circRNA, and ceRNA systems (Yang et al. 2011). These interactions were obtained using software and pipelines to analyze high-throughput sequencing data sets, including RNA-RNA interaction, CLIP-seq, and degradome sequencing.

StarBase offers a range of tools for integrated data, such as miRFunction and ceRNA functions (Li et al. 2014). These tools are designed to predict the function of noncoding RNAs (ncRNAs), and protein-coding genes based on the regulatory networks of ceRNAs. The database includes a wide-ranging collection of about 2.9 million ceRNAs, one million microRNA-messenger RNA (miRNA-mRNA) interactions, 3.5 million RNA-RNA interactions, and 1.5 million disease-related data that have been collected using the integrated tools. StarBase is known as the central functional database for miRNA networks because of its extensive collection of interactions and the availability of essential tools.

The processed interaction data in StarBase has been effectively combined with gene annotations, signaling pathways, cancer genomics, and transcriptomic data. This integration has increased our understanding of the functions and mechanisms of the regulatory networks mediated by RNAs and RNA-binding proteins in developmental biology (Kim et al. 2021).

Table 3.4 MiRNA network tools. WB is used for Web-based, and PK is used for Programming Language Package in Platform category

Name	Summary	Organism	Platform	Website	Reference
StarBase	ceRNA database	Human, mouse	WB	https://rnasysu.com/encori/	Yang et al. (2011)
HMDD	Human miRNA disease database	Human	WB	https://www.cuilab.cn/hmdd	Lu et al. (2008)
miRNet	microRNA network analysis	Animals	WB, PK	https://www.mirnet.ca/	Fan et al. (2016)
MIENTURNET	microRNA enrichment analysis	Animals	WB	http://userver.bio.uniroma1.it/apps/mienturnet/	Licursi et al. (2019)
miRPath	miRNA pathway analysis	Animals	WB	http://www.microrna.gr/miRPathv4	Tastsoglou et al. (2023)
miR2Disease	Human miRNA disease database	Human	WB	http://www.mir2disease.org/	Jiang et al. (2009)
dbDEMC	Differentially expressed miRNAs in cancer database	Human	WB	https://www.biosino.org/dbDEMC	Xu et al. (2022)
miTALOS	miRNA pathway analysis	Human, mouse	WB	http://mips.helmholtz-muenchen.de/mitalos	Preusse et al. (2016)
TissueAtlas	Tissue specificity miRNA database	Human, mouse	WB	https://ccb-web.cs.uni-saarland.de/tissueatlas2/	Keller et al. (2022)

3.6.2 *miRNet*

miRNet is a web-based tool designed to elucidate miRNA functions via network-based visual analytics, since its first release in 2016 (Fan et al. 2016). The primary focus of version 1.0 was on exploring interactions between miRNAs and target genes. Establishing upon this essential concept, miRNet version 2.0 contains additional data and analysis options, such as the inclusion of transcription factors (TFs) and single-nucleotide polymorphisms (SNPs) that impact the regulation of miRNAs (Chang and Xia 2023). This updated version (v2.0) integrates new datasets related to miRNA-binding sites and the associations between miRNAs and diseases. Moreover, significant improvements have been made, including the implementation of new functions for easier network formation and visual exploration, redesign of the web interface, and enhancement of the workflow efficiency. Collectively, these enhancements contribute to supporting high-performance and real-time data analysis (Chang et al. 2020). Furthermore, an R library of miRNet has been released for downstream analyses.

3.6.3 *MIENTURNET*

MIENTURNET (MicroRNA ENrichment TURned NETwork) is a web-based application designed to provide target enrichment and network-based analysis (Licursi et al. 2019). It utilizes two distinct target prediction databases, miRTarbase and TargetScan, to enhance target identification, network analysis, and functional enrichment analysis. Furthermore, functional enrichment analyses incorporate KEGG (Kanehisa and Goto 2000), REACTOME (Gillespie et al. 2022), WikiPathways (Martens et al. 2021), and Disease Ontology (Schriml et al. 2019). MIENTURNET performs these functions for six distinct organisms, (*Homo sapiens, Mus musculus, Rattus norvegicus,* Drosophila *melanogaster, Danio rerio, and Caenorhabditis elegans*), providing researchers with user-friendly, expedient, and visually comprehensive visualizations. The intuitive web interface of this tool has the potential to enable researchers without computational or informatics skills to efficiently explore miRNA activity in many cellular processes of interest.

3.6.4 *Human MicroRNA Disease Database (HMDD)*

The Human MicroRNA Disease Database (HMDD) serves as an integrative database for human miRNAs and their associated diseases, containing supporting experimentally validated evidence (Lu et al. 2008). The initial version of this database (HMDD v1.0, 2007) was pioneering in the literature and has four objects for annotation: the name of miRNA, the name of the disease, a PubMed ID for the reference source, and a paper that shows the relationship between miRNA and disease. Recent updates to HMDD include enhanced integrated interaction tools and improved information on the associations between human miRNAs and diseases (Huang et al. 2019).

The current version of the database (HMDD v4.0) contains manually curated annotations of 53,530 data points between miRNAs and diseases, consisting of 1817 human miRNA genes, 79 virus-derived miRNAs, and 2360 diseases from 37,090 reports (Cui et al. 2023). Noteworthy additions in the updates include exosomal miRNAs and virus-encoded miRNAs, which are two additional categories of miRNA-disease linkages. The latest version introduces disease network analysis, utilizing disease-associated miRNAs. HMDD is consistently and promptly updated with new entries of miRNA-disease information.

3.7 Conclusions

The exploration of miRNAs and their regulatory roles in gene expression has been greatly facilitated by the development of bioinformatics tools and databases. In this chapter, we have reviewed a plethora of resources designed to aid researchers in various aspects of miRNA analysis, from sequence repositories to target prediction tools and network analysis platforms. In navigating the vast landscape of miRNA research, bioinformatics tools serve as indispensable companions, aiding researchers in unraveling the intricate web of miRNA functions and interactions. Our exploration of these tools has shed light on both their capabilities and limitations, prompting the formulation of a practical workflow tailored to maximize their utility while mitigating potential pitfalls (Fig. 3.1). The most important considerations in this chapter include the following:

3.7.1 Sequence Annotation and Identification

- Commence by annotating sequence files with the organism of interest, leveraging repositories like miRBase and RNAcentral, which provide a wealth of miRNA sequences and annotations essential for initial exploration. These resources offer comprehensive collections of miRNA sequences, secondary structure information, and associated literature references, facilitating thorough sequence annotation.

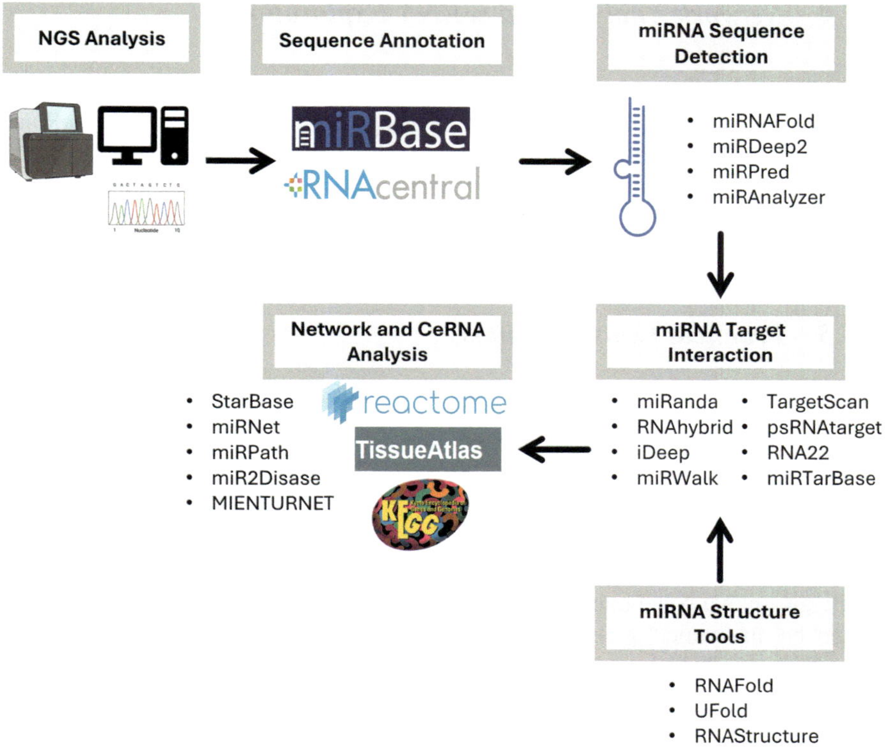

Fig. 3.1 Practical workflow for microRNA analysis. It presents a set of tools that can be used to analyze RNA-Seq data in a systematic manner

3.7.2 MiRNA Detection

- Proceed with miRNA detection using sophisticated tools such as miRDeep2 and miRNAFold, which employ algorithms to identify novel miRNAs from high-throughput sequencing data. By predicting miRNA precursors and secondary structures, these tools facilitate accurate annotation, enabling researchers to uncover previously unrecognized miRNA candidates.

3.7.3 Target Prediction and Validation

- Advance to miRNA targeting analysis, utilizing tools like TargetScan and miRTarBase to predict and validate miRNA-target interactions. These resources elucidate regulatory networks and offer insights into the biological significance of miRNA-mediated gene regulation, essential for understanding the functional implications of miRNA activity.

3.7.4 Network Analysis and ceRNA Exploration

- Delve deeper into miRNA regulatory networks and explore the competing endogenous RNA (ceRNA) hypothesis using platforms like StarBase, miRNet, and MIENTURNET. These tools unravel the complexities of miRNA interactions, providing valuable insights into gene expression regulation and facilitating the investigation of regulatory crosstalk within cellular systems.

3.7.5 Practical Considerations

- Remain mindful of species-specific differences in miRNA biogenesis. As such, the majority of available tools and databases are tailored and validated for specific organisms. Before using a tool, it is crucial to verify its suitability for the intended task and organism of interest. Researchers should check if the tool has been tested with samples relevant to their study to ensure accurate results.
- Be aware that some published tools designed for miRNA detection may no longer be maintained or regularly updated. To mitigate this issue, we recommend selecting tools and databases that receive regular updates, ensuring access to the most recent information. By utilizing continuously maintained resources, researchers can avoid potential discrepancies arising from outdated data and algorithms.
- It is important to keep in mind that many machine learning-based miRNA detection algorithms rely on the chemical or biological structures of molecules for identification (Gomes et al. 2013). Unlike known miRNA sequences, there is a lack of repositories containing sequences suitable for use as negative examples during training. However, a recent study has introduced a novel framework for generating negative data, offering a promising solution for training ab initio detection models (Orhan et al. 2023). Researchers can consider incorporating such frameworks into their analysis to enhance the robustness and accuracy of miRNA detection algorithms.

In conclusion, this chapter underscores the significance of bioinformatics tools in advancing our understanding of miRNA biology. By leveraging these resources, researchers can uncover novel insights into miRNA function, regulatory networks, and their implications in health and disease. However, it is essential to remain cognizant of the limitations and considerations associated with each tool to ensure accurate and reliable analysis of miRNA data.

Acknowledgments This work has been supported by the Abdullah Gul University Support Foundation (AGUV).

Compliance with Ethical Standards

Conflict of Interest Authors declare that they have no conflict of interest.

Ethical Approval This chapter contains no studies with human/animal participants performed by authors.

References

Agarwal V, Bell GW, Nam JW, Bartel DP (2015) Predicting effective microRNA target sites in mammalian mRNAs. elife 4. https://doi.org/10.7554/eLife.05005

Ala U (2020) Competing endogenous RNAs, non-coding RNAs and diseases: an intertwined story. Cells 9

Bartel DP (2004) MicroRNAs: genomics, biogenesis, mechanism, and function. Cell 116

Bateman A, Agrawal S, Birney E et al (2011) RNAcentral: a vision for an international database of RNA sequences. RNA 17. https://doi.org/10.1261/rna.2750811

Betel D, Koppal A, Agius P et al (2010) Comprehensive modeling of microRNA targets predicts functional non-conserved and non-canonical sites. Genome Biol 11. https://doi.org/10.1186/gb-2010-11-8-r90

Brameier M, Wiuf C (2007) Ab initio identification of human microRNAs based on structure motifs. BMC Bioinformatics 8. https://doi.org/10.1186/1471-2105-8-478

Chang L, Xia J (2023) MicroRNA Regulatory Network Analysis Using miRNet 2.0. In: Methods in Molecular Biology

Chang L, Zhou G, Soufan O, Xia J (2020) miRNet 2.0: network-based visual analytics for miRNA functional analysis and systems biology. Nucleic Acids Res 48. https://doi.org/10.1093/nar/gkaa467

Chen L, Heikkinen L, Wang C et al (2019) Trends in the development of miRNA bioinformatics tools. Brief Bioinform 20

Cheng S, Guo M, Wang C et al (2016) MiRTDL: a deep learning approach for miRNA target prediction. IEEE/ACM Trans Comput Biol Bioinform 13. https://doi.org/10.1109/TCBB.2015.2510002

Coronnello C, Benos PV (2013) ComiR: combinatorial microRNA target prediction tool. Nucleic Acids Res 41. https://doi.org/10.1093/nar/gkt379

Cui C, Zhong B, Fan R, Cui Q (2023) HMDD v4.0: a database for experimentally supported human microRNA-disease associations. Nucleic Acids Res:gkad717. https://doi.org/10.1093/nar/gkad717

Dai X, Zhuang Z, Zhao PX (2018) psRNATarget: a plant small RNA target analysis server (2017 release). Nucleic Acids Res 46:W49–W54. https://doi.org/10.1093/NAR/GKY316

Engreitz JM, Ollikainen N, Guttman M (2016) Long non-coding RNAs: spatial amplifiers that control nuclear structure and gene expression. Nat Rev Mol Cell Biol 17

Enright AJ, John B, Gaul U et al (2003) MicroRNA targets in drosophila. Genome Biol 5. https://doi.org/10.1186/gb-2003-5-1-r1

Fahlgren N, Carrington JC (2010) miRNA target prediction in plants. Methods Mol Biol 592. https://doi.org/10.1007/978-1-60327-005-2_4

Fan Y, Siklenka K, Arora SK et al (2016) miRNet - dissecting miRNA-target interactions and functional associations through network-based visual analysis. Nucleic Acids Res 44. https://doi.org/10.1093/nar/gkw288

Friedländer MR, Chen W, Adamidi C et al (2008) Discovering microRNAs from deep sequencing data using miRDeep. Nat Biotechnol 26. https://doi.org/10.1038/nbt1394

Friedländer MR, MacKowiak SD, Li N et al (2012) MiRDeep2 accurately identifies known and hundreds of novel microRNA genes in seven animal clades. Nucleic Acids Res 40. https://doi.org/10.1093/nar/gkr688
Gillespie M, Jassal B, Stephan R et al (2022) The reactome pathway knowledgebase 2022. Nucleic Acids Res 50. https://doi.org/10.1093/nar/gkab1028
Gomes CPC, Cho J, Hood L et al (2013) A review of computational tools in microRNA discovery. Front Genet. https://doi.org/10.3389/fgene.2013.00081
Griffiths-Jones S (2004) The microRNA registry. Nucleic Acids Res 32. https://doi.org/10.1093/nar/gkh023
Griffiths-Jones S, Grocock RJ, van Dongen S et al (2006) miRBase: microRNA sequences, targets and gene nomenclature. Nucleic Acids Res 34. https://doi.org/10.1093/nar/gkj112
Gruber AR, Lorenz R, Bernhart SH et al (2008) The Vienna RNA websuite. Nucleic Acids Res 36. https://doi.org/10.1093/nar/gkn188
Hackenberg M, Sturm M, Langenberger D et al (2009) miRanalyzer: a microRNA detection and analysis tool for next-generation sequencing experiments. Nucleic Acids Res 37. https://doi.org/10.1093/nar/gkp347
Hackenberg M, Rodríguez-Ezpeleta N, Aransay AM (2011) MiRanalyzer: an update on the detection and analysis of microRNAs in high-throughput sequencing experiments. Nucleic Acids Res 39. https://doi.org/10.1093/nar/gkr247
Hendrix D, Levine M, Shi W (2010) MiRTRAP, a computational method for the systematic identification of miRNAs from high throughput sequencing data. Genome Biol 11. https://doi.org/10.1186/gb-2010-11-4-r39
Hertel J, Stadler PF (2006) Hairpins in a Haystack: Recognizing microRNA precursors in comparative genomics data. In: Bioinformatics
Higashi S, Fournier C, Gautier C et al (2015) Mirinho: an efficient and general plant and animal pre-miRNA predictor for genomic and deep sequencing data. BMC Bioinformatics 16. https://doi.org/10.1186/s12859-015-0594-0
Hsu SD, Lin FM, Wu WY et al (2011) MiRTarBase: a database curates experimentally validated microRNA-target interactions. Nucleic Acids Res 39. https://doi.org/10.1093/nar/gkq1107
Huang JC, Babak T, Corson TW et al (2007a) Using expression profiling data to identify human microRNA targets. Nat Methods 4. https://doi.org/10.1038/nmeth1130
Huang TH, Fan B, Rothschild MF et al (2007b) MiRFinder: an improved approach and software implementation for genome-wide fast microRNA precursor scans. BMC Bioinformatics 8. https://doi.org/10.1186/1471-2105-8-341
Huang Z, Shi J, Gao Y et al (2019) HMDD v3.0: a database for experimentally supported human microRNA-disease associations. Nucleic Acids Res 47:10.1093/nar/gky1010
Huang HY, Lin YCD, Li J et al (2020) miRTarBase 2020: updates to the experimentally validated microRNA–target interaction database. Nucleic Acids Res 48:D148–D154. https://doi.org/10.1093/NAR/GKZ896
Huang HY, Lin YCD, Cui S et al (2022) MiRTarBase update 2022: an informative resource for experimentally validated miRNA-target interactions. Nucleic Acids Res 50. https://doi.org/10.1093/nar/gkab1079
Jiang Q, Wang Y, Hao Y et al (2009) miR2Disease: a manually curated database for microRNA deregulation in human disease. Nucleic Acids Res 37. https://doi.org/10.1093/nar/gkn714
Kalvari I, Nawrocki EP, Argasinska J et al (2018) Non-coding RNA analysis using the Rfam database. Curr Protoc Bioinformatics 62. https://doi.org/10.1002/cpbi.51
Kalvari I, Nawrocki EP, Ontiveros-Palacios N et al (2021) Rfam 14: expanded coverage of metagenomic, viral and microRNA families. Nucleic Acids Res 49. https://doi.org/10.1093/nar/gkaa1047
Kanehisa M, Goto S (2000) KEGG: Kyoto encyclopedia of genes and genomes. Nucleic Acids Res 28
Karagkouni D, Paraskevopoulou MD, Chatzopoulos S et al (2018) DIANA-TarBase v8: a decade-long collection of experimentally supported miRNA-gene interactions. Nucleic Acids Res 46. https://doi.org/10.1093/nar/gkx1141

Keller A, Gröger L, Tschernig T et al (2022) MiRNATissueAtlas2: an update to the human miRNA tissue atlas. Nucleic Acids Res 50. https://doi.org/10.1093/nar/gkab808
Kertesz M, Iovino N, Unnerstall U et al (2007) The role of site accessibility in microRNA target recognition. Nat Genet 39. https://doi.org/10.1038/ng2135
Kim S, Kim S, Chang HR et al (2021) The regulatory impact of RNA-binding proteins on microRNA targeting. Nat Commun 12. https://doi.org/10.1038/s41467-021-25078-5
Kozomara A, Griffiths-Jones S (2014) MiRBase: annotating high confidence microRNAs using deep sequencing data. Nucleic Acids Res 42. https://doi.org/10.1093/nar/gkt1181
Kozomara A, Birgaoanu M, Griffiths-Jones S (2019) miRBase: from microRNA sequences to function. Nucleic Acids Res 47:D155–D162. https://doi.org/10.1093/NAR/GKY1141
Krol J, Loedige I, Filipowicz W (2010) The widespread regulation of microRNA biogenesis, function and decay. Nat Rev Genet 11
Lee D, Shin C (2012) MicroRNA-target interactions: new insights from genome-wide approaches. Ann N Y Acad Sci 1271. https://doi.org/10.1111/j.1749-6632.2012.06745.x
Lee B, Park S, Baek J, Yoon S (2016) DeepTarget: End-to-end learning framework for MicroRNA target prediction using deep recurrent neural networks. In: ACM-BCB 2016 - 7th ACM Conference on Bioinformatics, Computational Biology, and Health Informatics
Lewis BP, Burge CB, Bartel DP (2005) Conserved seed pairing, often flanked by adenosines, indicates that thousands of human genes are microRNA targets. Cell 120
Li JH, Liu S, Zhou H et al (2014) StarBase v2.0: decoding miRNA-ceRNA, miRNA-ncRNA and protein-RNA interaction networks from large-scale CLIP-Seq data. Nucleic Acids Res 42:10.1093/nar/gkt1248
Licursi V, Conte F, Fiscon G, Paci P (2019) MIENTURNET: an interactive web tool for microRNA-target enrichment and network-based analysis. BMC Bioinformatics 20. https://doi.org/10.1186/s12859-019-3105-x
Lim LP, Lau NC, Weinstein EG et al (2003) The microRNAs of Caenorhabditis elegans. Genes Dev 17. https://doi.org/10.1101/gad.1074403
Liu B, Childs-Disney JL, Znosko BM et al (2016) Analysis of secondary structural elements in human microRNA hairpin precursors. BMC Bioinformatics 17. https://doi.org/10.1186/s12859-016-0960-6
Lorenz R, Bernhart SH, Höner zu Siederdissen C et al (2011) ViennaRNA Package 2.0. Algorithms Mol Biol 6:1–14. https://doi.org/10.1186/1748-7188-6-26/TABLES/2
Lorenz R, Hofacker IL, Stadler PF (2016) RNA folding with hard and soft constraints. Algorithms Mol Biol 11. https://doi.org/10.1186/S13015-016-0070-Z
Lu M, Zhang Q, Deng M et al (2008) An analysis of human microRNA and disease associations. PLoS One 3. https://doi.org/10.1371/journal.pone.0003420
Martens M, Ammar A, Riutta A et al (2021) WikiPathways: connecting communities. Nucleic Acids Res 49. https://doi.org/10.1093/nar/gkaa1024
Mathelier A, Carbone A, Hofacker I (2011) MIReNA: Finding microRNAs with high accuracy and no learning at genome scale and from deep sequencing data. In: Bioinformatics
Mathews DH (2014) RNA secondary structure analysis using RNAstructure. Curr Protoc Bioinformatics. https://doi.org/10.1002/0471250953.bi1206s46
McGeary SE, Lin KS, Shi CY et al (2019) The biochemical basis of microRNA targeting efficacy. Science 1979:366. https://doi.org/10.1126/science.aav1741
Miranda KC, Huynh T, Tay Y et al (2006) A pattern-based method for the identification of MicroRNA binding sites and their corresponding Heteroduplexes. Cell 126. https://doi.org/10.1016/j.cell.2006.07.031
Mückstein U, Tafer H, Hackermüller J et al (2006) Thermodynamics of RNA–RNA binding. Bioinformatics 22:1177–1182. https://doi.org/10.1093/BIOINFORMATICS/BTL024
O'Brien J, Hayder H, Zayed Y, Peng C (2018) Overview of microRNA biogenesis, mechanisms of actions, and circulation. Front Endocrinol 9
Orhan ME, Demirci YM, Saçar Demirci MD (2023) NeRNA: a negative data generation framework for machine learning applications of noncoding RNAs. Comput Biol Med 159:106861. https://doi.org/10.1016/j.compbiomed.2023.106861

Pan X, Shen HB (2017) RNA-protein binding motifs mining with a new hybrid deep learning based cross-domain knowledge integration approach. BMC Bioinformatics 18. https://doi.org/10.1186/s12859-017-1561-8

Pasquinelli AE (2012) MicroRNAs and their targets: recognition, regulation and an emerging reciprocal relationship. Nat Rev Genet 13

Peterson SM, Thompson JA, Ufkin ML et al (2014) Common features of microRNA target prediction tools. Front Genet 5. https://doi.org/10.3389/FGENE.2014.00023

Preusse M, Theis FJ, Mueller NS (2016) miTALOS v2: analyzing tissue specific microRNA function. PLoS One 11. https://doi.org/10.1371/journal.pone.0151771

Rehmsmeier M, Steffen P, Höchsmann M, Giegerich R (2004) Fast and effective prediction of microRNA/target duplexes. RNA 10. https://doi.org/10.1261/rna.5248604

Saçar Demirci MD, Baumbach J, Allmer J (2017) On the performance of pre-microRNA detection algorithms. Nat Commun 8:330. https://doi.org/10.1038/s41467-017-00403-z

Salmena L, Poliseno L, Tay Y et al (2011) A ceRNA hypothesis: the Rosetta stone of a hidden RNA language? Cell 146:353. https://doi.org/10.1016/J.CELL.2011.07.014

Sayed D, Abdellatif M (2011) Micrornas in development and disease. Physiol Rev 91. https://doi.org/10.1152/physrev.00006.2010

Schriml LM, Mitraka E, Munro J et al (2019) Human disease ontology 2018 update: classification, content and workflow expansion. Nucleic Acids Res 47. https://doi.org/10.1093/nar/gky1032

Srivastava PK, Moturu TR, Pandey P et al (2014) A comparison of performance of plant miRNA target prediction tools and the characterization of features for genome-wide target prediction. BMC Genomics 15. https://doi.org/10.1186/1471-2164-15-348

Sticht C, De La Torre C, Parveen A, Gretz N (2018) Mirwalk: an online resource for prediction of microrna binding sites. PLoS One 13. https://doi.org/10.1371/journal.pone.0206239

Sweeney BA, Tagmazian AA, Ribas CE et al (2020) Exploring non-coding RNAs in RNAcentral. Curr Protoc Bioinformatics 71. https://doi.org/10.1002/cpbi.104

Sweeney BA, Hoksza D, Nawrocki EP et al (2021a) R2DT is a framework for predicting and visualising RNA secondary structure using templates. Nat Commun 12. https://doi.org/10.1038/s41467-021-23555-5

Sweeney BA, Petrov AI, Ribas CE et al (2021b) RNAcentral 2021: secondary structure integration, improved sequence search and new member databases. Nucleic Acids Res 49:D212–D220. https://doi.org/10.1093/NAR/GKAA921

Tastsoglou S, Skoufos G, Miliotis M et al (2023) DIANA-miRPath v4.0: expanding target-based miRNA functional analysis in cell-type and tissue contexts. Nucleic Acids Res. https://doi.org/10.1093/nar/gkad431

Tav C, Tempel S, Poligny L, Tahi F (2016) miRNAFold: a web server for fast miRNA precursor prediction in genomes. Nucleic Acids Res 44. https://doi.org/10.1093/nar/gkw459

Tempel S, Tahi F (2012) A fast ab-initio method for predicting miRNA precursors in genomes. Nucleic Acids Res 40. https://doi.org/10.1093/nar/gks146

Wang S, Kim J, Jiang X et al (2014) GAMUT: GPU accelerated microRNA analysis to uncover target genes through CUDA-miRanda. BMC Med Genet 7. https://doi.org/10.1186/1755-8794-7-S1-S9

Wheeler TJ, Eddy SR (2013) Nhmmer: DNA homology search with profile HMMs. Bioinformatics 29. https://doi.org/10.1093/bioinformatics/btt403

Xu F, Wang Y, Ling Y et al (2022) dbDEMC 3.0: functional exploration of differentially expressed miRNAs in cancers of human and model organisms. Genomics Proteomics Bioinformatics 20:10.1016/j.gpb.2022.04.006

Xue C, Li F, He T et al (2005) Classification of real and pseudo microRNA precursors using local structure-sequence features and support vector machine. BMC Bioinformatics 6. https://doi.org/10.1186/1471-2105-6-310

Yang JH, Li JH, Shao P et al (2011) StarBase: a database for exploring microRNA-mRNA interaction maps from Argonaute CLIP-Seq and Degradome-Seq data. Nucleic Acids Res 39. https://doi.org/10.1093/nar/gkq1056

Chapter 4
MicroRNAs As Promising Therapeutic Targets

Suryaji Patil, Ye Tian, Zhihao Chen, Wei Bai, Xin Zhao, and Airong Qian

Abstract MicroRNAs (miRNA) represent a class of noncoding RNAs that regulate gene expression by binding and inhibiting mRNA expression. Various human disorders, such as cancer, cardiovascular diseases, bone diseases, autoimmune diseases, and inflammatory diseases, are influenced by numerous deregulated miRNAs, which inhibit genes necessary for normal phenotypes or promote the expression of those that are normally suppressed. This influence, in turn, determines the fate of the cells and disease progression. Many are either upregulated or downregulated in the disease state, and their levels in serum can be used to diagnose the state or progression of the disorder, allowing them to be targeted either by inhibitors or mimics, respectively. In some situations, the same miRNA is expressed in various diseased tissues and even displays targeting of the same mRNA or more than one mRNA. The advancements in chemical modifications and targeted delivery systems have paved the way for miRNAs to be used in clinical trials to establish their therapeutic potential. In this chapter, we have summarized various strategies that can be employed to target miRNAs in diverse diseases and the current and completed clinical trials involving miRNAs.

Keywords MiRNAs · Human disorders · Therapeutic targets · Delivery systems · Clinical trials

---

S. Patil · Y. Tian · Z. Chen · W. Bai · X. Zhao · A. Qian (✉)
Lab for Bone Metabolism, Xi'an Key Laboratory of Special Medicine and Health Engineering, Key Lab for Space Biosciences and Biotechnology, Research Center for Special Medicine and Health Systems Engineering, NPU-UAB Joint Laboratory for Bone Metabolism, School of Life Sciences, Northwestern Polytechnical University, Xi'an, Shaanxi, China
e-mail: suryaji@nwpu.edu.cn; tianye@nwpu.edu.cn; chenzhihao@nwpu.edu.cn; qianair@nwpu.edu.cn

L. M. Vaschetto (ed.), *miRNAs, Human Health and Diseases*, Epigenetics and Human Health 13, https://doi.org/10.1007/978-3-031-64788-8_4

List of Abbreviations

ADCY5	Adenylate Cyclase 5
ADPKD	Autosomal Dominant Polycystic Kidney Disease
AGO	Argonaut
ALP	Alkaline Phosphatase
AMD	Age-Related Macular Degeneration
cAMP	Cyclic Adenosine Monophosphate
ANKRD46	Ankyrin Repeat Domain 46
AR	Allergic Rhinitis
ASO	Antisense Oligonucleotide
ATP	Adenosine Triphosphate
BC	Breast Cancer
BMD	Bone Mineral Density
BMSC	Bone Marrow Mesenchymal Cells
BNA-2′	4’-Bridged Nucleic Acid
CADM1	Cell Adhesion Molecule 1
CAP	Chondrocyte-Affinity Peptide
CF	Cardiac Fibroblasts
eCIRP	Extracellular Cold-Inducible RNA Binding Protein
CNV	Choroidal Neovascularization
CXCL12	C-X-C Motif Chemokine Ligand 12
CYP	Cytochrome P450
DKK1	Dickkopf-1
DMM	Destabilization of the Medial Meniscus
DNMT1	DNA Methyltransferase I
EBV	Epstein-Barr Virus
EGFR	Epidermal Growth Factor Receptor
EZH2	Enhancer of Zeste Homolog 2
FOXA1	Forkhead Box A1
FOXO1	Forkhead Box O1
GC	Gastric Cancer
GRIA1	Glutamate AMPA Receptor Subunit 1
HBV	Hepatitis B Virus
HCC	Hepatocellular Carcinoma
HCV	Hepatitis C Virus
HDAC4	Histone Deacetylase 4
HF	Heart Failure
HSUR	Herpesvirus Saimiri U RNA
HVS	Herpesvirus Saimiri
IGFBP-3	Insulin Growth Factor Binding Protein-3
IL	Interleukin
IL13Rα1	Il-13 Receptor A1 Chain
ITGA5	Integrin Subunit Alpha 5

JAK	Janus Kinase
KCNQ1OT1	Potassium Voltage-Gated Channel Subfamily Q Member 1 Overlapping Transcript 1
LEP	Leptin
LNA	Locked Nucleic Acid
MACF1	Microtubule Actin Cross-Linking Factor 1
MAP2K4	Mitogen-Activated Protein Kinase Kinase 4
MAPK	Mitogen-Activated Protein Kinase
MEF2C	Myocyte Enhancer Factor 2c
MI	Myocardial Infarction
MIAT	Myocardial Infarction-Associated Transcript
MOE-2	O-Methoxyethyl
MRC2	Mannose Receptor C Type 2
MRE	MiRNA Recognition Elements
MSC	Mesenchymal Stem Cells
iNOS	Inducible Nitric Oxide Synthase
NPC	Nasopharyngeal Carcinoma
OA	Osteoarthritis
OCN	Osteocalcin
OP	Osteoporosis
OVA	Ovalbumin
OVX	Ovariectomy
PAA	Polyacrylic Acid
PAMAM	Polyamidoamine
PC1	Polycystin-1
PC2	Polycystin-2
PEI	Polyethylenimine
PGF	Placental Growth Factor
PLGA	Poly (Lactic-Co-Glycolic) Acid
PMOP	Postmenopausal Osteoporosis
PPARα	Peroxisome Proliferator-Activated Receptor-α
PS	Phosphorothioate
PTEN	Phosphatase and Tensin Homolog
PTENP1	Phosphatase and Tensin Homolog Pseudogene 1
PVAm	Polyvinylamine
RANKL	Receptor Activator Nuclear Factor-K Ligand
RISC	RNA-Induced Silencing Complex
RUNX2	Runt-Related Transcription Factor 2
RXR	Retinoid X Receptor
SCI	Spinal Cord Injury
SLE	Systemic Lupus Erythematosus
α-SMA	α-Small Muscle Actin
SOCS1	Suppressor Of Cytokine Signaling 1
STAT	Signal Transducers and Activators Of Transcription
TGF-β	Transforming Growth Factor β

TGFBR1	Transforming Growth Factor β Receptor 1
TNF-α	Tumor Necrosis Factor-A
mTOR	Mammalian Target of Rapamycin
TPM1	Tumor Suppressor Tropomyosin 1
TSLC1	Tumor Suppressor in Lung Cancer 1
TUG1	Taurine Upregulated 1
UTR	Untranslated Region
VCAM1	Vascular Cell Adhesion Molecule 1
VEGFA	Vascular Endothelial Growth Factor A
XIST	X-Inactive Specific Transcript

4.1 Introduction

MicroRNAs constitute a class of small, endogenously produced noncoding RNAs that are highly conserved throughout the species. These exhibit a length of about 18–22 nucleotides and regulate posttranscriptional gene expression by destabilizing mRNA through hybridization to their 3′ untranslated region (UTR) (Bushati and Cohen 2007). Though miRNAs share structural similarities with small interfering RNAs (siRNAs), the fact that they are produced endogenously and bind to partially complementary sites in the target mRNA's 3′ UTR sets them apart (Pauley and Chan 2008). The important characteristic of a miRNA is its capability to regulate multiple mRNAs and thus multiple gene targets, but at a concerted level, they can regulate as much as 30% of the genes in a cell (Qin and Zhang 2011; Dissanayake and Inoue 2016). Therefore, it is not surprising when microRNAs regulate the gene expression of thousands of genes in various organisms, not only in normal physiological scenarios but also in disease states. Because of their ability to influence various biological processes, miRNA expression profiling has gained popularity in identifying them as biomarkers for disease (Pritchard et al. 2012). It begins with miRNA isolation, profiling, detection of targets, and then modulation of miRNA levels in vitro and in vivo, which finally accumulates in identifying miRNAs as targets (Lu and Rothenberg 2018).

The transcription of the majority of the miRNA is carried out by RNA polymerase II (RNA pol II), either as an individual transcription unit or as part of a gene. Like the RNA polymerase II products, these miRNAs also possess 5′ methylated caps and 3′ polyadenylated tails (Finnegan and Pasquinelli 2013). Initially, it was assumed that intergenic regions code for the miRNAs, but extensive studies have shown that intronic regions of coding and noncoding transcription units are responsible for miRNA generation, which is concurrently transcribed with other transcripts (Pauley and Chan 2008). When miRNA is intronic, precursor miRNA is released before mRNA splicing can be finished (Finnegan and Pasquinelli 2013). The process of miRNA biogenesis is evolutionarily conserved and entails sequential cleavages mediated by two RNase III enzymes, Drosha and Dicer. It starts in the nucleus, where miRNAs are transcribed by RNA polymerase II as long primary transcripts (pri-miRNAs) containing one or two hairpin structures that contain the sequence of mature miRNA. Pri-miRNAs then undergo maturation through sequential processing (Lee et al. 2003). First, pri-miRNAs are processed in the nucleus into an approximately 70-nucleotide-long stem-loop structure called the precursor-miRNA (pre-miRNA) by a member of the RNase III superfamily nucleases, Drosha. After cleavage by Drosha, pre-miRNA is exported into the cytoplasm through Exportin-5, where it is processed by another RNase III family member, Dicer, to produce the mature, double-stranded, 22 nucleotide-long RNA in the cytoplasm. This duplex strand is unwounded by helicase into a mature miRNA (Yi et al. 2003; Davis-Dusenbery and Hata 2010; Ho et al. 2022). This duplex, together with a member of the Argonaut (AGO) protein family, assembles a complex known as the RNA-induced silencing complex (RISC), which contains only one strand of the duplex miRNA, which guides miRNA to silence target mRNA (Iwakawa and Tomari 2022; Fig. 4.1).

Since the discovery of the crucial role of miRNAs in the regulation of gene expression and their ability to be detected in the serum in various diseases, miRNAs have become viable candidates as biomarkers and prospective targets for the development of novel therapeutic drugs such as miRNA mimics and miRNA inhibitors (Lu and Rothenberg 2018).

Therefore, in this chapter, we have summarized various strategies that can be employed to regulate miRNA function and to present miRNAs as novel targets to either suppress or upregulate their expression to alleviate disease progression.

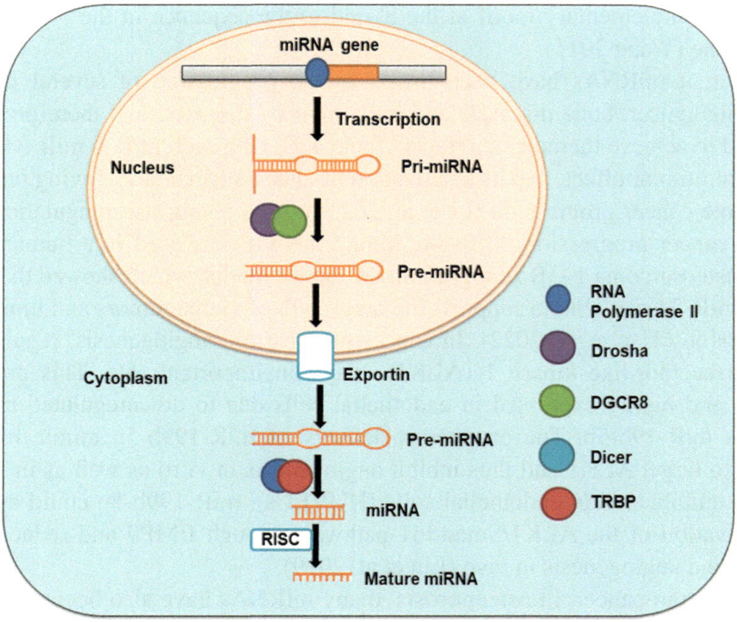

Fig. 4.1 Schematic representation of miRNA synthesis. (Adapted from (Gao et al. 2020) originally published under a Creative Commons Attribution (CC BY) license (http://creativecommons.org/licenses/by/4.0/)

4.2 Strategies to Regulate MicroRNA

MiRNome profiling has become a prevalent tool for various aspects of biological research, including the diagnosis and prognosis of various diseases such as cancer, providing an understanding of the role of miRNAs in gene regulation (White 2011; Lee and Dutta 2009). Despite the progress in identifying target expression-specific miRNAs, their small size and lack of suitable miRNA delivery systems remain limiting factors for their broader application as therapeutic agents. Nevertheless, these obstacles can be resolved by either modifying miRNAs and/or polymers to synthesize novel delivery systems (Patil et al. 2019). As for their targeting to achieve therapeutic targeting, two main approaches can be employed. The first is to reestablish the level and function of miRNAs that have been lost, and the second is to inhibit the levels and, thus, functions of aberrantly overexpressed miRNAs. For the former approach, miRNA mimics can be employed, while for the latter, a variety of interfering strategies, such as miRNA inhibitors, can be utilized (Ling 2016).

4.2.1 MiRNA Upregulation As a miRNA Regulation Strategy

This approach is categorized as a miRNA-gain-of-function strategy and is primarily used to restore the function of mutated or downregulated miRNAs. This strategy involves the generation of double-stranded miRNA-like RNA fragments having a partially complementary motif at the 5′ end to the sequence in the 3′UTR of the target gene (Wang 2011).

Different miRNAs have been linked to the progression of several diseases, including cancer, bone diseases, and inflammatory diseases, and therefore can be targeted to achieve therapeutic effects (Table 4.1). One such miR is miR-34, which has an antitumor effect, and its inactivation has been attributed to driving just about the whole cancer progression (Li et al. 2021c). As a result, its upregulation could inhibit cancer progression. MiR-34a mimic therapy was used in a human orthotropic osteosarcoma 143B xenograft tumor mouse model, which showed that delivery of miR-34 was able to suppress the outgrowth of viable tumors and limit tumor progression (Tian et al. 2022). In cancer progression, angiogenesis, regulated by activin receptor-like kinase 1 (ALK1), plays an important role. This protein is mainly and highly expressed in endothelial cells due to downregulated miRNAs such as miR-19b-5p. The exogenous delivery of miR-199b-5p mimic has been shown to target ALK1 and thus inhibit angiogenesis in vitro as well as in vivo. In human umbilical vein endothelial cells (HUVECs), miR-199b-5p could attenuate the activation of the ALK1/Smad/Id1 pathway through BMP9 and reduce tumor growth and angiogenesis in vivo (Lin et al. 2019).

Apart from cancer, in osteoporosis, many miRNAs have also been reported to be expressed in low concentrations, resulting in the reduced osteogenic capacity of bone cells and exacerbating the disease. One such miRNA is miR-27a, which, at a normal level, favors the osteoblastogenesis of mesenchymal stem cells

Table 4.1 Possible miRNA mimics and their targets in various diseases

MiRNA	Disease	Target	Relative expression	Regulation strategy	Delivery system/Route of administration	Model animals or cells	Effect of regulation strategy	Reference
MiR-34a	Cancer	Sirtuin (SIRT1)	Down	Up	Polyvinylamine	Osteosarcoma xenograft tumor mouse model	Antitumor	Li et al. (2021c) and Tian et al. (2022)
MiR-199b-5p		ALK1			Lipofectamine 2000	MDA-MB-231 xenograft tumor mouse model	Inhibit angiogenesis in HUVECs	Lin et al. (2019)
MiR-27a	Osteoporosis	Mef2c			Lipofectamine 2000	OVX model andMef2C knockout model mice	Promotes osteoblastogenesis	You et al. (2016) and Fujii et al. (2021)
MiR-29a		RANKL, CXCL12			Lipofectamine 2000	OVX model mice	Increases BMD	Lian et al. (2019)
MiR-19a-3p and miR-874-3p		HDAC4, LEP			Lipofectamine 2000	hBMSCs cells	Alleviates osteoporosis	Chen et al. (2019) and Mei et al. (2021)
MiR-203		DKK1			Lipofectamine 2000	OVX rats	Lowers adipogenic differentiation and increases osteogenesis	Qiao et al. (2018)
MiR-181c-5p and miR-497		–			Lipofectamine 2000	Aging and OVX model mice	Promotes the differentiation and mineralization of osteoblast cells	Ma et al. (2019)
MiR-143	Osteosarcoma	MAPK7			Lipofectamine 2000	Osteosarcoma tissue or cells	Promotes cell proliferation, migration, and invasion of U2OS	Hou et al. (2019) and Dong et al. (2017)
MiR-22	Myocardial infarction	TGFBR1			Lipofectamine 2000	Myocardial infarction model mice	Improves cardiac function	Hong et al. (2016)
MiR-29b-3p		FOS			Lipofectamine 2000	Cardiac fibroblasts	Inhibits cell proliferation, migration, and differentiation of CF	Xue et al. (2020)
MiR-143	Allergic rhinitis	IL13Rα1			Lentivirus	Human primary NEC cells	Reduces granulocyte-macrophage colony-stimulating factor, eotaxin, and mucin 5 AC	Teng et al. (2015)
MiR-133b		Nlrp3			Lipofectamine 2000	Allergic rhinitis mouse model	Lowers OVA-specific IgE and inflammatory cytokines	Xiao et al. (2017)
MiR-34/449	Asthma	IGFBP-3, Nur-77			Intraperitoneal	Asthma mouse model	Diminishes airway remodeling and autophagy	Yin et al. (2017)
MiR-20a-5p		ATG7			Intranasal	Asthma mouse model	Diminishes airway remodeling	(Yu et al. (2021)
MiR-34a	Spinal cord injury	Notch pathway			Intrathecal	SCI model rats	Suppresses neuronal apoptosis and alleviates microglia inflammation	Jian et al. (2020)
MiR-155	Systemic lupus erythematosus	SOCS1			Lentivirus	PBMSC cells	Increases STAT3 phosphorylation and IL-21	Rasmussen et al. (2015)

(MSCs). However, its low expression in osteoporosis reduces the osteogenic potential of MSCs but promotes adipogenesis. MiR-27a, when overexpressed, has been shown to promote osteoblastogenesis by targeting myocyte enhancer factor 2c (Mef2c), an important factor involved in the regulation of osteoclastogenesis and pathologic bone resorption (You et al. 2016; Fujii et al. 2021). Decreased expression of miR-29a has also been shown to reduce mineralized matrix production in osteogenic cells and increase osteoclast differentiation and pit formation. In OVX mice, overexpression of miR-29a in osteoblasts increased bone mineral density (BMD), bone parameters, and osteogenesis of bone marrow mesenchymal cells (BMSCs) through osteoclast inhibition by binding and inhibiting receptor activator nuclear factor-κ ligand (RANKL) and C-X-C motif chemokine ligand 12 (CXCL12) (Lian et al. 2019). Additionally, miR-19a-3p and miR-874-3p, when exogenously overexpressed, have also been shown to increase the levels of Runt-related transcription factor 2 (Runx2), osteocalcin (OCN), and alkaline phosphatase (ALP) activity by targeting the 3′UTR of histone deacetylase 4 (HDAC4) and leptin (LEP), respectively, to alleviate osteoporosis (Chen et al. 2019; Mei et al. 2021). MiR-203 is involved in regulating the differentiation of BMSCs in postmenopausal osteoporosis (PMOP). Its low levels in PMOP promote adipogenic differentiation of BMSCs, while upregulation has been reported to improve osteogenesis by targeting Dickkopf-1 (DKK1) (Qiao et al. 2018). Moreover, treatment with miR-181c-5p and miR-497 mimics has also been shown to promote the differentiation and mineralization of osteoblast cells (Ma et al. 2019). Therefore, employing miRNA mimics can be a promising strategy to regulate miRNA function in bone diseases.

Osteosarcoma (OS) is regarded as one of the most aggressive bone tumors. Despite numerous efforts, the prognosis for OS remains elusive. In many cellular mechanisms, mitogen-activated protein kinase (MAPK) pathways commonly convey signals, and MAPK7 has been identified as an oncogene. Studies have shown that the overexpression of MAPK7 in human OS cells can promote cell proliferation, migration, and invasion while silencing MAPK7 to effectively suppress cell proliferation, cell migration, and invasion (Tesser-Gamba et al. 2016). Among various miRNAs that are downregulated in osteosarcoma tissues and cell lines, miR-143 has been shown to promote cell proliferation, migration, and invasion capabilities of U2OS cells. The upregulation of miR-143 through a gain-of-function strategy showed that miR-143 targeted MAPK7 and inhibited cell migration and invasion of U2OS cells (Hou et al. 2019; Dong et al. 2017). Previous studies have detailed the various miRNAs that are downregulated and responsible for the progression of human osteosarcoma (Llobat and Gourbault 2021). Therefore, upregulating miRNA in OS can be a potential therapeutic approach.

In addition to cancer and bone disorders, inflammatory diseases also influence the progression of other diseases, such as diabetes, cardiovascular diseases, cancer, arthritis, and autoimmune diseases, through a cascade of protein signaling (Arulselvan et al. 2016; Vanaja et al. 2015). Noncoding RNAs, including microRNAs, play an important role in inflammation by regulating the expression of inflammatory genes in various diseases and therefore could potentially be used as biomarkers and therapeutic targets to control unusual inflammatory responses. As such, many have been reviewed for their role in inflammation in various diseases and their potential as target molecules (Das and Rao 2022; Tahamtan et al. 2018; Gaudet et al. 2018).

Cardiac fibrosis is commonly associated with myocardial infarction (MI), and low levels of miR-22 have been implicated in infarcted myocardium and hypoxic cardiac fibroblasts (CFs). Overexpression of miR-22 in mice has been shown to improve cardiac function in mice after MI due to the downregulation of TGF-β receptor 1 (TGFBR1) (Hong et al. 2016). In TGF-β1-treated cardiac CF, a low level of miR-29b-3p has also been reported to promote the proliferation, migration, and differentiation of CFs. However, overexpression led to the inhibition of cell proliferation, migration, and differentiation of CFs through targeted inhibition of FOS (Xue et al. 2020). Moreover, overexpression of miR-143 in IL-13-stimulated NECs from allergic rhinitis (AR) patients could also significantly reduce the expression of the IL-13 receptor α1 chain (IL13Rα1), an important receptor involved in the interleukin-3 signaling pathway, resulting in a reduction of granulocyte-macrophage colony-stimulating factor, eotaxin, and mucin 5AC and alleviating AR (Teng et al. 2015). Moreover, in the allergic rhinitis mouse model, upregulation of miR-133b has been shown to reduce the levels of OVA-specific IgE and the expression of inflammatory cytokines generated by cervical lymph node cells (Xiao et al. 2017).

Insulin growth factor binding protein-3 (IGFBP-3) is implicated in many human diseases and also plays an important function in the pathogenesis of asthma by regulating the tumor necrosis factor (TNF)-α-induced activities of the NF-κB signaling pathway that can cause apoptosis or autophagy. In asthma, miR-34/449 expression is downregulated with an increase in the expression of autophagy-related proteins. Forced expression of miR-34/449 has been shown to diminish airway remodeling and autophagy by targeting target IGFBP-3 and Nur-77 (Yin et al. 2017). MiR-20a-5p overexpression has also been reported to alleviate asthma symptoms by improving airway remodeling in model mice through suppression of apoptosis, inflammation, and autophagy (Yu et al. 2021). A spinal cord injury (SCI) significantly influences the mobility of the affected person, having social, financial, and psychological burdens. SCI can be treated with medication, surgery, and rehabilitation approaches, but the lack of definite treatment limits the improvement of neurological function in chronic SCI patients (Oh and Jeon 2016). Numerous studies have been conducted to understand the cause of SCI and many point to the role of miRNA. Following SCI, decreased miR-34a levels have been shown to lead to an increase in the expressions of Notch 1, Jagged 1, CD11b, inducible nitric oxide synthase (iNOS), IL-1β, and IL-6, but with miR-34a treatment, the expression levels of these proteins could be lowered (Jian et al. 2020). Moreover, exosomes overexpressing miR-709 could significantly reduce the inflammatory response and recover lost motor functions after spinal cord injury, by targeting NKAP, an active contributor to inflammation (Xiong et al. 2022).

Systemic lupus erythematosus (SLE) is a chronic autoimmune disease characterized by deregulated immune cells, such as dendritic cells, T and B cells, and autoantibodies. Dysregulated immune cells can damage various organs and tissues through the formation of multiple cellular components, autoantibodies, immunocomplexes, activation of the complement system, and cytokines (Fava and Petri 2019; Wu et al. 2020a). Previous studies have suggested that aberrant changes in miRNA expression are one of the epigenetic factors in SLE (Long et al. 2018; Hiramatsu-Asano and Wada 2022). The reduced expression of miR-4512 in monocytes and macrophages has been reported to exaggerate SLE by targeting the TLR4 pathway and promoting proinflammatory cytokine expression and neutrophil extracellular trap formations. Overexpression of miR-4512, on the other hand, was significantly able to reduce the release of myeloperoxidase and cell-free DNA by neutrophils, thus inhibiting neutrophil extracellular trap formations (Yang et al. 2021). The production of autoantibodies and inflammatory responses is regulated by cytokines, especially IL-21, and a higher level of suppressor of cytokine signaling 1 (SOCS1) has been reported in $CD4^{+}$ T cells from SLE patients in parallel with low levels of miR-155. Overexpression of miR-155 has been shown to increase STAT3 phosphorylation and IL-21 production in SLE patients through SOCS1 inhibition (Rasmussen et al. 2015). The study has demonstrated the implication of several miRNAs in the pathogenesis of SLE and has even highlighted the therapeutic potential of the manipulation of miRNA expression in vivo to achieve therapeutic effects in lupus mouse models. Moreover, miRNA expression patterns in peripheral blood cells, plasma, and urine can have potential as biomarkers and viable therapies for the treatment of SLE (Hong et al. 2020b).

Taken together, these results demonstrate that miRNA mimic delivery in vitro and in vivo can be a promising alternative for the development of prospective miRNA regulation strategies for a range of cancerous and noncancerous disorders.

4.2.2 MiRNA Inhibition As a miRNA Regulation Strategy

In various diseases, miRNAs are highly expressed, and therefore, inhibition of such miRNAs through miRNA inhibitors can mitigate the negative effects on target mRNA and restore normal expression and cell phenotype. MiRNA inhibitors can be classified into two categories: natural miRNA inhibitors and artificial miRNA inhibitors (Fig. 4.2). Natural miRNA inhibitors involve competing endogenous RNAs (ceRNAs) that compete to bind to miRNA recognition elements (MREs) to reduce miRNA availability and mitigate the repression of target RNAs. There are three main classes of naturally occurring miRNA inhibitors: long noncoding RNAs, RNAs encoded by pseudogenes, and viral RNA transcripts (Table 4.2). Additionally, excess levels of mRNA can also sequester and suppress the activities of miRNAs (Bak and Mikkelsen 2014). Artificial miRNA inhibitors include the synthesis of RNA transcripts through chemical and genetic methods and their delivery through gene vectors (Tang et al. 2017).

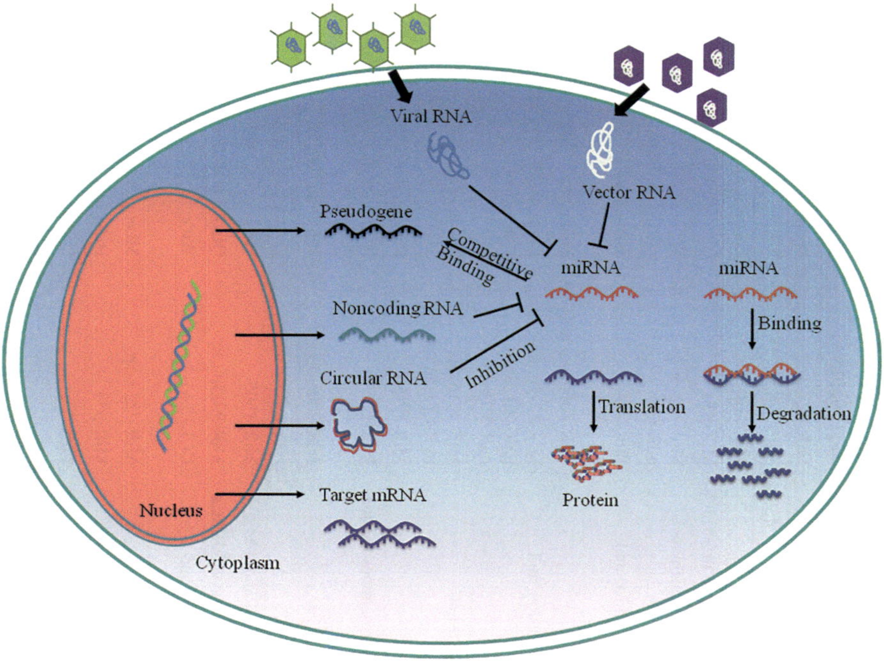

Fig. 4.2 Approaches for miRNA inhibition

4.2.2.1 Long Noncoding RNAs As miRNA Inhibitors

Long noncoding RNAs constitute a class of RNA transcripts with a length of more than 200 nucleotides. These ncRNAs are involved in most of the cellular activities ranging from transcription and RNA stability to translation, and some even act as ceRNA (Bridges et al. 2021; Napoli 2021).

LncRNAs have been linked to several disorders, including osteoporosis (Li et al. 2021a). LncRNA potassium voltage-gated channel subfamily Q member 1 overlapping transcript 1 (lncRNA KCNQ1OT1) has been shown to have an anti-osteoporotic effect in ovariectomy-induced (OVX) PMOP model mice. KCNQ1OT1 has been shown to target the pro-osteoporotic miR-421-3p, thus elevating the mammalian target of rapamycin (mTOR) to enhance MC3T3-E1 cell functions (Wang et al. 2023). In addition to KCNQ1OT1, various lncRNAs have also been reported to regulate the action of miRNAs that negatively impact bone metabolism due to their excessively high expression levels. LncRNAs such as NEAT1, TINCR, and MCF2L-AS1 have been shown to target miR-29b-3p to increase BMP1 expression (Zhang et al. 2019b), miR-761 to increase Wnt2 (Zheng et al. 2020), and miR-33a to positively regulate the expression of Runx2 (Chen et al. 2020a), respectively, while some have been reported to target more than one miRNA, such as H19. LncRNA H19 can target miR-188 (Wang et al. 2018), miR-138 (Wu et al. 2018),

Table 4.2 MiRNAs inhibitors in various diseases

Type	Name	Disease	Role in disease	Target (miRNA)	Reference
LncRNA	KCNQ1OT1	Osteoporosis	Anti-osteoporotic	MiR-421-3p	Wang et al. (2023)
	NEAT1		Promotes osteogenic differentiation of hBMSCs	MiR-29b-3p	Zhang et al. (2019b)
	TINCR		Promotes migration of rat MSCs	MiR-761	Zheng et al. (2020)
	MCF2L-AS1		Stimulates osteogenic differentiation in BMSCs	MiR-33a	Chen et al. (2020a)
	H19		Improve the function of bone mesenchymal stem cells	MiR-188, miR-138 and miR-675	Wang et al. (2018), Wu et al. (2018) and Huang et al. (2015)
Pseudogene	PTENP1	Osteoporosis	Suppresses osteoclast differentiation and attenuates osteoporosis	MiR-214	Wang et al. (2020)
		Tumors	Tumor suppression	MiR-106b and miR-93	Zhang et al. (2017)
	TUSC2P	Esophageal squamous cell carcinoma	Inhibits proliferation, and invasion, and promotes apoptosis	MiR-17-5p, miR-608, miR-661, and miR-520a-3p	Liu et al. (2018) and Rutnam et al. (2014)
	CHIAP2	Lung adenocarcinoma	Inhibits proliferation and invasion	MiR-3614-5p, miR-873-3p	Shang et al. (2019)
	INTS6P1	Hepatocellular carcinoma	Suppresses tumor growth	MiR-17-5p	Peng et al. (2015)
	CYP2A7	Nicotine metabolism	Activates tobacco-related nitrosamines and the metabolism of nicotine	MiR-126*	Nakano et al. (2015)
CircRNA	CircST6GALNAC6	Bladder cancer	Metastasis inhibition	MiR-200a-3p	Tan et al. (2021)
	RFWD2 and INO80	Bone disorders	Promotes osteogenesis of hASCs	Hsa-miR-6817-5p	Huang et al. (2019)
	0076906		Promotes osteogenic differentiation of hMSCs	MiR-1305	Wen et al. (2020)
	0048211		Increases osteogenic differentiation of hBMCs	MiR-93-5p	Qiao et al. (2020)
	003795		Promotes osteoblast differentiation and mineralization.	COL15A1	Wu et al. (2020b)
	AFF4		Promotes MC3T3-E1 proliferation.	MiR-7223-5p	Mi et al. (2019)
	cPWWP2A		Inhibits apoptosis in osteoblast	MiR-579	Hong et al. (2019)
Viral transcript	HSURs1	Virally transformed T cells	Causes aggressive leukemias and lymphomas in New World primates	MiR-27	Cazalla et al. (2010)
	HBV RNAs transcript	Hepatocellular carcinoma	Promotes apoptosis	MiR 15a/16	Liu et al. (2013, 2015)

and miR-675 (Huang et al. 2015) to improve the function of bone mesenchymal stem cells.

Thus, these studies show that miRNA expression can be targeted to regulate gene expression through lncRNAs, and their ability to target more than one miRNA makes them an ideal candidate to develop miRNA inhibition therapeutics.

4.2.2.2 Pseudogenes As miRNA Inhibitors

Pseudogenes comprise genomic DNA sequences similar to functional genes but lack functional characteristics. However, despite their nonfunctional characteristics, they can influence a significant portion of gene activity. Depending on how they are produced, pseudogenes are categorized into two groups: processed pseudogenes and unprocessed pseudogenes, with processed pseudogenes being the most prominent in mammals. Owing to their sequence complementarity, pseudogenes can bind to miRNA and regulate gene expression at the miRNA level (Di Sanzo et al. 2020). Various pseudogenes have been shown to function as miRNA sponges in various cancers.

Phosphatase and tensin homolog (PTEN) is a well-known tumor suppressor in various cancer types, and many miRNAs in various cancers have been shown to target PTEN to drive cancer progression. Its pseudogene, PTENP1, has been shown to posttranscriptionally regulate PTEN by sponging the action of miRNAs that target PTEN, thereby increasing PTEN levels and suppressing cancer (Travis et al. 2023). For example, PTENP1 has been shown to sponge (1) miR-214 to regulate osteoclast differentiation to attenuate osteoporosis and (2) miR-106b and miR-93 in gastric cancer (Wang et al. 2020; Zhang et al. 2017). The pseudogene of tumor suppressor candidate-2, TUSC2P, has been shown to share binding sites with various miRNAs such as miR-17-5p, miR-608, miR-661, and miR-520a-3p in esophageal squamous cell carcinoma. Its overexpression in EC109 and TE-1 cell lines altered the expression of TUSC2, inhibiting proliferation and invasion and promoting apoptosis (Liu et al. 2018; Rutnam et al. 2014). In lung adenocarcinoma, the pseudogene CHIAP2 has been shown to inhibit the proliferation and invasion of lung adenocarcinoma cells by regulating the WNT signal pathway through miR-3614-5p or miR-873-3p (Shang et al. 2019). Moreover, the expression of Foxo3, a member of the forkhead family of transcription factors, is regulated by the Foxo3 pseudogene (Foxo3P), which can bind to at least eight miRNAs. This binding would lead to the inhibition of tumor growth and cancer cell proliferation (Yang et al. 2016). A pseudogene of tumor suppressor INTS6, INTS6P1, in HCC has been shown to function as an antitumor pseudogene. INTS6P1 exerted the tumor suppressive effect by competing for oncogenic miR-17-5p (Peng et al. 2015).

In addition to diseases, pseudogenes are also involved in physiological processes, such as nicotine metabolism. Human cytochrome P450 (CYP)2A6 is involved in activating tobacco-related nitrosamines and the metabolism of nicotine. However, it is targeted by miR-126*. A study shows that the pseudogene of CYP2A7 could sponge the action of miR-126*, increasing the level of CYP2A6 (Nakano et al. 2015). Therefore, it can be assumed that a better understanding of pseudogenes can contribute to the development of novel molecular therapies targeting miRNAs for human diseases.

Moreover, pseudogenes can produce modulatory effects through various mechanisms, such as binding to transcription factors and producing miRNA/piRNA or proteins (Lou et al. 2020). Therefore, a better understanding of these mechanisms is needed to develop novel molecular therapies based on pseudogenes to target miRNAs for human diseases and have an effective translation into clinical application.

4.2.2.3 Circular RNAs (CircRNAs) As miRNA Inhibitors

CircRNAs are single-stranded, covalently closed-loop RNA molecules that are produced through back-splicing and, owing to their circular shape, are resistant to nuclease degradation. CircRNAs regulate biological functions through transcriptional regulation by acting as microRNA sponges and protein templates (Zhou et al. 2020; Cao et al. 2021). Most studies highlight the role of circRNAs in sponging miRNAs to exert an anti-disease effect, such as CircRNA ST6GALNAC6, which could suppress bladder cancer metastasis by sponging the action of miR-200a-3p (Tan et al. 2021). Various other circRNAs have also been reported to act as miRNA inhibitors, such as high levels of circRNAs RFWD2 and circINO80 in hASCs, which have been shown to inhibit hsa-miR 6817-5p (Huang et al. 2019), while circRNA 0076906 and circRNA 0048211 can sponge the action of miR-1305 and miR 93-5p, respectively (Wen et al. 2020; Qiao et al. 2020). In addition, circRNA 003795 has been shown to target miR-1249-5p to improve the expression of COL15A1 and OPN (Wu et al. 2020b), circRNA AFF4 to target miR-7223-5p to inhibit PI3KR1 (Mi et al. 2019), and circRNA cPWWP2A to attenuate cytotoxicity in human osteoblasts through inhibition of miR-579 (Hong et al. 2019). These studies show that circRNAs can be used to target the miRNAs involved in bone metabolism to improve bone homeostasis.

4.2.2.4 Viral RNA Transcripts As miRNA Inhibitors

Viral RNAs, during their infective process, disturb miRNA regulatory networks by competing with host RNAs (Arancio 2021). Primate herpesviruses express more noncoding RNAs (ncRNAs) than any other mammalian virus. T cells transformed by the monkey virus Herpesvirus saimiri (HVS) express seven viral U-rich ncRNAs termed Herpesvirus saimiri U RNAs (HSURs). HSURs 1 and 2 share a sequence complementary to three host-cell miRNAs: miR-27, miR-16, and miR-142-3p (Cazalla and Steitz 2010). In the presence of HVS-transformed cells, miR-27 has been shown to express at reduced levels because of its sequence-specific and binding-dependent degradation by HVS RNA (Cazalla et al. 2010). Therefore, this sequence-specificity and subsequent binding-dependent degradation abilities of viral RNA transcripts can be used to regulate miRNA function. Hepatitis B virus (HBV)-mediated chronic hepatitis has been known to eventually manifest into hepatocellular carcinoma (HCC) and inhibit etoposide-induced apoptosis of hepatoma cells. MiR-15a/16 has been shown to target the anti-apoptotic Bcl-2 gene. HBV mRNAs have been shown to contain miR-15a/16 complementary sites, and therefore, when introduced, HBV mRNAs were able to bind and sequester endogenous miR-15a/16 and increase Bcl-2 gene expression (Liu et al. 2013) to drive HCC. Smad 7, a transforming growth factor β (TGF-β) pathway inhibitor, is a known target of miR-15a. During the infection of HBV, miR-15a level is reduced, relinquishing its inhibitory controls on Smad 7, which abrogated TGF-β signaling, thus inhibiting apoptosis of HepG2 cells and promoting HCC (Liu et al. 2015). These studies demonstrate the potential of viral transcripts as a novel approach to regulating miRNA function. For further information regarding the interplay between miRNAs and viral infection, please refer to Chap. 11 in this book.

4.2.2.5 Antimirs As miRNA Inhibitors

Various studies have detailed the use of antimir RNAs as therapeutic strategies to alleviate various disorders (Table 4.3). In cancer, miR-21 is extensively studied due to its expression in various cancers and targeting of several mRNAs. MiRNA array profiling shows that higher expression of miR-21 in breast cancer (BC) tumors can target and downregulate the tumor suppressor tropomyosin 1 (TPM1), a cancer suppressor (Zhu et al. 2007). The introduction of anti-miR-21 oligonucleotides into MCF-7 cells has been shown to reduce cell growth and migration of MCF-7 and MDA-MB-231 cell lines in vitro. Moreover, anti-miR-21 inhibited the growth of tumors in nude xenograft mouse models. However, miR-21 inhibition led to a considerable increase in the mRNA and protein expression of ankyrin repeat domain 46 (ANKRD46), suggesting ANKRD46 as another direct target of miR-21 in BC (Si et al. 2007; Yan et al. 2011). In invasive breast cancers, miR-21 expression levels are significantly elevated compared to the noninvasive component group, suggesting its potential as an independent molecular marker to study breast cancer invasiveness (Petrović et al. 2014). In addition to miR-21, overexpression of miR-155 is also associated with significant lymph node positivity and an enhanced proliferation index in the advanced breast cancer stage (Zheng et al. 2012).

MiR-21 has also been reported to play a significant role in driving SLE by increasing DNA hypomethylation in $CD4^+$ T cells through DNA methyltransferase I (DNMT1) inhibition and stimulating autoimmune-associated methylation-sensitive gene overexpression (Pan et al. 2010). MiR-21 inhibition has been shown to reduce IL-6R, Janus kinase (JAK), and signal transducers and activators of transcription 3 (STAT3) levels in vivo and improve hindlimb movement and coordination in model rats (Ning et al. 2019). Moreover, high levels of miR-21 in the infarcted zone after MI in vivo have been shown to promote Col-1, α-SMA, and F-actin expression and activation of CFs, exaggerating MI. The downregulation of miR-21 expression could mitigate fibrosis through Smad7 and improve MI (Yuan et al. 2017).

Some miRNAs, such as miR-155 and miR-155-3p, have also been implicated in suppressing apoptosis in cancer cells. The inhibition of miR-155 in HS578T cells

Table 4.3 Possible miRNA targets for miRNA inhibitors in various diseases

Disease	MiRNA	Target	Level in disease	Role in disease	Regulation Strategy	Reference
Breast cancer	MiR 21	TPM1	Up	Promotes breast cancer	Inhibition or Downregulation	Zhu et al. (2007)
		ANKRD46				Yan et al. (2011)
	MiR 27a	CADM1				Zhang et al. (2019a)
SLE	MiR 21	DNMT1		Mediates SLE pathogenesis		Pan et al. (2010)
MI		Smad7		Promotes cardiac fibrosis		Yuan et al. (2017)
Osteosarcoma	MiR 27a	MAP2K4		Pro-metastatic		Pan et al. (2014)
Osteoarthritis	MiR 34a-5p	–		Cartilage degradation		(Endisha et al. 2021)
Osteoporosis	MiR18a-3p	ADCY5		Reduces osteogenic differentiation		Wang et al. (2021)
		GRIA1				Zhao et al. (2022)
	MiR 23b-3p	MRC2		Inhibits osteogenic differentiation of hMSCs		Li et al. (2021b)
	MiR 133a	Wnt/β-catenin		Promote osteoclastogenesis and differentiation		Li et al. (2018)
	MiR 205-5p	Runx2		Inhibits COL1A1 and ALP activity		Huang et al. (2020b)
	MiR 214	β-Catenin		Reduces osteoblast differentiation		Li et al. (2017)
	MiR 138–5	MACF1		Inhibits osteoblast differentiation		Chen et al. (2020b, 2022)
	MiR-129-5p	TCF4		Inhibits osteoblast differentiation in vitro and in vivo		Yin et al. (2020)
	MiR-12,200-5p	APC, TCF4, TCF7, WNT3A, WNT5A, LRP		Reduces osteoblast differentiation		Li et al. (2023)
	MiR-208a-3p	Activin A receptor type I		Inhibits osteoblast activity and bone formation		Arfat et al. (2018)

using miR-155 ASO has been shown to reduce cell proliferation and increase apoptosis. Similar results were also observed in the MDA-MB-157 cell line, where it reduced cell viability and significantly increased the number of apoptotic cells. Moreover, miR-155 ASO suppressed the growth of transplanted tumors in vivo while increasing the expression of caspase-3 (Zheng et al. 2013), suggesting its anticancer and pro-apoptotic functions. Cell adhesion molecule 1 (CADM1), or tumor suppressor in lung cancer 1 (TSLC1), plays a crucial role as a tumor suppressor. In breast cancer, it is often reported to be inactive due to upregulated miR-155-3p (Zhang et al. 2019a), providing another target to limit breast cancer progression.

Additionally, extensive work has been done on understanding the role of noncoding RNAs, including miRNAs, and their potential for therapeutics in bone-related diseases such as osteosarcoma (OS), osteoarthritis (OA), and osteoporosis (Qian and Tian 2021). Many miRNAs can be detected in the serum of normal and diseased patients, making them ideal biomarkers in the prognosis and diagnosis of various diseases. MiR-21, miR-23a, miR-24, miR-93, miR-100, miR-122a, miR-124a, miR-125b, and miR-148a have been shown to express abundantly in the serum of osteoporotic patients, while miR-21, miR-23a, miR-24, miR-25, miR-100, and miR-125b can be found in the bone tissue of osteoporotic patients (Seeliger et al. 2014). Upregulated miRNAs, such as miR-27a, have been shown to drive OS progression by acting as pro-metastatic miRNAs. A study has shown that miR-27a targets mitogen-activated protein kinase kinase 4 (MAP2K4), and when inhibited, miR-27a could increase MAP2K4 expression, resulting in the inhibition of cell proliferation and migration of MG63 cells (Pan et al. 2014). The upregulated expression of miR-34a-5p in the plasma, cartilage, and synovium has been associated with late-stage OA, and antisense oligonucleotide treatment has shown to improve cellular gene expression and cartilage-protective effects in the destabilization of the medial meniscus (DMM) and high-fat diet/DMM models (Endisha et al. 2021), providing miR-24-5p as a potential target for OA treatment. In hBMSCs and vertebral compression fracture tissues of osteoporosis patients, adenylate cyclase 5 (ADCY5), an enzyme responsible for converting adenosine triphosphate (ATP) into cyclic adenosine monophosphate (cAMP), was targeted by overexpressed miR-18a-3p, leading to a reduction in osteogenic differentiation (Wang et al. 2021). Moreover, miR18a-3p could also target glutamate AMPA receptor subunit 1 (GRIA1) in OP and osteoporotic vertebral fracture patients to reduce osteogenic differentiation (Zhao et al. 2022). Its knockdown, on the other hand, promoted osteogenic differentiation, indicating targeting of miR-18a-3p as a therapeutic strategy for OP. High expression of miR-23b-3p in the bone tissues of PMOP mice has also been reported to inhibit osteogenic differentiation of hMSCs through targeted inhibition of a receptor associated with collagen metabolism, mannose receptor C type 2 (MRC2), while higher expression of miR-133a in PMOP patients promotes osteoclastogenesis and RANKL-induced differentiation of RAW264.7 and THP-1 cells into osteoclasts. Silencing or knockdown of miR-23b-3p and miR-133a could alleviate PMOP by promoting the Wnt/β-catenin pathway and modulating the levels of osteoclastogenesis-related factors during osteogenic differentiation in hMSCs and in mice (Li et al. 2018, 2021b). Huang et al. have reported that increased levels of miR-205-5p inhibit COL1A1 and ALP activity, whereas downregulation of miR-205-5p can improve COL1A1 and ALP levels. This inhibitory effect of miR-205-5p on osteoblast differentiation was attributed to its direct targeting of Runx2 (Huang et al. 2020b). Moreover, increased levels of miR-214 in bone tissue from osteoporosis patients have also been associated with lower osteoblast differentiation due to β-catenin inhibition (Li et al. 2017). Mir 138-5p has been shown to act as a mechano-sensitive miRNA and target microtubule actin cross-linking factor 1 (MACF1) and β-catenin expression to inhibit osteoblast differentiation (Chen et al. 2021). Moreover, miR-138-5p has been shown to exacerbate bone loss in senile osteoporosis, and its inhibition through antimir 138-5p sensitizes bone anabolic action and stimulates bone formation in aged mice (Chen et al. 2020b, 2022). Additionally, miR-129-5p has been shown to inhibit osteoblast differentiation of MC3T3-E1 cells as well as bone formation in vivo by repressing the Wnt/β-catenin pathway via targeted inhibition of TCF4. Notably, recombinant inhibitor-129-5p (nCAR/anti129) could reverse the anti-osteogenic effect by inhibiting miR-129-5p in vitro and in vivo (Yin et al. 2020). MiR-12,200-5p has also been demonstrated to target multiple members of the Wnt signaling pathway, such as APC, TCF4, TCF7, WNT3A, WNT5A, and LRP6, reducing osteoblast differentiation. The study has shown that suppression of miR-12,200-5p could alleviate PMOP (Li et al. 2023). MiR-208a-3p inhibition using antagomir 208a-3p could also promote osteoblast activity, in vitro matrix mineralization, and in vivo bone formation under mechanical unloading conditions (Arfat et al. 2018).

Taken together, these studies show that inhibition of miRNAs could be a novel therapeutic strategy for the development of various therapeutics to treat different diseases.

4.2.3 RNA Modifications

Antagomirs, also known as anti-miRNA ASOs or blockmirs, are chemically synthesized oligonucleotide antagonists of miRNAs used to silence the target miRNAs. Owing to their specificities and complementarities to target miRNAs, antagomirs sequester specific miRNAs, inducing miRNA repression and inhabiting target mRNA degradation (Chang 2012; Atri et al. 2019). For example, antagomirs against miR-16, miR-122, miR-192, and miR-194 have been shown to significantly reduce the levels of corresponding miRNAs in vivo (Krützfeldt et al. 2005) and therefore could be employed as a therapeutic strategy to silence overexpressed miRNAs in various diseases. Given their broader range of influence on gene expression and single-stranded nature, RNAs need protection from nuclease degradation or alterations to enhance their binding properties. This can be achieved either by chemical modifications of the nucleoside backbone of the RNA through phosphoramidite precursors or by the enzymatic polymerization of nucleoside triphosphates (Flamme et al. 2019).

Chemical modifications of the phosphate backbone and ribose ring can improve nuclease resistance and biocompatibility and reduce the immune response. The most common chemical modifications that are carried out are on phosphorothioate (PS) backbones by replacing one of the non-bridging oxygen atoms in the internucleotide phosphate group with sulfur, resulting in an improvement in cellular uptake, bioavailability, and resistance to phosphodiesterase enzymes (Zhu et al. 2022). PS-OMe modification of miRNA 130b-3p has been shown to improve the stability of the miRNA and reduce extracellular cold-inducible RNA-binding protein (eCIRP), a mediator in sepsis, to inhibit inflammation and lung injury and improve survival in murine sepsis (Borjas et al. 2023).

Oligonucleotide analogs modified at the 2′-position of the furanose sugar improve the drug-like properties of antisense oligonucleotides (ASOs). These modifications predispose the sugar pucker toward the 3′-endo-conformation; improve ASO affinity for its biological target, pharmacokinetic, and toxicological properties; and enhance metabolic stability (Prakash and Bhat 2007). The modifications at the 2′- and 4′-positions of the furanose ring can produce the 2′, 4′-bridged nucleic acid (BNA) or locked nucleic acid (LNA) analogs. ASOs containing such analogs exhibit improved nuclease resistance in vitro as well as in vivo, with exceptionally increased binding affinity for target RNA (Prakash 2011). As a result, LNA-modified oligonucleotides have been exploited to improve the hybridization properties of LNA with their target RNA molecules with high efficiency and could be useful for miRNA in situ hybridization and miRNA expression profiling (Válóczi et al. 2004). Moreover, seed-targeting 8-mer LNA oligonucleotides called tiny LNAs could simultaneously inhibit miRNAs within families having the same seed with associated upregulation of direct targets. Interestingly, unconjugated tiny LNAs could be taken up by normal tissues as well as breast tumors in mice, with long-term miRNA silencing having low off-target effects (Obad et al. 2011). Further modifications of ribose sugar at 2′ include 2′-methoxyethyl (MOE), 2′-O-methyl (O-Me), and 2′-fluoro (F) (Fig. 4.3). These modifications can improve the binding affinity and increase the base-pairing melting temperature (Zhu et al. 2022). These modifications, though they provide stability to RNAs, and their inability to successfully enter the cells and produce therapeutic effects remain a significant obstacle.

4.2.4 Delivery Systems

Most of the antagomirs or mimic RNAs are single-stranded, and to protect them from degradation and achieve maximum therapeutic potential in vitro *and* in vivo, they need to be delivered safely and effectively to the organ or tissue of interest (Feng et al. 2021). For this purpose, various gene delivery systems, both viral and nonviral, have been developed. Viral vectors, though they have high transfection efficiency, and their immunogenicity and toxicity can limit their broader application. Nonviral delivery systems, on the other hand, such as cationic lipids, quantum dots, and polymers such as chitosan, polyethyleneimine (PEI), polyacrylic acid

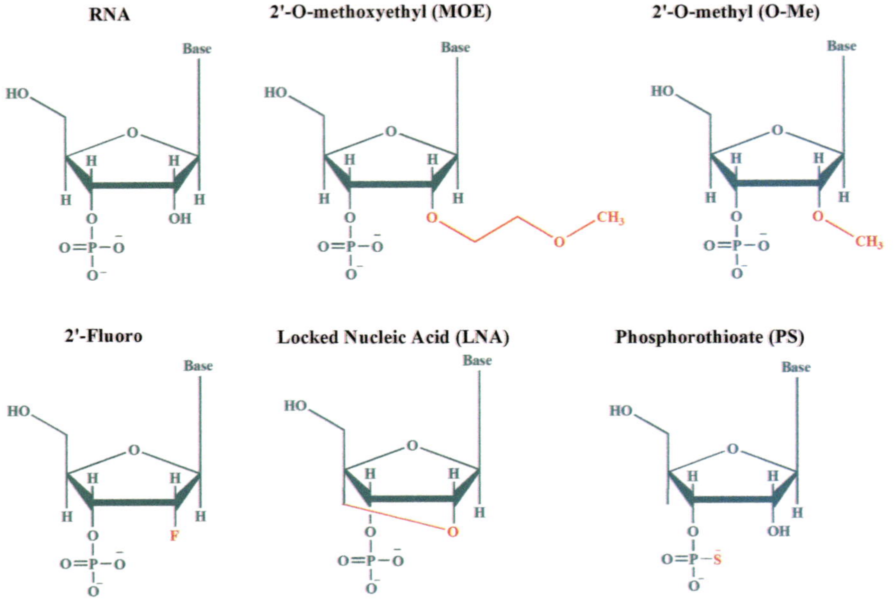

Fig. 4.3 RNA modifications at the 2′ position of the ribose ring. (Adapted from (Feng et al. 2021) originally published under a Creative Commons Attribution (CC BY) license (http://creativecommons.org/licenses/by/4.0/)

(PAA), polyamidoamine (PAMAM), and polyvinylamine (PVAm), are relatively non-immunogenic and have a larger surface area for nucleic acid binding, showing promising in vitro and in vivo results. However, they have low transfection efficiency (Patil et al. 2019). Modification to the delivery systems can improve their transfection abilities and the targeted delivery of nucleic acids. In addition, modifications can endow them with additional features such as fluorescence capabilities and improved stability. One such modification is the hydrophobic modification, which can enhance endosomal escape and improve the nucleic acid transfection abilities of the cationic lipids (Gao et al. 2016, 2018; Qiang Jin et al. 2016).

Dendrimers, owing to their structurally controllable nature, have become versatile candidates as efficient gene vectors (Jiepin Yang et al. 2015). Polyamidoamine (PAMAM) is a classic example of the dendrimer, which has been reported as a novel RNA delivery carrier. Its modification with fluorescent molecules has been shown to lower immunogenic detection of the polymer-nucleic acid complex in vitro as well as in vivo (Guan et al. 2015; Yoo and Juliano 2000). Polyethyleneimine (PEI) is considered a gold standard among gene delivery vectors owing to its high transfection efficiency (Chunping Ma et al. 2016) and ability to encapsulate and deliver RNA with high transfection efficiency (Yu et al. 2015). Additional modifications, such as poly (1,8-octanedio-citric acid)-co-polyethylene glycol, have been shown to further improve the transfection efficiency, photostability, and real-time tracking ability of developed vectors (Min Wang et al. 2017).

Despite their high transfection efficiency and high cationic surface charge, cationic polymers tend to display cytotoxicity and are often not cell- or tissue-specific. This can be resolved by further modification through targeting moieties such as peptides. For example, the nuclear targeting of nucleic acids can be improved by conjugating polymers with NLS-peptides (Dean et al. 2005), or mitochondrial delivery with high efficiency by conjugating triphenylphosphonium (TPP) to dendrimers (Wang et al. 2014), or modification of cationic polymers such as PAMAM- or PVAm-targeting peptides, such as SDSSD or CAP, to increase the targeting efficiency of these polymers toward bone and cartilage. Such modifications have been shown to improve the delivery of miRNA inhibitors or mimics to bones and cartilage without any off-targeting effect and improve cellular function more efficiently than unmodified polymers (Zhao et al. 2023; Patil et al. 2023). The study shows that vector conjugation to lipids, such as mPPS-FA (methyl PEG-PEI-sebacoyl chloride-folic acid), can improve the targeting of RNA toward cancer cells, (Yoshiyuki Hattori and Maitani 2006). Moreover, linking folate to polymers can effectively deliver synthetic RNA into nasopharyngeal tumor KB cells and significantly suppress KB xenograft growth in mice (Yoshizawa et al. 2008), folic acid-linked methyl PEG-2000, PEI-600, and sebacoyl chloride to improve high GFP transfection efficiency (Xu et al. 2012), or folate-linked nanoparticles designed for suicide gene therapy have been shown to inhibit cell growth (Hattori and Maitani 2005). In addition to folic acid, galactose can also be used to improve gene delivery to tumor cells. Lipids modified with galactose have been shown to successfully deliver nucleic acids to HepG2 cells and could induce high gene expression (Sakashita et al. 2014). Additionally, polyethyleneimine (PEI) cross-linked by myo-inositol (INO) and a galactose-grafted PEG could enhance the stability of the vector and transfection efficiency in the asialoglycoprotein receptor (ASGPR)-positive liver cells in vitro, effectively suppress tumor growth, and improve the survival time in tumor-bearing mice (Ling Liu et al. 2018).

The modification can provide an alternative to improve targeting and lower host response-related concerns; however, their long-term suitability for in vivo applications needs further investigation.

4.3 MicroRNA-Targeting Drugs in Clinical Trials

With advancements in the development of targeted delivery systems, many preclinical and clinical trials in various diseases have been carried out to establish the therapeutic potential of antagomir and miRNA (Table 4.4). Among other systemic diseases, autosomal dominant polycystic kidney disease (ADPKD) is the most common hereditary systemic disease characterized by enlarged kidneys with growing cysts and hypertension due to a mutation in the *PKD1* or *PKD2* genes, which encode polycystin-1 (PC1) or polycystin-2 (PC2), respectively, that are involved in mechano-transduction (Cornec-Le Gall et al. 2019). A short oligonucleotide inhibitor of miR-17, RGLS4326, was designed as a potential treatment for ADPKD. The

Table 4.4 miRNA therapeutics in clinical trials

Disease	Drug	Function	Drug target	Route of administration	Clinical Trial Identifier	Phase	Status	Reference
Alport syndrome	RG012	Inhibitor	MiR-21	Subcutaneous	NCT03373786	I	Completed (2019)	Li et al. (2013)
Hepatitis C	Miravirsen	Inhibitor	MiR-122	Subcutaneous	NCT01200420	II	Completed (2011)	Janssen et al. (2013)
Malignant pleural mesothelioma, non-small cell lung cancer	Mesomir	MiR-16 mimic	BCL2, CDK1, ETS1, and JUN	Intravenous	NCT02369198	I	Completed (2017)	van Zandwijk et al. (2017)
Advanced solid tumors	MRX34	MiR-34a mimic	MET, MYC	Intravenous	NCT01829971	I	Terminated (2017)	Hong et al. (2020a)
Lymphomas and leukemias	Cobomarsen (MRG-106)	Inhibitor	MiR-155	Intratumoral	NCT02580552	I	Completed (2020)	Querfeld et al. (2018)
Healthy volunteers	MRG-110	Inhibitor	MiR-92	Intradermal	NCT03603431	I	Completed (2019)	Huang et al. (2020a)
Heart failure	CDR132L	Inhibitor	MiR-132	Intravenous	NCT04045405	Ib	Completed (2020)	Täubel et al. (2021)

result of the preclinical study showed that RGLS4326 was preferentially distributed to kidneys and cysts derived from collecting ducts and targeted miR-17 to revive *Pkd1* and *Pkd2 levels*. Moreover, it showed an encouraging preclinical safety profile and assuaged cyst growth in human in vitro models and PKD mouse models (Lee et al. 2019). MiR-17 has been shown to induce kidney cysts in mouse and human ADPKD by inhibiting oxidative phosphorylation and stimulating proliferation by suppressing peroxisome proliferator-activated receptor-α (PPARα) (Hajarnis et al. 2017). A miR-21-specific oligonucleotide has been shown to inhibit miR-21 function without any adverse effects, resulting in improvements in kidney disease and dysfunction in mice by enhancing PPARα/retinoid X receptor (PPARα/RXR) activity and signaling pathways (Gomez et al. 2015). Moreover, RG-012, a miR-21 inhibitor in clinical trial (NCT03373786), is being studied for its safety, pharmacodynamics, and pharmacokinetics in Alport syndrome, caused by mutations in the collagen IV genes COL4A3, COL4A4, and COL4A5, predisposing a person to the risk of kidney failure among other abnormalities. MiR-122 has been shown to stabilize hepatitis C virus (HCV) viral RNA by recruiting Ago-2 to the genomic RNA and slowing its degradation in infected cells by cytoplasmic 5′ exonuclease Xrn1. Therefore, it has been considered a target molecule for the treatment of hepatitis C virus (HCV) infection (Li et al. 2013). Miravirsen is a locked nucleic acid-modified DNA phosphorothioate antisense oligonucleotide that can bind to miR-122, forming a stable heteroduplex to inhibit its function. The Phase 2a study (NCT01200420) demonstrated that Miravirsen was able to reduce the levels of HCV RNA in a dose-dependent manner with no adverse events (Janssen et al. 2013). A clinical study (NCT02369198) has shown that the miR-16 mimic delivered against malignant pleural mesothelioma and non-small cell lung cancer can effectively limit cancer progression without producing any safety concerns (van Zandwijk et al. 2017). In preclinical studies, Cobomarsen (MRG-106), an oligonucleotide inhibitor of miR-155, was able to reduce ABC-DLBCL cell proliferation and promote apoptosis. In addition, MRG-106 was effective in reducing tumor volume in a xenograft NSG mouse model (Anastasiadou et al. 2021). Moreover, Cobomarsen was also tested in clinical trials (NCT02580552, NCT03713320, and NCT03837457) for its safety, tolerability, pharmacokinetics, and potential efficacy against various lymphomas and leukemias (Querfeld et al. 2018).

MiR-92a, due to its anti-angiogenic activity, has been the focus of many studies, and as such, miR-92a inhibitors have been explored for their therapeutic potential. For the prevention of keloid scars, a miR-92 mimic formulation, Remlarsen (MRG-201), was studied for its safety and tolerability. Both phase I (NCT02603224) and phase II (NCT03601052) studies have shown MRG-110 to have therapeutic potential. In a preclinical study, a locked nucleic acid-modified miR-92a inhibitor, MRG-110, was investigated in excisional wounds in db/db mice and normal pigs. In both models, MRG-110 facilitated tissue formation, angiogenesis, and wound healing by promoting the expression of the pro-angiogenic gene, integrin alpha 5, in vitro and in vivo without raising safety concerns (Gallant-Behm et al. 2018; Huang et al. 2020a). Further study in humans showed that MRG-110 effectively targeted and inhibited miR-92a and increased integrin subunit alpha 5 (ITGA5) and

CD93 expression, the target genes of miR-92a (Abplanalp et al. 2020), suggesting its effectiveness in wound healing.

In myocardial infarction (MI) and heart failure (HF), miR-132 activation has been linked to adverse remodeling and pathological hypertrophy in the heart, providing miR-132 as a potential target to avoid MI and HF. To that effect, CDR132L, a synthetic lead-optimized oligonucleotide inhibitor, was designed. Its administration in chronic post-MI HF model pigs was able to improve cardiac function by reversing cardiac remodeling (Batkai et al. 2021). Moreover, it was assessed for safety, pharmacokinetics, and exploratory pharmacodynamic effects in a phase 1b clinical study (NCT04045405) in chronic ischemic HF patients. The results demonstrated that CDR132L was safe, well-tolerated, and effective (Täubel et al. 2021). Therefore, a phase II study (NCT05350969) with a large number of participants is set to be conducted to further evaluate the efficacy and safety of CDR132L.

However, there are some studies in which, despite their effectiveness, miRNA therapeutics demonstrated serious safety concerns, such as a liposome containing a mimic of miR-34a (MRX34) (NCT01829971). It was able to penetrate advanced solid tumors to modulate target gene expression but was terminated early due to severe immune-mediated adverse events (Hong et al. 2020a). Therefore, it is necessary to pay as much attention to developing a safe therapeutic as it is to developing an effective therapeutic.

4.4 Conclusion

MiRNA is a versatile molecule with a significant impact on several cellular functions by way of its influence on the genotypic regulation of cells and tissues. Differential miRNA expression in diseases makes it a potential therapeutic target or even a therapeutic drug molecule. The inhibitors, due to their complementarity and endogenous competitiveness due to the shared MREs, or miRNA mimics, can be used to regulate miRNA expression. That being said, the clinical application of miRNA as miRNA targeting therapy is limited by various factors; chief among them are its single-stranded nature, the fact that some miRNAs have multiple targets, and the different expression profiles of certain miRNAs among cell tissues within a single patient. Nevertheless, improvements in delivery methods to facilitate targeted delivery and chemical modification have made it possible to successfully deliver miRNA therapeutics to the desired organs and even to cells with notable success. Given the optimistic results of various in vitro and in vivo studies and a deeper understanding of miRNA function, many clinical trials involving miRNA as targets or therapeutic agents have been conducted. Some of them have even been completed with satisfactory results, while others have faced immunogenic problems. Therefore, with additional optimization in delivery systems, chemical modifications, in vitro and in vivo models, and ex vivo studies with human cells and tissues, miRNAs can emerge as next-generation therapeutics for the treatment of various diseases.

Compliance with Ethical Standards

Funding This work was funded by the Natural Science Foundation of China (grant numbers 82072106 and 31570940), the Key R&D Projects in Shaanxi Province (grant numbers 2021SF-242, 2021GXLH-01-02, 2022KWZ-13), and the Fundamental Research Funds for the Central Universities (grant number D5000210746 and D5000230122).

Conflict of Interest Authors declare they have no conflict of interest.

Ethical Approval This chapter does not contain any studies with human/animal participants performed by any of the authors.

References

Abplanalp WT, Fischer A, John D et al (2020) Efficiency and target Derepression of anti-miR-92a: results of a first in human study. Nucleic Acid Therapeutics 30(6):335–345. https://doi.org/10.1089/nat.2020.0871

Anastasiadou E, Seto AG, Beatty X et al (2021) Cobomarsen, an oligonucleotide inhibitor of miR-155, slows DLBCL tumor cell growth in vitro and in vivo. Clin Cancer Res 27(4):1139–1149. https://doi.org/10.1158/1078-0432.Ccr-20-3139

Arancio W (2021) ceRNA analysis of SARS-CoV-2. Arch Virol 166(1):271–274. https://doi.org/10.1007/s00705-020-04856-4

Arfat Y, Basra MAR, Shahzad M et al (2018) miR-208a-3p suppresses osteoblast differentiation and inhibits bone formation by targeting ACVR1. Mol Therapy Nucleic Acids 11:323–336. https://doi.org/10.1016/j.omtn.2017.11.009

Arulselvan P, Fard MT, Tan WS et al (2016) Role of antioxidants and natural products in inflammation. Oxidative Med Cell Longev 2016:5276130. https://doi.org/10.1155/2016/5276130

Atri C, Guerfali FZ, Laouini D (2019) Chapter 6 - MicroRNAs in diagnosis and therapeutics. In: Mallick B (ed) AGO-driven non-coding RNAs. Academic Press, Cambridge, pp 137–177. https://doi.org/10.1016/B978-0-12-815669-8.00006-3

Bak RO, Mikkelsen JG (2014) miRNA sponges: soaking up miRNAs for regulation of gene expression. WIREs RNA 5(3):317–333. https://doi.org/10.1002/wrna.1213

Batkai S, Genschel C, Viereck J et al (2021) CDR132L improves systolic and diastolic function in a large animal model of chronic heart failure. Eur Heart J 42(2):192–201. https://doi.org/10.1093/eurheartj/ehaa791

Borjas T, Jacob A, Kobritz M et al (2023) An engineered miRNA PS-OMe miR130 inhibits acute lung injury by targeting eCIRP in sepsis. Mol Med 29(1):21. https://doi.org/10.1186/s10020-023-00607-8

Bridges MC, Daulagala AC, Kourtidis A (2021) LNCcation: lncRNA localization and function. J Cell Biol 220(2). https://doi.org/10.1083/jcb.202009045

Bushati N, Cohen SM (2007) microRNA functions. Annu Rev Cell Dev Biol 23:175–205. https://doi.org/10.1146/annurev.cellbio.23.090506.123406

Cao YZ, Sun JY, Chen YX et al (2021) The roles of circRNAs in cancers: perspectives from molecular functions. Gene 767:145182. https://doi.org/10.1016/j.gene.2020.145182

Cazalla D, Steitz JA (2010) Down-regulation of a host microRNA by a viral noncoding RNA. Cold Spring Harb Symp Quant Biol 75:321–324. https://doi.org/10.1101/sqb.2010.75.009

Cazalla D, Yario T, Steitz JA (2010) Down-regulation of a host microRNA by a Herpesvirus saimiri noncoding RNA. Science (New York, NY) 328(5985):1563–1566. https://doi.org/10.1126/science.1187197

Chang C (2012) Chapter 12 - approaches to autoimmune diseases using epigenetic therapy. In: Tollefsbol TO (ed) Epigenetics in human disease. Academic Press, San Diego, pp 225–251. https://doi.org/10.1016/B978-0-12-388415-2.00012-3
Chen R, Qiu H, Tong Y et al (2019) MiRNA-19a-3p alleviates the progression of osteoporosis by targeting HDAC4 to promote the osteogenic differentiation of hMSCs. Biochem Biophys Res Commun 516(3):666–672. https://doi.org/10.1016/j.bbrc.2019.06.083
Chen Q, Wang M, Wu S (2020a) The lncRNA MCF2L-AS1 controls osteogenic differentiation by regulating miR-33a. Cell Cycle 19(9):1059–1065. https://doi.org/10.1080/15384101.2020.1747776
Chen Z, Zhao F, Liang C et al (2020b) Silencing of miR-138-5p sensitizes bone anabolic action to mechanical stimuli. Theranostics 10(26):12263–12278. https://doi.org/10.7150/thno.53009
Chen Z, Zhang Y, Zhao F et al (2021) miR-138–5p negatively regulates osteoblast differentiation through inhibiting β-catenin under simulated microgravity in MC3T3-E1 cells. Acta Astronaut 182:240–250. https://doi.org/10.1016/j.actaastro.2021.01.052
Chen Z, Huai Y, Chen G et al (2022) MiR-138-5p targets MACF1 to aggravate aging-related bone loss. Int J Biol Sci 18(13):4837–4852. https://doi.org/10.7150/ijbs.71411
Chunping Ma XZ, Yang L, Yujiao W, Liu H, Zhang X, Wei Y (2016) Preparation of fluorescent organic nanoparticles from polyethylenimine and sucrose for cell imaging. Mater Sci Eng C 68:37–42
Cornec-Le Gall E, Alam A, Perrone RD (2019) Autosomal dominant polycystic kidney disease. Lancet 393(10174):919–935. https://doi.org/10.1016/s0140-6736(18)32782-x
Das K, Rao LVM (2022) The role of microRNAs in inflammation. Int J Mol Sci 23(24). https://doi.org/10.3390/ijms232415479
Davis-Dusenbery BN, Hata A (2010) Mechanisms of control of microRNA biogenesis. J Biochem 148(4):381–392. https://doi.org/10.1093/jb/mvq096
Dean DA, Strong DD, Zimmer WE (2005) Nuclear entry of nonviral vectors. Gene Ther 12:881–890
Di Sanzo M, Quaresima B, Biamonte F et al (2020) FTH1 Pseudogenes in cancer and cell metabolism. Cells 9(12):2554
Dissanayake E, Inoue Y (2016) MicroRNAs in allergic disease. Curr Allergy Asthma Rep 16(9):67. https://doi.org/10.1007/s11882-016-0648-z
Dong X, Lv B, Li Y et al (2017) MiR-143 regulates the proliferation and migration of osteosarcoma cells through targeting MAPK7. Arch Biochem Biophys 630:47–53. https://doi.org/10.1016/j.abb.2017.07.011
Endisha H, Datta P, Sharma A et al (2021) MicroRNA-34a-5p promotes joint destruction during osteoarthritis. Arthritis Rheumatol 73(3):426–439. https://doi.org/10.1002/art.41552
Fava A, Petri M (2019) Systemic lupus erythematosus: diagnosis and clinical management. J Autoimmun 96:1–13. https://doi.org/10.1016/j.jaut.2018.11.001
Feng R, Patil S, Zhao X et al (2021) RNA therapeutics - research and clinical advancements. Front Mol Biosci 8:710738. https://doi.org/10.3389/fmolb.2021.710738
Finnegan EF, Pasquinelli AE (2013) MicroRNA biogenesis: regulating the regulators. Crit Rev Biochem Mol Biol 48(1):51–68. https://doi.org/10.3109/10409238.2012.738643
Flamme M, McKenzie LK, Sarac I et al (2019) Chemical methods for the modification of RNA. Methods 161:64–82. https://doi.org/10.1016/j.ymeth.2019.03.018
Fujii T, Murata K, Mun S-H et al (2021) MEF2C regulates osteoclastogenesis and pathologic bone resorption via c-FOS. Bone Res 9(1):4. https://doi.org/10.1038/s41413-020-00120-2
Gallant-Behm CL, Piper J, Dickinson BA et al (2018) A synthetic microRNA-92a inhibitor (MRG-110) accelerates angiogenesis and wound healing in diabetic and nondiabetic wounds. Wound Repair Regeneration 26(4):311–323. https://doi.org/10.1111/wrr.12660
Gao Y, Alam U, Tang Q et al (2016) Functional lipids based on [12]aneN3 and naphthalimide as efficient non-viral gene vectors. Org Biomol Chem 14:6346–6354
Gao Y, Alam U, Ding A-X et al (2018) [12]aneN3-based lipid with naphthalimide moiety for enhanced gene transfection efficiency. Bioorg Chem 79:334–340
Gao Y, Patil S, Qian A (2020) The role of MicroRNAs in bone metabolism and disease. Int J Mol Sci 21(17):6081

Gaudet AD, Fonken LK, Watkins LR et al (2018) MicroRNAs: roles in regulating Neuroinflammation. Neuroscient Rev J Bringing Neurobiol Neurol Psychiat 24(3):221–245. https://doi.org/10.1177/1073858417721150

Gomez IG, MacKenna DA, Johnson BG et al (2015) Anti-microRNA-21 oligonucleotides prevent Alport nephropathy progression by stimulating metabolic pathways. J Clin Invest 125(1):141–156. https://doi.org/10.1172/jci75852

Guan L, Huang S, Chen Z et al (2015) Low cytotoxicity fluorescent PAMAM dendrimer as gene carriers for monitoring the delivery of siRNA. J Nanopart Res 17:385

Hajarnis S, Lakhia R, Yheskel M et al (2017) microRNA-17 family promotes polycystic kidney disease progression through modulation of mitochondrial metabolism. Nat Commun 8:14395. https://doi.org/10.1038/ncomms14395

Hattori Y, Maitani Y (2005) Folate-linked nanoparticle-mediated suicide gene therapy in human prostate cancer and nasopharyngeal cancer with herpes simplex virus thymidine kinase. Cancer Gene Ther 12:796–809

Hiramatsu-Asano S, Wada J (2022) Therapeutic approaches targeting miRNA in systemic lupus Erythematosus. Acta Med Okayama 76(4):359–371. https://doi.org/10.18926/amo/63887

Ho PTB, Clark IM, Le LTT (2022) MicroRNA-based diagnosis and therapy. Int J Mol Sci 23(13). https://doi.org/10.3390/ijms23137167

Hong Y, Cao H, Wang Q et al (2016) MiR-22 may suppress Fibrogenesis by targeting TGFβR I in cardiac fibroblasts. Cell Physiol Biochem 40(6):1345–1353. https://doi.org/10.1159/000453187

Hong H, Sun Y, Deng H et al (2019) Dysregulation of cPWWP2A-miR-579 axis mediates dexamethasone-induced cytotoxicity in human osteoblasts. Biochem Biophys Res Commun 517(3):491–498. https://doi.org/10.1016/j.bbrc.2019.07.095

Hong DS, Kang YK, Borad M et al (2020a) Phase 1 study of MRX34, a liposomal miR-34a mimic, in patients with advanced solid tumours. Br J Cancer 122(11):1630–1637. https://doi.org/10.1038/s41416-020-0802-1

Hong SM, Liu C, Yin Z et al (2020b) MicroRNAs in systemic lupus Erythematosus: a perspective on the path from biological discoveries to clinical practice. Curr Rheumatol Rep 22(6):17. https://doi.org/10.1007/s11926-020-00895-7

Hou Y, Feng H, Jiao J et al (2019) Mechanism of miR-143-3p inhibiting proliferation, migration and invasion of osteosarcoma cells by targeting MAPK7. Artif Cells Nanomed Biotechnol 47(1):2065–2071. https://doi.org/10.1080/21691401.2019.1620252

Huang Y, Zheng Y, Jia L et al (2015) Long noncoding RNA H19 promotes osteoblast differentiation via TGF-β1/Smad3/HDAC signaling pathway by deriving miR-675. Stem Cells 33(12):3481–3492. https://doi.org/10.1002/stem.2225

Huang X, Cen X, Zhang B et al (2019) The roles of circRFWD2 and circINO80 during NELL-1-induced osteogenesis. J Cell Mol Med 23(12):8432–8441. https://doi.org/10.1111/jcmm.14726

Huang CK, Kafert-Kasting S, Thum T (2020a) Preclinical and clinical development of noncoding RNA therapeutics for cardiovascular disease. Circ Res 126(5):663–678. https://doi.org/10.1161/circresaha.119.315856

Huang M, Li X, Zhou C et al (2020b) Noncoding RNA miR-205-5p mediates osteoporosis pathogenesis and osteoblast differentiation by regulating RUNX2. J Cell Biochem 121(10):4196–4203. https://doi.org/10.1002/jcb.29599

Iwakawa H-o, Tomari Y (2022) Life of RISC: formation, action, and degradation of RNA-induced silencing complex. Mol Cell 82(1):30–43. https://doi.org/10.1016/j.molcel.2021.11.026

Janssen HLA, Reesink HW, Lawitz EJ et al (2013) Treatment of HCV infection by targeting MicroRNA. N Engl J Med 368(18):1685–1694. https://doi.org/10.1056/NEJMoa1209026

Jian YP, Dong SJ, Xu SS et al (2020) MicroRNA-34a suppresses neuronal apoptosis and alleviates microglia inflammation by negatively targeting the notch pathway in spinal cord injury. Eur Rev Med Pharmacol Sci 24(3):1420–1427. https://doi.org/10.26355/eurrev_202002_20199

Jiepin Yang QZ, Chang H, Cheng Y (2015) Surface-engineered Dendrimers in gene delivery. Chem Rev 115:5274–5300

Krützfeldt J, Rajewsky N, Braich R et al (2005) Silencing of microRNAs in vivo with 'antagomirs'. Nature 438(7068):685–689. https://doi.org/10.1038/nature04303
Lee YS, Dutta A (2009) MicroRNAs in cancer. Annu Rev Pathol 4:199–227. https://doi.org/10.1146/annurev.pathol.4.110807.092222
Lee Y, Ahn C, Han J et al (2003) The nuclear RNase III Drosha initiates microRNA processing. Nature 425(6956):415–419. https://doi.org/10.1038/nature01957
Lee EC, Valencia T, Allerson C et al (2019) Discovery and preclinical evaluation of anti-miR-17 oligonucleotide RGLS4326 for the treatment of polycystic kidney disease. Nat Commun 10(1):4148. https://doi.org/10.1038/s41467-019-11918-y
Li Y, Masaki T, Lemon SM (2013) miR-122 and the hepatitis C RNA genome: more than just stability. RNA Biol 10(6):919–923. https://doi.org/10.4161/rna.25137
Li JP, Zhuang HT, Xin MY et al (2017) MiR-214 inhibits human mesenchymal stem cells differentiating into osteoblasts through targeting β-catenin. Eur Rev Med Pharmacol Sci 21(21):4777–4783
Li Z, Zhang W, Huang Y (2018) MiRNA-133a is involved in the regulation of postmenopausal osteoporosis through promoting osteoclast differentiation. Acta Biochim Biophys Sin 50(3):273–280. https://doi.org/10.1093/abbs/gmy006
Li D, Yang C, Tian Y et al (2021a) Roles and mechanism of Long noncoding RNAs in bone diseases. In: Qian A, Tian Y (eds) Noncoding RNAs and bone. Springer, Singapore, pp 95–128. https://doi.org/10.1007/978-981-16-2402-5_5
Li R, Ruan Q, Yin F et al (2021b) MiR-23b-3p promotes postmenopausal osteoporosis by targeting MRC2 and regulating the Wnt/β-catenin signaling pathway. J Pharmacol Sci 145(1):69–78. https://doi.org/10.1016/j.jphs.2020.11.004
Li S, Wei X, He J et al (2021c) The comprehensive landscape of miR-34a in cancer research. Cancer Metastasis Rev 40(3):925–948. https://doi.org/10.1007/s10555-021-09973-3
Li H, Yin C, Li J et al (2023) MiR-12200-5p targets multiple members of Wnt signaling pathway to inhibit osteoblast differentiation and bone formation. Endocr Metab Immune Disord Drug Targets. https://doi.org/10.2174/1871530323666230301150350
Lian WS, Ko JY, Chen YS et al (2019) MicroRNA-29a represses osteoclast formation and protects against osteoporosis by regulating PCAF-mediated RANKL and CXCL12. Cell Death Dis 10(10):705. https://doi.org/10.1038/s41419-019-1942-1
Lin X, Qiu W, Xiao Y et al (2019) MiR-199b-5p suppresses tumor angiogenesis mediated by vascular endothelial cells in breast cancer by targeting ALK1. Front Genet 10:1397. https://doi.org/10.3389/fgene.2019.01397
Ling H (2016) Non-coding RNAs: therapeutic strategies and delivery systems. In: Slaby O, Calin G (eds) Non-coding RNAs in colorectal cancer. Advances in experimental medicine and biology, vol 937. Springer, Cham. https://doi.org/10.1007/978-3-319-42059-2_12
Ling Liu Z-MZ, Liuc Q, Jiangc S-S, Zhang Q, Cen L-Q, Gao J, Gao X-G, Huang J-D, Liu Y, Yao H (2018) A novel galactose-PEG-conjugated biodegradable copolymer is an efficient gene delivery vector for immunotherapy of hepatocellular carcinoma. Biomaterials 184:20–30
Liu N, Zhang J, Jiao T et al (2013) Hepatitis B virus inhibits apoptosis of hepatoma cells by sponging the MicroRNA 15a/16 cluster. J Virol 87(24):13370–13378. https://doi.org/10.1128/jvi.02130-13
Liu N, Jiao T, Huang Y et al (2015) Hepatitis B virus regulates apoptosis and tumorigenesis through the microRNA-15a-Smad7-transforming growth factor beta pathway. J Virol 89(5):2739–2749. https://doi.org/10.1128/jvi.02784-14
Liu F, Gong R, He B et al (2018) TUSC2P suppresses the tumor function of esophageal squamous cell carcinoma by regulating TUSC2 expression and correlates with disease prognosis. BMC Cancer 18(1):894. https://doi.org/10.1186/s12885-018-4804-9
Llobat L, Gourbault O (2021) Role of MicroRNAs in human osteosarcoma: future perspectives. Biomedicines 9(5). https://doi.org/10.3390/biomedicines9050463
Long H, Wang X, Chen Y et al (2018) Dysregulation of microRNAs in autoimmune diseases: pathogenesis, biomarkers and potential therapeutic targets. Cancer Lett 428:90–103. https://doi.org/10.1016/j.canlet.2018.04.016

Lou W, Ding B, Fu P (2020) Pseudogene-derived lncRNAs and their miRNA sponging mechanism in human cancer. Front Cell Dev Biol 8:85. https://doi.org/10.3389/fcell.2020.00085
Lu TX, Rothenberg ME (2018) MicroRNA. J Allergy Clin Immunol 141(4):1202–1207. https://doi.org/10.1016/j.jaci.2017.08.034
Ma J, Lin X, Chen C et al (2019) Circulating miR-181c-5p and miR-497-5p are potential biomarkers for prognosis and diagnosis of osteoporosis. J Clin Endocrinol Metabol 105(5):1445–1460. https://doi.org/10.1210/clinem/dgz300
Mei L, Li M, Zhang T (2021) MicroRNA miR-874-3p inhibits osteoporosis by targeting leptin (LEP). Bioengineered 12(2):11756–11767. https://doi.org/10.1080/21655979.2021.2009618
Mi B, Xiong Y, Chen L et al (2019) CircRNA AFF4 promotes osteoblast cells proliferation and inhibits apoptosis via the Mir-7223-5p/PIK3R1 axis. Aging 11(24):11988–12001. https://doi.org/10.18632/aging.102524
Min Wang YG, Meng Y, Ma PX, Mao C, Lei B (2017) Photoluminescent and biodegradable polycitrate-polyethylene glycol-polyethyleneimine polymers as highly biocompatible and efficient vectors for bioimaging-guided siRNA and miRNA delivery. Acta Biomater 54:69–80
Nakano M, Fukushima Y, Yokota S et al (2015) CYP2A7 pseudogene transcript affects CYP2A6 expression in human liver by acting as a decoy for miR-126. Drug Metab Dispos 43(5):703–712. https://doi.org/10.1124/dmd.115.063255
Napoli S (2021) LncRNAs and available databases. Methods Mol Biol 2348:3–26. https://doi.org/10.1007/978-1-0716-1581-2_1
Ning SL, Zhu H, Shao J et al (2019) MiR-21 inhibitor improves locomotor function recovery by inhibiting IL-6R/JAK-STAT pathway-mediated inflammation after spinal cord injury in model of rat. Eur Rev Med Pharmacol Sci 23(2):433–440. https://doi.org/10.26355/eurrev_201901_16852
Obad S, dos Santos CO, Petri A et al (2011) Silencing of microRNA families by seed-targeting tiny LNAs. Nat Genet 43(4):371–378. https://doi.org/10.1038/ng.786
Oh SK, Jeon SR (2016) Current concept of stem cell therapy for spinal cord injury: a review. Korean J Neurotrauma 12(2):40–46. https://doi.org/10.13004/kjnt.2016.12.2.40
Pan W, Zhu S, Yuan M et al (2010) MicroRNA-21 and MicroRNA-148a contribute to DNA Hypomethylation in lupus CD4+ T cells by directly and indirectly targeting DNA Methyltransferase 1. J Immunol 184(12):6773–6781. https://doi.org/10.4049/jimmunol.0904060
Pan W, Wang H, Jianwei R et al (2014) MicroRNA-27a promotes proliferation, migration and invasion by targeting MAP2K4 in human osteosarcoma cells. Cell Physiol Biochem 33(2):402–412. https://doi.org/10.1159/000356679
Patil S, Gao Y, Lin X et al (2019) The development of functional non-viral vectors for gene delivery. Int J Mol Sci 20(21):5491. https://doi.org/10.3390/ijms20215491
Patil S, Gao Y, Qian A (2023) Peptide-modified PAMAM-based bone-targeting RNA delivery system. Future J Pharmaceutical Sci 9(1):109. https://doi.org/10.1186/s43094-023-00560-7
Pauley KM, Chan EKL (2008) MicroRNAs and their emerging roles in immunology. Ann N Y Acad Sci 1143(1):226–239. https://doi.org/10.1196/annals.1443.009
Peng H, Ishida M, Li L et al (2015) Pseudogene INTS6P1 regulates its cognate gene INTS6 through competitive binding of miR-17-5p in hepatocellular carcinoma. Oncotarget 6(8):5666–5677. https://doi.org/10.18632/oncotarget.3290
Petrović N, Mandušić V, Stanojević B et al (2014) The difference in miR-21 expression levels between invasive and non-invasive breast cancers emphasizes its role in breast cancer invasion. Med Oncol 31(3):867. https://doi.org/10.1007/s12032-014-0867-x
Prakash TP (2011) An overview of sugar-modified oligonucleotides for antisense therapeutics. Chem Biodivers 8(9):1616–1641. https://doi.org/10.1002/cbdv.201100081
Prakash TP, Bhat B (2007) 2'-modified oligonucleotides for antisense therapeutics. Curr Top Med Chem 7(7):641–649. https://doi.org/10.2174/156802607780487713
Pritchard CC, Cheng HH, Tewari M (2012) MicroRNA profiling: approaches and considerations. Nat Rev Genet 13(5):358–369. https://doi.org/10.1038/nrg3198

Qian A, Tian Y (2021) Noncoding RNAs and bone, 1st edn. Springer, Singapore. https://doi.org/10.1007/978-981-16-2402-5

Qiang Jin LF, Wang D-D, Jing-Jing W, Hou J, Dai Z-R, Sun S-G, Wang J-Y, Ge G-B, Cui J-N, Yang L (2016) A highly selective near-infrared fluorescent probe for carboxylesterase 2 and its bioimaging applications in living cells and animals. Biosens Bioelectron 83:193–199

Qiao L, Liu D, Li CG et al (2018) MiR-203 is essential for the shift from osteogenic differentiation to adipogenic differentiation of mesenchymal stem cells in postmenopausal osteoporosis. Eur Rev Med Pharmacol Sci 22(18):5804–5814. https://doi.org/10.26355/eurrev_201809_15906

Qiao L, Li CG, Liu D (2020) CircRNA_0048211 protects postmenopausal osteoporosis through targeting miRNA-93-5p to regulate BMP2. Eur Rev Med Pharmacol Sci 24(7):3459–3466. https://doi.org/10.26355/eurrev_202004_20804

Qin S, Zhang C (2011) MicroRNAs in vascular disease. J Cardiovasc Pharmacol 57(1):8–12. https://doi.org/10.1097/FJC.0b013e318203759b

Querfeld C, Foss FM, Kim YH et al (2018) Phase 1 trial of Cobomarsen, an inhibitor of Mir-155, in cutaneous T cell lymphoma. Blood 132(Supplement 1):2903–2903. https://doi.org/10.1182/blood-2018-99-119861

Rasmussen TK, Andersen T, Bak RO et al (2015) Overexpression of microRNA-155 increases IL-21 mediated STAT3 signaling and IL-21 production in systemic lupus erythematosus. Arthritis Res Ther 17(1):154. https://doi.org/10.1186/s13075-015-0660-z

Rutnam ZJ, Du WW, Yang W et al (2014) The pseudogene TUSC2P promotes TUSC2 function by binding multiple microRNAs. Nat Commun 5:2914. https://doi.org/10.1038/ncomms3914

Sakashita M, Mochizuki S, Sakurai K (2014) Hepatocyte-targeting gene delivery using a lipoplex composed of galactose-modified aromatic lipid synthesized with click chemistry. Bioorg Med Chem 22:5212–5219

Seeliger C, Karpinski K, Haug AT et al (2014) Five freely circulating miRNAs and bone tissue miRNAs are associated with osteoporotic fractures. J Bone Miner Res 29(8):1718–1728. https://doi.org/10.1002/jbmr.2175

Shang J, Wang Z, Chen W et al (2019) Pseudogene CHIAP2 inhibits proliferation and invasion of lung adenocarcinoma cells by means of the WNT pathway. J Cell Physiol 234(8):13735–13746. https://doi.org/10.1002/jcp.28053

Si ML, Zhu S, Wu H et al (2007) miR-21-mediated tumor growth. Oncogene 26(19):2799–2803. https://doi.org/10.1038/sj.onc.1210083

Tahamtan A, Teymoori-Rad M, Nakstad B et al (2018) Anti-inflammatory MicroRNAs and their potential for inflammatory diseases treatment. Front Immunol 9:1377. https://doi.org/10.3389/fimmu.2018.01377

Tan S, Kang Y, Li H et al (2021) circST6GALNAC6 suppresses bladder cancer metastasis by sponging miR-200a-3p to modulate the STMN1/EMT axis. Cell Death Dis 12(2):168. https://doi.org/10.1038/s41419-021-03459-4

Tang L, Chen HY, Hao NB et al (2017) microRNA inhibitors: natural and artificial sequestration of microRNA. Cancer Lett 407:139–147. https://doi.org/10.1016/j.canlet.2017.05.025

Täubel J, Hauke W, Rump S et al (2021) Novel antisense therapy targeting microRNA-132 in patients with heart failure: results of a first-in-human phase 1b randomized, double-blind, placebo-controlled study. Eur Heart J 42(2):178–188. https://doi.org/10.1093/eurheartj/ehaa898

Teng Y, Zhang R, Liu C et al (2015) miR-143 inhibits interleukin-13-induced inflammatory cytokine and mucus production in nasal epithelial cells from allergic rhinitis patients by targeting IL13Rα1. Biochem Biophys Res Commun 457(1):58–64. https://doi.org/10.1016/j.bbrc.2014.12.058

Tesser-Gamba F, Lopes LJ, Petrilli AS et al (2016) MAPK7 gene controls proliferation, migration and cell invasion in osteosarcoma. Mol Carcinog 55(11):1700–1713. https://doi.org/10.1002/mc.22420

Tian Y, Zhao Y, Yin C et al (2022) Polyvinylamine with moderate binding affinity as a highly effective vehicle for RNA delivery. J Control Release 345:20–37. https://doi.org/10.1016/j.jconrel.2022.03.003

Travis G, McGowan EM, Simpson AM et al (2023) PTEN, PTENP1, microRNAs, and ceRNA networks: precision targeting in cancer therapeutics. Cancers 15(20). https://doi.org/10.3390/cancers15204954

Válóczi A, Hornyik C, Varga N et al (2004) Sensitive and specific detection of microRNAs by northern blot analysis using LNA-modified oligonucleotide probes. Nucleic Acids Res 32(22):e175–e175. https://doi.org/10.1093/nar/gnh171

van Zandwijk N, Pavlakis N, Kao SC et al (2017) Safety and activity of microRNA-loaded minicells in patients with recurrent malignant pleural mesothelioma: a first-in-man, phase 1, open-label, dose-escalation study. Lancet Oncol 18(10):1386–1396. https://doi.org/10.1016/s1470-2045(17)30621-6

Vanaja SK, Rathinam VA, Fitzgerald KA (2015) Mechanisms of inflammasome activation: recent advances and novel insights. Trends Cell Biol 25(5):308–315. https://doi.org/10.1016/j.tcb.2014.12.009

Wang Z (2011) The guideline of the design and validation of MiRNA mimics. In: Wu W (ed) MicroRNA and cancer: methods and protocols. Humana Press, Totowa, pp 211–223. https://doi.org/10.1007/978-1-60761-863-8_15

Wang X, Shao N, Zhang Q et al (2014) Mitochondrial targeting dendrimer allows efficient and safe gene delivery. J Mater Chem B 2:2546–2553

Wang Y, Liu W, Liu Y et al (2018) Long noncoding RNA H19 mediates LCoR to impact the osteogenic and adipogenic differentiation of mBMSCs in mice through sponging miR-188. J Cell Physiol 233(9):7435–7446. https://doi.org/10.1002/jcp.26589

Wang CG, Wang L, Yang T et al (2020) Pseudogene PTENP1 sponges miR-214 to regulate the expression of PTEN to modulate osteoclast differentiation and attenuate osteoporosis. Cytotherapy 22(8):412–423. https://doi.org/10.1016/j.jcyt.2020.04.090

Wang L, Dong J, Ma J et al (2021) By inhibiting ADCY5, miR-18a-3p promotes osteoporosis and possibly contributes to spinal fracture. Biochem Biophys Res Commun 550:49–55. https://doi.org/10.1016/j.bbrc.2021.02.118

Wang Z, Zhang H, Li Q et al (2023) Long non-coding RNA KCNQ1OT1 alleviates postmenopausal osteoporosis by modulating miR-421-3p/mTOR axis. Sci Rep 13(1):2333. https://doi.org/10.1038/s41598-023-29546-4

Wen J, Guan Z, Yu B et al (2020) Circular RNA hsa_circ_0076906 competes with OGN for miR-1305 biding site to alleviate the progression of osteoporosis. Int J Biochem Cell Biol 122:105719. https://doi.org/10.1016/j.biocel.2020.105719

White P (2011) Profiling the miRNome: detecting global miRNA expression levels with DNA microarrays. In: Harper SQ (ed) RNA Interference Techniques. Humana Press, Totowa, pp 91–111. https://doi.org/10.1007/978-1-61779-114-7_6

Wu J, Zhao J, Sun L et al (2018) Long non-coding RNA H19 mediates mechanical tension-induced osteogenesis of bone marrow mesenchymal stem cells via FAK by sponging miR-138. Bone 108:62–70. https://doi.org/10.1016/j.bone.2017.12.013

Wu H, Chang C, Lu Q (2020a) The epigenetics of lupus Erythematosus. Adv Exp Med Biol 1253:185–207. https://doi.org/10.1007/978-981-15-3449-2_7

Wu J, Ren W, Zheng Z et al (2020b) Mmu_circ_003795 regulates osteoblast differentiation and mineralization in MC3T3-E1 and MDPC23 by targeting COL15A1. Mol Med Rep. https://doi.org/10.3892/mmr.2020.11264

Xiao L, Jiang L, Hu Q et al (2017) MicroRNA-133b ameliorates allergic inflammation and symptom in murine model of allergic rhinitis by targeting Nlrp3. Cell Physiol Biochem 42(3):901–912. https://doi.org/10.1159/000478645

Xiong W, Li C, Kong G et al (2022) Treg cell-derived exosomes miR-709 attenuates microglia pyroptosis and promotes motor function recovery after spinal cord injury. J Nanobiotechnol 20(1):529. https://doi.org/10.1186/s12951-022-01724-y

Xu Z, Jin J, Siu LKS et al (2012) Folic acid conjugated mPEG-PEI600 as an efficient non-viral vector for targeted nucleic acid delivery. Int J Pharm 426:182–192

Xue Y, Fan X, Yang R et al (2020) miR-29b-3p inhibits post-infarct cardiac fibrosis by targeting FOS. Biosci Rep 40(9). https://doi.org/10.1042/bsr20201227

Yan LX, Wu QN, Zhang Y et al (2011) Knockdown of miR-21 in human breast cancer cell lines inhibits proliferation, in vitro migration and in vivo tumor growth. Breast Cancer Res 13(1):R2. https://doi.org/10.1186/bcr2803

Yang W, Du WW, Li X et al (2016) Foxo3 activity promoted by non-coding effects of circular RNA and Foxo3 pseudogene in the inhibition of tumor growth and angiogenesis. Oncogene 35(30):3919–3931. https://doi.org/10.1038/onc.2015.460

Yang B, Huang X, Xu S et al (2021) Decreased miR-4512 levels in monocytes and macrophages of individuals with systemic lupus Erythematosus contribute to innate immune activation and Neutrsophil NETosis by targeting TLR4 and CXCL2. Front Immunol 12:756825. https://doi.org/10.3389/fimmu.2021.756825

Yi R, Qin Y, Macara IG et al (2003) Exportin-5 mediates the nuclear export of pre-microRNAs and short hairpin RNAs. Genes Dev 17(24):3011–3016. https://doi.org/10.1101/gad.1158803

Yin H, Zhang S, Sun Y et al (2017) MicroRNA-34/449 targets IGFBP-3 and attenuates airway remodeling by suppressing Nur77-mediated autophagy. Cell Death Dis 8(8):e2998. https://doi.org/10.1038/cddis.2017.357

Yin C, Tian Y, Yu Y et al (2020) miR-129-5p inhibits bone formation through TCF4. Front Cell Dev Biol 8:600641. https://doi.org/10.3389/fcell.2020.600641

Yoo H, Juliano RL (2000) Enhanced delivery of antisense oligonucleotides with fluorophore-conjugated PAMAM dendrimers. Nucleic Acids Res 28(21):4225–4231

Yoshiyuki Hattori MS, Maitani Y (2006) Folate-linked lipid-based nanoparticles deliver a NFk B decoy into activated murine macrophage-like RAW264.7 cells. Biol Pharm Bull 29(7):1516–1520

Yoshizawa T, Hattori Y, Hakoshima M et al (2008) Folate-linked lipid-based nanoparticles for synthetic siRNA delivery in KB tumor xenografts. Eur J Pharm Biopharm 70:718–725

You L, Pan L, Chen L et al (2016) MiR-27a is essential for the shift from Osteogenic differentiation to Adipogenic differentiation of Mesenchymal stem cells in postmenopausal osteoporosis. Cell Physiol Biochem 39(1):253–265. https://doi.org/10.1159/000445621

Yu J-C, Zhu S, Feng P-J et al (2015) Cationic fluorescent polymer core–shell nanoparticles for encapsulation, delivery, and non-invasively tracking the intracellular release of siRNA. Chem Commun 51:2976–2979. https://doi.org/10.1039/c4cc09685a

Yu Y, Men S, Zhang Y (2021) miR-20a-5p ameliorates ovalbumin (OVA)-induced mouse model of allergic asthma through targeting ATG7-regulated cell death, fibrosis and inflammation. Int Immunopharmacol 95:107342. https://doi.org/10.1016/j.intimp.2020.107342

Yuan J, Chen H, Ge D et al (2017) Mir-21 promotes cardiac fibrosis after myocardial infarction via targeting Smad7. Cell Physiol Biochem 42(6):2207–2219. https://doi.org/10.1159/000479995

Zhang R, Guo Y, Ma Z et al (2017) Long non-coding RNA PTENP1 functions as a ceRNA to modulate PTEN level by decoying miR-106b and miR-93 in gastric cancer. Oncotarget 8(16):26079–26089. https://doi.org/10.18632/oncotarget.15317

Zhang G, Zhong L, Luo H et al (2019a) MicroRNA-155-3p promotes breast cancer progression through down-regulating CADM1. Onco Targets Ther 12:7993–8002. https://doi.org/10.2147/ott.S206180

Zhang Y, Chen B, Li D et al (2019b) LncRNA NEAT1/miR-29b-3p/BMP1 axis promotes osteogenic differentiation in human bone marrow-derived mesenchymal stem cells. Pathol Res Pract 215(3):525–531. https://doi.org/10.1016/j.prp.2018.12.034

Zhao M, Dong J, Liao Y et al (2022) MicroRNA miR-18a-3p promotes osteoporosis and possibly contributes to spinal fracture by inhibiting the glutamate AMPA receptor subunit 1 gene (GRIA1). Bioengineered 13(1):370–382. https://doi.org/10.1080/21655979.2021.2005743

Zhao Y, Deng X, Tan S et al (2023) Co-polymer carrier with dual advantages of cartilage-penetrating and targeting improves delivery and efficacy of MicroRNA treatment of osteoarthritis. Adv Healthc Mater 12(6):2202143. https://doi.org/10.1002/adhm.202202143

Zheng SR, Guo GL, Zhang W et al (2012) Clinical significance of miR-155 expression in breast cancer and effects of miR-155 ASO on cell viability and apoptosis. Oncol Rep 27(4):1149–1155. https://doi.org/10.3892/or.2012.1634

Zheng SR, Guo GL, Zhai Q et al (2013) Effects of miR-155 antisense oligonucleotide on breast carcinoma cell line MDA-MB-157 and implanted tumors. Asian Pac J Cancer Prev 14(4):2361–2366. https://doi.org/10.7314/apjcp.2013.14.4.2361
Zheng J, Huang Y, Li Y et al (2020) lncRNA-TINCR functions as a competitive endogenous RNA to regulate the migration of Mesenchymal stem cells by sponging miR-761. Biomed Res Int 2020:9578730. https://doi.org/10.1155/2020/9578730
Zhou WY, Cai ZR, Liu J et al (2020) Circular RNA: metabolism, functions and interactions with proteins. Mol Cancer 19(1):172. https://doi.org/10.1186/s12943-020-01286-3
Zhu S, Si ML, Wu H et al (2007) MicroRNA-21 targets the tumor suppressor gene tropomyosin 1 (TPM1). J Biol Chem 282(19):14328–14336. https://doi.org/10.1074/jbc.M611393200
Zhu Y, Zhu L, Wang X et al (2022) RNA-based therapeutics: an overview and prospectus. Cell Death Dis 13(7):644. https://doi.org/10.1038/s41419-022-05075-2

Chapter 5
The Interplay Between MicroRNAs and Genetic Diseases

Sakshi Kushwaha, Anto George, Harsimrat Kaur, Ajay Elangovan, Sindduja Muthukumar, Mahalaxmi Iyer, Mukesh Kumar Yadav, Jyoti Parkash, Dhivya Venkatesan, and Balachandar Vellingiri

Abstract miRNA is a class of noncoding RNA believed to have no significance in human health but proven to have greater implication in disease pathophysiology. miRNAs have a substantial role in diseases such as cancers, neurological diseases, and heart diseases. Furthermore, the impact of miRNAs has also been extended to genetic diseases such as β-Thalassemia. Both point and large mutations and mutations in 3′UTR of miRNAs may have significant effects on disease occurrence. Mutations in the 3′UTR of mRNAs can disrupt miRNA binding, impairing normal gene regulation, which can lead to improper protein formation and deposition, contributing to disease pathology. MiRNAs play an important role in disease mechanisms, and they have the potential to be used in biomarker and therapeutic applications. The current review focuses on the miRNAs involved in genetic diseases, Mendelian disorders, and its epigenetic variations, highlighting the promi-

---

Sakshi Kushwaha, Anto George, Harsimrat Kaur and Ajay Elangovan contributed equally with all other contributors.

S. Kushwaha · A. George · H. Kaur · A. Elangovan · S. Muthukumar · B. Vellingiri (✉)
Human Cytogenetics and Stem Cell Laboratory, Department of Zoology, School of Basic Sciences, Central University of Punjab, Bathinda, Punjab, India
e-mail: balachandar.vellingiri@cup.edu.in; http://cup.edu.in/Dr_Balachandar_Vellingiri.php

M. Iyer
Department of Biotechnology, Karpagam Academy of Higher Education (Deemed to be University), Coimbatore, Tamil Nadu, India

Department of Microbiology, Central University of Punjab, Bathinda, Punjab, India

M. K. Yadav
Department of Microbiology, Central University of Punjab, Bathinda, Punjab, India

J. Parkash
Neurochemistry and Neuroendocrinology lab, Department of Zoology, Central University of Punjab, Bathinda, Punjab, India

D. Venkatesan
Department of Biotechnology, Karpagam Academy of Higher Education (Deemed to be University), Coimbatore, Tamil Nadu, India

L. M. Vaschetto (ed.), *miRNAs, Human Health and Diseases*, Epigenetics and Human Health 13, https://doi.org/10.1007/978-3-031-64788-8_5

nent role of miRNA in hemoglobinopathies. We have also discussed the role of miRNAs as diagnostic markers and as a therapeutic option in genetic diseases. Additionally, we highlight literature evidencing the manner in which miRNAs function as reliable diagnostic and therapeutic markers in various genetic diseases.

Keywords miRNAs · Genetic diseases · Hemoglobinopathies · Biomarker · Gene therapy

List of Abbreviations

AA	Adeno-associated viral vectors
AG	Argonaute
ALS	Amyotrophic lateral sclerosis
ASOs	Antisense oligonucleotides
BRCA1	BReast CAncer gene 1
CDC25A	Cell division cycle 25A
CF	Cystic fibrosis
CFTR	Cystic fibrosis transmembrane conductance regulator
CMT1A	Charcot-Marie-Tooth disease type 1A
CVD	Cardiovascular disease
DCAMKL-1	Doublecortin and CaM kinase-like 1 protein
ddPCR	Droplet Digital PCR
DGCR8	DiGeorge Critical Region 8
DICER1	Ribonuclease III endoribonuclease
DMD	Duchenne Muscular Dystrophy
DNMT	DNA methyltransferase
dsRNA	Double stranded RNA
EV	Extracellular vesicles
EZH2	Enhancer of zeste homolog 2
FRM1	Fragile X Messenger Ribonucleoprotein 1
GCLC/M	Glutamate-cysteine ligase, catalytic/modifier subunit
GFPT1	Glutamine-fructose-6-phosphate transaminase 1
HBB	Human β-globin gene
HbF	Fetal hemoglobin
HCC	Hepatitis B virus-infected hepatocellular carcinoma
HDAC	Histone deacetylation
HDL	High-density lipoprotein
HHD	Hailey-Haley disease
JNK/SAPK	Jun N-terminal kinase/stress-activated protein kinase
KLFD	Kruppel-like transcription factor D
LDL	Low-density lipoprotein
LNAs	Locked nucleic acids
LSD-1	Lysine-specific demethylase 1
MCF-7	Michigan Cancer Foundation-7

MDA-MB-237	Metastasis breast cancer cell line
miR	micro RNA
miRISC	mi-RNA-induced silencing complexes
miRNA	micro RNA
mRNA	Messenger RNA
MYC	Myelocytomatosis oncogene
ncRNA	Noncoding RNA
NF-κB	Nuclear factor kappa B
NMP1	Nucleophosmin
NRF2	Nuclear-related factor 2
PEAMT	Phosphoethanolamine N-methyltransferase
PKA-R	cAMP-dependent protein kinase
PMP22	Peripheral myelin protein 22
Pre-DM	Pre-diabetic stage
Pre-miRNA	Precursor micro RNA
Pri-miRNA	Primary micro RNA
PTEN	Phosphatase and TENsin
RAN	Ras-related nuclear protein
REEP1	Receptor expression-enhancing protein 1
RT-PCR	Reverse transcriptase-polymerase chain reaction
RTT	Rett Syndrome
SCD	Sickle cell disease
SiRNA	Small interfering RNA
SLITRK1	Slit and Trk-like family member 1
SOD1	Superoxide dismutase 1
SPG	Spastic paraplegia
UMTD	Ultrasound-targeted microbubble distraction
UTR	Untranslated region
VHL	Von Hippel-Lindau syndrome
WES	Whole-exome sequencing
WGS	Whole-genome sequencing
XPO5	Exportin 5

5.1 Introduction

Numerous new findings have been made since the initial draft of the human genome was released in February 2001, providing insight into the complexity of both the human genome and the human proteome. Research focus has shifted from messenger RNAs (mRNAs) to noncoding RNAs (ncRNAs) as primary regulators of the human genome. These regulatory sequences have the ability to generate noncoding transcripts that control gene expression and protein activity but do not encode proteins. They are significant mediators that need to be understood in medical research due to their presence and role in numerous human diseases (Ardekani and Naeini

2010). For years, the significance of ncRNAs has been underestimated. Recently, however, microRNAs (miRNAs) have emerged as the most extensively studied among them. As our understanding of the human genome and its functions has deepened, the roles of miRNAs have become increasingly evident (Esteller 2011). According to Urbich et al., miRNAs are thought to regulate about one-third of the human genome (Urbich et al. 2008). Although the exact processes underlying miRNA transcription are unknown, their expression is thought to be influenced by their locations in exons, noncoding genes, and protein-coding gene introns as well as by their proximity to other genes in the genome (Baskerville and Bartel 2005). If miRNAs have distinct promoters, they express themselves on their own and are arranged in clusters within the genome that share the same transcriptional regulatory units (Lee et al. 2004; Altuvia et al. 2005). An estimated 50% of the miRNAs expressed in the genome are thought to be transcribed from genes that do not code for proteins, with the remaining miRNAs being coded in the introns of coding genes (Saini et al. 2007). Ever since its discovery, miRNAs have been suspected to have pathogenic roles in various diseases, especially in humans, which has been confirmed by their altered expression patterns in cancers (Bonci et al. 2008; Visone and Croce 2009; Kavitha et al. 2022) and neurodegenerative diseases (Balachandar et al. 2022; Vellingiri et al. 2022; Vellingiri 2023; Elangovan et al. 2023b). Dysregulation of miRNAs can lead to two contrasting scenarios. Firstly, given their well-established role in regulating gene expression, significant alterations in miRNA activity could ultimately have a profound impact on cell survival. Secondly, if the changes in miRNA activity are minor, they might result in negligible perturbations to the cellular proteome, potentially leading to no observable consequences (Meola et al. 2009). Given their potential roles in various diseases, miRNAs warrant significant research attention. Several online databases compile valuable experimental evidence linking miRNAs to disease development (Jiang et al. 2009; Xie et al. 2013; Li and Rana 2014). Moreover, the multifaceted involvement of miRNAs in disease processes underscores their high potential as biomarkers (Zhang and Farwell 2008). Beyond mutations in protein-coding genes, alterations in miRNAs, including large deletions, point mutations, and target site mutations, can also contribute to the complexities of disease pathologies (Georges et al. 2003; Duan et al. 2007). This highlights the crucial role of miRNAs in regulating the cellular transcriptome (Nowakowski et al. 2018). These miRNA-mediated regulatory processes hold significant molecular implications for the theranostics of various diseases. This review has comprehensively explored the multifaceted role of miRNAs in genetic diseases, specifically Mendelian disorders. We have highlighted their significant contribution to the pathogenesis of hemoglobinopathies. Additionally, the potential of miRNAs as valuable diagnostic markers and promising tools for gene therapy has been discussed.

5.2 The miRNA World

MicroRNAs (miRNAs) are often encoded within the genome's introns and intergenic regions. These short, noncoding RNA molecules typically form hairpin structures with double-stranded RNA (dsRNA) stem loops of approximately 70 nucleotides in length (Morey and Avner 2004). Within the nucleus, miRNA biogenesis commences with the transcription of long primary transcripts known as pri-miRNAs. These pri-miRNAs are subsequently cleaved into precursor pre-miRNA hairpins by the RNAase III endonuclease Dicer, resulting in mature miRNAs of approximately 21-23 nucleotides (Bilen et al. 2006). Prior to cytoplasmic translocation, the export of pre-miRNAs is facilitated by Exportin 5 (XPO5) and RAN-GTP. Pre-miRNAs are converted by DICER1, a protein with an additional RNase III domain, into mature, double-stranded miRNA duplexes with 2-nt 3′-overhang structures that are recognized by Argonaute (AGO). The passenger strand is destroyed, whereas the guide strand is chosen by AGO to generate operational miRNA-induced silencing complexes (miRISCs) to silence target mRNAs along with other proteins (Bartel 2018; Michlewski and Cáceres 2019). Processing of pri-miRNAs and pre-miRNAs occurs rapidly in seconds or minutes, generating miRNA duplexes. AGO loading is crucial for miRNA production, with slow degradation as it has a median half-life of over 10 h (Kingston and Bartel 2019). miRNA can be localized in intergenic or intragenic regions, but most of them are located within intronic regions of the host genes. Intronic miRNAs and host genes often co-regulate and co-transcribe from a common promoter. The seed region, which is six to seven nucleotides long, plays an important role in miRNA recognition of targeted mRNA (Rodriguez et al. 2004; Baskerville and Bartel 2005; Kim and Kim 2007). Targeted seed site of more than 45,000 miRNA is conserved in mRNA 3′UTR, signifying its regulatory function (Friedman et al. 2009). High-throughput analyses revealed sequence features of pri-miRNA variants, which are crucial for efficient processing. It includes a UG motif at the base, a UGU motif in the apical loop, a CNNC motif, a mismatched GHG motif in the basal stem region, and stem length of 35 ± 1 base pairs (Auyeung et al. 2013; Fang and Bartel 2015). MiRNAs form multiple-to-multiple regulatory networks, that is, multiple miRNAs can target one single mRNA and multiple mRNAs can be regulated by a unique miRNA (Plaisier et al. 2012). Thus, it has been proposed that the global expression of miRNAs governs mammalian biological processes, acting as a "cellular buffer" (Cui et al. 2023).

More than 1800 miRNA genes in humans and 1100 miRNA genes in mice have been identified (Kozomara and Griffiths-Jones 2014). For humans, approximately 1900 miRNA precursors have been annotated in the miRBase v.22.1 database (Kozomara et al. 2019). Mutations affecting miRNA function and biogenesis have been seen to have a pathological role. Point or large mutations like deletions or duplications in the miRNA locus, 3′UTR associated mutations, and mutations in the genes involved in processing and functioning of miRNA lead to changes in global miRNA functions (Meola et al. 2009). RNA modifications can change the stability and efficiency of miRNAs, thus controlling miRNA dosage (Jaffrey 2014).

RNA-binding proteins can also alter miRNA biogenesis, controlling their expression by binding to them (Sin-Chan et al. 2019). Missense mutations in Drosha have been linked to microcephaly and intellectual disabilities (Barish et al. 2022). Cell-specific miRNAs are synthesized and loaded into exosomes in response to specific stimuli to mediate intercellular communication (Valadi et al. 2007). The miRNA-loaded circulating extracellular vesicles are a sign of cellular stress and can act as a disease biomarker for diagnostics (Abels and Breakefield 2016; De Miguel et al. 2020). miRNAs also regulate liver diseases (Mukherjee et al. 2022). To summarize, miRNAs regulate target mRNAs in the body and the intricate role played directly or indirectly influence a wide range of physiological activities; thereby, any dysfunction of these RNA structures has a potential to give rise to pathological effects in the body.

5.3 miRNA Potential As a Biomarker for Genetic Diseases

Given their extensive research and well-established roles, miRNAs hold significant promise as biomarkers for various diseases. This includes not only cancers but also prominent neurodegenerative disorders like Parkinson's disease (PD) and Alzheimer's disease (AD), which already have established genetic biomarkers (Elangovan et al. 2023a, b; Subramaniam et al. 2023; Vellingiri et al. 2023). Recent research suggests the potential use of heavy metals within the body as biomarkers for diseases, gaining momentum in scientific investigations (Iyer et al. 2023; Vellingiri 2023). The diverse roles of miRNAs in various developmental processes range from metabolism and cell division to differentiation, apoptosis, and even the development and differentiation of neuronal cells. The function of miRNAs in cells can be affected by various mechanisms, such as single point mutations within the miRNA or its target, or through epigenetic processes that silence miRNA transcription.

The past few decades have witnessed an explosion of discoveries and advancements in miRNA research, suggesting a bright future for these molecules in disease diagnosis, prognosis, and treatment (Correia et al. 2017; Dwivedi et al. 2019). This section will focus on their potential as biomarkers for genetic diseases. For further information regarding the utility of miRNAs as biomarkers for disease diagnosis please refer to Chap. 6. Notably, the stability and tissue-specific nature of circulating cell-free miRNAs within the extracellular space make them a promising avenue for noninvasive diagnostics in humans (Hong et al. 2021). A growing body of evidence from various studies links the expression levels of circulating miRNAs, particularly those encapsulated within protective exosomes, to a diverse range of diseases. This includes cancer, liver damage, infectious diseases, cardiovascular issues, and neurodegenerative disorders (Bhardwaj et al. 2013; Zeng et al. 2017; Aghili et al. 2018; Biswas et al. 2019; Coban et al. 2020; Teksoy et al. 2020). Given their crucial role in regulating numerous biological processes, altered expression of circulating miRNAs emerges as a promising avenue for the development of biomarkers capable of detecting a wide spectrum of genetic disease anomalies and pathologies (Correia et al. 2017).

MiRNAs are found not only within cells but also outside of them, circulating in various bodily fluids like tears, breast milk, saliva, colostrum, urine, amniotic fluid, cerebrospinal fluid, bronchial lavage fluid, pleural fluid, peritoneal fluid, seminal fluid, and gastric juices. The remarkable stability of these extracellular miRNAs, combined with the ease of noninvasive collection from bodily fluids, positions them as highly attractive candidates for diagnostic and prognostic biomarkers across various disease states (Ortiz-Quintero 2016; Prajzlerová et al. 2023).

A growing body of evidence suggests a paradigm shift in our understanding of miRNA release. This viewpoint emphasizes the systematic and active secretion of miRNAs by both immune and nonimmune cells into the extracellular space (Robbins and Morelli 2014; Fernández-Messina et al. 2015; de Candia et al. 2016) within exosomes, heterogeneous microvesicles, or in association with low-density lipoprotein (LDL), high-density lipoprotein (HDL) (Correia et al. 2017), or other RNA-binding proteins such as NPM1-nucleophosmin and AGO2 (Tijsen et al. 2012), which accounts for their stability and hence can be detected using advanced technologies such as RNA sequencing, next-generation sequencing, and microarray analysis, alongside the use of bioinformatic tools. Also, some miRNAs have been confirmed using more precise and cost-efficient techniques like RT-PCR or ddPCR (Constâncio et al. 2023) to analyze their dysregulation. These methods are detailed in Table 5.1.

5.4 Genetic Diseases Associated with Dysfunctional miRNAs

There are diverse mechanisms by which mutations can influence global miRNA function. These include point mutations (single-nucleotide polymorphisms), large-scale deletions and duplications, alterations within the 3′ untranslated region (UTR) of target mRNAs, and mutations in genes involved in miRNA processing (Meola et al. 2009). Notably, large mutations can disrupt both protein-coding genes and intragenic miRNAs residing within host genes. Furthermore, chromosomal abnormalities associated with aberrant gene copy number can lead to deletions or duplications of miRNA-encoding genes, potentially contributing to disease phenotypes. Studies investigating the mutational landscape of disease genes within the Human Gene Mutation Database (HGMD) reveal a spectrum of mutations that can significantly impact miRNA expression. A prime example is Dent disease, an X-linked recessive disorder characterized by the complete deletion of a cluster of miRNAs at Xp11.22, including miR-500, miR-600, miR-188, miR-362-5p, miR-362-3p, miR-501, miR-502, miR-532-5p, and miR-532-3p (Delabar et al. 1993; Antonarakis 1993; Fisher et al. 1994; Brown 2000; Stenson et al. 2008; Meola et al. 2009). Conversely, chromosomal abnormalities like trisomy 21 in Down syndrome lead to miRNA triplication, as observed with miR-99, let-7c, miR-125b, miR-155, and

Table 5.1 Circulating noninvasive miRNA biomarkers for genetic diseases

Disease	Biological Sample	Altered circulating miRNAs compared to HCs	Technique	Reference
DLBCL	Serum	miR-210 (↑) miR-21 (↑) miR -155 (↑)	qRT-PCR	Lawrie et al. (2008)
LABC	Whole blood	miR-331 (↑) mir-195 (↓) mir-181a (↓)	qRT-PCR	McAnena et al. (2019)
Pompe disease	Plasma	miR-127-3p (↑) miR-136-3p (↑) miR-133a-3p (↑) miR-99b-5p (↑) miR-329-3p (↑) miR-144-5p (↓) miR-410-3p (↓) miR-15b-5p (↓) miR-182-5p (↓)	Small-RNA sequencing	Tarallo et al. (2019)
DM1	Serum	miR-1 (↑) miR-133a/b (↑*) miR-206 (↑)	qRT-PCR	Koehorst et al. (2020)
GNCA	Serum	miR-16 (↑) miR -25 (↑) miR -92a (↑) miR -451 (↑) miR-486-5p (↑)	TaqMan low-density array, qRT-PCR	Zhu et al. (2014)
ADPKD	Urine	mir-143 (↑) mir-3619 (↑) mir-1 (↓) mir-133b (↓) mir-205(↓)	RNA sequencing	Ben-Dov et al. (2014)
Type -1 diabetes	Serum	miR-152 (↑) miR-30a-5p (↑) miR-181a (↑) miR-24 (↑) miR-148a (↑) miR-210 (↑) miR-27a (↑) miR-29a (↑) miR-26a (↑) miR-27b (↑) miR-25 (↑) miR-200a (↑)	miRNA purification, Solexa sequencing, and qRT-PCR	Nielsen et al. (2012)
Familial atrial fibrillation	Plasma	miR-1 (↓) miR-21 (↓) miR-328 (↑) miR-208 (↑)	qRT-PCR	Komal et al. (2019)
AMI	Plasma	*miR-1* (↑)	Real-time RT-PCR	Ai et al. (2010)

(continued)

Table 5.1 (continued)

Disease	Biological Sample	Altered circulating miRNAs compared to HCs	Technique	Reference
β-Thalassemia	Plama	miR-609 (↑) miR-15a/16-1 (↑) miR-486-3p (↑) miR-26b (↑) miR-199b-5p (↑) miR-210 (↑) miR-34a (↑) miR-138 (↑) miR-326 (↑) miR-96 (↓) miR-146a (↓) miR-223-3p (↓) miR-144 i (↓)	RT-PCR	Wang et al. (2015, 2021)

Arrow direction of expression of miRNA, * only in a subset of patients, *HC* healthy controls, *DLBCL* Diffuse large B-cell lymphoma, *LABC* locally advanced breast cancer, *DM1* Myotonic dystrophy type I, *GNCA* Gastric non-cardia adenocarcinoma, *ADPKD* Autosomal dominant polycystic kidney disease, *AMI* acute myocardial infarction

miR-802 (Delabar et al. 1993; Antonarakis 1993; Fisher et al. 1994; Brown 2000; Stenson et al. 2008; Meola et al. 2009).

Single-nucleotide polymorphisms (SNPs) can significantly influence miRNA function, as exemplified by a mutation within the miR-96 seed region in the diminuendo mouse mutant. This mutation, associated with progressive hearing loss in both humans and mice, resulted in hearing impairment even in heterozygous conditions and profound deafness in the homozygous state (Lewis et al. 2009). Additionally, the mutation was linked to hair cell defects. Bioinformatic analyses revealed that the mutation directly impacts gene expression, particularly those encoding transcription factors critical for hair cell development and survival (Lewis et al. 2009). Mutations within the 3′ untranslated region (UTR) of mRNAs can contribute to disease pathogenesis by creating novel miRNA binding sites. These newly formed sites, predicted targets for specific miRNAs, function as seed regions for miRNA-mRNA interactions and subsequent gene regulation. Supporting this concept, mutations within the REEP1 gene's 3′UTR, a predicted miR-140 target site, have been linked to autosomal dominant spastic paraplegia (SPG31) (Züchner et al. 2006; Beetz et al. 2008). Similarly, mutations in the 3′UTR of SLITRK1 mRNA in familial Tourette's syndrome are proposed to enhance repression by miR-24 (Abelson et al. 2005).

Mutations in genes whose products are involved in miRNA processing led to global miRNA dysfunction and are proposed to be linked with various pathological conditions, including cardiovascular, central nervous system, and immunological diseases. In vivo studies utilizing Dicer, an RNase III enzyme critical for miRNA biogenesis, have provided valuable insights into miRNA function. Chen et al. (2008) demonstrated for the first time the essential role of Dicer in cardiac muscle contraction, suggesting the importance of miRNAs in both healthy and pathological cardiac conditions. Further supporting this concept, preliminary findings from in vivo investigations using Dicer-deficient mouse models have shed light on the crucial role of miRNAs within the mammalian central nervous system (Schaefer et al. 2007; Cuellar et al. 2008; Damiani et al. 2008; De Pietri et al. 2008). Studies have demonstrated that Dicer loss of function disrupts the functionality of T lymphocytes, regulatory T-cell differentiation, and B-cell activity (Muljo et al. 2005; Cobb et al. 2006; Koralov et al. 2008; Liston et al. 2008). Furthermore, mutations in genes encoding proteins involved in miRNA processing, such as FRM1 gene loss in Fragile X syndrome and DGCR8 gene deletion in DiGeorge syndrome, have also been implicated in disease development (Meola et al. 2009; Fig. 5.1).

Mencía et al. (2009) identified two distinct nucleotide changes within the seed region of human miR-96 in two unrelated Spanish families suffering from deafness. Interestingly, both families harbored the mutations, miR-96 (+13G > A) and

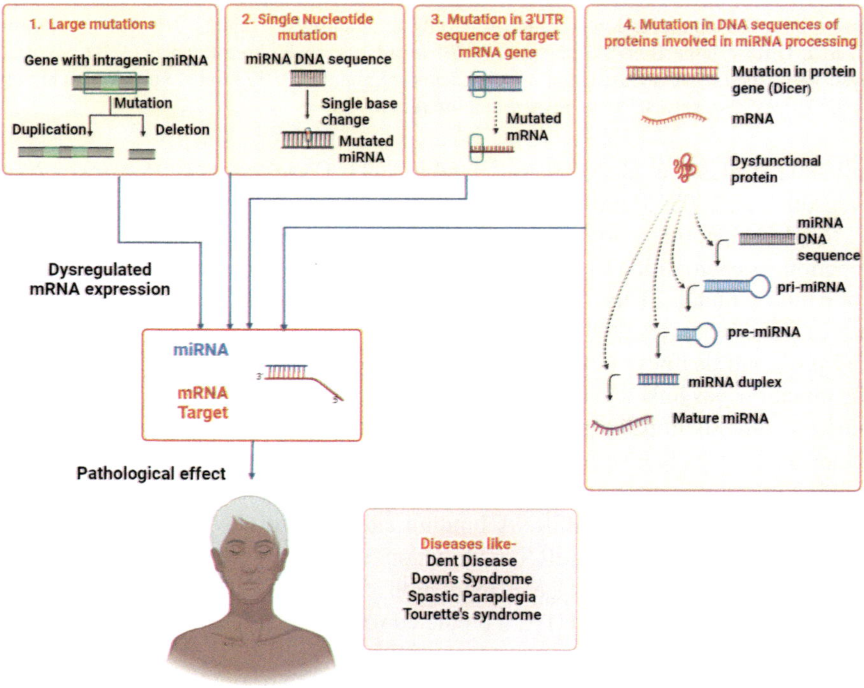

Fig. 5.1 Mutations may be associated with dysfunctional miRNA and dysregulated mRNA expression. Various diseases and syndromes (e.g., Dent disease, Down's syndrome, spastic paraplegia) have been linked to mutations causing dysfunctional miRNA. Large mutations like deletions and duplications can impact intragenic miRNA-containing host genes; single-nucleotide mutations such as those occurred in the seed region of miRNAs, mutations in the 3′UTR of mRNAs that are predicted target sites of certain miRNA, and mutations in genes whose products are involved in miRNA processing (e.g., dicer), lead to global miRNA dysfunction and impact the mRNA expression of target and, at times, nontarget genes

(+14C > A), further supporting their potential role in hearing loss. These mutations were absent in unrelated healthy controls. Studies have revealed significant alterations in muscle-specific miRNA expression in Duchenne muscular dystrophy (DMD) patients. miR-1, miR-206, miR-133, miR-499, miR-208a, and miR-208b levels were reported to be upregulated, while miR-29b and miR-29c levels were downregulated in DMD condition (Zaharieva et al. 2013; Li et al. 2014; Zanotti et al. 2015). Moreover, Matsuzaka et al. (2014) identified miR-1, miR-133a, and miR-206 as potential biomarkers for limb-girdle muscular dystrophy, Becker's muscular dystrophy, and facioscapulohumeral muscular dystrophy (Matsuzaka et al. 2014). These findings suggest the potential utility of miRNAs as biomarkers for monitoring disease progression, assessing severity, and evaluating therapeutic response.

In the case of amyotrophic lateral sclerosis (ALS) patients, skeletal muscles show variations in the expression of miR-206, miR-208b, and miR-499 as it progresses and miR-206, miR-143-3p, and miR-374b-5p act as prospective ALS indicators (Waller et al. 2017; Di Pietro et al. 2017). Cystic fibrosis (CF) is an autosomal recessive genetic condition caused by mutations in the CF transmembrane conductance regulator (CFTR) gene (Kerem et al. 2014; Fabbri et al. 2017). Studies have shown that miR-509-3p, miR-494, miR-223, and miR-145 can directly and indirectly influence the production of CFTR protein in primary bronchial epithelial cells obtained from patients with CF (Wang et al. 2000).

Rett syndrome (RTT), a complex neurological disorder affecting the X chromosome, arises from mutations in the MeCP2 gene. miR-199a plays a crucial role in connecting MeCP2 to the mTOR signaling pathway, and disruptions in this pathway contribute to RTT development. These disruptions are characterized by decreased levels of several miRNAs (miR-122a, miR-342, miR-409, miR-146a, miR-146b, and miR-130) and increased levels of others (miR-382, miR-296, miR-221, miR-92, miR-29b, miR-329, and miR-199b) (Lyst and Bird 2015; Tsujimura et al. 2015; Mohan et al. 2021).

Sézary syndrome (SS), a rare and serious form of skin cancer (cutaneous T-cell lymphoma or CTCL), is associated with unusually high levels of a specific group of miRNAs known as the miR-199a2/214 cluster (Qin et al. 2012; Benoit et al. 2017). Researchers identified eight specific miRNAs (miR-21, miR-140, miR-145, miR-214, miR-195, miR-199a, miR-451, and miR-483) associated with multiple osteochondromas (MO). These miRNAs could potentially help distinguish individuals with MO from healthy individuals. A study suggests that these differentially expressed miRNAs play a role in both the development and progression of MO by influencing the molecular signals crucial for healthy growth and development of chondrocytes (Zuntini et al. 2010). Mutations in the β-globin gene chain lead to β-thalassemia. Interestingly, certain miRNAs, including miR-486-3p, miR-23a, miR-15a/16-1, and miR-27a, seem to play a role in regulating proteins that suppress the production of γ-globin proteins (Gasparello et al. 2017; Thein 2018), and miR-210 is involved in γ-globin gene upregulation (Bavelloni et al. 2014; Vardiman et al. 2009).

miR-660, miR-510, miR-509-3, miR-325, and miR-188 have been associated with X-chromosome-linked schizophrenia (Feng et al. 2009). The age of cancer onset in people with Li-Fraumeni syndrome is influenced by miR-605 SNP (Id Said and Malkin 2015). miR-206, miR-130, miR-124, and miR-9 were found to have a role in pulmonary arterial hypertension (Kim et al. 2015). A study revealed high levels of miR-30c and miR-181a in the blood serum of patients with Ullrich congenital muscular dystrophy (Paco et al. 2015). In Hailey-Haley disease (HHD), oxidative stress triggers the activation of miR-125b, which plays a specific role in the development of the disease by influencing the production of proteins responsible for the growth and maturation of keratinocytes (Manca et al. 2011). Mutation in miR-140 causes skeletal dysplasia (Grigelioniene et al. 2019). A large British family spanning several generations was found to have a genetic mutation in miR-204-5p, specifically at the fourth position of its seed sequence. This mutation caused an eye disease known as hereditary retinal dystrophy (Conte et al. 2015). These examples exemplify the reason for which miRNA research holds immense potential to become a critical tool for future healthcare. This technology may potentially help doctors diagnose and predict the course of diseases, ultimately enabling personalized treatment plans based on individual patients' specific miRNA profiles (Finotti et al. 2019; Table 5.2).

5.5 Epigenetic Variations of miRNA and Their Implication in Genetic Diseases

Epigenetic mechanisms, such as DNA methylation, histone modifications, and non-coding RNA regulation, can induce heritable alterations in gene expression without modifying the underlying DNA sequence, resulting in phenotypic changes that do not correspond to changes in the genotype (Morales et al. 2017; Iyer et al. 2023) and regulatory pathways (Liu et al. 2013).

DNA methylation is a common modification where methyl groups are added to cytosine bases within CpG dinucleotides, which are concentrated in CpG islands found in the proximal promoter regions of about half of human genes. CpG islands have a high proportion of CpG sites with at least 55% GC content and a minimum length of 0.5 kb (Bird 1986, 2002). Methylation of CpG islands within the promoters of miRNA genes leads to gene silencing and the activation of tumor suppressor properties. This occurs by preventing the binding of transcription factors and RNA polymerase to the DNA through methylation-binding proteins that interact with methylated CpG islands. CpG island demethylation can activate gene expression and promote cancer progression (Jaenisch and Bird 2003), while hypermethylation in promoter CpG islands can silence tumor suppressor genes (Jin et al. 2017). For instance, promoter methylation in gliomas has been found to downregulate tumor-suppressive miRNAs, including miR-153, miR-126, miR-124a, and miR-200b/a/429 (Ning et al. 2015). It is also responsible for the silencing of miR-137, an important

Table 5.2 miRNAs playing roles in various genetic diseases

Genetic disease	miRNAs	References
Duchenne muscular dystrophy	miR-1 (↑) miR-206 (↑) miR-133 (↑) miR-499 (↑) miR-181 (↑) miR-208a (↑) miR-208b (↑) miR-29b (↓) miR-29c (↓)	Li et al. (2014), Zaharieva et al. (2013), Zanotti et al. (2015) and Salvatore et al. (2011)
Amyotrophic lateral sclerosis	miR-499 (↑) miR-206 (↑) miR-208b (↑) miR-143 (↑) miR-374b-5p (↓)	Di Pietro et al. (2017), Waller et al. (2017)
DFNA 50	miR-96	Mencía et al. (2009)
Cystic fibrosis	miR-145 (↑) miR-223 (↑) miR-9 (↑) miR-93 (↓) miR-494 (↑) miR-509-3p (↑)	Wang et al. (2000) and De Palma et al. (2020)
Rett syndrome	miR-199a miR-29b (↑) miR-329 (↑) miR-199b (↑) miR-382 (↑) miR-296 (↑) miR-221 (↑) miR-92 (↑) miR-146a (↓) miR-146b (↓) miR-130 (↓) miR-122a (↓) miR-342 (↓) miR-409 (↓)	Lyst and Bird (2015) and Tsujimura et al. (2015)

(continued)

Table 5.2 (continued)

Genetic disease	miRNAs	References
Sézary syndrome	miR-342 (↓) miR-31 (↑) miR- 424 (↑) miR-21 (↑) miR-214 (↑) miR-617 (↓) miR-133a (↓) miR-199a (↓)	Benoit et al. (2017), Qin et al. (2012), Salvatore et al. (2011)
Multiple osteochondromas	miR-21 (↑) miR-140 miR-145 (↓) miR-214 miR-195 miR-199a (↓) miR-451 (↑) miR-483	Zuntini et al. (2010)
β-Thalassemia	miR-486-3p (↑) miR-23a miR-15a (↑) miR-16-1 (↑) miR-27a miR-210 (↑)	Thein et al. (2018), Gasparello et al. (2017), Bavelloni et al. (2014) and Vardiman et al. (2009)
X-linked schizophrenia	miR-188 miR-325 (↑) miR-509-3 (↑) miR-510 miR-660	Feng et al. (2009)
Li-Fraumeni syndrome	miR-605 (↑)	Id Said and Malkin (2015)
Pulmonary arterial hypertension	miR-9 (↑) miR-124 (↑) miR-130 miR-206	Kim et al. (2015)
Ullrich congenital muscular dystrophy	miR-30c (↑) miR-181a (↑)	Paco et al. (2015)

(continued)

Table 5.2 (continued)

Genetic disease	miRNAs	References
Hailey-Haley disease	miR-125b (↑) miR-181a (↑) miR-99a (↑) miR-106a (↓)	Manca et al. (2011) and Salvatore et al. (2011)
Skeletal dysplasia	miR-140 (↑)	Grigelioniene et al. (2019)
Hereditary retinal dystrophy	miR-204(↑)	Conte et al. (2015)

DFNA-50 DeaFNess, Autosomal dominant 50; Arrow, direction of expression of miRNA

mediator that targets lysine-specific demethylase 1 (LSD1) in colorectal adenomas (Balaguer et al. 2010) and is observed in various cancers, including lung cancer (involving p16ink4A) (Belinsky et al. 1998), renal carcinoma (involving VHL) (Herman et al. 1994), and breast cancer (involving BRCA1) (Esteller et al. 2000).

In addition, histone modifications such as phosphorylation, methylation, acetylation, sumoylation, biotinylation, ubiquitylation, or ADP-ribosylation can occur on the amino-terminal regions of histones (Falkenberg and Johnstone 2014) where these modifications can modulate the structure and function of chromatin, thereby influencing the accessibility of the transcriptional machinery to miRNA genes (Miller and Grant 2013). A study comparing the gene expression profiles of HCC (hepatitis B virus-infected hepatocellular carcinoma) with a control group found that 14 miRNAs were overexpressed, and 16 miRNAs were downregulated, indicating their involvement in regulating H3K9 methylation and their potential role in the development of HCC (Ding et al. 2017). The miR-29 family regulated by histone modifications plays a crucial role in epigenetic regulation, with MYC and NF-κB being key transcription factors that induce histone deacetylation and methylation to repress miR-29 expression in various cancers, while EZH2 and HDACs are recruited to the miR-29 promoter to further enhance the epigenetic repression.

MicroRNAs also have the ability to affect the activity of enzymes involved in epigenetic processes. They can control the expression of enzymes responsible for DNA methylation and histone modification like histone deacetylase (HDAC), histone methyltransferase (EZH2), and DNA methyltransferase (DNMT) by targeting them, which plays a crucial role in regulating epigenetic changes and cell fate. Furthermore, certain miRNAs themselves can be influenced by epigenetic factors, creating a feedback loop between miRNAs and the machinery that controls epigenetic modifications (Sato et al. 2016; Yao et al. 2019). These findings have expanded our knowledge of the relationship between miRNAs and epigenetic mechanisms (Fig. 5.2).

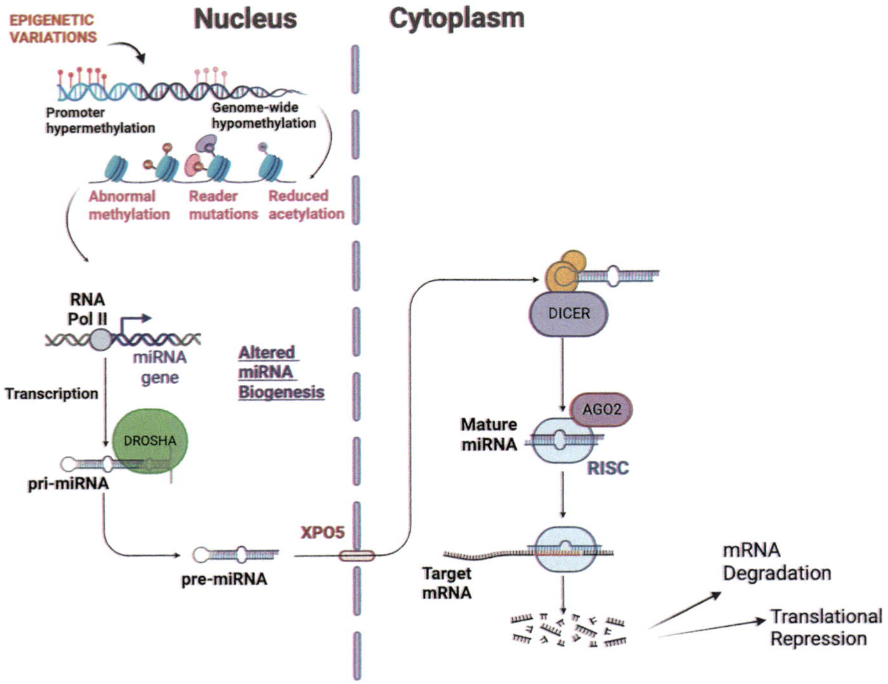

Fig. 5.2 Epigenetic effects on miRNA biogenesis and function. This illustration shows the impact of epigenetic variations on microRNA (miRNA) biogenesis, where alterations in DNA methylation and histone modifications can disrupt the normal processing of miRNA precursors. These epigenetic changes contribute to the degradation of mature miRNAs and translational repression of target mRNAs, highlighting the intricate regulatory mechanisms underlying gene expression

5.6 Mutations in miRNA and Mendelian Disorders

Genetics and genomics have recently undergone a revolution as a result of the development of high-throughput next-generation sequencing techniques, which have helped to pinpoint the etiology of many Mendelian disorders. The use of whole-genome sequencing (WGS) and whole-exome sequencing (WES) on large populations has generated a number of publicly available sequence datasets that have shown the extent of human genetics (Ombrello et al. 2014).

Numerous fundamental biological processes are thought to be influenced by miRNAs (Bartel 2004). There are three main types of mutation mechanisms that affect miRNA function, namely, (1) mutations that primarily affect miRNAs, either larger mutations or point mutations in the mature sequence; (2) mutations in genes that are predicted to have an effect on global health because they are involved in the general processes of miRNA processing and function; and (3) mutations in the 3′UTR of mRNAs that can result in the removal or in the de novo generation of a target recognition site for a specific miRNA (Meola et al. 2009).

A growing body of evidence suggests that mutations within genes can disrupt miRNA biogenesis, potentially contributing to human genetic diseases (Stenson et al. 2008). This is exemplified by intragenic deletions observed in Mendelian disorders like Duchenne muscular dystrophy (DMD) (Venugopal et al. 2018), choroideremia, and Dent disease (Sethupathy et al. 2007), where such deletions can significantly impact the expression of one or more miRNAs. Furthermore, point mutations can also influence miRNA biogenesis. A compelling example is a single-nucleotide polymorphism (SNP) identified within miR-125a, which disrupts the processing pathway from pri-miRNA to pre-miRNA (Duan et al. 2007). This finding highlights the potential of SNPs to not only impede miRNA synthesis but also to affect target selection, potentially leading to significant biological consequences.

Mendelian disorders can also occur by miRNA alterations, protein misfolding, and mitochondrial effects due to environmental factors (Tukker et al. 2021). Some miRNAs, for example, can regulate neurotoxicant-induced neurodegeneration in the brain in spinocerebellar ataxias and Parkinson's disease (Wallace et al. 2020; Singh and Yadav 2020). Given their stability, tissue specificity, and ease of quantification, miRNAs are currently being investigated as prospective biomarkers for environmental exposures (Kotsyfakis and Patelarou 2019; Wallace et al. 2020; Tukker et al. 2021).

Limb-girdle congenital myasthenic syndrome is a neuromuscular condition caused by mutations in the gene encoding glutamine-fructose-6-phosphate transaminase 1 (GFPT1). A mutation in the GFPT1 gene c.*22C>A region causes illegitimate miRNA binding, resulting in low protein expression. The causal mutation c.*22C>A has been identified, and the creation of such miRNA target sites is thought to be a relevant pathomechanism in Mendelian disorders (Dusl et al. 2015). The C allele of rs2910164, an SNP found in the third nucleotide of the miR146A-3p seed sequence, has been linked to an increased susceptibility to psoriasis (Zhang et al. 2014), multiple sclerosis in women (Li et al. 2015), leprosy (Cezar-de-Mello et al. 2014), and a lower risk of ankylosing spondylitis (Xu et al. 2015) disorders (Machowska et al. 2022). Charcot-Marie-Tooth disease type 1A (CMT1A), the most common inherited peripheral neuropathy, is caused by a genome duplication that includes the PMP22 gene, which is dosage sensitive. miRNAs operate as repressors of posttranscriptional gene expression, and overexpression of one such miRNA (miR-29a) in CMT1A rodent models has been demonstrated to diminish PMP22 transcript and protein levels (Pipis et al. 2023). Thus, mutations in miRNA act as a driving force in the occurrence of Mendelian disorders (Fig. 5.3).

5.7 Role of miRNA in Hemoglobinopathies

MicroRNAs (miRNAs) are emerging as key regulators in diseases affecting the endometrium, blood system (hematologic disorders), and infectious processes. Hemoglobinopathies, the most prevalent blood disorders, arise from genetic

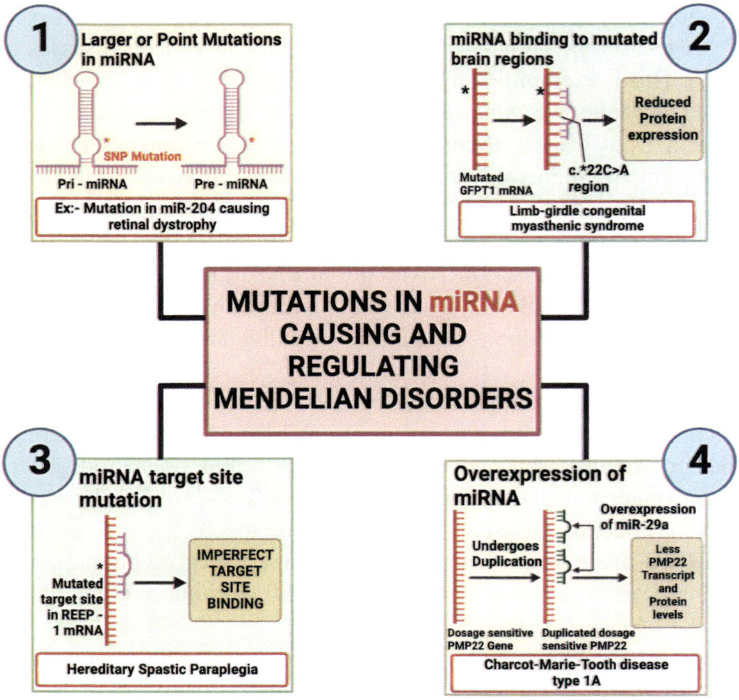

Fig. 5.3 Mendelian disorders can be caused and regulated by different types of mutations in miRNA. Firstly, large or point mutations in miRNAs (e.g., mutation in miR-204 causing retinal dystrophy) can lead to disorders. Secondly, the binding of miRNA to mutated gene regions increases the pathogenicity of disorders, as seen in limb-girdle congenital myasthenic syndrome, where the binding of miRNAs to the miRNA binding sites of the mutated GFPT1 gene's c.*22C > A region causes a reduction in protein expression. Thirdly, mutations in miRNA target sites, like those in REEP-1, can result in disorders such as hereditary spastic paraplegia. Lastly, overexpression of miRNAs, such as miR-29a, diminishes PMP22 transcript and protein levels in Charcot-Marie-Tooth disease type 1A

mutations in the human beta-globin gene (HBB), leading to abnormal hemoglobin structure. Two common examples are beta-thalassemia and sickle cell disease (SCD) (Ahmadvand et al. 2014). A crucial therapeutic strategy for these diseases involves increasing fetal hemoglobin (HbF) production, as HbF expression is high in the fetus but low in adults (Hagh et al. 2011). Enhancing fetal hemoglobin (HbF) is an important therapeutic strategy for treating ineffective hematopoiesis and anemia. Reducing the accumulation of α-globin chains in erythroid precursors and inhibiting ineffective erythropoiesis improve the oxygen supply to tissues and alleviates clinical symptoms by balancing the ratio of pathological chains of α/β (Fathallah et al. 2009; Fard et al. 2013).

A variety of methods, such as hydroxyurea, epigenetic modifications, and miR-based regulation, are used to induce γ-globin, which may be used therapeutically in patients with sickle cell disease (SCD) and β-thalassemia (Atweh and Loukopoulos

2001; Sankaran 2011; Sankaran et al. 2011; Fard et al. 2013; Lulli et al. 2013). In the following sections, we explore the importance of miRNA in hemoglobinopathies like β-thalassemia and sickle cell disease (SCD).

5.8 miRNAs in β-Thalassemia

Over 200 autosomal mutations in the globin gene cause thalassemia, an inherited blood disorder that manifests as a failure to produce normal globin chains and a variety of phenotypes depending on the severity of anemia and clinical complications (Khanahmad et al. 2006; Saki et al. 2012). Patients with α- and β-thalassemia, caused by mutations in the α- and β-globin genes, suffer from persistent anemia and ineffective red blood cell production (erythropoiesis) (Dehghani-Fard et al. 2012). In β-thalassemia, a lack of β-globin chain production leads to excess α-chains accumulating in red blood cells, causing hemolysis and further hampering erythropoiesis. Current treatment strategies for these patients focus on stimulating fetal hemoglobin (HbF) synthesis and regular blood transfusions (Atweh and Loukopoulos 2001). It is possible that modifications in the expression of these small RNAs can help thalassemic patients to experience fewer clinical complications, since miRs are involved in the expression of globin genes and also transcriptionally regulate erythroid-specific genes like KLFD (Kruppel-like transcription factor D) (Fornari et al. 2013).

In thalassemic cells, the erythrocyte membrane is destroyed by the accumulation of globin chains (Bilen et al. 2006). The majority of miRs that block the expression of the α-gene alleviate hemolytic anemia. By directly targeting erythroid-specific KLFD, miR-144, an erythroid-specific miR, inhibits cell lysis (Fu et al. 2009). KLFD functions as a co-regulator of both genes through its interaction with the α-globin gene promoter and the CACCC sites in miR-144 (Wijgerde et al. 1996; Fu et al. 2009). In children with β-thalassemia major, it has been demonstrated that the expression of the α/β globin gene is negatively regulated by the level of miR-144. The main treatment for β-thalassemia is based on this regulation, wherein patients with thalassemia may experience fewer clinical complications by avoiding the accumulation of excess α-globin (Wu et al. 2010). Another potential inhibitor of α-globin gene expression is miR-150. This miR functions in different cell types, including megakaryocytes, lymphocytes, and erythroids. It is expressed during erythroid differentiation, but in polycythemia vera, it is overexpressed. Apart from aimed at the myeloblastosis oncogene (MYB), another function of this miR is to regulate the fate of progenitor cells of erythroid and reduce the expression of the α-globin gene (Lee et al. 2004; Altuvia et al. 2005).

In cells with β-thalassemia mutations, there was decreased expression of miR-503, which controls cell cycle arrest and apoptosis (Roy et al. 2012). The target gene for this miR, cell division cycle 25A (CDC25A), is involved in DNA damage responses as well as the cell cycle (Mailand et al. 2000). Due to a lack of this miR, thalassemic cells exhibit 1000-fold overexpression of CDC25A, which is a significant contributing factor to the ineffective hematopoiesis that occurs in these patients (Roy et al. 2012). The induction of miR-503, miR322/424, or other hypoxia-dependent factors like miR-21 and cyclin-dependent kinase inhibitor p21 may suppress CDC25A and inhibit the growth of erythroid progenitor thalassemic cells and cancer cells, respectively (Roy et al. 2012; Queiroz de Oliveira et al. 2009; Sarkar et al. 2010). Another miR specific to erythroids is miR-451, which also is overexpressed during erythropoiesis and induces erythroid differentiation from CD133+ cells (Zhan et al. 2007; Kouhkan et al. 2013). Yet, in β-thalassemia/HbE cells, miR-451 is substantially overexpressed and linked to ineffective hematopoiesis, chronic hemolytic anemia, and overall severity of thalassemia. Overexpression of miR-451 is linked to declining levels of GATA1, α-chain, and glycophorin-A transcripts (Svasti et al. 2010). It is also seen in thalassemic cells, which have higher reticulocyte counts and lower hemoglobin levels. Zhan et al. (2007) found that overexpression of miR-451 is stronger than γ-globin in increasing the expression of α- and β-globin genes. Other hemoglobinopathies and polycythemia vera, where erythropoiesis occurs at a greater level, also exhibit elevated levels of this miR (Bruchova et al. 2007; Chen et al. 2008).

Emerging evidence suggests that miRNAs contribute to the regulation of globin gene expression, influencing the development of red blood cells. Recent studies identified distinct miRNA profiles that regulate α-, β-, and γ-globin expression (Wang et al. 2021). Several miRNAs, including miR-15a/16-1, miR-486-3p, and miR-26b, have been shown to promote γ-globin expression, potentially offering therapeutic targets for β-thalassemia. Furthermore, Roy et al. (2012) observed downregulation of miR-503, a cell cycle and apoptosis regulator, in β-thalassemia major cells, suggesting its potential contribution to the disease pathology. Studies exploring miRNA expression profiles in β-thalassemia are shedding light on potential mechanisms underlying the disease. Levin et al. (2021) reported significant alterations in miRNA levels within extracellular vesicles (EVs) isolated from β-thalassemia major patients. Specifically, they observed a twofold increase in 21 miRNAs and a threefold decrease in 17 miRNAs. Furthermore, differential expression of six miRNAs and the involvement of EVs in erythroid cell survival via the JNK/SAPK pathway were identified. Additionally, miR-144-loaded EVs were hypothesized to play a role in triggering apoptosis. Supporting these findings, Wang et al. (2021) observed decreased levels of hsa-miR-210, hsa-let-7b-5p, miR-15a, and miR-16-1, along with abnormal dysregulation of let-7 miRNAs in β-thalassemia pediatric patients. Moreover, miR-451 is upregulated in β-thalassemia/HbE cells and has been associated with severity of thalassemia (Zhan et al. 2007; Svasti et al. 2010). miR-148b3p, miR146a-5p, and miR-146b-5p also showed increased expression in HbE/β-thalassemia patients, whereas miR-320a exhibited downregulation (Zhai et al. 2014; Das et al. 2021).

Overall, the importance of miRs in regulating globin gene expression and reactivation of these genes could enable the use of these small ncRNA as novel therapeutic targets by altering the state of thalassemic cells and enhancing the physiology and symptoms of hemoglobinopathies.

5.9 Sickle Cell Disease (SCD)

A single amino acid replacement in the sixth position of the β-globin gene causes SCD, a common HBB disorder. Inside the cell, in an oxygen-free environment, the generated HbS polymerizes (Schaeffer et al. 2014). Researchers compared miRNA expression in reticulocytes, normal, and sickle mature erythrocytes. The observed differences could serve to understand anemia and related conditions (Byon and Papayannopoulou 2012). A study investigating how miRNAs affect the resistance of HbSS red blood cells to *Plasmodium falciparum* found that HbSS cells have higher levels of miR-451, let-7i, and miR-223 compared to HbA cells. This increase in specific miRNAs is believed to play a role in reducing the chance of the cells becoming infected by the parasite (LaMonte et al. 2012). Transfecting these miRNAs successfully diminished parasite infection rates. This is likely because they induce the production of molecules (such as PKA-R and PEAMT) that inhibit the parasite's protein production, and they also directly reduce parasite growth (Chen et al. 2008; Duraisingh and Lodish 2012). Inhibiting the activity of miR-451 and miR-223 using 2′-O-methyl antisense oligonucleotides reduced sickle cell resistance to malaria. This finding indicates that treatments capable of increasing these microRNAs in infected cells could potentially strengthen the immune response (Duraisingh and Lodish 2012).

MiR22, miR-200a, and miR-320 play a key role in regulating the production of the CD71 transferrin receptor. This receptor is vital for the final stages of red blood cell maturation and the body's ability to absorb iron (Schaar et al. 2009). Research indicates that miR-320 directly binds to the CD71 mRNA, preventing its translation into protein. CD71 plays a role in cell development, and its overexpression is associated with some cancers. Therefore, targeting CD71 with miR-320 could potentially slow cancer growth, especially in cancers that rely on rapid cell division and high iron needs (Chen et al. 2008). Reduced miR-320 levels in HbSS lead to high levels of CD71 on their surface. This indicates abnormal erythropoiesis caused by hemolysis and is further supported by the impaired maturation and shortened lifespan of these sickle cells (Byon and Papayannopoulou 2012).

Nuclear-related factor 2 (NRF2) is the primary regulator of the cellular antioxidant defense system and the oxidative stress response (Lee et al. 2004). Normally, miR-144 helps cells resist oxidative stress by targeting NRF2, which leads to low NRF2 and increased cellular glutathione levels. However, in HbSS cells, the increased expression of miR-144 leads to a decrease in NRF2, paradoxically making these cells more susceptible to oxidative stress. High levels of miR-144 are associated with severe anemia, characterized by a significant drop in red blood cell count (hemoglobin/hematocrit) and a lack of essential antioxidant proteins like glutamate-cysteine ligase, catalytic/modifier subunit (GCLC/M), and superoxide dismutase 1 (SOD1) (Sangokoya et al. 2010). The specific way miR-144 regulates NRF2 renders HbSS cells susceptible to oxidative stress, hemolysis, and ultimately more severe anemia. This finding highlights the potential of manipulating miRNA expression in red blood cells as a new approach to alleviate the clinical and pathological symptoms experienced by patients with SCD (Byon and Papayannopoulou 2012). 5-azacytidine, decitabine, hydroxyurea, and butyrate are agents used to induce HbF synthesis in patients with SCD (Bunn 1999; Fathallah and Atweh 2006). According to the previous discussion, certain miR has the ability to stimulate HbF synthesis in primary erythroid cells, thereby displacing drug-based methods and improving patient clinical signs without adverse effects.

5.10 miRNA As a New Target for Gene Therapy

The ability of miRNA to regulate the expression of numerous genes simultaneously makes it a promising candidate for therapeutic applications, either as miRNA mimics or as targets for therapeutic intervention in the form of antimiRs (Lecellier et al. 2005; Iorio and Croce 2012). A single miRNA can successfully control whole cellular pathways, leading to major shifts in cellular function, despite the comparatively minimal impact on specific targeted genes (Gebert and MacRae 2019).

MicroRNA mimics are artificially created double-stranded RNA molecules designed to mimic the sequence of specific miRNAs. These molecules are intended to compensate for the reduced expression of miRNAs in diseases (Rupaimoole and Slack 2017) by specifically targeting and binding to mRNA molecules. One example of this is the miR-34 family, which is frequently found to be defective in human cancers and plays a critical role in suppressing tumor development (Jansson and Lund 2012; Shah et al. 2016).

Viral vectors, such as lentiviral, retroviral, and adeno-associated viral (AAV) vectors, are widely employed for miRNA delivery due to their ability to transduce both dividing and nondividing cells (Zhang and Godbey 2006; Liu and Berkhout 2011). Among these vectors, AAV vectors hold particular promise for miRNA delivery owing to their minimal toxicity and capacity for targeted expression, minimizing off-target effects (Zhang and Godbey 2006). This advantage is exemplified by a Phase II clinical trial for Huntington's disease, where researchers successfully utilized an AAV vector system to deliver the therapeutic miRNA drug, adeno-associated virus-mediated micro-RNA-130 (AMT-130) (Miniarikova et al. 2016; Samaranch et al. 2017; Keskin et al. 2019). While viral vectors have proven effective in delivering miRNA antagonists or mimics to tumor tissues, their potential for triggering an immune response remains a significant concern in clinical applications. To address this limitation, various nonviral delivery methods have been developed as safe alternatives for miRNA delivery (Yin et al. 2014). One such approach involves encapsulating miRNA mimics within NPs (nanoparticles). NPs serve as carriers for miRNA and have the advantage of slow degradation by endonucleases, thus avoiding nonspecific binding with other protein targets. This chemical-based concept for miRNA delivery offers a promising solution to enhance the efficacy of in vivo delivery (Rupaimoole and Slack 2017; Kobayashi et al. 2018).

Various nonviral delivery methods have also been developed for miRNA delivery, including lipid conjugation, antibody-dependent targeting, physical methods (e.g., gene gun and electroporation), and noninvasive methods (e.g., UMTD, ultrasound-targeted microbubble distraction). These methods have shown efficacy in delivering miRNAs to different cell types both in vitro and in vivo (Saiyed et al. 2022). Ji et al. (2016) investigated the therapeutic potential of ultrasound microbubble treatment (UMTD) in regulating miRNA-133a for breast cancer treatment. They employed MCF-7 and MDA-MB-231 cell lines, along with an in vivo model, and observed that UMTD exposure at a low frequency of 10 MHz significantly decreased cell proliferation and suppressed tumor growth in the in vivo model, potentially through the upregulation of miRNA-133a. Anti-miRs, such as antisense oligonucleotides (ASOs) or locked nucleic acids (LNAs) hinder cancer-related pathways by blocking oncomiRs. An example of this is miR-21, which prevents apoptosis by activating the PI3K pathway and inhibiting PTEN. By using anti-miR-21 therapy, breast cancer cells experienced increased activation of apoptotic factors, effectively controlling their growth (Yan et al. 2011). Natural substances also provide a rich source of miRNA interfering molecules (Zhang et al. 2020). Curcumin, for instance, has been found to exert inhibitory effects on breast cancer cell proliferation by influencing the expression of multiple miRNAs (Norouzi et al. 2018).

Lastly, the coadministration of miRNAs and siRNAs presents a potential strategy to enhance the therapeutic efficacy of miRNAs (Lam et al. 2015). Recent studies have demonstrated the successful co-expression of miRNAs and siRNAs in recombinant plasmids, showcasing their synergistic effects in human lung cancer cells (Petrek et al. 2019). In colorectal cancer xenografts, the intratumoral delivery of siRNA targeting DCAMKL-1 via nanoparticles led to the upregulation of let-7a and miR-144. This, in turn, resulted in the suppression of proto-oncogenes c-Myc and Notch-1. Consequently, tumor inhibition growth was observed (Sureban et al. 2011). Thus, future research should focus on optimizing delivery methods, addressing immune responses, and exploring the synergistic effects of miRNAs and siRNAs to fully harness the therapeutic potential of miRNA-based therapies.

5.11 Future of miRNA Research

miRNAs have the potential to be used as novel therapies because they are essential in the development and prognosis of disease. The security and efficiency of miRNA distribution have recently drawn attention because of the instability of miRNAs and their complicated environment, including in vivo nuclease degradation (Fu et al. 2019). Recent research on the expression of miRNAs in many tumor types has shown that aberrantly expressed miRNAs are helpful for disease categorization, diagnosis, staging, and prognosis (Cheng et al. 2017). The targeted inhibition of miRNAs plays a significant role in many facets of human disorders with a prominent therapeutic effect. There are many ways to inhibit miRNAs, but chemically altered antisense oligonucleotides (anti-miRs) have produced the most noticeable results (Li and Rana 2014).

MiRNA masks hold promise as future research tools to elucidate the functional impact of miRNAs on specific pathways involving target genes. They represent a valuable complement to anti-miRNA oligonucleotides (AMOs) and are designed as single-stranded 22-nucleotide oligoribonucleotides with 2′-O-methyl modifications (Wang 2011; Fu et al. 2019). miRNA masks function through perfect complementarity, specifically interacting with miRNA binding sites located within the 3′ untranslated region (3′UTR) of target gene mRNA. More research has revealed how specific regulatory miRNAs known as "miRNA sponges" are regulated by other biological molecules. To stop miRNAs from interacting with their natural targets, these aggressive sponges attach to and sequester them. Potential applications for miRNA sponges include illness detection, prognosis, and treatment. Our knowledge of how different RNA classes communicate with one another in the transcriptome can be greatly improved thanks to the miRSponge database. This resource will also be helpful for future study on miRNAs and their use in medicine (Wang et al. 2015). However, no commercially available therapeutic nano delivery systems based on miRNAs have yet been authorized for use in treating diseases and genetic disorders. Future research should concentrate on creating methods that effectively silence genes (Mehta et al. 2021). miRNAs offer promising potential as therapeutic targets and indicators of drug responses and treatment strategies for CVD. This can be achieved by assessing circulating miRNAs and examining their expression levels in serum or plasma samples (Sessa et al. 2023).

Vasu and colleagues investigated the levels of miRNAs in the bloodstream among individuals with various forms of diabetes, including recent-onset or long-standing type 1 diabetes, obesity, pre-diabetes, type 2 diabetes, and gestational diabetes (Vasu et al. 2019). During the pre-diabetic stage (pre-DM), islets experience stress and release miRNAs. The authors summarized their findings regarding miRNA profiles that have the potential to indicate the progression of different diabetes types. In the future, advancements in the identification of specific biomarker signatures may contribute to the early detection and classification of diabetic conditions, as well as the development of treatments aimed at preventing and reversing diabetes (Vasu et al. 2019). Future research may show miRNAs as valuable tools for management or diagnostic purposes; however, before miRNA can be used in clinical practice, the delivery of miRNA targeting medicines must be resolved (Oliviero et al. 2019).

5.12 Conclusion

This review has highlighted the importance of miRNAs in gene regulation and their significance in disease formation. We have also discussed the significance of miRNA in hemoglobinopathies. Mutations not only in genes but also in ncRNAs can impact the disease's origin or pathophysiology. Epigenetic modifications in miRNAs may be associated with Mendelian disorders, paving a way to the specificity of miRNA in the genetic diseases. By exploring the disease mechanisms, we can explore therapeutic approaches based on miRNAs.

Acknowledgments Authors are thankful to the Central University of Punjab, Bathinda–151401, Punjab, and Karpagam Academy of Higher Education, Coimbatore, for providing the necessary facilities to conduct this manuscript.

Compliance with Ethical Standards

Conflict of Interest Authors declares that no conflict of interest.

Ethical Approval This chapter is a review of previously published accounts; as such, no animal or human studies were performed.

References

Abels ER, Breakefield XO (2016) Introduction to extracellular vesicles: biogenesis, RNA cargo selection, content, release, and uptake. Cell Mol Neurobiol 36(3):301–312. https://doi.org/10.1007/s10571-016-0366-z

Abelson JF, Kwan KY, O'Roak BJ et al (2005) Sequence variants in *SLITRK1* are associated with Tourette's syndrome. Science 310(5746):317–320. https://doi.org/10.1126/science.1116502

Aghili Z, Nasirizadeh N, Divsalar A et al (2018) A highly sensitive miR-195 nanobiosensor for early detection of Parkinson's disease. Artificial cells Nanomed Biotechnol 46(sup1):32–40

Ahmadvand M, Noruzinia M, Fard AD et al (2014) The role of epigenetics in the induction of fetal hemoglobin: a combination therapy approach. Int J Hematol Oncol Stem Cell Res 8(1):9

Ai J, Zhang R, Li Y, Pu J, Lu Y, Jiao J et al (2010) Circulating microRNA-1 as a potential novel biomarker for acute myocardial infarction. Biochem Biophys Res Commun 391(1):73–77

Altuvia Y, Landgraf P, Lithwick G et al (2005) Clustering and conservation patterns of human microRNAs. Nucleic Acids Res 33(8):2697–2706

Antonarakis SE (1993) Human chromosome 21: genome mapping and exploration, circa 1993. Trends Genet 9(4):142–148. https://doi.org/10.1016/0168-9525(93)90210-9

Ardekani AM, Naeini MM (2010) The role of microRNAs in human diseases. Avicenna J Med Biotechnol 2(4):161

Atweh GF, Loukopoulos D (2001) Pharmacological induction of fetal hemoglobin in sickle cell disease and β-thalassemia. Elsevier, pp 367–373

Auyeung VC, Ulitsky I, McGeary SE et al (2013) Beyond secondary structure: primary-sequence determinants license pri-miRNA hairpins for processing. Cell 152(4):844–858

Balachandar V, Mahalaxmi I, Neethu R et al (2022) New insights into epigenetics as an influencer: an associative study between maternal prenatal factors in autism Spectrum disorder (ASD). Neurol Perspect 2(2):78–86

Balaguer F, Link A, Lozano JJ et al (2010) Epigenetic silencing of miR-137 is an early event in colorectal carcinogenesis. Cancer Res 70(16):6609–6618

Barish S, Senturk M, Schoch K, Minogue AL et al (2022) The microRNA processor DROSHA is a candidate gene for a severe progressive neurological disorder. Hum Mol Genet 31(17):2934–2950

Bartel DP (2004) MicroRNAs: genomics, biogenesis, mechanism, and function. Cell 116(2):281–297

Bartel DP (2018) Metazoan micrornas. Cell 173(1):20–51

Baskerville S, Bartel DP (2005) Microarray profiling of microRNAs reveals frequent coexpression with neighboring miRNAs and host genes. RNA 11(3):241–247
Bavelloni A, Poli A, Fiume R et al (2014) PLC-beta 1 regulates the expression of miR-210 during mithramycin-mediated erythroid differentiation in K562 cells. Oncotarget 5(12):4222–4231. https://doi.org/10.18632/oncotarget.1972
Beetz C, Schüle R, Deconinck T et al (2008) REEP1 mutation spectrum and genotype/phenotype correlation in hereditary spastic paraplegia type 31. Brain 131(4):1078–1086. https://doi.org/10.1093/brain/awn026
Belinsky SA, Nikula KJ, Palmisano WA et al (1998) Aberrant methylation of p16 INK4a is an early event in lung cancer and a potential biomarker for early diagnosis. Proc Natl Acad Sci USA 95(20):11891–11896. https://doi.org/10.1073/pnas.95.20.11891
Ben-Dov IZ, Tan YC, Morozov P, Wilson PD, Rennert H, Blumenfeld JD, Tuschl T (2014) Urine microRNA as potential biomarkers of autosomal dominant polycystic kidney disease progression: description of miRNA profiles at baseline. PLoS One 9(1):e86856
Benoit BM, Jariwala N, O'Connor G et al (2017) CD164 identifies CD4+ T cells highly expressing genes associated with malignancy in Sézary syndrome: the Sézary signature genes, FCRL3, Tox, and miR-214. Arch Dermatol Res 309(1):11–19. https://doi.org/10.1007/s00403-016-1698-8
Bhardwaj A, Arora S, Prajapati K et al (2013) Cancer "stemness"-regulating microRNAs: role, mechanisms and therapeutic potential. Curr Drug Targets 14(10):1175–1184
Bilen J, Liu N, Bonini NM (2006) A new role for microRNA pathways: modulation of degeneration induced by pathogenic human disease proteins. Cell Cycle 5(24):2835–2838
Bird AP (1986) CpG-rich islands and the function of DNA methylation. Nature 321(6067):209–213
Bird A (2002) DNA methylation patterns and epigenetic memory. Genes Dev 16(1):6–21
Biswas S, Haleyurgirisetty M, Lee S et al (2019) Development and validation of plasma miRNA biomarker signature panel for the detection of early HIV-1 infection. EBioMedicine 43:307–316
Bonci D, Coppola V, Musumeci M et al (2008) The miR-15a–miR-16-1 cluster controls prostate cancer by targeting multiple oncogenic activities. Nat Med 14(11):1271–1277
Brown AS (2000) Genome-wide variation in recombination in female meiosis: a risk factor for non-disjunction of chromosome 21. Hum Mol Genet 9(4):515–523. https://doi.org/10.1093/hmg/9.4.515
Bruchova H, Yoon D, Agarwal AM et al (2007) Regulated expression of microRNAs in normal and polycythemia vera erythropoiesis. Exp Hematol 35(11):1657–1667
Bunn HF (1999) Induction of fetal hemoglobin in sickle cell disease. Blood J Am Soc Hematol 93(6):1787–1789
Byon JC, Papayannopoulou T (2012) MicroRNAs: allies or foes in erythropoiesis? J Cell Physiol 227(1):7–13
Cezar-de-Mello PFT, Toledo-Pinto TG, Marques CS et al (2014) Pre-miR-146a (rs2910164 G>C) single nucleotide polymorphism is genetically and functionally associated with leprosy. PLoS Negl Trop Dis 8(9):e3099. https://doi.org/10.1371/journal.pntd.0003099
Chen J-F, Murchison EP, Tang R et al (2008) Targeted deletion of dicer in the heart leads to dilated cardiomyopathy and heart failure. Proc Natl Acad Sci USA 105(6):2111–2116. https://doi.org/10.1073/pnas.0710228105
Cheng W-T, Rosario R, Muthukaruppan A et al (2017) MicroRNA profiling of ovarian granulosa cell tumours reveals novel diagnostic and prognostic markers. Clin Epigenetics 9(1):72. https://doi.org/10.1186/s13148-017-0372-0
Coban N, Pirim D, Erkan AF et al (2020) Hsa-miR-584-5p as a novel candidate biomarker in Turkish men with severe coronary artery disease. Mol Biol Rep 47:1361–1369
Cobb BS, Hertweck A, Smith J et al (2006) A role for dicer in immune regulation. J Exp Med 203(11):2519–2527. https://doi.org/10.1084/jem.20061692
Constâncio V, Tavares NT, Henrique R et al (2023) MiRNA biomarkers in cancers of the male reproductive system: are we approaching clinical application? Andrology 11(4):651–667. https://doi.org/10.1111/andr.13258

Conte I, Hadfield KD, Barbato S et al (2015) MiR-204 is responsible for inherited retinal dystrophy associated with ocular coloboma. Proc Natl Acad Sci USA 112(25). https://doi.org/10.1073/pnas.1401464112

Correia CN, Nalpas NC, McLoughlin KE et al (2017) Circulating microRNAs as potential biomarkers of infectious disease. Front Immunol 8:118

Cuellar TL, Davis TH, Nelson PT et al (2008) Dicer loss in striatal neurons produces behavioral and neuroanatomical phenotypes in the absence of neurodegeneration. Proc Natl Acad Sci USA 105(14):5614–5619. https://doi.org/10.1073/pnas.0801689105

Cui Y, Qi Y, Ding L et al (2023) miRNA dosage control in development and human disease. Trends Cell Biol

Damiani D, Alexander JJ, O'Rourke JR et al (2008) Dicer inactivation leads to progressive functional and structural degeneration of the mouse retina. J Neurosci 28(19):4878–4887. https://doi.org/10.1523/JNEUROSCI.0828-08.2008

Das SS, Das S, Byram PK, Rahaman M et al (2021) MicroRNA expression patterns in HbE/β-thalassemia patients: the passwords to unlock fetal hemoglobin expression in β-hemoglobinopathies. Blood Cell Mol Dis 87:102523. https://doi.org/10.1016/j.bcmd.2020.102523

de Candia P, De Rosa V, Casiraghi M et al (2016) Extracellular RNAs: a secret arm of immune system regulation. J Biol Chem 291(14):7221–7228

De Miguel Pérez D, Rodriguez Martínez A, Ortigosa Palomo A et al (2020) Extracellular vesicle-miRNAs as liquid biopsy biomarkers for disease identification and prognosis in metastatic colorectal cancer patients. Sci Rep 10(1):3974. https://doi.org/10.1038/s41598-020-60212-1

De Pietri TD, Pulvers JN, Haffner C et al (2008) miRNAs are essential for survival and differentiation of newborn neurons but not for expansion of neural progenitors during early neurogenesis in the mouse embryonic neocortex. Development 135(23):3911–3921. https://doi.org/10.1242/dev.025080

Dehghani-Fard A, Kaviani S, Saki N et al (2012) The emerging role of immunomodulatory agents in fetal hemoglobin induction. Int J Hematol Oncol Stem Cell Res:35–36

Delabar J-M, Theophile D, Rahmani Z et al (1993) Molecular mapping of twenty-four features of down syndrome on chromosome 21. Eur J Hum Genet 1(2):114–124. https://doi.org/10.1159/000472398

De Palma FDE, Raia V, Kroemer G, Maiuri MC (2020) The multifaceted roles of microRNAs in cystic fibrosis. Diagnostics 10(12):1102

Di Pietro L, Baranzini M, Berardinelli MG et al (2017) Potential therapeutic targets for ALS: MIR206, MIR208b and MIR499 are modulated during disease progression in the skeletal muscle of patients. Sci Rep 7(1):9538. https://doi.org/10.1038/s41598-017-10161-z

Ding W, Yang H, Gong S et al (2017) Candidate miRNAs and pathogenesis investigation for hepatocellular carcinoma based on bioinformatics analysis. Oncol Lett 13(5):3409–3414

Duan R, Pak C, Jin P (2007) Single nucleotide polymorphism associated with mature miR-125a alters the processing of pri-miRNA. Hum Mol Genet 16(9):1124–1131

Duraisingh MT, Lodish HF (2012) Sickle cell MicroRNAs inhibit the malaria parasite. Cell Host Microbe 12(2):127–128

Dusl M, Senderek J, Muller JS et al (2015) A 3'-UTR mutation creates a microRNA target site in the GFPT1 gene of patients with congenital myasthenic syndrome. Hum Mol Genet 24(12):3418–3426. https://doi.org/10.1093/hmg/ddv090

Dwivedi S, Purohit P, Sharma P (2019) MicroRNAs and diseases: promising biomarkers for diagnosis and therapeutics. Indian J Clin Biochem 34:243–245

Elangovan A, Babu HWS, Iyer M et al (2023a) Untangle the mystery behind DS-associated AD—is APP the main protagonist? Ageing Res Rev 87:101930. https://doi.org/10.1016/j.arr.2023.101930

Elangovan A, Venkatesan D, Selvaraj P et al (2023b) miRNA in Parkinson's disease: from pathogenesis to theranostic approaches. J Cell Physiol 238(2):329–354. https://doi.org/10.1002/jcp.30932

Esteller M (2011) Non-coding RNAs in human disease. Nat Rev Genet 12(12):861–874
Esteller M, Silva JM, Dominguez G et al (2000) Promoter hypermethylation and BRCA1 inactivation in sporadic breast and ovarian tumors. JNCI J Natl Cancer Inst 92(7):564–569
Fabbri E, Tamanini A, Jakova T et al (2017) A peptide nucleic acid against MicroRNA miR-145-5p enhances the expression of the cystic fibrosis Transmembrane conductance regulator (CFTR) in Calu-3 cells. Molecules 23(1):71. https://doi.org/10.3390/molecules23010071
Falkenberg KJ, Johnstone RW (2014) Histone deacetylases and their inhibitors in cancer, neurological diseases and immune disorders. Nat Rev Drug Discov 13(9):673–691
Fang W, Bartel DP (2015) The menu of features that define primary microRNAs and enable de novo design of microRNA genes. Mol Cell 60(1):131–145
Fard AD, Kaviani S, Noruzinia M et al (2013) Evaluation of H3 histone methylation and colony formation in erythroid progenitors treated with thalidomide and sodium butyrate. Lab Hematol 19(1):1–5
Fathallah H, Atweh GF (2006) Induction of fetal hemoglobin in the treatment of sickle cell disease. ASH Education Program Book 1:58–62
Fathallah H, Taher A, Bazarbachi A et al (2009) Differences in response to fetal hemoglobin induction therapy in β-thalassemia and sickle cell disease. Blood Cell Mol Dis 43(1):58–62
Feng J, Sun G, Yan J et al (2009) Evidence for X-chromosomal schizophrenia associated with microRNA alterations. PLoS One 4(7):e6121. https://doi.org/10.1371/journal.pone.0006121
Fernández-Messina L, Gutiérrez-Vázquez C, Rivas-García E et al (2015) Immunomodulatory role of microRNAs transferred by extracellular vesicles. Biol Cell 107(3):61–77. https://doi.org/10.1111/boc.201400081
Finotti A, Fabbri E, Lampronti I et al (2019) MicroRNAs and long non-coding RNAs in genetic diseases. Mol Diagn Ther 23(2):155–171. https://doi.org/10.1007/s40291-018-0380-6
Fisher SE, Black G, Lloyd SE et al (1994) Isolation and partial characterization of a chloride channel gene which is expressed in kidney and is a candidate for Dent's disease (an X-linked hereditary nephrolithiasis). Hum Mol Genet 3:2053–2059
Fornari TA, Lanaro C, Ferreira R et al (2013) HPFH and Delta-Beta thalassemia have different profiles of Micrornas and transcription factors. Blood 122(21):186–186. https://doi.org/10.1182/blood.V122.21.186.186
Friedman RC, Farh KK-H, Burge CB et al (2009) Most mammalian mRNAs are conserved targets of microRNAs. Genome Res 19(1):92–105. https://doi.org/10.1101/gr.082701.108
Fu Y-F, Du T-T, Dong M et al (2009) Mir-144 selectively regulates embryonic α-hemoglobin synthesis during primitive erythropoiesis. Blood 113(6):1340–1349. https://doi.org/10.1182/blood-2008-08-174854
Fu Y, Chen J, Huang Z (2019) Recent progress in microRNA-based delivery systems for the treatment of human disease. ExRNA 1(1):24. https://doi.org/10.1186/s41544-019-0024-y
Gasparello J, Fabbri E, Bianchi N et al (2017) BCL11A mRNA targeting by miR-210: a possible network regulating γ-globin gene expression. IJMS 18(12):2530. https://doi.org/10.3390/ijms18122530
Gebert LF, MacRae IJ (2019) Regulation of microRNA function in animals. Nat Rev Mol Cell Biol 20(1):21–37
Georges M, Charlier C, Cockett N (2003) The callipyge locus: evidence for the trans interaction of reciprocally imprinted genes. Trends Genet 19(5):248–252
Grigelioniene G, Suzuki HI, Taylan F et al (2019) Gain-of-function mutation of microRNA-140 in human skeletal dysplasia. Nat Med 25(4):583–590. https://doi.org/10.1038/s41591-019-0353-2
Hagh MF, Fard AD, Saki N et al (2011) Molecular mechanisms of hemoglobin F induction. International journal of hematology-oncology and stem. Cell Res:5–9
Herman JG, Latif F, Weng Y et al (1994) Silencing of the VHL tumor-suppressor gene by DNA methylation in renal carcinoma. Proc Natl Acad Sci USA 91(21):9700–9704. https://doi.org/10.1073/pnas.91.21.9700

Hong LZ, Zhou L, Zou R et al (2021) Systematic evaluation of multiple qPCR platforms, NanoString and miRNA-Seq for microRNA biomarker discovery in human biofluids. Sci Rep 11(1):4435
Id Said B, Malkin D (2015) A functional variant in miR-605 modifies the age of onset in Li-Fraumeni syndrome. Cancer Genet 208(1–2):47–51. https://doi.org/10.1016/j.cancergen.2014.12.003
Iorio MV, Croce CM (2012) MicroRNA dysregulation in cancer: diagnostics, monitoring and therapeutics. A comprehensive review. EMBO Mol Med 4(3):143–159. https://doi.org/10.1002/emmm.201100209
Iyer M, Anand U, Thiruvenkataswamy S et al (2023) A review of chromium (Cr) epigenetic toxicity and health hazards. Sci Total Environ 882:163483
Jaenisch R, Bird A (2003) Epigenetic regulation of gene expression: how the genome integrates intrinsic and environmental signals. Nat Genet 33(3):245–254
Jaffrey SR (2014) An expanding universe of mRNA modifications. Nat Struct Mol Biol 21(11):945–946. https://doi.org/10.1038/nsmb.2911
Jansson MD, Lund AH (2012) MicroRNA and cancer. Mol Oncol 6(6):590–610
Ji Y, Han Z, Shao L, Zhao Y (2016) Evaluation of in vivo antitumor effects of low-frequency ultrasound-mediated miRNA-133a microbubble delivery in breast cancer. Cancer Med 5(9):2534–2543. https://doi.org/10.1002/cam4.840
Jiang Q, Wang Y, Hao Y et al (2009) miR2Disease: a manually curated database for microRNA deregulation in human disease. Nucleic Acids Res 37(suppl_1):D98–D104
Jin C, Li M, Ouyang Y, Tan Z et al (2017) MiR-424 functions as a tumor suppressor in glioma cells and is down-regulated by DNA methylation. J Neuro-Oncol 133(2):247–255. https://doi.org/10.1007/s11060-017-2438-4
Kavitha M, Jayachandran D, Aishwarya SY et al (2022) A new insight into the diverse facets of microRNA-31 in oral squamous cell carcinoma. Egypt J Med Hum Genet 23(1):149. https://doi.org/10.1186/s43042-022-00361-2
Kerem E, Konstan MW, De Boeck K et al (2014) Ataluren for the treatment of nonsense-mutation cystic fibrosis: a randomised, double-blind, placebo-controlled phase 3 trial. Lancet Respir Med 2(7):539–547. https://doi.org/10.1016/S2213-2600(14)70100-6
Keskin S, Brouwers CC, Sogorb-Gonzalez M et al (2019) AAV5-miHTT lowers huntingtin mRNA and protein without off-target effects in patient-derived neuronal cultures and astrocytes. Mol Therapy-Meth Clin Dev 15:275–284
Khanahmad H, Daloii MN, Shokrgozar M et al (2006) A novel single step double positive double negative selection strategy for β-globin gene replacement. Biochem Biophys Res Commun 345(1):14–20
Kim Y-K, Kim VN (2007) Processing of intronic microRNAs. EMBO J 26(3):775–783. https://doi.org/10.1038/sj.emboj.7601512
Kim J-D, Lee A, Choi J et al (2015) Epigenetic modulation as a therapeutic approach for pulmonary arterial hypertension. Exp Mol Med 47(7):e175–e175. https://doi.org/10.1038/emm.2015.45
Kingston ER, Bartel DP (2019) Global analyses of the dynamics of mammalian microRNA metabolism. Genome Res 29(11):1777–1790
Kobayashi M, Sawada K, Kimura T (2018) Is microRNA replacement therapy promising treatment for cancer. Non-Coding RNA Investig 2:56
Koehorst E, Ballester-Lopez A, Arechavala-Gomeza V, Martínez-Piñeiro A, Nogales-Gadea G (2020) The biomarker potential of miRNAs in myotonic dystrophy type I. J Clin Med 9(12):3939
Komal S, Yin JJ, Wang SH, Huang CZ, Tao HL, Dong JZ et al (2019) MicroRNAs: Emerging biomarkers for atrial fibrillation. J Cardiol 74(6):475–482
Koralov SB, Muljo SA, Galler GR et al (2008) Dicer ablation affects antibody diversity and cell survival in the B lymphocyte lineage. Cell 132(5):860–874. https://doi.org/10.1016/j.cell.2008.02.020

Kotsyfakis M, Patelarou E (2019) MicroRNAs as biomarkers of harmful environmental and occupational exposures: a systematic review. Biomarkers 24(7):623–630. https://doi.org/10.1080/1354750X.2019.1652348
Kouhkan F, Soleimani M, Daliri M et al (2013) miR-451 up-regulation, induce erythroid differentiation of CD133+ cells independent of cytokine cocktails. Iran J Basic Med Sci 16(6):756
Kozomara A, Griffiths-Jones S (2014) miRBase: annotating high confidence microRNAs using deep sequencing data. Nucleic Acids Res 42(D1):D68–D73
Kozomara A, Birgaoanu M, Griffiths-Jones S (2019) miRBase: from microRNA sequences to function. Nucleic Acids Res 47(D1):D155–D162
Lam JK, Chow MY, Zhang Y et al (2015) siRNA versus miRNA as therapeutics for gene silencing. Mol Therapy-Nucleic Acids 4
LaMonte G, Philip N, Reardon J et al (2012) Translocation of sickle cell erythrocyte microRNAs into plasmodium falciparum inhibits parasite translation and contributes to malaria resistance. Cell Host Microbe 12(2):187–199
Lawrie CH, Saunders NJ, Soneji S, Palazzo S, Dunlop HM, Cooper CDO et al (2008) MicroRNA expression in lymphocyte development and malignancy. Leukemia 22(7):1440–1446
Lecellier C-H, Dunoyer P, Arar K et al (2005) A cellular MicroRNA mediates antiviral defense in human cells. Science 308(5721):557–560. https://doi.org/10.1126/science.1108784
Lee Y, Kim M, Han J et al (2004) MicroRNA genes are transcribed by RNA polymerase II. EMBO J 23(20):4051–4060
Levin C, Koren A, Rebibo-Sabbah A et al (2021) Extracellular vesicle MicroRNA that are involved in β-thalassemia complications. IJMS 22(18):9760. https://doi.org/10.3390/ijms22189760
Lewis MA, Quint E, Glazier AM et al (2009) An ENU-induced mutation of miR-96 associated with progressive hearing loss in mice. Nat Genet 41(5):614–618. https://doi.org/10.1038/ng.369
Li Z, Rana TM (2014) Therapeutic targeting of microRNAs: current status and future challenges. Nat Rev Drug Discov 13(8):622–638. https://doi.org/10.1038/nrd4359
Li X, Li Y, Zhao L et al (2014) Circulating muscle-specific miRNAs in Duchenne muscular dystrophy patients. Mol Therapy Nucleic Acids 3:e177. https://doi.org/10.1038/mtna.2014.29
Li Y, Du C, Wang W et al (2015) Genetic association of MiR-146a with multiple sclerosis susceptibility in the Chinese population. Cell Physiol Biochem 35(1):281–291. https://doi.org/10.1159/000369695
Liston A, Lu L-F, O'Carroll D et al (2008) Dicer-dependent microRNA pathway safeguards regulatory T cell function. J Exp Med 205(9):1993–2004. https://doi.org/10.1084/jem.20081062
Liu YP, Berkhout B (2011) miRNA cassettes in viral vectors: problems and solutions. Biochimica et Biophysica Acta (BBA)-Gene Regulatory Mechanisms 1809(11–12):732–745
Liu X, Chen X, Yu X et al (2013) Regulation of microRNAs by epigenetics and their interplay involved in cancer. J Exp Clin Cancer Res 32(1):96. https://doi.org/10.1186/1756-9966-32-96
Lulli V, Romania P, Morsilli O et al (2013) MicroRNA-486-3p regulates γ-globin expression in human erythroid cells by directly modulating BCL11A. PLoS One 8(4):e60436
Lyst MJ, Bird A (2015) Rett syndrome: a complex disorder with simple roots. Nat Rev Genet 16(5):261–275. https://doi.org/10.1038/nrg3897
Machowska M, Galka-Marciniak P, Kozlowski P (2022) Consequences of genetic variants in miRNA genes. Comput Struct Biotechnol J 20:6443–6457. https://doi.org/10.1016/j.csbj.2022.11.036
Mailand N, Falck J, Lukas C et al (2000) Rapid destruction of human Cdc25A in response to DNA damage. Science 288(5470):1425–1429
Manca S, Magrelli A, Cialfi S et al (2011) Oxidative stress activation of miR-125b is part of the molecular switch for Hailey-Hailey disease manifestation: molecular mechanisms in Hailey-Hailey disease. Exp Dermatol 20(11):932–937. https://doi.org/10.1111/j.1600-0625.2011.01359.x
Matsuzaka Y, Kishi S, Aoki Y et al (2014) Three novel serum biomarkers, miR-1, miR-133a, and miR-206 for limb-girdle muscular dystrophy, Facioscapulohumeral muscular dystrophy, and Becker muscular dystrophy. Environ Health Prev Med 19(6):452–458. https://doi.org/10.1007/s12199-014-0405-7

McAnena P, Tanriverdi K, Curran C, Gilligan K, Freedman JE, Brown JA, Kerin MJ (2019) Circulating microRNAs miR-331 and miR-195 differentiate local luminal a from metastatic breast cancer. BMC Cancer 19:1–10

Mehta M, Satija S, Paudel KR et al (2021) Targeting respiratory diseases using miRNA inhibitor based nanotherapeutics: current status and future perspectives. Nanomedicine 31:102303. https://doi.org/10.1016/j.nano.2020.102303

Mencía Á, Modamio-Høybjør S, Redshaw N et al (2009) Mutations in the seed region of human miR-96 are responsible for nonsyndromic progressive hearing loss. Nat Genet 41(5):609–613. https://doi.org/10.1038/ng.355

Meola N, Gennarino VA, Banfi S (2009) microRNAs and genetic diseases. PathoGenetics 2(1):7. https://doi.org/10.1186/1755-8417-2-7

Michlewski G, Cáceres JF (2019) Post-transcriptional control of miRNA biogenesis. Rna 25(1):1–16

Miller JL, Grant PA (2013) The role of DNA methylation and histone modifications in transcriptional regulation in humans. In: Kundu TK (ed) Epigenetics: development and disease. Springer, Dordrecht, pp 289–317

Miniarikova J, Zanella I, Huseinovic A et al (2016) Design, characterization, and lead selection of therapeutic miRNAs targeting huntingtin for development of gene therapy for Huntington's disease. Mol Therapy Nucleic Acids 5

Mohan G, Bharathi G, Vellingiri B (2021) A middle-aged woman with severe scoliosis and encephalopathy. JAMA Neurol 78(2):251–252

Morales S, Monzo M, Navarro A (2017) Epigenetic regulation mechanisms of microRNA expression. Biomol Concepts 8(5–6):203–212. https://doi.org/10.1515/bmc-2017-0024

Morey C, Avner P (2004) Employment opportunities for non-coding RNAs. FEBS Lett 567(1):27–34

Mukherjee AG, Wanjari UR, Gopalakrishnan AV et al (2022) Exploring the regulatory role of ncRNA in NAFLD: a particular focus on PPARs. Cells 11(24):3959. https://doi.org/10.3390/cells11243959

Muljo SA, Ansel KM, Kanellopoulou C et al (2005) Aberrant T cell differentiation in the absence of dicer. J Exp Med 202(2):261–269. https://doi.org/10.1084/jem.20050678

Ning X, Shi Z, Liu X et al (2015) DNMT1 and EZH2 mediated methylation silences the microRNA-200b/a/429 gene and promotes tumor progression. Cancer Lett 359(2):198–205

Norouzi S, Majeed M, Pirro M et al (2018) Curcumin as an adjunct therapy and microRNA modulator in breast cancer. Curr Pharm Des 24(2):171–177

Nowakowski TJ, Rani N, Golkaram et al (2018) Regulation of cell-type-specific transcriptomes by microRNA networks during human brain development. Nat Neurosci 21(12):1784–1792. https://doi.org/10.1038/s41593-018-0265-3

Oliviero A, Della Porta G, Peretti GM et al (2019) MicroRNA in osteoarthritis: physiopathology, diagnosis and therapeutic challenge. Br Med Bull 130(1):137–147. https://doi.org/10.1093/bmb/ldz015

Ombrello MJ, Sikora KA, Kastner DL (2014) Genetics, genomics, and their relevance to pathology and therapy. Best Pract Res Clin Rheumatol 28(2):175–189. https://doi.org/10.1016/j.berh.2014.05.001

Ortiz-Quintero B (2016) Cell-free microRNAs in blood and other body fluids, as cancer biomarkers. Cell Prolif 49(3):281–303. https://doi.org/10.1111/cpr.12262

Paco S, Casserras T, Rodríguez MA et al (2015) Transcriptome analysis of Ullrich congenital muscular dystrophy fibroblasts reveals a disease extracellular matrix signature and key molecular regulators. PLoS One 10(12):e0145107. https://doi.org/10.1371/journal.pone.0145107

Petrek H, Batra N, Ho PY et al (2019) Bioengineering of a single long noncoding RNA molecule that carries multiple small RNAs. Appl Microbiol Biotechnol 103:6107–6117

Pipis M, Won S, Poh R et al (2023) Post-transcriptional microRNA repression of PMP22 dose in severe Charcot-Marie-tooth disease type 1. Brain 146(10):4025–4032. https://doi.org/10.1093/brain/awad203

Plaisier CL, Pan M, Baliga NS (2012) A miRNA-regulatory network explains how dysregulated miRNAs perturb oncogenic processes across diverse cancers. Genome Res 22(11):2302–2314. https://doi.org/10.1101/gr.133991.111

Prajzlerová K, Šenolt L, Filková M (2023) Is there a potential of circulating miRNAs as biomarkers in rheumatic diseases? Genes & Diseases 10(4):1263–1278

Qin Y, Buermans HP, Van Kester MS et al (2012) Deep-sequencing analysis reveals that the miR-199a2/214 cluster within DNM3os represents the vast majority of aberrantly expressed microRNAs in Sezary syndrome. J Invest Dermatol 132(5):1520

Queiroz de Oliveira PE, Zhang L, Wang Z et al (2009) Hypoxia-mediated regulation of Cdc25A phosphatase by p21 and miR-21. Cell Cycle 8(19):3157–3164

Robbins PD, Morelli AE (2014) Regulation of immune responses by extracellular vesicles. Nat Rev Immunol 14(3):195–208

Rodriguez A, Griffiths-Jones S, Ashurst JL et al (2004) Identification of mammalian microRNA host genes and transcription units. Genome Res 14(10a):1902–1910

Roy P, Bhattacharya G, Lahiri A et al (2012) hsa-miR-503 is Downregulated in β thalassemia major. Acta Haematol 128(3):187–189. https://doi.org/10.1159/000339492

Rupaimoole R, Slack FJ (2017) MicroRNA therapeutics: towards a new era for the management of cancer and other diseases. Nat Rev Drug Discov 16(3):203–222

Saini HK, Griffiths-Jones S, Enright AJ (2007) Genomic analysis of human microRNA transcripts. Proc Natl Acad Sci 104(45):17719–17724. https://doi.org/10.1073/pnas.0703890104

Saiyed AN, Vasavada AR, Johar SRK (2022) Recent trends in miRNA therapeutics and the application of plant miRNA for prevention and treatment of human diseases. Futur J Pharm Sci 8(1):24. https://doi.org/10.1186/s43094-022-00413-9

Saki N, Dehghani Fard A, Kaviani S et al (2012) Beta thalassemia: epidemiology and diagnostic and treatment approaches in Iran. Genetics in the 3rd Millennium 10(1):2674–2683

Salvatore M, Magrelli A, Taruscio D (2011) The role of microRNAs in the biology of rare diseases. Int J Mol Sci 12(10):6733–6742

Samaranch L, Blits B, San Sebastian W et al (2017) MR-guided parenchymal delivery of adeno-associated viral vector serotype 5 in non-human primate brain. Gene Ther 24(4):253–261

Sangokoya C, Telen MJ, Chi J-T (2010) microRNA miR-144 modulates oxidative stress tolerance and associates with anemia severity in sickle cell disease. Blood J Am Soc Hematol 116(20):4338–4348

Sankaran VG (2011) Targeted therapeutic strategies for fetal hemoglobin induction. Hematol 2010 Am Soc Hematol Education Program Book 2011(1):459–465

Sankaran VG, Menne TF, Šćepanović D et al (2011) MicroRNA-15a and-16-1 act via MYB to elevate fetal hemoglobin expression in human trisomy 13. Proc Natl Acad Sci 108(4):1519–1524

Sarkar S, Dey BK, Dutta A (2010) MiR-322/424 and-503 are induced during muscle differentiation and promote cell cycle quiescence and differentiation by down-regulation of Cdc25A. Mol Biol Cell 21(13):2138–2149

Sato S, Katsushima K, Shinjo K et al (2016) Histone deacetylase inhibition in prostate cancer triggers miR-320–mediated suppression of the androgen receptor. Cancer Res 76(14):4192–4204

Schaar DG, Medina DJ, Moore DF et al (2009) miR-320 targets transferrin receptor 1 (CD71) and inhibits cell proliferation. Exp Hematol 37(2):245–255

Schaefer A, O'Carroll D, Tan CL et al (2007) Cerebellar neurodegeneration in the absence of microRNAs. J Exp Med 204(7):1553–1558. https://doi.org/10.1084/jem.20070823

Schaeffer EK, West RJ, Conine SJ et al (2014) Multiple physical stresses induce γ-globin gene expression and fetal hemoglobin production in erythroid cells. Blood Cell Mol Dis 52(4):214–224

Sessa F, Salerno M, Esposito M et al (2023) miRNA Dysregulation in cardiovascular diseases: current opinion and future perspectives. IJMS 24(6):5192. https://doi.org/10.3390/ijms24065192

Sethupathy P, Borel C, Gagnebin M et al (2007) Human microRNA-155 on chromosome 21 differentially interacts with its polymorphic target in the AGTR1 3′ Untranslated region: a mecha-

nism for functional single-nucleotide polymorphisms related to phenotypes. Am J Hum Genet 81(2):405–413. https://doi.org/10.1086/519979
Shah MY, Ferrajoli A, Sood AK et al (2016) microRNA therapeutics in cancer—an emerging concept. EBioMedicine 12:34–42
Sin-Chan P, Mumal I, Suwal T et al (2019) A C19MC-LIN28A-MYCN oncogenic circuit driven by hijacked super-enhancers is a distinct therapeutic vulnerability in ETMRs: a lethal brain tumor. Cancer Cell 36(1):51–67
Singh T, Yadav S (2020) Role of microRNAs in neurodegeneration induced by environmental neurotoxicants and aging. Ageing Res Rev 60:101068. https://doi.org/10.1016/j.arr.2020.101068
Stenson P, Mort M, Ball E et al (2008) The human gene mutation database. Genome Med 1
Subramaniam MD, Aishwarya Janaki P, Abishek Kumar B et al (2023) Retinal changes in Parkinson's disease: a non-invasive biomarker for early diagnosis. Cell Mol Neurobiol 43(8):3983–3996. https://doi.org/10.1007/s10571-023-01419-4
Sureban SM, May R, Mondalek FG et al (2011) Nanoparticle-based delivery of siDCAMKL-1 increases microRNA-144 and inhibits colorectal cancer tumor growth via a Notch-1 dependent mechanism. J Nanobiotechnol 9(1):40. https://doi.org/10.1186/1477-3155-9-40
Svasti S, Masaki S, Penglong T et al (2010) Expression of microRNA-451 in normal and thalassemic erythropoiesis. Ann Hematol 89(10):953–958. https://doi.org/10.1007/s00277-010-0980-7
Tarallo A, Carissimo A, Gatto F, Nusco E, Toscano A, Musumeci O et al (2019) microRNAs as biomarkers in Pompe disease. Genet Med 21(3):591–600
Teksoy O, Sahinturk V, Cengiz M et al (2020) The protective effects of silymarin on thioacetamide-induced liver damage: measurement of miR-122, miR-192, and miR-194 levels. Appl Biochem Biotechnol 191:528–539
Thein SL (2018) Molecular basis of β thalassemia and potential therapeutic targets. Blood Cell Mol Dis 70:54–65. https://doi.org/10.1016/j.bcmd.2017.06.001
Tijsen AJ, Pinto YM, Creemers EE (2012) Circulating microRNAs as diagnostic biomarkers for cardiovascular diseases. Am J Phys Heart Circ Phys 303(9):H1085–H1095. https://doi.org/10.1152/ajpheart.00191.2012
Tsujimura K, Irie K, Nakashima H et al (2015) miR-199a links MeCP2 with mTOR signaling and its Dysregulation leads to Rett syndrome phenotypes. Cell Rep 12(11):1887–1901. https://doi.org/10.1016/j.celrep.2015.08.028
Tukker AM, Royal CD, Bowman AB et al (2021) The impact of environmental factors on monogenic Mendelian diseases. Toxicol Sci 181(1):3–12. https://doi.org/10.1093/toxsci/kfab022
Urbich C, Kuehbacher A, Dimmeler S (2008) Role of microRNAs in vascular diseases, inflammation, and angiogenesis. Cardiovasc Res 79(4):581–588
Valadi H, Ekström K, Bossios A et al (2007) Exosome-mediated transfer of mRNAs and microRNAs is a novel mechanism of genetic exchange between cells. Nat Cell Biol 9(6):654–659
Vardiman JW, Thiele J, Arber DA et al (2009) The 2008 revision of the World Health Organization (WHO) classification of myeloid neoplasms and acute leukemia: rationale and important changes. Blood 114(5):937–951. https://doi.org/10.1182/blood-2009-03-209262
Vasu S, Kumano K, Darden CM et al (2019) MicroRNA signatures as future biomarkers for diagnosis of diabetes states. Cells 8(12):1533. https://doi.org/10.3390/cells8121533
Vellingiri B (2023) A deeper understanding about the role of uranium toxicity in neurodegeneration. Environ Res 116430
Vellingiri B, Suriyanarayanan A, Selvaraj P et al (2022) Role of heavy metals (copper (Cu), arsenic (As), cadmium (Cd), iron (Fe) and lithium (Li)) induced neurotoxicity. Chemosphere 301:134625
Vellingiri B, Balasubramani K, Iyer M et al (2023) Role of telomeres and telomerase in Parkinson's disease—a new Theranostics? Adv Biol 7(12):2300097
Venugopal A, Chandran M, Eruppakotte N et al (2018) Monogenic diseases in India. Mutation Research Rev Mutation Res 776:23–31
Visone R, Croce CM (2009) MiRNAs and cancer. Am J Pathol 174(4):1131–1138

Wallace DR, Taalab YM, Heinze S et al (2020) Toxic-metal-induced alteration in miRNA expression profile as a proposed mechanism for disease development. Cells 9(4):901. https://doi.org/10.3390/cells9040901

Waller R, Goodall EF, Milo M et al (2017) Serum miRNAs miR-206, 143-3p and 374b-5p as potential biomarkers for amyotrophic lateral sclerosis (ALS). Neurobiol Aging 55:123–131. https://doi.org/10.1016/j.neurobiolaging.2017.03.027

Wang Z (2011) The principles of MiRNA-masking antisense oligonucleotides technology. In: Wu W (ed) MicroRNA and cancer. Humana Press, Totowa, pp 43–49

Wang S, Yue H, Derin RB et al (2000) Accessory protein facilitated CFTR-CFTR interaction, a molecular mechanism to potentiate the Chloride Channel activity. Cell 103(1):169–179. https://doi.org/10.1016/S0092-8674(00)00096-9

Wang P, Zhi H, Zhang Y et al (2015) miRSponge: a manually curated database for experimentally supported miRNA sponges and ceRNAs. Database 2015:bav098

Wang F, Ling L, Yu D (2021) MicroRNAs in β-thalassemia. Am J Med Sci 362(1):5–12. https://doi.org/10.1016/j.amjms.2021.02.011

Wijgerde M, Gribnau J, Trimborn T et al (1996) The role of EKLF in human beta-globin gene competition. Genes Dev 10(22):2894–2902

Wu C, Xue J, Dang X (2010) Detection of miR-144 gene in peripheral blood of children with β-thalassemia major and its significance. China Trop Med 10(3):285–286

Xie B, Ding Q, Han H et al (2013) miRCancer: a microRNA–cancer association database constructed by text mining on literature. Bioinformatics 29(5):638–644

Xu HY, Wang ZY, Chen JF et al (2015) Association between Ankylosing spondylitis and the miR-146a and miR-499 polymorphisms. PLoS One 10(4):e0122055. https://doi.org/10.1371/journal.pone.0122055

Yan LX, Wu QN, Zhang Y et al (2011) Knockdown of miR-21 in human breast cancer cell lines inhibits proliferation, in vitro migration and in vivotumor growth. Breast Cancer Res 13(1):R2. https://doi.org/10.1186/bcr2803

Yao Q, Chen Y, Zhou X (2019) The roles of microRNAs in epigenetic regulation. Curr Opin Chem Biol 51:11–17

Yin H, Kanasty RL, Eltoukhy AA et al (2014) Non-viral vectors for gene-based therapy. Nat Rev Genet 15(8):541–555

Zaharieva IT, Calissano M, Scoto M et al (2013) Dystromirs as serum biomarkers for monitoring the disease severity in Duchenne muscular dystrophy. PLoS One 8(11):e80263. https://doi.org/10.1371/journal.pone.0080263

Zanotti S, Gibertini S, Curcio M et al (2015) Opposing roles of miR-21 and miR-29 in the progression of fibrosis in Duchenne muscular dystrophy. Biochim Biophys Acta (BBA) - Mol Basis Dis 1852(7):1451–1464. https://doi.org/10.1016/j.bbadis.2015.04.013

Zeng D, Wang Z, Meng Z et al (2017) DNA tetrahedral nanostructure-based electrochemical miRNA biosensor for simultaneous detection of multiple miRNAs in pancreatic carcinoma. ACS Appl Mater Interfaces 9(28):24118–24125

Zhai P-F, Wang F, Su R et al (2014) The regulatory roles of MicroRNA-146b-5p and its target platelet-derived growth factor receptor α (PDGFRA) in erythropoiesis and Megakaryocytopoiesis. J Biol Chem 289(33):22600–22613. https://doi.org/10.1074/jbc.M114.547380

Zhan M, Miller CP, Papayannopoulou T et al (2007) MicroRNA expression dynamics during murine and human erythroid differentiation. Exp Hematol 35(7):1015–1025. https://doi.org/10.1016/j.exphem.2007.03.014

Zhang B, Farwell MA (2008) microRNAs: a new emerging class of players for disease diagnostics and gene therapy. J Cell Mol Med 12(1):3–21. https://doi.org/10.1111/j.1582-4934.2007.00196.x

Zhang X, Godbey WT (2006) Viral vectors for gene delivery in tissue engineering. Adv Drug Deliv Rev 58(4):515–534

Zhang W, Yi X, Guo S et al (2014) A single-nucleotide polymorphism of miR-146a and psoriasis: an association and functional study. J Cell Mol Med 18(11):2225–2234. https://doi.org/10.1111/jcmm.12359
Zhang B, Tian L, Xie J et al (2020) Targeting miRNAs by natural products: a new way for cancer therapy. Biomed Pharmacother 130:110546
Zhu C, Ren C, Han J, Ding Y, Du J, Dai N et al (2014) A five-microRNA panel in plasma was identified as potential biomarker for early detection of gastric cancer. Br J Cancer 110(9):2291–2299
Züchner S, Wang G, Tran-Viet K-N et al (2006) Mutations in the novel mitochondrial protein REEP1 cause hereditary spastic paraplegia type 31. Am J Hum Genet 79(2):365–369
Zuntini M, Salvatore M, Pedrini E et al (2010) MicroRNA profiling of multiple osteochondromas: identification of disease-specific and normal cartilage signatures. Clin Genet 78(6):507–516. https://doi.org/10.1111/j.1399-0004.2010.01490.x

Chapter 6
Circulating miRNAs As Biomarkers for Early Disease Diagnosis

Jonas Holdmann, Andreas Savelsbergh, and Lukas Markert

Abstract In today's medical world, early detection of disease is a crucial factor in ensuring that patients receive appropriate and adequate treatment. Particularly in the context of cancer, early diagnosis can have a significant impact on prognosis. The search is therefore intensifying for suitable biomarkers that can improve or even replace our modern screening programs. MicroRNAs, as stable molecules in various body fluids, are ideal for liquid biopsy and have been widely studied in epigenetic research over the last decade. This chapter highlights the current state of research in the most common cancers and common diseases such as type 2 diabetes mellitus and coronary artery disease and covers the difficulties and challenges in recent approaches up to clinical implementation.

Keywords MicroRNA · Liquid biopsy · Biomarker · Cancer · Diagnostic · Clinical application

---

J. Holdmann (✉)
St. Marien-Hospital, Department of Internal Medicine, Cologne, Germany

Division of Functional Genomics, Chair of Biochemistry and Molecular Medicine, Witten/Herdecke University, Witten, Germany
e-mail: Jonas.holdmann@uni-wh.de

A. Savelsbergh
Division of Functional Genomics, Chair of Biochemistry and Molecular Medicine, Witten/Herdecke University, Witten, Germany

L. Markert
Division of Functional Genomics, Chair of Biochemistry and Molecular Medicine, Center for Biomedical Education and Research (ZBAF), Witten/Herdecke University, Witten, Germany

Helios Park Hospital Leipzig, Department of Anesthesiology and Intensive Care, Leipzig, Germany

L. M. Vaschetto (ed.), *miRNAs, Human Health and Diseases*, Epigenetics and Human Health 13, https://doi.org/10.1007/978-3-031-64788-8_6

List of Abbreviations

AGO2	Protein Argonaute 2
AMI	Acute myocardial infarction
AUC	Area under the curve
CAD	Coronary arterial disease
c-miRNA	Circulating microRNA
DNA	Deoxynucleic acid
ds-miRNA	Double-stranded microRNA
GC	Gastric cancer
HbA1c	Glycated hemoglobin
HCC	Hepatocellular carcinoma
LDCT	Low-dose computed tomography
NGS	Next-generation sequencing
NSCLC	Non-small-cell lung cancer
mRNA	Messenger RNA
miRNA	microRNA
MRI	Magnetic resonance imaging
MVB	Multivesicular body
pre-miRNA	Precursor microRNA
PET	Positron emission tomography
RISC	RNA-induced silencing complex
RNA	Ribonucleic acid
ROC	Receiver operating characteristic
RT-qPCR	Real-time quantitative polymerase chain reaction
TRBP	Trans-activation response RNA binding protein
T2D	Type 2 diabetes

6.1 Circulating MiRNAs As Biomarkers

MicroRNAs (miRNAs) are single-stranded, small non-coding RNA molecules consisting of ~22 nucleotides (Bartel 2004). First discovered in 1993 (Lee et al. 1993), the latest release of miRBase (v22.1) contains 1917 mature miRNAs for the human genome (Kozomara et al. 2019).

MiRNAs are involved in the regulation of more than 60% of human protein-coding genes (Friedman et al. 2009) and are thereby implicated in a variety of biological processes, including cell differentiation, development, proliferation, stress response, and apoptosis (He and Hannon 2004). Research nowadays shows their essential role in RNA silencing and the posttranscriptional regulation of genes in plants, animals, and DNA viruses (Sullivan and Ganem 2005).

The biogenesis and release of miRNAs are shown schematically in Fig. 6.1. MiRNA genes are part of the genome, transcribed by RNA polymerase II (Lee et al. 2004) or III (Borchert et al. 2006). The generated primary miRNA is a several

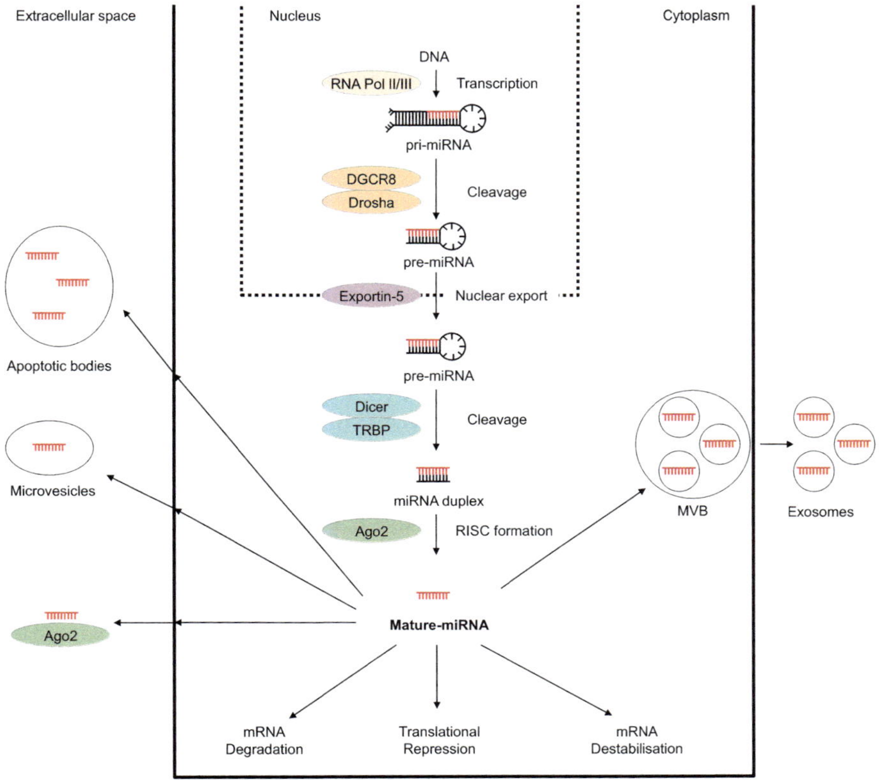

Fig. 6.1 Biogenesis and release of miRNAs. MiRNA genes are transcribed by RNA polymerase II/III. The pri-miRNA is a stem-loop formed structure, subsequently processed within the nucleus. The resulting pre-miRNA shows a characteristic hairpin structure. After being exported to the cytoplasm, the pre-miRNA hairpin is cleaved by the RNase-III-like enzyme Dicer, producing imperfect ds-miRNA. The ds-miRNA is divided, only one miRNA strand is incorporated into RISC, and the other strand is degraded. RISC is a multiprotein complex composed of a miRNA, the Dicer, TRBP, and AGO2. Within RISC, the miRNA is able to repress translation, and degrade or destabilize mRNAs. MiRNAs leave their cell via exosomes, microvesicles, protein bound (AGO2) or during apoptosis

kilobase long, stem-loop formed structure, which is subsequently processed by RNase III Drosha and the microprocessor complex subunit DGCR8 (DiGeorge syndrome critical region 8) within the nucleus (Cai et al. 2004). The resulting product is a 70–80 nt long precursor miRNA (pre-miRNA) with a characteristic hairpin structure. After being exported to the cytoplasm, the pre-miRNA hairpin is cleaved by the RNase-III-like enzyme Dicer, producing imperfect ~22-nt long, double-stranded miRNAs (ds-miRNA) (Lund and Dahlberg 2006). Next, the ds-miRNA is divided, and only one mature miRNA strand is incorporated into the RNA-induced silencing complex (RISC) (Schwarz et al. 2003), and the other strand is degraded by RISC. RISC is a multiprotein complex composed of a miRNA, the Dicer, trans-activation response RNA-binding protein (TRBP), and the protein Argonaute2 (AGO2) (Gregory et al. 2005). Within this complex, the mature miRNA is able to repress translation, degrade, or destabilize messenger RNAs (mRNAs) playing an important role in posttranscriptional gene regulation (McManus and Sharp 2002). MiRNAs leave their cell of origin via exosomes, microvesicles, protein bound (AGO2), or during apoptosis (Turchinovich et al. 2012). Exosomes are the result of exocytosis of multivesicular bodies (MVBs) and thought to be involved in cell-cell communication by being taken up by other cells (Mulcahy et al. 2014). Whether this communication is targeted or random is the subject of current research.

The ubiquitous occurrence of miRNAs in human body fluids and tissues has led to increased research into their role in cancer development and early detection, as well as the aim to early detect other common diseases, including diabetes, coronary artery disease, or neurological diseases, for example, Alzheimer's or Parkinson's disease (Malhotra et al. 2023; Fazmin et al. 2020; Sebastiani et al. 2017). Several studies have investigated the connection between miRNA expression profiles and certain types of cancer (Valihrach et al. 2020). In 2007, an association between chronic lymphatic leukemia and miRNA deregulation was shown for the first time (Marton et al. 2007). Subsequently, various miRNAs have been associated with different types of cancer. In colorectal cancer, dysregulated miRNAs appear to play a role in progression and metastasis as tumor suppressors (Huang et al. 2021), whereas other miRNAs seem to have oncogenic effects, for example, in breast cancer by promoting proliferation and metastasis (Mandal et al. 2012).

Moreover, miRNAs are detectable in various body fluids, including urine, saliva, ascites, breast milk, semen, cerebrospinal fluid, and blood (Weber et al. 2010). The remarkable stability of circulating miRNAs (c-miRNAs) from solid tumor tissues in serum and plasma samples in extreme conditions, that is, low or high pH, extended storage, boiling, or freeze-thaw cycles were demonstrated early on (Mitchell et al. 2008; Chen et al. 2008). The high stability during circulation in blood, for example, is achieved in part by binding to AGO2 within the RISC; moreover, miRNAs also show higher stability than mRNAs (Li et al. 2021).

Compared to protein-based biomarkers, the high stability of miRNAs represents a major advantage for diagnostics in the processing of samples. In addition, cell-specific miRNAs allow conclusions to be drawn about a specific tissue, whereas in the diagnosis of acute coronary syndrome, an elevated protein-based biomarker such as the highly sensitive troponin can also be confused with various other diseases, such as chronic renal failure, stroke, or subarachnoid hemorrhage, making precise classification difficult (Patil et al. 2011). In addition, research into new, improved protein-based biomarkers is currently not particularly productive due to the small number of clinically relevant proteins, their complex structure, and the establishment of suitable measurement methods (Condrat et al. 2020). Since nucleotides have the disadvantages of extremely high polarity, low concentrations, and chromatography problems, which currently make sample processing much more difficult, miRNAs are currently favored in detecting new biomarkers from human plasma or blood (Zhang 2017).

Since quantification of dysregulated miRNAs is well-feasible using real-time quantitative polymerase chain reaction (RT-qPCR) or high-throughput techniques such as microarrays or next-generation sequencing (NGS), they are classified as a potential new form of biomarkers for the early detection of diverse diseases (Ho et al. 2022). The use of various body fluids as liquid biopsies to detect miRNAs to diagnose or predict diseases is shown in Fig. 6.2.

6.2 MiRNAs As Diagnostic Biomarkers in Common Diseases

Research into the use of suitable biomarkers in diagnosis is showing progress, particularly in chronic diseases that have a slow progression or show changes in miRNA expression patterns before the onset of symptoms or metabolic imbalances. Here, we review the current state of research on miRNAs as biomarkers exemplary for coronary artery diseases (CAD) and type 2 diabetes (T2D), two of the most common diseases worldwide. It should be noted that in addition to the two diseases mentioned above, progress has also been made in recent years in the field of neurodegenerative diseases such as Alzheimer's disease (Swarbrick et al. 2019), multiple sclerosis (Zailaie et al. 2022), Parkinson's disease (Roser et al. 2018), and myasthenia gravis (Sabre et al. 2020) regarding circulating miRNAs as biomarkers. As these diseases are mostly chronic and, unlike CAD or T2D, the prevention of secondary damage plays a minor role, there is no in-depth evaluation of miRNAs as biomarkers for nervous system disorders.

In the following, the miRNAs with the highest sensitivity, specificity, and differentiation potential are identified based on receiver operating characteristic (ROC) analysis. This type of analysis allows the comparison of the diagnostic ability of different biomarkers based on the area under the ROC curve (AUC), which is calculated based on sensitivity and specificity (Florkowski 2008).

Fig. 6.2 Schematic illustration of the use of body fluid miRNAs as disease biomarkers. Body fluids serve as the source of miRNAs, which are obtained as liquid biopsies. The samples are then analyzed using different detection methods such as quantitative real-time PCR, microarrays, or next-generation sequencing. In addition, the use of circulating miRNAs as biomarkers in the diagnosis and prognosis of cancer and other common diseases is the subject of current research

6.2.1 Coronary Artery Disease

Cardiovascular diseases are the leading cause of mortality worldwide. These can be subdivided into coronary arterial diseases (CAD) and stroke. Although some advances in the detection and treatment of CAD have reduced CAD-associated mortality in recent years, particularly in the United States and Europe, there is still a clinical need for new biomarkers for early detection. In countries such as China, where Western lifestyles are being adapted in terms of malnutrition and physical inactivity, there is an increasing trend in the incidence of CAD and its associated consequences (Roth et al. 2020). Currently, cardiac troponin is the main serum-based biomarker used to detect acute myocardial infarction (AMI). Still, this biomarker is limited by several factors, such as its early absence or interference with the presence of renal impairment (Suzuki et al. 2018). MiRNAs appear to play an important role in the pathogenesis of CAD by affecting atherogenesis regarding endothelial dysfunction or plaque stability (Gao et al. 2015). MiRNAs can regulate the development of CAD through several mechanisms. Modulation of angiogenesis is the most important, but they are also involved in leukocyte adhesion (Nariman-Saleh-Fam et al. 2019), inflammatory responses (Yang et al. 2018), and modulation of vascular smooth muscle cell activity (Lai et al. 2018). Remarkably, several miRNAs affect different aspects of this process or different targets within it. For example, miR-206 regulates several signaling pathways by regulating the expression of its key proteins (e.g., vascular endothelial growth factor or endothelial nitric oxide synthase), all of which are involved in angiogenic processes (Wang et al. 2016; Tang et al. 2015).

Several dysregulated miRNAs have been described in serum or plasma from patients with CAD or acute coronary syndrome, with the most promising ones described below.

For example, the analysis revealed that circulating miR-133a had a diagnostic accuracy with an AUC of 0.918 in the comparison of 154 patients with CAD versus a control group of 92 individuals without coronary artery stenosis (Wang et al. 2013). A study validated a miRNA panel consisting of miR-132, miR-150, and miR-186, showing a high discriminatory potential with an AUC of 0.91 (Zeller et al. 2014).

An interesting study investigating the discriminatory potential of two cardiomyocyte-derived miRNAs (miR-208b and miR-499) showed an AUC of 0.99 in separating patients with AMI from a healthy control group, highlighting their high potential as clinical biomarkers (Agiannitopoulos et al. 2018). The reliability and associated biomarker potential of miR-499 were confirmed with an AUC of 0.98 in a meta-analysis that included over 3000 participants from a total of 14 studies (Zhao et al. 2019). A recent meta-analysis based on a total of 2069 patients with CAD and 1338 controls from 16 articles showed that miR-155 is significantly decreased in the plasma of patients with CAD or AMI, making it suitable for the diagnosis and monitoring of CAD (Ran et al. 2023). For further information regarding the importance of miRNAs in cardiovascular disease please refer to Chap. 9.

6.2.2 Type 2 Diabetes Mellitus

Type 2 diabetes (T2D) is a major and worrying health challenge worldwide. It is currently estimated to affect nearly 6.5% of the world's population, with an increasing incidence, especially in Western countries. The annual mortality rate is about one million, making type 2 diabetes the ninth leading cause of death worldwide (Khan et al. 2020). According to recent estimates, T2D accounts for more than 90% of all diabetes diseases and is therefore primarily considered in this section (Henning 2018). To avoid secondary complications associated with diabetes, such as micro- and macrovascular complications, peripheral neuropathy, or neonatal complications (Cade 2008), screening is carried out on risk patients using glycated hemoglobin (HbA1c), fasting plasma glucose levels, and an oral glucose tolerance test (Brar 2019). Despite this, almost half of all patients are unaware of their diabetic condition, which highlights the urgent demand for better screening programs (Ogurtsova et al. 2022).

The use of alternative biomarkers like C-reactive protein, cytokines, adiponectin, ferritin, etc. has been early proposed (Urdea et al. 2009). As these are not disease-specific, there is a need to explore new biomarkers that can help identify individuals at risk of developing T2D, preferably before glucose metabolism becomes dysregulated.

One of the first important studies established a miRNA panel (miR-15a, miR-28-3p, miR-126, miR-223, and miR-320) that was able to distinguish patients with T2D from healthy controls and was furthermore able to predict T2D up to 10 years before diagnosis due to dysregulated miRNA expression in normoglycemic patients (Zampetaki et al. 2010).

Type 2 diabetes is a multifactorial disease. The most important factors in the development of this type of diabetes are insulin resistance with secondary hyperinsulinemia and impaired insulin secretion. Insulin is secreted by pancreatic ß-cells whose apoptosis leads to progressive (postprandial) insulin deficiency. MiRNAs appear to have an impact on the pathogenesis of diabetes. There has been early evidence that overexpression of miR-375 suppresses glucose-induced insulin secretion (Poy et al. 2004). Recently, it has been shown that mir-204 targets key ß-cell genes, as the glucagon-like peptide-1 receptor (Jo et al. 2018), and thus regulates important processes in ß-cell biology including insulin production and incretin-mediated insulin secretion (Xu et al. 2019).

In a meta-analysis, the most frequently studied and reported differentially expressed miRNAs in type 2 diabetes mellitus were miR-122, −126b, −144, −146a, −210, and -375, while interestingly miR-144 was also differentially expressed in the context of prediabetes (González-Sánchez et al. 2021). In particular, endothelial-derived miR-126 is a potential biomarker, and studies have shown its high discriminatory potential with an AUC of 0.97 in the diagnosis of T2D (Benko et al. 2023) and suitability as an independent predictor of long-term all-cause mortality in patients with T2D (Pordzik et al. 2021). Another meta-analysis showed that increased expression of miR-126 alone is associated with a reduced risk of developing T2D, highlighting its potential as a biomarker (He et al. 2022).

6.3 Circulating MiRNAs in Cancer Diagnosis

The search for suitable biomarkers in cancer diagnosis is a subject of intense research, as early detection of cancer plays a crucial role in the treatment and prognosis of cancer patients. If cancer is diagnosed at an advanced, possibly already metastasized stage, a curative therapy approach is no longer possible in most cases. For this reason, mass screenings for the early detection of certain cancers have been and are being implemented in national health guidelines around the world. However, directly after cardiovascular diseases, cancer is the second leading cause of death worldwide and a global health burden. The most common causes of cancer-related deaths are cancer of the lung, colon and rectum, stomach, breast, liver, and prostate (Kocarnik et al. 2022).

The oncogenesis of different types of cancer is a highly complex genetic event in which oncogenes and tumor suppressor genes play an important role. In recent years, epigenetic alterations have increasingly become the focus of research, considering DNA methylation and the expression of noncoding RNAs such as miRNAs. The influence of miRNAs on cancer development occurs at different levels. As such, miRNAs have been described to affect the interaction of the tumor with the immune system (Cubillos-Ruiz et al. 2012; Zonari et al. 2013). They can exert an effect on other cells, in particular on cancer-associated fibroblasts, which lead to the progression of cancer growth (Mitra et al. 2012). The exact biological understanding of individual miRNAs and their role in the human body is extremely complex, as miRNAs exert great influence on specific signaling pathways, but can also be the target of other regulatory processes themselves. Current research is investigating the potential of liquid biopsies taken from various body fluids considering both miRNA-based single biomarkers and biomarker panels in the diagnosis of cancer (Markert et al. 2021). MiRNAs are not always tissue-specific, but during oncogenesis, frequently dysregulated miRNA expression forms a tissue-specific pattern that is associated with the cancer microenvironment, stage of progression, therapeutic response, and patient survival (Wang et al. 2018).

In the following, the most frequent causes of cancer-related deaths are explored regarding the use of miRNAs as biomarkers. The most promising miRNAs are summarized based on study quality, case numbers, discrimination values (sensitivity, specificity, and area under the curve (AUC)), and potential for clinical implementation (Table 6.1).

Table 6.1 Dysregulated circulating miRNAs as diagnostic biomarkers from different body fluids in human cancer

Disease	miRNAs	Regulation	Reference
Lung cancer (NSCLC)	miRNA-96 (serum, tissue)	↑	Wu et al. (2017b)
Lung cancer (Adenocarcinoma)	miR-486 miR-21, miR-375, miR-200b (sputum panel)	↓ ↑	Yu et al. (2010)
Prostate cancer	miR-181a miR-141, miR-221, miR-21, miR-375 (plasma)	↓ ↑	Fabris et al. (2016)
	miR-H9, miR-3659	↑	Ran et al. (2023)
Breast cancer	miR-155 (serum)	↑	Sun et al. (2012); Zhao et al. (2012)
	miR-1246, miR-206, miR-24, miR-373 (plasma panel)	↑	Jang et al. (2021)
Colorectal cancer	miR-21 (serum, tissue)	↑	(Weng et al. 2018)
	miR-202-3p, miR-21-3p, miR-1229-3p, miR-1246, miR-532-3p (serum)	↓ ↑	Guo et al. (2018)
	miR-135b-5p (stool)	↑	(Li et al. 2020)
Liver cancer	miR-148a (plasma)	↓	Dawood et al. (2021) and Han et al. (2019b)

6.3.1 *Lung Cancer*

Lung cancer remains as the leading cancer-associated cause of death worldwide. The current gold standard in early detection of lung cancer is the radiological examination of high-risk groups using low-dose computed tomography (LDCT) (Manser et al. 2013). Depending on the diagnostic results, these are subsequently verified by an invasive sample biopsy, or a follow-up is performed at a defined time interval (Kuijvenhoven et al. 2020).

Since radiological options involve exposure to ionizing radiation and invasive sample biopsy is also associated with adverse risks to health, the use of noninvasive biomarkers has been a subject of recent investigation.

Lung cancer has various subtypes, which differ in their development, treatment, and prognosis. The search for suitable biomarkers is therefore usually performed for one subtype compared to a healthy control group. It was shown that exosomal miRNAs have comparable levels in plasma from patients with non-small-cell lung cancer (NSCLC) and tumor tissue in NSCLC, suggesting that they can be used for liquid biopsy (Rabinowits et al. 2009). Since then, a variety of exosomal miRNAs have been described for early detection of lung cancer (Hu et al. 2020). For example, increased levels of exosomal miRNA-96 in lung cancer tissue and serum of NSCLC have shown to be correlated. In addition, miR-96 from lung cancer tissue and serum differs significantly from lung tissue and serum of the healthy control group. Elevated miRNA-96 levels in lung cancer tissues are associated with increased cell growth, migration, and cisplatin resistance. Additionally, it is positively associated with metastasis and high-grade cancer status (Wu et al. 2017b).

Sputum was also searched for possible biomarkers for lung cancer. A biomarker panel consisting of 4 sputum-derived miRNAs (miR-21, miR-486, miR-375, and miR-200b) achieved sensitivity and specificity of 80.6% and 92.5%, respectively, in distinguishing 64 patients with NSCLC from a healthy control group consisting of 58 patients (Yu et al. 2010).

So far, miRNA-analysis has not been included in the guidelines for the detection of lung cancer, although the combination of current diagnostics with individual miRNAs would be conceivable. An interesting study with 939 patients retrospectively investigated the diagnostic potential of a plasma-based miRNA panel. It showed a high negative predictive value of over 99%, a sensitivity of 87%, and a specificity of 81% for the panel consisting of 24 miRNAs. The combination of LDCT and the miRNA panel resulted in a fivefold reduction to 3.7% in false-positive findings (Sozzi et al. 2014). From today's perspective, miR-96 is a promising single biomarker that could play an important role in the diagnosis of lung cancer. Nevertheless, further studies with high sample sizes are needed to validate the results reported so far.

6.3.2 Prostate Cancer

Prostate cancer is the second-most diagnosed cancer and the fifth leading cause of cancer-related death worldwide (Sung et al. 2021; Jemal et al. 2011). To date, the gold standard in the diagnosis of prostate cancer remains the targeted and systematic transrectal or transperineal biopsy. The biopsies carry a certain risk of post procedural complications such as infection or significant bleeding (Loeb et al. 2013; Wei et al. 2015), leading to an ongoing need for other early and effective diagnostic tools (Rafikova et al. 2023).

In 2008, it was shown for the first time that the tumor-associated miRNA miR-141 is present and measurable in the plasma of patients with prostate cancer, making it a potential biomarker (Mitchell et al. 2008). Currently, the most reliable miRNAs in plasma are miR-141, miR-181a, miR-221, miR-21, and miR-375, as they showed consistent deregulation in different studies (Fabris et al. 2016). Moreover, several studies proposed that exosome-encapsulated miRNAs may be used as biomarkers in early stages of prostate cancer (Nilsson et al. 2009). A recent study showed that the ratio of miR-H9 to miR-3659 is significantly higher in patients with prostate cancer compared to the control group and the discrimination reached an AUC of 0.80 (Kang et al. 2022). In comparing the results published to date on urine-based biomarkers, one challenge is whether and how different urine fractions alter the results. Some studies examined urine sediment (Stuopelyte et al. 2016), while others used whole urine (Ghorbanmehr et al. 2019), cell-free urine (Bryant et al. 2012), or urinary exosomes (Holdmann et al. 2022).

To date, miRNA-based biomarkers have not been included in prostate cancer guidelines. However, five liquid biopsy assays for the detection of prostate cancer have already been approved by the FDA in the United States, namely, Oncotype DX AR-V7 Nucleus Detect, Foundation One Liquid, Sangia Total PSA Test, Progensa PCA 3 Assay, 4 K Score Test (Munteanu et al. 2020). This is an example that liquid biopsy will play an important role in the future of personalized medicine for diagnostics and especially for possible therapy decisions.

6.3.3 Breast Cancer

Breast cancer is the most common cancer and the second leading cause of cancer-associated death in women worldwide (Sung et al. 2021). It is estimated that one in four women will develop breast cancer in her lifetime. As with all types of cancer, early detection increases the long-term survival probability of patients (Rivera-Franco and Leon-Rodriguez 2018). The current standard of care in breast cancer screening is mammography for women between the ages of 40 and 59. In addition, ultrasound, magnetic resonance imaging (MRI), or positron emission tomography (PET) can be used for verification, although the last mentioned are not used for standard screening due to their high costs (Roganovic et al. 2015).

The survival probability of patients with metastases decreases significantly, so these patients would benefit from biomarkers for early detection of breast cancer.

A special patient cohort within breast cancer patients is the group of so-called triple negative breast cancer patients. Normally, the presence of hormone receptors for estrogen and progesterone as well as the HER2/neu oncoprotein is determined in the primary diagnosis and, among other things, influences the choice of therapy. Patients with a triple negative result are classified as triple negative and are associated with a poor response to drug therapy and increased mortality (Dent et al. 2007). For this common premenopausal breast cancer, established screening programs are often too late due to early onset. For this reason, there is a particular need for suitable biomarkers for early detection (Sukumar et al. 2021).

The miRNA with the greatest potential for a breast-specific biomarker for early diagnosis is miR-155. Several studies describe its function as an oncogenic miRNA, promoting the proliferation of cancer cells (Liu et al. 2019) and evasion of apoptosis (Shen et al. 2015) and also regulating tumor suppressors (Cheng et al. 2015). Several studies have shown that the expression of miR-155 is significantly increased in the serum of patients with breast cancer (Sun et al. 2012; Zhao et al. 2012). Interestingly, the expression in the blood decreased significantly after treatment with surgery and chemotherapy (Anwar et al. 2020). A meta-analysis concluded that the differentiation of healthy and diseased patients by miR-155 has an AUC of 0.95 (Garrido-Palacios et al. 2023).

A biomarker panel consisting of four plasma-derived miRNAs (miR-1246, miR-206, miR-24, miR-373) showed a sensitivity of 98% and a specificity of 96%. Compared with other biomarker studies, this study had a relatively large number of cases with 226 breast cancer patients and 146 healthy patients (Jang et al. 2021). A comparable result with an AUC of 0.95 was obtained with a serum-derived biomarker panel consisting of eight miRNAs (Loke et al. 2019). For liquid biopsy, blood-based miRNA assays appear to be suitable for diagnosing breast cancer, though other studies have shown possible discrimination in urine (Hirschfeld et al. 2020; Ando et al. 2019). Due to the lack of large-scale, multicenter studies, miRNAs are not yet included in current guidelines despite their great potential for early detection of breast cancer.

6.3.4 Colorectal Cancer

Colorectal cancer is the third most common cancer and the second leading cause of cancer-associated deaths worldwide (Sung et al. 2021). Most colorectal cancers develop slowly over many years from precancerous stages to malignant tumors. Therefore, the chances of early detection through screening programs are higher than other cancers. Screening programs for early cancer detection have been effectively implemented, based on two columns: colonoscopy and fecal diagnostic tests (Brenner et al. 2014; Davidson et al. 2021). In the diagnosis of colorectal carcinoma and its precursors, the gold standard remains colonoscopy, although it may be associated with procedural and postprocedural complications (Kim et al. 2019; Paszat et al. 2021). Therefore, the search for appropriate biomarkers for early detection continues to be the subject of current research.

When considering single biomarkers for colorectal cancer, miR-21, which is also frequently described in the context of other cancers, is the most studied and validated miRNA. It is often referred to as "oncomiR" in the literature as it targets various tumor suppressor genes and thus may be involved in carcinogenesis, including colorectal cancer (Feng and Tsao 2016; Li et al. 2016). A significantly higher concentration of miR-21 was found in tissue samples and serum from patients with colorectal carcinoma compared to tissue and serum from healthy individuals. A validation of 239 serum samples showed a sensitivity of 91.9% and a specificity of 81.1% with an AUC of 0.92 (Toiyama et al. 2013).

Current studies often use biomarker panels to achieve better discriminatory values. A serum-derived biomarker panel consisting of five miRNAs (miR-1246, miR-202-3p, miR-21-3p, miR-1229-3p, and miR-532-3p) showed a sensitivity and specificity of 91% in a study size of 575 samples (Guo et al. 2018). Plasma-based miRNA panels also showed great potential in discriminating healthy patients from those with colorectal cancer, with AUC values exceeding 0.9 (Herreros-Villanueva et al. 2019; Fang et al. 2015).

Recently, the investigation of miRNA from feces attracted attention since fecal miRNA is also very stable and therefore suitable for analysis. Among individual stool-based biomarkers, miR-135b-5p showed good discriminatory potential with a sensitivity of 97% and an AUC of 0.87 (Li et al. 2020). Panels consisting of two to four miRNAs achieved comparable results with AUC values ranging from 0.81 to 0.89 (Wu et al. 2017a; Epstein et al. 2016).

To implement the use of miRNA for colorectal cancer detection from the field of experimental research into clinical practice, a large-scale clinical trial is currently conducted to validate a microRNA-based fecal test for colorectal cancer screening.

6.3.5 Liver Cancer

Hepatocellular carcinoma (HCC) is the third leading cause of cancer-related mortality worldwide and a global health challenge, with an estimated annual occurrence of approximately 1.4 million diagnosed cases by 2040 (Rumgay et al. 2022). HCC accounts for more than 80% of primary liver cancer cases, and the prognosis of patients depends on the staging of the disease, highlighting the importance of early diagnosis (Sung et al. 2021). Radiological approaches such as ultrasound, computed tomography, and magnetic resonance imaging are commonly used to diagnose HCC; nevertheless, it is difficult to detect small lesions in dense liver tissue (Simmons et al. 2017). Screening for other serological tumor markers, such as alpha-fetoprotein, is associated with insufficient sensitivity and specificity to accurately detect HCC (Seo et al. 2015).

Various studies have investigated the differential expression of miRNAs between patients with hepatocellular carcinoma and healthy individuals (Zhao et al. 2021; Conti et al. 2021). One study demonstrated that circulating miRNAs could serve as biomarkers for the accurate detection of HCC by analyzing serum samples from 345 patients with HCC and 1033 healthy individuals using miRNA expression microarrays. A panel of eight miRNAs was tested on a validation set with promising results. The diagnostic values for discriminating between cancer and HCC risk control samples resulted in an AUC of 0.99, a sensitivity of 97.7%, and a specificity of 94.7% (Yamamoto et al. 2020). Regarding single biomarkers, several studies have suggested miR-148a as a potential biomarker for detecting HCC. It was found to discriminate HCC from healthy controls with an AUC of 0.99, a sensitivity of 95.8%, and a specificity of 85.1% (Dawood et al. 2021). Another study found that in HCC patients with negative or low AFP, plasma miR-148a discriminated with an AUC of 0.95, reaching a sensitivity of 90.6% and a specificity of 92.6% (Han et al. 2019a).

Although miRNAs are a potential biological target for the diagnosis of HCC, further clinical validation studies are needed before they can be used as biomarkers in clinical practice.

6.3.6 Gastric Cancer

Gastric cancer (GC) is the fourth most common cancer-related cause of death worldwide, and current estimates suggest that its incidence will continue to rise in the future (Sung et al. 2021). Upper gastrointestinal endoscopy remains the gold standard for GC diagnosis (Karimi et al. 2014). In contrast to Asian countries such as Japan, where population-wide screening is performed due to very high incidences, this approach would not be justified in European countries with low incidences (Park et al. 2014). Early diagnosis is crucial for prognosis but extremely difficult due to the long absence of specific symptoms in gastric cancer (Machlowska et al. 2020).

Considering that miRNAs appear to promote cell proliferation and migration in gastric cancer (Duan et al. 2016), a high-quality study developed a clinical assay for their detection based on a 12-miRNA biomarker panel. The 12-miRNA panel had an AUC of 0.93, with an overall sensitivity and specificity of 87% and 68.4%, respectively (So et al. 2021).

In addition to the classical search for biomarkers, clinical trials are currently investigating miRNA markers regulated by probiotics. Probiotic supplementation with *Helicobacter pylori* eradication promotes regression of intestinal metaplasia by regulating the Wnt/beta-catenin pathway (Ji and Yang 2020). Since miRNAs play a critical role in regulating the Wnt/beta-catenin pathway in gastric carcinogenesis, probiotics may modulate miRNAs involved in the Wnt/beta-catenin pathway and gastric carcinogenesis (Wu et al. 2016). A clinical trial has been designed to identify miRNAs regulated by probiotics in gastric carcinogenesis. Currently, there are no miRNA-based biomarkers that can be used for the early detection of gastric cancer. However, developments in recent years and the results show the potential of miRNAs as diagnostic biomarkers and highlight the need for further investigation.

6.4 Challenges and Difficulties in Clinical Implementing

MiRNAs offer great potential as biomarkers in many different areas of personalized medicine regarding diagnosis, therapy, and prognosis. However, there are still major hurdles for the clinical implementation of new miRNA-based biomarkers, as many current studies work with different extraction procedures, body fluids, analytical methods, or examination protocols. Moreover, the sample size remains often small, and there is generally a high degree of heterogeneity in the studies in terms of possible concomitant diseases, medications, age differences, and gender (de Boer et al. 2013). In addition to the already large natural variance of miRNA expression patterns, these factors lead to poor reproducibility of miRNA studies (Ameling et al. 2015).

6.4.1 Extraction and Quality Control

There are different commercial kits and procedures for the extraction of cellular or circulating microRNAs that may influence the outcome of studies (Brunet-Vega et al. 2015; Masotti et al. 2009). Depending on cell density, some extraction procedures favor long RNA fragments over small RNAs such as miRNAs (El-Khoury et al. 2016). It must be taken into account that the range of miRNA concentration in serum and plasma of healthy individuals is noteworthy large (Kroh et al. 2010). In addition, the handling of the samples (centrifugation influences the spectrum of miRNAs detected (Page et al. 2013)), their storage temperature (freeze thaw cycles reduce miRNA concentration (Glinge et al. 2017)), pretreatment (anticoagulants including heparin, citrate and EDTA influence the measurable miRNA expression profiles (Fichtlscherer et al. 2010)), and the use of a small, enriched RNA fraction compared to total miRNA can also have a decisive influence on the results generated (Poel et al. 2018).

It has been shown that the level of circulating miRNAs in plasma is significantly influenced by the processing of blood samples. When examining miRNA expression patterns in plasma as a cell-free fraction of blood, the presence of platelets leads to a bias of the results. Therefore and in contrast to serum, pretreatment of plasma samples prior to extraction is important (Cheng et al. 2013). Another body fluid suitable for liquid biopsy is urine, whereas urine samples also vary greatly in composition depending on age, diet, the time the urine remains in the bladder, and how long it is stored (Parekh et al. 2007). Furthermore, there have been differences in terms of protein concentration or the presence of blood cells among urine samples and an effect on small RNA concentration is also conceivable (Thomas et al. 2010).

Quality control is essential in the validation of potential miRNA-based biomarkers (Bustin et al. 2009). Exogenous spike-in controls are foreign small RNA molecules of the same length as native miRNAs that may be derived from unrelated species such as Caenorhabditis elegans. They are added to the samples before extraction and allow intermediate measurements to be made, giving an indication of relevant miRNA losses (Li and Kowdley 2012). Endogenous control methods are currently based only on published data listing sequences that are consistently present or absent in, for example, plasma samples. To assess miRNA recovery, both blood cell-derived and circulating miRNAs are selected (Duy et al. 2015). The state-of-the-art approach is a quality control panel based on endogenous microRNAs and synthetic spike-in molecules quantified by RT-qPCR technology (Androvic et al. 2019).

6.4.2 Quantification and Analysis

Basically, three different methods are used for miRNA quantification: microarrays, RT-qPCR, and RNA-Seq such as NGS, each of which has specific advantages and disadvantages.

Microarrays were the first available high-throughput procedures that usually needed a relatively high RNA input, possibly limiting the investigation of body liquids with low RNA concentrations (Ono et al. 2015; Mestdagh et al. 2014). NGS is considered the most important technology in miRNA research, as it works without predefined target miRNAs or specific probes, enabling the discovery of yet unknown miRNA sequences. The disadvantages are the relatively time-consuming laboratory preparation, the lack of standardized protocols, and the comparatively high costs. Both high-throughput methods do not allow exact quantification and require a partly complex bioinformatic processing of the results (Pritchard et al. 2012).

The state-of-the-art method for accurately quantifying of miRNAs is still RT-qPCR, with ease of execution in molecular genetics laboratories, standardized protocols and analyses, and relatively low cost. However, like microarrays, only known miRNAs can be quantified (Chugh and Dittmer 2012). In summary, microarrays and NGS are suitable for screening, while RT-qPCR is helpful in validating the results obtained, but convenience, dynamic range, variability, or experience with a particular technique, together with local availability and cost, may ultimately determine the best microRNA profiling platform.

Data analysis is also an important step in the study of miRNAs and poses certain challenges depending on the technology chosen. While the more established methods, such as microarrays and RT-qPCR, are often supplied with appropriate software that does not require advanced bioinformatics skills, the analysis of NGS data is very complex due to the large amount of data generated and the lack of quality standards for statistical processing (Cristiano and Veltri 2016).

Primary analysis includes quality control, data normalization, and, in the case of NGS, comparison with a database such as miRbase or a known genome. Normalization is important for the comparability of samples within a study to reduce differences due to technical variations in the preparation process. If the normalization of the data is not done properly, the validity of the reported results should be questioned as they may lead to negligent misconceptions (Schwarzenbach et al. 2015).

In the secondary analysis, the differentially expressed miRNAs are tested and compared using different statistical approaches, and the most interesting candidates are presented in terms of fold change, sensitivity, specificity, and diagnostic accuracy. But, as well, there are certain challenges associated with analyzing the expression patterns of miRNAs. Changes in expression patterns do not necessarily correspond to the exact number of miRNAs in the sample because, for example, they are dependent on the oligonucleotides present (Linsen et al. 2009), and some cDNA may be amplified more efficiently during PCR. One way to counteract this is to use unique molecular identifiers (UMI) that are assigned to the miRNAs during their preparation for quantification. UMIs are unique sequences that are used to accurately identify the input miRNA after sequencing, using software to avoid a possible amplification bias (Fu et al. 2018).

6.4.3 Comparability and Clinical Implementation

The comparability of studies investigating miRNAs as biomarkers of diseases is limited by different quantification technologies, different protocols for performing procedures such as miRNA extraction and potential pitfalls in analysis and interpretation as previously described (Yamamoto et al. 2020).

The implementation of these molecules as biomarkers in clinical practice faces several challenges. Firstly, standardized miRNA detection technologies, efficient normalization, and validation methods must be developed to avoid the frequent poor reproducibility; it is not uncommon for follow-up studies with the same question to show deviating miRNA expression profiles. Secondly, the appropriate biological material (whole blood, plasma, serum, cerebrospinal fluid, urine, or saliva) for miRNA measurement of specific diseases should be defined, as miRNAs show unique and stable patterns in different body fluids (Mitchell et al. 2008). Once these standards have been established, randomized controlled trials with large patient cohorts will be needed to validate the results to date and enable clinical translation.

In assessing the results and the interpretations derived thereof, it must be emphasized that in many studies the correlation of expression patterns of miRNAs with certain diseases is investigated. It would be an incorrect conclusion to draw a causal relationship between the described correlations, since the findings could also be due to other, possibly yet unknown, alterations of the human body. In order to derive a causal theory from the results, it is first necessary to confirm that the miRNAs of potential interest originate from the tissue that is essentially linked to a disease, as it has already been described for several types of cancer (Liu et al. 2019). Considering single dysregulated miRNAs as potential biomarkers, the understanding of its function toward a certain mRNA has to be seen in context of the other miRNAs targeting that mRNA (Andl et al. 2020). Conclusions about the causality of its dysregulation cannot be drawn, as the dysregulated miRNAs can either be the cause or the effect of cancer or disease development.

We are currently in a process of continuous expansion of our technical and analytical capabilities, making it difficult to establish globally standardized procedures. However, to finally use circulating biomarkers clinically, standardized validation of existing results is essential to meet the diagnostic criteria in human medicine and therefore one of the key objectives for further research.

6.5 Conclusion

In summary, miRNAs have great potential as biomarkers for certain diseases, including common diseases such as TD2 or CVD and especially in the field of cancer medicine. In recent years, the databases of known miRNAs have grown steadily, and our understanding of the genetic pathways that miRNAs affect is also expanding. While the sheer number of miRNAs found and their potential targets already requires a multitude of huge databases, an uncertain part of the network of epigenetic-genetic interactions remains to be understood and offers room for future research. In addition, the study methods, their application and interpretation are very complex, and the further development and establishment of standardized detection and validation procedures is still essential for the comparability of studies. Nevertheless, the prospects for miRNA biomarkers are exciting. Their chemical stability in tissues, blood, and other body fluids and their ability to regulate target networks are technically and biologically attractive. Given the rapid development of current molecular techniques coupled with the growth of multiomics, it is likely that circulating miRNAs will be widely used in personalized medicine and will also enable targeted therapeutic interventions.

Acknowledgments We would like to thank Gerhard Ludwig for providing the illustration showing the biogenesis of miRNAs.

Compliance with Ethical Standards

Conflict of Interest JH, AS, and LM do hereby declare no conflict of interest.

Ethical Approval This chapter contains no studies with human/animal participants performed by authors.

References

Agiannitopoulos K, Pavlopoulou P, Tsamis K et al (2018) Expression of miR-208b and miR-499 in Greek patients with acute myocardial infarction. In Vivo 32(2):313–318. https://doi.org/10.21873/invivo.11239

Ameling S, Kacprowski T, Chilukoti RK et al (2015) Associations of circulating plasma microRNAs with age, body mass index and sex in a population-based study. BMC Med Genet 8:61. https://doi.org/10.1186/s12920-015-0136-7

Andl T, Ganapathy K, Bossan A et al (2020) MicroRNAs as guardians of the prostate: those who stand before cancer. What do we really know about the role of microRNAs in prostate biology? Int J Mol Sci 21(13):4796. https://doi.org/10.3390/ijms21134796

Ando W, Kikuchi K, Uematsu T et al (2019) Novel breast cancer screening: combined expression of miR-21 and MMP-1 in urinary exosomes detects 95% of breast cancer without metastasis. Sci Rep 9(1):13595. https://doi.org/10.1038/s41598-019-50084-5

Androvic P, Romanyuk N, Urdzikova-Machova L et al (2019) Two-tailed RT-qPCR panel for quality control of circulating microRNA studies. Sci Rep 9(1):4255. https://doi.org/10.1038/s41598-019-40513-w

Anwar SL, Tanjung DS, Fitria MS et al (2020) Dynamic changes of circulating Mir-155 expression and the potential application as a non-invasive biomarker in breast cancer. Asian Pac J Cancer Prev 21(2):491–497. https://doi.org/10.31557/apjcp.2020.21.2.491

Bartel DP (2004) MicroRNAs: genomics, biogenesis, mechanism, and function. Cell 116(2):281–297. https://doi.org/10.1016/s0092-8674(04)00045-5

Benko J, Sarlinova M, Mikusova V et al (2023) MiR-126 and miR-146a as markers of type 2 diabetes mellitus: a pilot study. Bratisl Lek Listy 124(7):527–533. https://doi.org/10.4149/bll_2023_081

Borchert GM, Lanier W, Davidson BL (2006) RNA polymerase III transcribes human microRNAs. Nat Struct Mol Biol 13(12):1097–1101. https://doi.org/10.1038/nsmb1167

Brar PC (2019) Update on the current modalities used to screen high risk youth for prediabetes and/or type 2 diabetes mellitus. Ann Pediatr Endocrinol Metab 24(2):71–77. https://doi.org/10.6065/apem.2019.24.2.71

Brenner H, Stock C, Hoffmeister M (2014) Effect of screening sigmoidoscopy and screening colonoscopy on colorectal cancer incidence and mortality: systematic review and meta-analysis of randomised controlled trials and observational studies. BMJ 348:g2467. https://doi.org/10.1136/bmj.g2467

Brunet-Vega A, Pericay C, Quílez ME et al (2015) Variability in microRNA recovery from plasma: comparison of five commercial kits. Anal Biochem 488:28–35. https://doi.org/10.1016/j.ab.2015.07.018

Bryant RJ, Pawlowski T, Catto JW et al (2012) Changes in circulating microRNA levels associated with prostate cancer. Br J Cancer 106(4):768–774. https://doi.org/10.1038/bjc.2011.595

Bustin SA, Benes V, Garson JA et al (2009) The MIQE guidelines: minimum information for publication of quantitative real-time PCR experiments. Clin Chem 55(4):611–622. https://doi.org/10.1373/clinchem.2008.112797

Cade WT (2008) Diabetes-related microvascular and macrovascular diseases in the physical therapy setting. Phys Ther 88(11):1322–1335. https://doi.org/10.2522/ptj.20080008

Cai X, Hagedorn CH, Cullen BR (2004) Human microRNAs are processed from capped, polyadenylated transcripts that can also function as mRNAs. RNA (New York, NY) 10(12):1957–1966. https://doi.org/10.1261/rna.7135204

Chen X, Ba Y, Ma L et al (2008) Characterization of microRNAs in serum: a novel class of biomarkers for diagnosis of cancer and other diseases. Cell Res 18(10):997–1006. https://doi.org/10.1038/cr.2008.282

Cheng HH, Yi HS, Kim Y et al (2013) Plasma processing conditions substantially influence circulating microRNA biomarker levels. PLoS One 8(6):e64795. https://doi.org/10.1371/journal.pone.0064795

Cheng CJ, Bahal R, Babar IA et al (2015) MicroRNA silencing for cancer therapy targeted to the tumour microenvironment. Nature 518(7537):107–110. https://doi.org/10.1038/nature13905

Chugh P, Dittmer DP (2012) Potential pitfalls in microRNA profiling. Wiley Interdiscip Rev RNA 3(5):601–616. https://doi.org/10.1002/wrna.1120

Condrat CE, Thompson DC, Barbu MG et al (2020) miRNAs as biomarkers in disease: latest findings regarding their role in diagnosis and prognosis. Cells 9(2). https://doi.org/10.3390/cells9020276

Conti I, Simioni C, Varano G et al (2021) MicroRNAs patterns as potential tools for diagnostic and prognostic follow-up in cancer survivorship. Cells 10(8):2069

Cristiano F, Veltri P (2016) Methods and techniques for miRNA data analysis. Methods Mol Biol 1375:11–23. https://doi.org/10.1007/7651_2015_238

Cubillos-Ruiz JR, Baird JR, Tesone AJ et al (2012) Reprogramming tumor-associated dendritic cells in vivo using miRNA mimetics triggers protective immunity against ovarian cancer. Cancer Res 72(7):1683–1693. https://doi.org/10.1158/0008-5472.Can-11-3160

Davidson KW, Barry MJ, Mangione CM et al (2021) Screening for colorectal cancer: US preventive services task force recommendation statement. JAMA 325(19):1965–1977. https://doi.org/10.1001/jama.2021.6238

Dawood AA, Saleh AA, Elbahr O et al (2021) Inverse relationship between the level of miRNA 148a-3p and both TGF-β1 and FIB-4 in hepatocellular carcinoma. Biochem Biophys Rep 27:101082. https://doi.org/10.1016/j.bbrep.2021.101082

de Boer HC, van Solingen C, Prins J et al (2013) Aspirin treatment hampers the use of plasma microRNA-126 as a biomarker for the progression of vascular disease. Eur Heart J 34(44):3451–3457. https://doi.org/10.1093/eurheartj/eht007

Dent R, Trudeau M, Pritchard KI et al (2007) Triple-negative breast cancer: clinical features and patterns of recurrence. Clin Cancer Res 13(15 Pt 1):4429–4434. https://doi.org/10.1158/1078-0432.Ccr-06-3045

Duan J, Zhang H, Qu Y et al (2016) Onco-miR-130 promotes cell proliferation and migration by targeting TGFβR2 in gastric cancer. Oncotarget 7(28):44522–44533. https://doi.org/10.18632/oncotarget.9936

Duy J, Koehler JW, Honko AN et al (2015) Optimized microRNA purification from TRIzol-treated plasma. BMC Genomics 16(1):95. https://doi.org/10.1186/s12864-015-1299-5

El-Khoury V, Pierson S, Kaoma T et al (2016) Assessing cellular and circulating miRNA recovery: the impact of the RNA isolation method and the quantity of input material. Sci Rep 6(1):19529. https://doi.org/10.1038/srep19529

Epstein JI, Egevad L, Amin MB et al (2016) The 2014 International Society of Urological Pathology (ISUP) consensus conference on Gleason grading of prostatic carcinoma: definition of grading patterns and proposal for a new grading system. Am J Surg Pathol 40(2):244–252. https://doi.org/10.1097/pas.0000000000000530

Fabris L, Ceder Y, Chinnaiyan AM et al (2016) The potential of MicroRNAs as prostate cancer biomarkers. Eur Urol 70(2):312–322. https://doi.org/10.1016/j.eururo.2015.12.054

Fang Z, Tang J, Bai Y et al (2015) Plasma levels of microRNA-24, microRNA-320a, and microRNA-423-5p are potential biomarkers for colorectal carcinoma. J Exp Clin Cancer Res 34(1):86. https://doi.org/10.1186/s13046-015-0198-6

Fazmin IT, Achercouk Z, Edling CE et al (2020) Circulating microRNA as a biomarker for coronary artery disease. Biomol Ther 10(10). https://doi.org/10.3390/biom10101354

Feng YH, Tsao CJ (2016) Emerging role of microRNA-21 in cancer. Biomed Rep 5(4):395–402. https://doi.org/10.3892/br.2016.747

Fichtlscherer S, De Rosa S, Fox H et al (2010) Circulating microRNAs in patients with coronary artery disease. Circ Res 107(5):677–684. https://doi.org/10.1161/circresaha.109.215566

Florkowski CM (2008) Sensitivity, specificity, receiver-operating characteristic (ROC) curves and likelihood ratios: communicating the performance of diagnostic tests. Clin Biochem Rev 29(Suppl 1):S83–S87

Friedman RC, Farh KK-H, Burge CB et al (2009) Most mammalian mRNAs are conserved targets of microRNAs. Genome Res 19(1):92–105. https://doi.org/10.1101/gr.082701.108

Fu Y, Wu PH, Beane T et al (2018) Elimination of PCR duplicates in RNA-seq and small RNA-seq using unique molecular identifiers. BMC Genomics 19(1):531. https://doi.org/10.1186/s12864-018-4933-1

Gao H, Guddeti RR, Matsuzawa Y et al (2015) Plasma levels of microRNA-145 are associated with severity of coronary artery disease. PLoS One 10(5):e0123477. https://doi.org/10.1371/journal.pone.0123477

Garrido-Palacios A, Rojas Carvajal AM, Núñez-Negrillo AM et al (2023) MicroRNA Dysregulation in early breast cancer diagnosis: a systematic review and meta-analysis. Int J Mol Sci 24(9):8270. https://doi.org/10.3390/ijms24098270

Ghorbanmehr N, Gharbi S, Korsching E et al (2019) miR-21-5p, miR-141-3p, and miR-205-5p levels in urine-promising biomarkers for the identification of prostate and bladder cancer. Prostate 79(1):88–95. https://doi.org/10.1002/pros.23714

Glinge C, Clauss S, Boddum K et al (2017) Stability of circulating blood-based MicroRNAs—pre-analytic methodological considerations. PLoS One 12(2):e0167969. https://doi.org/10.1371/journal.pone.0167969

González-Sánchez LE, Ortega-Camarillo C, Contreras-Ramos A et al (2021) miRNAs as biomarkers for diagnosis of type 2 diabetes: a systematic review. J Diabetes 13(10):792–816. https://doi.org/10.1111/1753-0407.13166

Gregory RI, Chendrimada TP, Cooch N et al (2005) Human RISC couples MicroRNA biogenesis and posttranscriptional gene silencing. Cell 123(4):631–640. https://doi.org/10.1016/j.cell.2005.10.022

Guo S, Zhang J, Wang B et al (2018) A 5-serum miRNA panel for the early detection of colorectal cancer. Onco Targets Ther 11:2603–2614. https://doi.org/10.2147/ott.S153535

Han J, Li J, Qian Y et al (2019a) Identification of plasma miR-148a as a noninvasive biomarker for hepatocellular carcinoma. Clin Res Hepatol Gastroenterol 43(5):585–593. https://doi.org/10.1016/j.clinre.2018.12.008

Han J, Wang F, Lan Y et al (2019b) KIFC1 regulated by miR-532-3p promotes epithelial-to-mesenchymal transition and metastasis of hepatocellular carcinoma via gankyrin/AKT signaling. Oncogene 38(3):406–420. https://doi.org/10.1038/s41388-018-0440-8

He L, Hannon GJ (2004) MicroRNAs: small RNAs with a big role in gene regulation. Nat Rev Genet 5(7):522–531. https://doi.org/10.1038/nrg1379

He Y, Li Y, Zhai Z et al (2022) Relationship of miRNA-126 and miRNA-122 expression with type 2 diabetes mellitus and related glucose metabolism parameters: a systematic review and meta-analysis. Exp Ther Med 24(5):652. https://doi.org/10.3892/etm.2022.11589

Henning RJ (2018) Type-2 diabetes mellitus and cardiovascular disease. Futur Cardiol 14(6):491–509. https://doi.org/10.2217/fca-2018-0045

Herreros-Villanueva M, Duran-Sanchon S, Martín AC et al (2019) Plasma MicroRNA signature validation for early detection of colorectal cancer. Clin Transl Gastroenterol 10(1):e00003. https://doi.org/10.14309/ctg.0000000000000003

Hirschfeld M, Rücker G, Weiß D et al (2020) Urinary Exosomal MicroRNAs as potential non-invasive biomarkers in breast cancer detection. Mol Diagn Ther 24(2):215–232. https://doi.org/10.1007/s40291-020-00453-y

Ho PTB, Clark IM, Le LTT (2022) MicroRNA-based diagnosis and therapy. Int J Mol Sci 23(13). https://doi.org/10.3390/ijms23137167

Holdmann J, Markert L, Klinger C et al (2022) MicroRNAs from urinary exosomes as alternative biomarkers in the differentiation of benign and malignant prostate diseases. J Circ Biomark 11:5–13. https://doi.org/10.33393/jcb.2022.2317

Hu C, Meiners S, Lukas C et al (2020) Role of exosomal microRNAs in lung cancer biology and clinical applications. Cell Prolif 53(6):e12828. https://doi.org/10.1111/cpr.12828

Huang X, Zhu X, Yu Y et al (2021) Dissecting miRNA signature in colorectal cancer progression and metastasis. Cancer Lett 501:66–82. https://doi.org/10.1016/j.canlet.2020.12.025

Jang JY, Kim YS, Kang KN et al (2021) Multiple microRNAs as biomarkers for early breast cancer diagnosis. Mol Clin Oncol 14(2):31. https://doi.org/10.3892/mco.2020.2193

Jemal A, Bray F, Center MM et al (2011) Global cancer statistics. CA Cancer J Clin 61(2):69–90. https://doi.org/10.3322/caac.20107

Ji J, Yang H (2020) Using probiotics as supplementation for helicobacter pylori antibiotic therapy. Int J Mol Sci 21(3). https://doi.org/10.3390/ijms21031136

Jo S, Chen J, Xu G et al (2018) miR-204 controls glucagon-like peptide 1 receptor expression and agonist function. Diabetes 67(2):256–264. https://doi.org/10.2337/db17-0506

Kang HW, Byun YJ, Moon SM et al (2022) Urinary hsv2-miR-H9 to hsa-miR-3659 ratio is an effective marker for discriminating prostate cancer from benign prostate hyperplasia in patients within the prostate-specific antigen grey zone. Investig Clin Urol 63(2):238–244. https://doi.org/10.4111/icu.20210493

Karimi P, Islami F, Anandasabapathy S et al (2014) Gastric cancer: descriptive epidemiology, risk factors, screening, and prevention. Cancer Epidemiol Biomarkers Prev 23(5):700–713. https://doi.org/10.1158/1055-9965.Epi-13-1057
Khan MAB, Hashim MJ, King JK et al (2020) Epidemiology of type 2 diabetes - global burden of disease and forecasted trends. J Epidemiol Glob Health 10(1):107–111. https://doi.org/10.2991/jegh.k.191028.001
Kim SY, Kim HS, Park HJ (2019) Adverse events related to colonoscopy: global trends and future challenges. World J Gastroenterol 25(2):190–204. https://doi.org/10.3748/wjg.v25.i2.190
Kocarnik JM, Compton K, Dean FE et al (2022) Cancer incidence, mortality, years of life lost, years lived with disability, and disability-adjusted life years for 29 cancer groups from 2010 to 2019: a systematic analysis for the global burden of disease study 2019. JAMA Oncol 8(3):420–444. https://doi.org/10.1001/jamaoncol.2021.6987
Kozomara A, Birgaoanu M, Griffiths-Jones S (2019) miRBase: from microRNA sequences to function. Nucleic Acids Res 47(D1):D155–d162. https://doi.org/10.1093/nar/gky1141
Kroh EM, Parkin RK, Mitchell PS et al (2010) Analysis of circulating microRNA biomarkers in plasma and serum using quantitative reverse transcription-PCR (qRT-PCR). Methods 50(4):298–301. https://doi.org/10.1016/j.ymeth.2010.01.032
Kuijvenhoven JC, Leoncini F, Crombag LC et al (2020) Endobronchial ultrasound for the diagnosis of centrally located lung tumors: a systematic review and meta-analysis. Respiration 99(5):441–450. https://doi.org/10.1159/000500363
Lai Z, Lin P, Weng X et al (2018) MicroRNA-574-5p promotes cell growth of vascular smooth muscle cells in the progression of coronary artery disease. Biomed Pharmacother 97:162–167. https://doi.org/10.1016/j.biopha.2017.10.062
Lee RC, Feinbaum RL, Ambros V (1993) The C. Elegans heterochronic gene lin-4 encodes small RNAs with antisense complementarity to lin-14. Cell 75(5):843–854
Lee Y, Kim M, Han J et al (2004) MicroRNA genes are transcribed by RNA polymerase II. EMBO J 23(20):4051–4060. https://doi.org/10.1038/sj.emboj.7600385
Li Y, Kowdley KV (2012) Method for microRNA isolation from clinical serum samples. Anal Biochem 431(1):69–75. https://doi.org/10.1016/j.ab.2012.09.007
Li C, Zhao L, Chen Y et al (2016) MicroRNA-21 promotes proliferation, migration, and invasion of colorectal cancer, and tumor growth associated with down-regulation of sec23a expression. BMC Cancer 16:605. https://doi.org/10.1186/s12885-016-2628-z
Li L, Wang A, Cai M et al (2020) Identification of stool miR-135b-5p as a non-invasive diaognostic biomarker in later tumor stage of colorectal cancer. Life Sci 260:118417. https://doi.org/10.1016/j.lfs.2020.118417
Li Z, Chen D, Wang Q et al (2021) mRNA and microRNA stability validation of blood samples under different environmental conditions. Forensic Sci Int Genet 55:102567. https://doi.org/10.1016/j.fsigen.2021.102567
Linsen SE, de Wit E, Janssens G et al (2009) Limitations and possibilities of small RNA digital gene expression profiling. Nat Methods 6(7):474–476. https://doi.org/10.1038/nmeth0709-474
Liu Q, Peng F, Chen J (2019) The role of Exosomal MicroRNAs in the tumor microenvironment of breast cancer. Int J Mol Sci 20(16). https://doi.org/10.3390/ijms20163884
Loeb S, Vellekoop A, Ahmed HU et al (2013) Systematic review of complications of prostate biopsy. Eur Urol 64(6):876–892. https://doi.org/10.1016/j.eururo.2013.05.049
Loke SY, Munusamy P, Koh GL et al (2019) A circulating miRNA signature for stratification of breast lesions among women with abnormal screening mammograms. Cancers (Basel) 11(12). https://doi.org/10.3390/cancers11121872
Lund E, Dahlberg JE (2006) Substrate selectivity of exportin 5 and dicer in the biogenesis of microRNAs. Cold Spring Harb Symp Quant Biol 71:59–66. https://doi.org/10.1101/sqb.2006.71.050
Machlowska J, Baj J, Sitarz M et al (2020) Gastric cancer: epidemiology, risk factors, classification, genomic characteristics and treatment strategies. Int J Mol Sci 21(11). https://doi.org/10.3390/ijms21114012

Malhotra S, Miras MCM, Pappolla A et al (2023) Liquid biopsy in neurological diseases. Cells 12(14). https://doi.org/10.3390/cells12141911
Mandal CC, Ghosh-Choudhury T, Dey N et al (2012) miR-21 is targeted by omega-3 polyunsaturated fatty acid to regulate breast tumor CSF-1 expression. Carcinogenesis 33(10):1897–1908. https://doi.org/10.1093/carcin/bgs198
Manser R, Lethaby A, Irving LB et al (2013) Screening for lung cancer. Cochrane Database Syst Rev 2013(6):Cd001991. https://doi.org/10.1002/14651858.CD001991.pub3
Markert L, Holdmann J, Klinger C et al (2021) Small RNAs as biomarkers to differentiate benign and malign prostate diseases: an alternative for transrectal punch biopsy of the prostate? PLoS One 16(3):e0247930. https://doi.org/10.1371/journal.pone.0247930
Marton S, Garcia MR, Robello C et al (2007) Small RNAs analysis in CLL reveals a deregulation of miRNA expression and novel miRNA candidates of putative relevance in CLL pathogenesis. Leukemia 22:330. https://doi.org/10.1038/sj.leu.2405022. https://www.nature.com/articles/2405022#supplementary-information
Masotti A, Caputo V, Da Sacco L et al (2009) Quantification of small non-coding RNAs allows an accurate comparison of miRNA expression profiles. J Biomed Biotechnol 2009:659028. https://doi.org/10.1155/2009/659028
McManus MT, Sharp PA (2002) Gene silencing in mammals by small interfering RNAs. Nat Rev Genet 3(10):737–747. https://doi.org/10.1038/nrg908
Mestdagh P, Hartmann N, Baeriswyl L et al (2014) Evaluation of quantitative miRNA expression platforms in the microRNA quality control (miRQC) study. Nat Methods 11(8):809–815. https://doi.org/10.1038/nmeth.3014
Mitchell PS, Parkin RK, Kroh EM et al (2008) Circulating microRNAs as stable blood-based markers for cancer detection. Proc Natl Acad Sci USA 105(30):10513–10518. https://doi.org/10.1073/pnas.0804549105
Mitra AK, Zillhardt M, Hua Y et al (2012) MicroRNAs reprogram normal fibroblasts into cancer-associated fibroblasts in ovarian cancer. Cancer Discov 2(12):1100–1108. https://doi.org/10.1158/2159-8290.Cd-12-0206
Mulcahy LA, Pink RC, Carter DR (2014) Routes and mechanisms of extracellular vesicle uptake. J Extracell Vesicles 3. https://doi.org/10.3402/jev.v3.24641
Munteanu VC, Munteanu RA, Onaciu A et al (2020) MiRNA-based inspired approach in diagnosis of prostate cancer. Medicina (Kaunas) 56(2). https://doi.org/10.3390/medicina56020094
Nariman-Saleh-Fam Z, Vahed SZ, Aghaee-Bakhtiari SH et al (2019) Expression pattern of miR-21, miR-25 and PTEN in peripheral blood mononuclear cells of patients with significant or insignificant coronary stenosis. Gene 698:170–178. https://doi.org/10.1016/j.gene.2019.02.074
Nilsson J, Skog J, Nordstrand A et al (2009) Prostate cancer-derived urine exosomes: a novel approach to biomarkers for prostate cancer. Br J Cancer 100(10):1603–1607. https://doi.org/10.1038/sj.bjc.6605058
Ogurtsova K, Guariguata L, Barengo NC et al (2022) IDF diabetes atlas: global estimates of undiagnosed diabetes in adults for 2021. Diabetes Res Clin Pract 183:109118. https://doi.org/10.1016/j.diabres.2021.109118
Ono S, Lam S, Nagahara M et al (2015) Circulating microRNA biomarkers as liquid biopsy for cancer patients: pros and cons of current assays. J Clin Med 4(10):1890–1907. https://doi.org/10.3390/jcm4101890
Page K, Guttery DS, Zahra N et al (2013) Influence of plasma processing on recovery and analysis of circulating nucleic acids. PLoS One 8(10):e77963. https://doi.org/10.1371/journal.pone.0077963
Parekh RS, Kao WH, Meoni LA et al (2007) Reliability of urinary albumin, total protein, and creatinine assays after prolonged storage: the family investigation of nephropathy and diabetes. Clin J Am Soc Nephrol 2(6):1156–1162. https://doi.org/10.2215/cjn.01030207
Park JY, von Karsa L, Herrero R (2014) Prevention strategies for gastric cancer: a global perspective. Clin Endosc 47(6):478–489. https://doi.org/10.5946/ce.2014.47.6.478

Paszat LF, Sutradhar R, Luo J et al (2021) Perforation and post-polypectomy bleeding complicating colonoscopy in a population-based screening program. Endosc Int Open 9(4):E637–e645. https://doi.org/10.1055/a-1381-7149

Patil H, Vaidya O, Bogart D (2011) A review of causes and systemic approach to cardiac troponin elevation. Clin Cardiol 34(12):723–728. https://doi.org/10.1002/clc.20983

Poel D, Buffart TE, Oosterling-Jansen J et al (2018) Evaluation of several methodological challenges in circulating miRNA qPCR studies in patients with head and neck cancer. Exp Mol Med 50(3):e454. https://doi.org/10.1038/emm.2017.288

Pordzik J, Eyileten-Postuła C, Jakubik D et al (2021) MiR-126 is an independent predictor of Long-term all-cause mortality in patients with type 2 diabetes mellitus. J Clin Med 10(11). https://doi.org/10.3390/jcm10112371

Poy MN, Eliasson L, Krutzfeldt J et al (2004) A pancreatic islet-specific microRNA regulates insulin secretion. Nature 432(7014):226–230. https://doi.org/10.1038/nature03076

Pritchard CC, Cheng HH, Tewari M (2012) MicroRNA profiling: approaches and considerations. Nat Rev Genet 13(5):358–369. https://doi.org/10.1038/nrg3198

Rabinowits G, Gerçel-Taylor C, Day JM et al (2009) Exosomal microRNA: a diagnostic marker for lung cancer. Clin Lung Cancer 10(1):42–46. https://doi.org/10.3816/CLC.2009.n.006

Rafikova G, Gilyazova I, Enikeeva K et al (2023) Prostate cancer: genetics, epigenetics and the need for immunological biomarkers. Int J Mol Sci 24(16). https://doi.org/10.3390/ijms241612797

Ran T, Chen J, Cheng Y et al (2023) A meta-analysis of the relationship between circulating microRNA-155 and coronary artery disease. PLoS One 18(4):e0274277. https://doi.org/10.1371/journal.pone.0274277

Rivera-Franco MM, Leon-Rodriguez E (2018) Delays in breast cancer detection and treatment in developing countries. Breast Cancer (Auckl) 12:1178223417752677. https://doi.org/10.1177/1178223417752677

Roganovic D, Djilas D, Vujnovic S et al (2015) Breast MRI, digital mammography and breast tomosynthesis: comparison of three methods for early detection of breast cancer. Bosn J Basic Med Sci 15(4):64–68. https://doi.org/10.17305/bjbms.2015.616

Roser AE, Caldi Gomes L, Schünemann J et al (2018) Circulating miRNAs as diagnostic biomarkers for Parkinson's disease. Front Neurosci 12:625. https://doi.org/10.3389/fnins.2018.00625

Roth GA, Mensah GA, Johnson CO et al (2020) Global burden of cardiovascular diseases and risk factors, 1990-2019: update from the GBD 2019 study. J Am Coll Cardiol 76(25):2982–3021. https://doi.org/10.1016/j.jacc.2020.11.010

Rumgay H, Arnold M, Ferlay J et al (2022) Global burden of primary liver cancer in 2020 and predictions to 2040. J Hepatol 77(6):1598–1606. https://doi.org/10.1016/j.jhep.2022.08.021

Sabre L, Punga T, Punga AR (2020) Circulating miRNAs as potential biomarkers in myasthenia gravis: tools for personalized medicine. Front Immunol 11:213. https://doi.org/10.3389/fimmu.2020.00213

Schwarz DS, Hutvágner G, Du T et al (2003) Asymmetry in the assembly of the RNAi enzyme complex. Cell 115(2):199–208. https://doi.org/10.1016/S0092-8674(03)00759-1

Schwarzenbach H, da Silva AM, Calin G et al (2015) Data normalization strategies for MicroRNA quantification. Clin Chem 61(11):1333–1342. https://doi.org/10.1373/clinchem.2015.239459

Sebastiani G, Nigi L, Grieco GE et al (2017) Circulating microRNAs and diabetes mellitus: a novel tool for disease prediction, diagnosis, and staging? J Endocrinol Investig 40(6):591–610. https://doi.org/10.1007/s40618-017-0611-4

Seo SI, Kim HS, Kim WJ et al (2015) Diagnostic value of PIVKA-II and alpha-fetoprotein in hepatitis B virus-associated hepatocellular carcinoma. World J Gastroenterol 21(13):3928–3935. https://doi.org/10.3748/wjg.v21.i13.3928

Shen R, Wang Y, Wang CX et al (2015) MiRNA-155 mediates TAM resistance by modulating SOCS6-STAT3 signalling pathway in breast cancer. Am J Transl Res 7(10):2115–2126

Simmons O, Fetzer DT, Yokoo T et al (2017) Predictors of adequate ultrasound quality for hepatocellular carcinoma surveillance in patients with cirrhosis. Aliment Pharmacol Ther 45(1):169–177. https://doi.org/10.1111/apt.13841

So JBY, Kapoor R, Zhu F et al (2021) Development and validation of a serum microRNA biomarker panel for detecting gastric cancer in a high-risk population. Gut 70(5):829–837. https://doi.org/10.1136/gutjnl-2020-322065

Sozzi G, Boeri M, Rossi M et al (2014) Clinical utility of a plasma-based miRNA signature classifier within computed tomography lung cancer screening: a correlative MILD trial study. J Clin Oncol 32(8):768–773. https://doi.org/10.1200/jco.2013.50.4357

Stuopelyte K, Daniunaite K, Bakavicius A et al (2016) The utility of urine-circulating miRNAs for detection of prostate cancer. Br J Cancer 115(6):707–715. https://doi.org/10.1038/bjc.2016.233

Sukumar J, Gast K, Quiroga D et al (2021) Triple-negative breast cancer: promising prognostic biomarkers currently in development. Expert Rev Anticancer Ther 21(2):135–148. https://doi.org/10.1080/14737140.2021.1840984

Sullivan CS, Ganem D (2005) MicroRNAs and viral infection. Mol Cell 20(1):3–7. https://doi.org/10.1016/j.molcel.2005.09.012

Sun Y, Wang M, Lin G et al (2012) Serum microRNA-155 as a potential biomarker to track disease in breast cancer. PLoS One 7(10):e47003. https://doi.org/10.1371/journal.pone.0047003

Sung H, Ferlay J, Siegel RL et al (2021) Global cancer statistics 2020: GLOBOCAN estimates of incidence and mortality worldwide for 36 cancers in 185 countries. CA Cancer J Clin. https://doi.org/10.3322/caac.21660

Suzuki K, Komukai K, Nakata K et al (2018) The usefulness and limitations of point-of-care cardiac troponin measurement in the emergency department. Intern Med 57(12):1673–1680. https://doi.org/10.2169/internalmedicine.0098-17

Swarbrick S, Wragg N, Ghosh S et al (2019) Systematic review of miRNA as biomarkers in Alzheimer's disease. Mol Neurobiol 56(9):6156–6167. https://doi.org/10.1007/s12035-019-1500-y

Tang Y, Zhang Y, Chen Y et al (2015) Role of the microRNA, miR-206, and its target PIK3C2α in endothelial progenitor cell function—potential link with coronary artery disease. FEBS J 282(19):3758–3772. https://doi.org/10.1111/febs.13372

Thomas CE, Sexton W, Benson K et al (2010) Urine collection and processing for protein biomarker discovery and quantification. Cancer Epidemiol Biomarkers Prev 19(4):953–959. https://doi.org/10.1158/1055-9965.Epi-10-0069

Toiyama Y, Takahashi M, Hur K et al (2013) Serum miR-21 as a diagnostic and prognostic biomarker in colorectal cancer. J Natl Cancer Inst 105(12):849–859. https://doi.org/10.1093/jnci/djt101

Turchinovich A, Weiz L, Burwinkel B (2012) Extracellular miRNAs: the mystery of their origin and function. Trends Biochem Sci 37(11):460–465. https://doi.org/10.1016/j.tibs.2012.08.003

Urdea M, Kolberg J, Wilber J et al (2009) Validation of a multimarker model for assessing risk of type 2 diabetes from a five-year prospective study of 6784 Danish people (Inter99). J Diabetes Sci Technol 3(4):748–755. https://doi.org/10.1177/193229680900300422

Valihrach L, Androvic P, Kubista M (2020) Circulating miRNA analysis for cancer diagnostics and therapy. Mol Asp Med 72:100825. https://doi.org/10.1016/j.mam.2019.10.002

Wang F, Long G, Zhao C et al (2013) Plasma microRNA-133a is a new marker for both acute myocardial infarction and underlying coronary artery stenosis. J Transl Med 11:222. https://doi.org/10.1186/1479-5876-11-222

Wang M, Ji Y, Cai S et al (2016) MiR-206 suppresses the progression of coronary artery disease by modulating vascular endothelial growth factor (VEGF) expression. Med Sci Monit 22:5011–5020. https://doi.org/10.12659/msm.898883

Wang H, Peng R, Wang J et al (2018) Circulating microRNAs as potential cancer biomarkers: the advantage and disadvantage. Clin Epigenetics 10:59. https://doi.org/10.1186/s13148-018-0492-1

Weber JA, Baxter DH, Zhang S et al (2010) The MicroRNA Spectrum in 12 body fluids. Clin Chem 56(11):1733–1741. https://doi.org/10.1373/clinchem.2010.147405

Wei T-C, Lin T-P, Chang Y-H et al (2015) Transrectal ultrasound-guided prostate biopsy in Taiwan: a nationwide database study. J Chin Med Assoc 78(11):662–665. https://doi.org/10.1016/j.jcma.2015.04.011

Weng W, Liu N, Toiyama Y et al (2018) Novel evidence for a PIWI-interacting RNA (piRNA) as an oncogenic mediator of disease progression, and a potential prognostic biomarker in colorectal cancer. Mol Cancer 17(1):16–16. https://doi.org/10.1186/s12943-018-0767-3

Wu K, Ma L, Zhu J (2016) miR-483-5p promotes growth, invasion and self-renewal of gastric cancer stem cells by Wnt/β-catenin signaling. Mol Med Rep 14(4):3421–3428. https://doi.org/10.3892/mmr.2016.5603

Wu CW, Cao X, Berger CK et al (2017a) Novel approach to fecal occult blood testing by assay of erythrocyte-specific microRNA markers. Dig Dis Sci 62(8):1985–1994. https://doi.org/10.1007/s10620-017-4627-6

Wu H, Zhou J, Mei S et al (2017b) Circulating exosomal microRNA-96 promotes cell proliferation, migration and drug resistance by targeting LMO7. J Cell Mol Med 21(6):1228–1236. https://doi.org/10.1111/jcmm.13056

Xu G, Thielen LA, Chen J et al (2019) Serum miR-204 is an early biomarker of type 1 diabetes-associated pancreatic beta-cell loss. Am J Physiol Endocrinol Metab 317(4):E723–e730. https://doi.org/10.1152/ajpendo.00122.2019

Yamamoto Y, Kondo S, Matsuzaki J et al (2020) Highly sensitive circulating MicroRNA panel for accurate detection of hepatocellular carcinoma in patients with liver disease. Hepatol Commun 4(2):284–297. https://doi.org/10.1002/hep4.1451

Yang S, Mi X, Chen Y et al (2018) MicroRNA-216a induces endothelial senescence and inflammation via Smad3/IκBα pathway. J Cell Mol Med 22(5):2739–2749. https://doi.org/10.1111/jcmm.13567

Yu L, Todd NW, Xing L et al (2010) Early detection of lung adenocarcinoma in sputum by a panel of microRNA markers. Int J Cancer 127(12):2870–2878. https://doi.org/10.1002/ijc.25289

Zailaie SA, Siddiqui JJ, Al Saadi RM et al (2022) Serum based miRNA as a diagnostic biomarker for multiple sclerosis: a systematic review and meta-analysis. Immunol Investig 51(4):947–962. https://doi.org/10.1080/08820139.2021.1887888

Zampetaki A, Kiechl S, Drozdov I et al (2010) Plasma microRNA profiling reveals loss of endothelial miR-126 and other microRNAs in type 2 diabetes. Circ Res 107(6):810–817. https://doi.org/10.1161/circresaha.110.226357

Zeller T, Keller T, Ojeda F et al (2014) Assessment of microRNAs in patients with unstable angina pectoris. Eur Heart J 35(31):2106–2114. https://doi.org/10.1093/eurheartj/ehu151

Zhang G (2017) Nucleoside/Nucleotide Biomarkers. In: Targeted Biomarker Quantitation by LC–MS. Wiley, New York, pp 389–406. https://doi.org/10.1002/9781119413073.ch25

Zhao S, Wu Q, Gao F et al (2012) Serum microRNA-155 as a potential biomarker for breast cancer screening. Chin Sci Bull 57(26):3466–3468. https://doi.org/10.1007/s11434-012-5362-1

Zhao J, Yu H, Yan P et al (2019) Circulating MicroRNA-499 as a diagnostic biomarker for acute myocardial infarction: a meta-analysis. Dis Markers 2019:6121696. https://doi.org/10.1155/2019/6121696

Zhao J, Wang Y, Su H et al (2021) Non-coding RNAs as biomarkers for hepatocellular carcinoma-a systematic review. Clin Res Hepatol Gastroenterol 45(6):101736. https://doi.org/10.1016/j.clinre.2021.101736

Zonari E, Pucci F, Saini M et al (2013) A role for miR-155 in enabling tumor-infiltrating innate immune cells to mount effective antitumor responses in mice. Blood 122(2):243–252. https://doi.org/10.1182/blood-2012-08-449306

Chapter 7
miRNAs As Biomarkers for Neurodegenerative Diseases

Moisés Manuel Gallardo-Pérez and Solón Javier Garcés-Eisele

Abstract microRNAs (miRNAs) are small noncoding RNAs that regulate mRNA stability and/or protein biosynthesis and, therefore, gene expression. They have been studied and linked to different neurodegenerative disorders. In this chapter, we review miRNAs with strong potential as biomarkers for neurodegenerative disease. We analyzed the available studies on neurodegenerative diseases that report miRNA alterations, considered the reproducibility of the results, and limited the report to studies that analyzed miRNAs in human body fluids of easy access (blood, CSF). Nevertheless, occasionally, we chose to mention miRNAs, which although not analyzed in readily accessible human body fluids, have shown significant dysregulation in human brain and eye tissues. Of the studies consulted for this work, only a few have used large-scale methods such as NGS. In general, real-time PCR has been used to analyze miRNA expression levels. Therefore, we have employed the guidelines for "minimum information for publication of quantitative real-time PCR experiments" (MIQE guidelines) to evaluate the different studies.

Keywords microRNAs · Biomakers · Neurodegenerative disorders

List of Abbreviations

AD	Alzheimer's disease
ALS	Amyotrophic lateral sclerosis
AMD	Age-related macular degeneration

M. M. Gallardo-Pérez (✉)
Universidad Popular Autónoma del Estado de Puebla, Puebla, Mexico

Centro de Hematología y Medicina Interna, Clínica Ruiz, Puebla, Mexico

S. J. Garcés-Eisele
Universidad Popular Autónoma del Estado de Puebla, Puebla, Mexico

Laboratorios Ruiz, SYNLAB, Puebla, Mexico

L. M. Vaschetto (ed.), *miRNAs, Human Health and Diseases*, Epigenetics and Human Health 13, https://doi.org/10.1007/978-3-031-64788-8_7

CSF	Cerebrospinal fluid
DLB	Dementia with Lewy bodies
FRDA	Autosomal recessive Friedreich ataxia
FTD	Frontotemporal Dementia
HD	Huntington's disease
miRNAs	micro RNAs
MJD	Machado-Joseph disease
PCR	Polymerase chain reaction
PD	Parkinson's disease
PDD	Parkinson disease with dementia
polyQ	Polyglutamine
SCA	Spinocerebellar ataxia

7.1 Introduction

Neurodegenerative diseases (ND) are a heterogeneous group of chronic central nervous system disorders, characterized by a progressive loss of neurons in defined brain regions. There is a broad number of ND including Alzheimer's disease (AD), Parkinson's disease (PD), dementia with Lewy bodies (DLB), and Huntington's disease (HD). The prevalence of these illnesses leads to a significant number of fatalities and extensive financial expenditures on healthcare across the globe, imposing a substantial weight on individuals, their loved ones, and the community.

Although all these diseases show a vast variety of different clinical courses and different deteriorated brain areas, all ND share alterations that lead to an altered conformation of proteins that form deposits like extracellular plaques or intracellular inclusions. These lesions are associated with neuroaxonal damage and neuron death and thus are the main cause of the initiation and progression of ND. Amyloid, tau, and synuclein are the types of protein deposits most related to the degenerative diseases of the nervous system.

Another condition to consider as a cause of ND are aging related changes. Most of the above mentioned diseases manifest themselves in later life. Therefore, a correct diagnosis may be hampered because a clinician may be attributing the neurological signs and symptoms from a patient to aging instead of seeking an explanation that leads to a diagnostic of ND.

7.1.1 Protein Aggregation

The protein aggregation is a key pathological hallmark of a variety of ND, and this information work for diagnosis and disease classification. The characteristic aggregating proteins in ND include Aβ peptides formed by cleavage of amyloid precursor protein (APP), tau, α-synuclein, TAR DNA-binding protein 43, superoxide dismutase, dipeptide repeat proteins (DPRs) (C9ORF72 gene), huntingtin (PolyQ)

(HTT gene), polyQ proteins (PolyQ), and cellular prion protein (PrPC) (PRNP gene). We discuss briefly each of the mentioned proteins.

7.1.1.1 Aβ Peptides

Aβ are peptides of between 39 and 43 amino acids derived from progressive processing of the Aβ precursor protein (APP). Neurotoxic Aβ is produced by the cleavage of APP, and the aggregation of soluble oligomers leads to the formation of senile plaques, a significant neuropathological marker of AD. APP is used as a substrate by α-secretase and β-secretase, resulting in the generation of soluble N-terminal peptides (APPsα and APPsβ) and C-terminal fragments (CTFα and CTFβ) that are associated with the cell membrane. γ-Secretase then cleaves the transmembrane peptides CTFα and CTFβ, causing the release of the p3 peptide from CTFα and β-amyloid from CTFβ. While p3 remains soluble, β-amyloid tends to aggregate (Abeysinghe et al. 2020).

At the synaptic level, Aβ hinders synaptic transmission by inhibiting neurotransmitter receptors and disrupting the sodium-potassium pump, resulting in impaired electrical and chemical transmission and the blocking of interneuronal connections, ultimately leading to synaptic dysfunction and the loss of neuronal synapses (García-Morales et al. 2021).

7.1.1.2 Tau Protein

Tau protein is a structural protein that normally stabilizes microtubules and promotes vesicle transport. Six tau protein isoforms are produced through alternative splicing of tau pre-mRNA exons 2, 3, and 10. Specifically, exon 10 contains the genetic information for the sequence that comprises the microtubule-binding domain. This sequence is responsible for generating either 3-repeat tau or 4-repeat tau, denoted as 3R and 4R tau, respectively. In healthy neurons and in cases of Alzheimer's disease (AD) and frontotemporal lobar degeneration (FTLD) caused by genetic mutations in the microtubule-associated protein tau gene (MAPT), the balance between the quantities of 3R and 4R tau is maintained. (Younes and Miller 2020). The initiation of hyperphosphorylation of tau protein is triggered by the deposition of Aβ plaques. This event subsequently leads to the misfolding of tau protein and the subsequent formation of aggregates known as neurofibrillary tangles (NFTs). Ultimately, these NFTs disrupt the communication between neurons, ultimately resulting in the death of the neuron (Abeysinghe et al. 2020; Sehar et al. 2022).

7.1.1.3 α-Synuclein

The function of α-synuclein (αSyn) is not clear, but it is thought to function in cell membrane remodeling at neuronal terminals. When aggregates of misfolded, overexpressed αSyn accumulate in neurons or surrounding glial cells, Lewy bodies are formed. There are three main types of "α-synucleinopathies": PD, Lewy body dementia (LBD), and multiple system atrophy (Sanford 2018).

Lewy body disease exists in multiple types, including brain stem-predominant, transitional (limbic), and diffuse (neocortical) forms associated with increasing degrees of pathological burden. Individuals with Lewy body spectrum disorders tend to have gray matter atrophy in the posterior parietal cortices with relative sparing of medial temporal lobe structures (Armstrong 2021).

7.1.1.4 TAR-DNA Binding Protein 43

TAR-DNA binding protein 43 (TARDBP) is a member of the ribonucleoprotein family and predominantly exists within the cellular nucleus. Its interactions with DNA and RNA molecules play a crucial role in modulating the processes of transcription, splicing, and the subsequent transportation of RNA. The majority of mutations found within the TARDBP gene are situated in exon 6, giving rise to a collection of nonfunctional mutations (Lee et al. 2011). The mechanisms by which TARDBP mutations lead to the development of ALS are yet to be determined. Mutations associated with the development of the disease increase the protein's translocation from the nucleus to the cytoplasm and promote aggregation.

7.1.1.5 Superoxide Dismutase 1

The SOD1 gene, located on chromosome 9q21q22, encodes the superoxide dismutase 1 protein. Different mutations that occur on chromosome 9 lead to significant misfolding of the SOD1 protein, resulting in its degradation by the ubiquitin proteasome system (UPS). However, a considerable fraction of misfolded proteins evade degradation by the UPS, thereby disrupting proteostasis and triggering secondary autophagic activation. The progressive accumulation of the misfolded mutant protein, in conjunction with other stressors such as aging, initiates a cellular stress response. It is important to note that SOD1 misfolding is not limited to mutant variants of the protein. SOD1 may also contribute to the pathogenesis of sporadic forms of ALS, particularly once cellular stress has been established through other mechanisms (Riancho et al. 2019).

7.1.1.6 Dipeptide Repeat Proteins

A mutation consisting of a GGGGCC hexanucleotide repeat is found on chromosome 9q21-q22 in a noncoding region of the open reading frame 72 of chromosome 9 (C9ORF72) located between two transcriptional start sites. Expansion of the C9ORF72 hexanucleotide repeat has been found in up to 40% of ALS families and up to 11% of FTD families (Riancho et al. 2019). While the repeat is located in a noncoding region, it has been demonstrated that the expansion is translated anyway in a repeat associated non-AUG translation (Freibaum et al. 2015).

The repeat expansions of C9ORF72 are thought to cause loss of function in transcription as well as possible gain of function due to toxic RNA foci. C9ORF72 repeat expansions can also produce neurotoxic dipeptide repeat proteins that contribute to neurodegeneration (Olney et al. 2017).

7.1.1.7 polyQ Proteins

Numerous neurodegenerative diseases in humans are caused by CAG repeats that give rise to abnormally long tracts in the translated exons of different genes. This results in the production of elongated stretches of polyglutamine in the proteins that are formed consequently. The most extensively studied mechanism by which polyglutamine disorders lead to disease is through a gain-of-function mechanism. This involves the enhanced ability of poly-Q proteins to attract host proteins, leading to the formation of aggregates and large inclusions that are positive for ubiquitin. Initially, it was strongly believed that these protein bodies had a pathogenic effect (Wojciechowska and Krzyzosiak 2011).

7.1.2 MicroRNAs and Neurodegenerative Disorders

microRNAs (miRNAs) are small noncoding RNAs that regulate mRNA stability and/or protein biosynthesis and, therefore, gene expression. miRNAs mediate post-transcriptional gene silencing of target genes by targeting the 3′ untranslated region of mRNA, nucleotides 2–7 of the 5′end of miRNA being the crucial sequence (Fig. 7.1). From a transient double-stranded miRNA duplex, the guide strand (mature miRNA strand) is incorporated into the RNA-induced silencing complex to mediate gene silencing, and the passenger strand is degraded (Lu and Rothenberg 2018).

Circulating miRNAs are stable and resistant to degradation from endogenous RNAs activity. Furthermore, they can withstand severe environmental conditions like long-term storage and multiple freeze-thaw cycles in multiple different types of bodily fluids.

The available techniques for the quantification of miRNAs are next-generation sequencing (NGS), quantitative real-time polymerase chain reaction (qRT-PCR), and microarrays. The most widely applied method is qRT-PCR due to its simplicity, high sensitivity, and capability to detect low levels of miRNAs, but as the sample is limited, it can only be applied for a reduced number of selected miRNAs. Microarrays enable measurement of about 2000 selected miRNAs, but analysis requires larger amounts of RNA, has lower reproducibility, and presents problems with cross

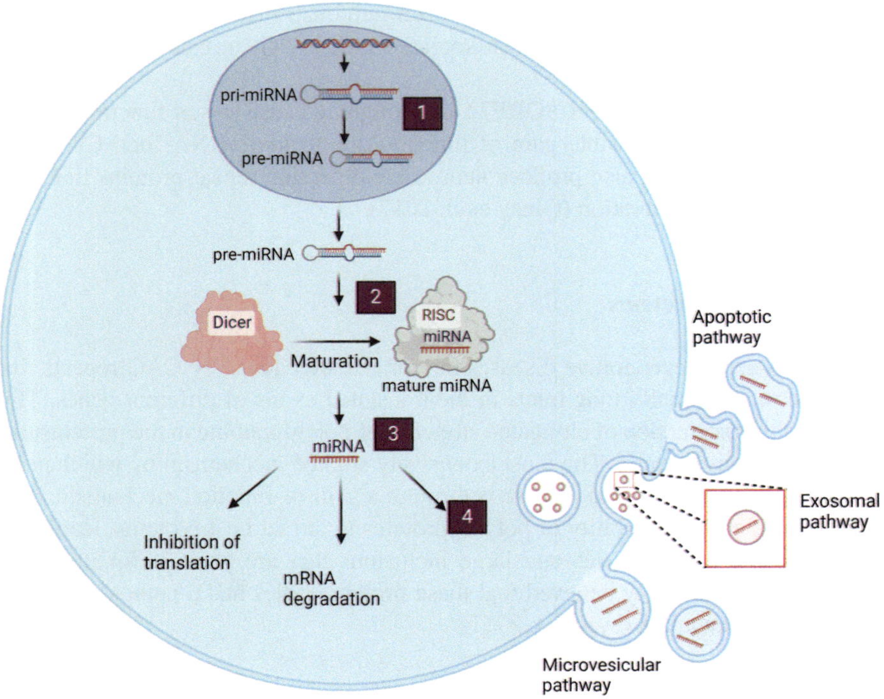

Fig. 7.1 Diagram of miRNA processing and extracellular vesicle transport. Following the generation of the pri-miRNA transcript, transcribed pri-miRNAs are processed by Drosha/DGCR8 (**1**). The pri-miRNAs are exported from the nucleus to the cytoplasm by exportin 5. Dicer cleaves the pre-microRNAs near the terminal loop, generating a mature miRNA duplex of approximately 22 nucleotides size (**2**). The mature miRNA is paired with the target mRNA to direct gene silencing through mRNA excision or translation repression (**3**). The transport of miRNAs out of the cell is carried out by microvesicles (**4**). Created with BioRender.com

hybridization. NGS allows high-throughput quantification of global miRNA levels, enabling the detection of novel miRNAs, but is yet an expensive method in comparison to qRT-PCR and microarrays (Van Den Berg et al. 2020).

The main limitation of the three methodologies is that currently there is no consensus about standardized methodologies and normalization, which makes it difficult to use miRNAs as routinary biomarkers. Nonetheless, miRNAs may have the potential to be used to accurately diagnose ND. The eventual use of miRNAs for diagnosis or prognosis may help doctors offer patients early and personalized treatment. The first barrier for the analysis of the role of miRNAs in ND is the inaccessibility of brain tissue; however, their study in animal and in vitro models allowed to understand the interactions between miRNAs and CNS function.

The regulation of the neuroinflammatory process by miRNAs has been extensively studied in the context of neuroinflammation. These molecules play a crucial role in modulating both innate and adaptive immunity. One of the ways in which they exert their regulatory function is by controlling the activation of microglia, the immune cells of the central nervous system. Additionally, miRNAs also regulate astrocyte reactivity, which further contributes to the overall immune response in the brain. Furthermore, miRNAs are involved in the regulation of peripheral immune cells such as neutrophils, macrophages, and leucocytes, as well as the activities of T and B cells. Their intricate involvement in these various aspects of immune regulation highlights the complexity of the neuroinflammatory process and underscores the significance of miRNAs as potential therapeutic targets in neuroinflammatory disorders (Khan et al. 2022).

In this review, we examine the existing research studies that specifically investigate the alterations of miRNA expression levels in ND. We considered the reproducibility of the reported results and narrowed down our focus to studies that have analyzed miRNAs in human body fluids. We believe that both blood and cerebrospinal fluid (CSF) are the most accessible body fluids in which miRNAs have been found and shown to be relatively stable, so the potential of miRNAs as biomarkers is greatest in these fluids. In order to indicate the biomarker potential of some microRNAs, we also make occasional references to them if they present significant dysregulation in human brain or eye tissues, while they have not yet been extensively studied in easily accessible human body fluids. Among the various studies we have consulted for this work, only a limited number have utilized large-scale methods such as next-generation sequencing (NGS). In most cases, researchers have relied on real-time polymerase chain reaction (PCR) to analyze the expression levels of miRNAs. Consequently, we have employed the established guidelines known as "minimum information for publication of quantitative real-time PCR experiments" (MIQE guidelines) to critically evaluate the different studies and their methodologies (Bustin et al. 2009). Table 7.1 shows a summary of the miRNAs analyzed in this chapter, sorted by neurodegenerative disorder.

Table 7.1 microRNAs (miRNAS) and neurodegenerative disorders

Disease	microRNA	Sample type	Upregulated (+) Downregulated (−)	Function/Target	Reference
AMD	miR-25	Plasma	–	Inhibits ITGAV and PEDF expressions, resulting in RPE apoptosis	Ertekin et al. (2014)
		Serum	–		Elshelmani et al. (2020)
	miR-155	PBNCs	+	Exerts its regulatory effect on mini-αA in oxidative stress-induced retinal pigment epithelial cell apoptosis through the regulation of CDK2 gene, an important regulator of G1/S-phase conversion	Litwińska et al. (2019)
		Plasma	–		Ulańczyk et al. (2019)
		Serum	–		Romano et al. (2017)
	miR-126a	PBNCs	+	Inhibits negative regulators of the VEGF pathway	Litwińska et al. (2019)
		Serum	+		Romano et al. (2017)
		Serum	+		Elshelmani et al. (2020)
	miR-146a	Vitreous humour and plasma	+	Has the ability to influence inflammatory cascades by interfering with IRAK and TRAF6, has been shown to bind the CFH gene in human neural and vascular endothelial cells	Ménard et al. (2016)
		Serum	+		Romano et al. (2017)
		Plasma	+		Ulańczyk et al. (2019)
		PBNCs	+		Litwińska et al. (2019)
	miR-184	Human retinal pigment epithelium	–	Downregulates LAMP1 gene and protein expression and reduced phagocytic activity in RPE	Murad et al. (2014)
SCA1	miR-144	Cerebellar and cortical samples from human brains	–	Regulating the expression of ATXN1	Persengiev et al. (2011)
	miR-101	Cerebellar and cortical samples from human brains	–		
SCA3	miR-25	Plasma	–	Regulating the expression of ATXN3	Shi et al. (2014)
		Lymphoblastoid cells	+		Krauss et al. (2019)
FD	miR-886-3p	Lymphoblasts	+	Regulating the expression of FXN	Mahishi et al. (2012)

FTD	miR-127-3p	Plasma	–	Regulating the expression of CISD1	Piscopo et al. (2018)
	miR-132	Human cerebral cortex	–	Regulating the expression of TMEM106B	Chen-Plotkin et al. (2012)
					Hébert et al. (2013)
		Plasma	–		Sheinerman et al. (2012)
	miR-632	CSF	–	Regulating the expression of HRK	Schneider et al. (2018)
DLB	miR-19b	Serum	–	Inhibits PTEN expression and FGF	Soto et al. (2022)
AD	miR-34a	Hippocampus and medial frontal gyrus	+	Downregulates ATG4B, triggering receptor expressed on myeloid cells 2 (TREM2), MMP2 and HSP70	Cogswell et al. (2008)
		Temporal cortex	+		Sarkar et al. (2016)
		Blood mononuclear cells	+		Schipper et al. (2007)
		Plasma	–		Cosín-Tomás et al. (2017)
		Serum	–		Maldonado-Lasuncion et al. (2019)
		CSF	–		Kiko et al. (2014)
	miR-107	Whole blood	–	Regulates BACE1 expression	Yılmaz et al. (2016)
	miR-125b	Serum	–	Regulates CDKN2A, SYN2 and ALOX15	Tan et al. (2014)
		CSF	+		Jin et al. (2018)
	miR-146a	Brain temporal lobe neocortex	+	Regulates of TLR2 which a primary receptor for encodes for beta-amyloid, CXCR4, FADD, MAPT, and downregulate VDAC1	Sethi and Lukiw (2009)
		CSF	+		Alexandrov et al. (2012) and Denk et al. (2015)
		CSF	–		Kiko et al. (2014)
		Serum	–		Dong et al. (2015)
		Plasma	–		Kiko et al. (2014)
		Hippocampus	+		Wang et al. 2016a
	miR-181c	Serum	–	Regulates DPYSL2 protein expression	Tan et al. (2014) and Abuelezz et al. (2022)
		Plasma	+		Barbagallo et al. (2020)

(continued)

Table 7.1 (continued)

Disease	microRNA	Sample type	Upregulated (+) Downregulated (−)	Function/Target	Reference
PD	miR-7	Peripheral blood lymphocytes	+	Regulates α-Synuclein expression and regulates BDNF expression	Alieva et al. (2015)
		CSF	+		Starhof et al. (2019)
		Serum	+		Citterio et al. (2023)
	miR-9	Peripheral blood lymphocites	+	miR-9-5p inhibition attenuated MPP+ -induced cell apoptosis, inflammation, and oxidative stress via directly targeting SIRT1	Alieva et al. (2015)
	miR-29a	Serum	−	miR-29 family members are associated with neuronal function: miR-29a inhibit mitochondrial antiviral signaling protein (MAVS) expression, targeting genes in the pro-apoptotic BH3-only family to inhibit apoptosis, miR-29c targets SP1	Bai et al. (2017)
		Saliva	−		Jiang et al. (2021)
		Whole blood	−		Margis et al. (2011)
	miR-29b	Serum	−		Bai et al. (2017)
		PBMCs	−		Martins et al. (2011)
	miR-29c	Serum	−		Bai et al. (2017)
		Saliva	−		Jiang et al. (2021)
		PBMCs	−		Martins et al. (2011)
		Serum	+		Ozdilek and Demircan (2021)
	miR-34a	Plasmatic extracellular vesicles	+	Might promote the axonal growth of dopaminergic neurons	Grossi et al. (2020)
		PBMCs	−		Yang et al. (2022a)
	miR-133b	Plasma	−	Regulates PITX3	Zhang et al. (2017b)
		Plasma	+		Chen et al. (2021)
	miR-221	Serum	−	Regulates transferrin receptor type 2 in PD model	Ma et al. (2016)
	miR-443	Plasma	−	Binds to the promoter region of FGF20	Zhang et al. (2017b)

HD	miR-9	Peripheral blood lymphocytes	–	Regulates the expression of co-repressor of repressor element 1-silencing transcription factor (RCOR1)	Chang et al. (2017)
	miR-10b	Prefrontal cortex	+	Targets brain-derived neurotrophic factor (BDNF)	Hoss et al. (2015)
	miR-34b	Plasma	+	Regulated TP53 expression	Gaughwin et al. (2011)
		Amygdala, substantia nigra, frontal cortex, cerebellum	–		Miñones-Moyano et al. (2011)
	miR-323b	Plasma	+	Might regulate mutant HTT	Ferraldeschi et al. (2021)
ALS	miR-206	Serum	+	Regulates HDAC4 mRNA, a regulatory factor that controls gene expression in nerves and muscles	Dobrowolny et al. (2021)
		Serum	+		Toivonen et al. (2014)
		Peripheral blood lymphocytes	+		Vrabec et al. (2018)
		Muscle samples	+		Pegoraro et al. (2017)
		Muscle samples	–		Jensen et al. (2016)
	miR-338	Leukocytes	+	Inhibits PYGB expression	De Felice et al. (2014)

7.2 Neurodegenerative Disorders

7.2.1 Age-Related Macular Degeneration

Age-related macular degeneration (AMD) is a chronic macular ailment that is progressive in nature and often affects those who are above the age of 75 (Mody and Joshi 2023). Progressive death of retinal pigment epithelium or photoreceptor cells is the hallmark of this degenerative retinal disease. The process initiates with the impairment of retina integrity followed by disintegration of retinal cell lining and death of photoreceptors. These photoreceptors or light-sensitive cells send visual signals and input to the brain. The disintegration of retinal cell lining and photoreceptor death results in visual deterioration, blindness, or complete vision loss (Kaur and Singh 2023). There are two types of AMD, nonexudative or non-neovascular (dry) and exudative or neovascular (wet). The "Dry" AMD refers to the presence of small yellow extracellular deposits of lipids, proteins, and cellular debris within the layers of the retina (drusen) in either early or intermediate disease, and the atrophic variant of dry AMD characterizes by the presence of atrophy in the macula; if the atrophy involves the fovea, it is categorized as advanced dry AMD and may lead to progressive permanent vision loss. "Wet" AMD is the most common form of advanced AMD, and the characteristic feature of this condition is the neovascularization in the macula, which leads to choroidal neovascularization (CNV). CNV is an unstable new blood vessel that arises from the choriocapillaris through a break in Bruch's membrane and extends into the subretinal pigment epithelium or subretinal space (Li et al. 2020); this neovascularization causes hemorrhaging and leakage of fluid in the inner retinal layers or subretinal space, ultimately causing fibrosis and permanent vision loss in the absence of treatment. The early or intermediate forms of dry AMD is associated with a more favorable prognosis if the disease course remains stable. It is important to be aware that for all forms of dry AMD, there is a 10–15% risk of progression to wet AMD at some point throughout the disease course (Thomas et al. 2021; Mody and Joshi 2023).

Several microRNAs have been associated with AMD. miR-146a-5p is a major regulator of the NF-κB pathway by inhibiting the expression of the interleukin-1-receptor-associated kinase-1 (IRAK1) and TNF receptor-associated factor 6 (TRAF6). IRAK1 and TRAF6 are known modulators of the NF-κB pathway, and their translational repression inhibits proinflammatory cytokine signaling (Berber et al. 2017; Juźwik et al. 2019). Likewise, NF-κB plays important roles in the early stages of angiogenesis and CNV. A disease-associated increase in miR-146a has been demonstrated in the vitreous humor, plasma, and peripheral blood cells of patients with neovascular AMD (Ménard et al. 2016; Romano et al. 2017; Litwińska et al. 2019; Ulańczyk et al. 2019).

miR-155-5p is a regulator of inflammation and microglial activation state in response to stress, potentially promoting the antiapoptotic role of mini-αA, a 19 amino acid peptide sequence in αA-crystallin that is a molecular chaperone and an antiapoptotic protein. This miRNA could exert its regulatory effect on mini-αA in oxidative stress-induced retinal pigment epithelial cell apoptosis through the regulation of the CDK2 gene, an important regulator of G1/S-phase conversion (Chen et al. 2023). It has been shown that miR-155 levels are decreased in serum and plasma of patients with neovascular AMD (Romano et al. 2017; Ulańczyk et al. 2019) and increased in peripheral blood cells (Litwińska et al. 2019).

miR-126 is encoded in the intron of EGF-like domain 7 and has been shown to be involved in tumor neovascularization. In an animal model of laser-induced CNV, the expression of miR-126 was decreased. Moreover, restoration of miR-126 decreased VEGFA levels, ultimately influencing its downstream pathway and reducing CNV severity (Wang et al. 2016b). Levels of miR-126 have been found to be increased in serum and peripheral blood cells of patients with neovascular AMD (Romano et al. 2017; Litwińska et al. 2019; ElShelmani et al. 2020).

Oxidative stress upregulates miR-25 in retinal pigment epithelium cells in very early stage, which is in turn accompanied by decreased phagocytosis and reduced growth factor secretion in those cells. Such changes precedes retinal pigment epithelium apoptosis and visual impairment in the sodium iodate-treated rats (Zhang et al. 2017a). A decrease in miR-25 has been demonstrated in plasma (Ertekin et al. 2014) and serum (ElShelmani et al. 2020) of patients with neovascular AMD.

miR-184 regulates the EZR (ezrin) gene known to encode a factor that binds to the lysosomal-associated membrane protein 1 (LAMP1) during the formation of phagocytic vacuoles. This means that miR-184 affects phagocytosis in human retinal pigment epithelium and is downregulated in AMD (Murad et al. 2014). It further promotes differentiation of the retinal pigment epithelium by targeting the AKT2/mTOR signaling pathway (Jiang et al. 2016). Yang et al. (2023) found that miR-184 is significantly upregulated in patients with central serous chorioretinopathy, a retinal disease related to AMD.

7.2.2 Hereditary Ataxia

Hereditary ataxia is a heterogeneous group of rare neurodegenerative disorders affecting cerebellar function, which may lead to significant disability. Cerebellar ataxia is usually diagnosed in a patient complaining about gait and balance difficulties. In addition to gait ataxia, neurological examination may reveal cerebellar dysarthria, dysmetria during nose-to-finger and heel-and-chin tests, and hypotonia. Ocular signs of cerebellar disease are very common and may appear as the first clinical features suggestive of cerebellar motor syndrome (Kuo 2019; Karamazovova et al. 2023). The polyglutamine (polyQ) diseases are a group of hereditary neurodegenerative pathologies caused by the expansion of CAG repeats in the coding part of certain genes. They include spinocerebellar ataxia (SCA) type 1, 2, 3, 6, 7, and 17; Huntington's disease; and dentatorubral-pallidoluysian atrophy. Also, hereditary ataxia can be subdivided by mode of inheritance, the most common being autosomal dominant spinocerebellar ataxia (SCA) and autosomal recessive Friedreich ataxia (FRDA).

SCA1 is a condition commonly caused by the expansion of CAG repeats in the Ataxin1 (ATXN1) gene. The disease develops if the number of CAG triplets exceeds 39 and no CAT inserts between them exist (Volovikov et al. 2020). ATXN1 interacts with transcription regulators, RNA splicing factors, and other nuclear receptors to drive cerebellar Purkinje cell dysfunction. The combined interactions between mutant ATXN1 and its binding partners alter the expression of surface receptors and ion channels, leading to alterations in membrane excitability; these alterations could result in motor impairment (Srinivasan and Shakkottai 2019). It has been proposed that miR-144 and miR-101 may potentially act as paracrine factors on Purkinje neurons (Lee et al. 2008; Persengiev et al. 2011). Downregulation of miR-144 and miR-101 in cerebellar and cortical samples from human brains have been reported (Persengiev et al. 2011).

Machado-Joseph disease (MJD) or spinocerebellar ataxia type 3 (SCA3) is caused by an unstable expansion of a CAG repeat sequence in the ATXN3 gene, and the pathogenic mutation results in an expanded polyQ tract at the C-terminus of ataxin-3. The alterations of the CNS associated with MJD/SCA3 consist of cell degeneration affecting multiple neuronal systems, especially the cerebellum, the brain stem, basal ganglia, some cranial nerves, and the spinal cord. Although the genetic cause of MJD/SCA3 has been clearly defined for decades, the mechanisms mediating the neurodegeneration observed in selected regions of MJD/SCA3 patients' brains are not yet fully elucidated (Matos et al. 2019). miR-25 could directly target the 3′UTR of the ATXN3 gene and its overexpression reduce polyQ induced toxicity and neuronal intranuclear inclusions formation in SCA3/MJD model cells (Huang et al. 2014). Downregulation of miR-25 in plasma (Shi et al. 2014) and upregulation in lymphoblastoid cells (Krauss et al. 2019) have been reported.

Friedrich's ataxia is an autosomal recessive disease resulting only from mutations in the Frataxin (FXN) gene, all of which lead to decreased levels of the functional frataxin protein. In 96% of the patients, mutations correspond to expanded GAA repeats in the first intron of the gene; such mutations decrease transcription of the cognate mRNA to approximately 10% of normal levels (Lynch et al. 2021). A study found that the levels of miR-886-3p in lymphoblasts are upregulated, and transfection with anti-miR-886-3p leads to differential increased frataxin levels in patient cells (Mahishi et al. 2012).

7.2.3 Frontotemporal Dementia

Frontotemporal dementia (FTD) is a heterogeneous disorder with distinct clinical phenotypes that include changes in behavior, language, psychiatric, executive control, and often motor symptoms (Olney et al. 2017). Clinical syndromes of FTD are associated with macroscopic changes characterized by frontal and anterior temporal lobe atrophy and microscopic changes marked by gliosis, microvacuolations, and synaptic and neuronal loss. Almost all FTD cases show illness-associated mutations in one of the following three genes: microtubule-associated protein tau (MAPT), progranulin (GRN), and chromosome 9 open reading frame 72 (C9ORF72) (Olney et al. 2017; Younes and Miller 2020).

There are several studies that have analyzed the levels of different miRNAs in patients with FTD. Some studies have reported significant downregulation of miR-132, which has been reported in studies performing miRNA isolation from human cerebral cortex (Chen-Plotkin et al. 2012; Hébert et al. 2013). In a study in blood plasma, miR-132 showed to be a useful biomarker to differentiate patients with mild cognitive impairment from healthy controls (Sheinerman et al. 2012). Chen-Plotkin et al. (2012) showed that miR-132 and miR-212 are decreased in frontotemporal lobar degeneration with TDP-43 inclusions (FTLD-TDP), and they can repress the so-far uncharacterized gene TMEM106B coding for a transmembrane protein located in endosome and lysosomal membranes, ultimately implicated in hypomyelinating leukodystrophy. This gene is highly expressed in human brain tissue, with particularly increased expression in GRN(+) FTLD-TDP. Increased levels of TMEM106B may likely cause intracellular retention of the granulin precursor. Moreover, deficiency of granulins due to homozygous mutations in the GRN gene lead to ceroid lipofuscinosis or Batten disease, while haploinsufficiency almost always are associated with neuronal FTD (Boland et al. 2022).

Some miRNA levels that have been evaluated in patients with FTD and have been compared with healthy controls are miR-632 and miR-127-3p. Schneider et al. showed that miR-632 expression is significantly decreased in sporadic FTD compared with healthy controls in samples from cerebrospinal fluid (CSF) (Schneider et al. 2018). Their in silico analysis revealed that HRK, a central mediator of apoptosis, is a potential target of miR-632 in the human frontal and temporal lobes. Low miR-632 levels could result in pathologically increased HRK and apoptosis, ultimately leading to degenerative changes within the frontal and temporal lobes of patients with FTD. miR-127-3p has been shown to be downregulated in plasma samples from FTD patients compared to healthy controls (Piscopo et al. 2018). Experimental data showed that miR-127-3p affects the induction of neuronal apoptosis by regulating the expression of CISD1 (mitoNEET), an iron-sulfur cluster transfer protein that binds to the mitochondrial outer membrane (He et al. 2016).

Magen et al. (2023) used next-generation sequencing to analyze circulating miRNAs in 293 subjects with FTD, developing a highly accurate prediction model from 13 miRNAs. Most of the selected miRNAs are widely distributed in CNS tissue (let-7c-5p, miR-26a-5p, miR-107, miR-125b-5p), suggesting that they might be directly involved in the disease state of the CNS. miR-26a, miR-326, miR-484, and miR-361 were strongly associated with FTD diagnosis.

7.2.4 Dementia with Lewy Bodies

Dementia with Lewy bodies (DLB) is the second most common neurodegenerative dementia after Alzheimer's disease. DLB and Parkinson's disease dementia (PDD) share many clinical, neurochemical, and morphological features. Both disorders result from the aggregation of α-synuclein (αSyn) into Lewy bodies, and dense αSyn pathology appears in the inferior regions of the hippocampus in both conditions, however, with different involvement of its subregions (Jellinger 2018). For example, DLB does not show the characteristic selective degeneration of ventrolateral substantia nigra neurons seen in Parkinson's disease, which could explain some of the clinical differences between both disorders (Walker et al. 2004). Risk factors associated with DLB include old age, mutations in the glucocerebrosidase (GBA1) and αSyn (SNCA) genes, and carrying the H1 haplotype of the microtubule-associated protein tau (MAPT). Other genes have been associated with DLB, including BCL7C and GABRB3 genes (Sezgin et al. 2019). The cognitive profile of DLB is characterized by particularly severe deficits in attention as well as executive and visuospatial functions (Jellinger 2018; Sezgin et al. 2019). Rapid eye movement sleep behavior disorder (RBD) is a prevalent symptom characterized by vivid, often frightening dreams accompanied by simple or complex motor behaviors. This unusual activity occurs due to the loss of muscle paralysis during rapid eye movement sleep. Additionally, the revised criteria for clinical diagnosis of DLB now recognize RBD as a core clinical feature (Armstrong 2021).

In individuals with idiopathic rapid eye movement behavior disorder (IRBD), serum miR-19b levels are downregulated even 5 years before the diagnosis of PD and DLB (Fernández-Santiago et al. 2015). This miRNA is significantly downregulated in IRBD subjects who converted to PD and DLB but not in those who remained disease-free. Soto et al. (2022) identified a serum miRNA biosignature of 12 miRNAs, including miR-19b, which could be useful in discriminating healthy controls, patients with IRBD, and patients with PD/DLB. This signature also holds potential as a disease progression biomarker. Moreover, in vitro studies suggest that upregulation of miR-19b may suppress the proliferation of hippocampal neural progenitor cells in mice, a process potentially linked to neurodegenerative disorders like dementia and schizophrenia (Horai et al. 2020; Chen et al. 2022).

7.2.5 Alzheimer's Disease

Alzheimer's disease (AD), the leading cause of dementia, remains enigmatic in its etiology. Although a definitive cause eludes us, the β-amyloid (Aβ) plaque formation and aggregation hypothesis stand as the most robust explanation for its pathogenesis (Abeysinghe et al. 2020). The accumulation of Aβ in the medial parietal cortex constitutes the first stage in AD (Solis et al. 2020; García-Morales et al. 2021). Another pathological cause of AD may be the formation of neurofibrillary tangles (NFTs) made up of an abnormal accumulation of hyperphosphorylated Tau protein that normally stabilizes microtubules and promotes vesicle transport. The abnormal tangles, however, block the transport process, which harms the synaptic interneuronal communication (Sehar et al. 2022). Some of the earliest symptoms of AD manifest years before receiving a clinical diagnosis of dementia include mood changes, anxiety, and modified sleep patterns. Nonetheless, alterations in memory, planning or problem-solving, familiar task completion, recognition of time or place, comprehension of visual images and spatial relationships, communication, use and recall of words and names during talking and writing, ability to retrace steps, judgment and interpersonal interactions, work/social activities, mood, and personality can be undetected until diagnosis (Atri 2019).

One of the microRNAs most strongly related to AD is miR-146a. This miRNA has been reported to be increased in human brain temporal lobe neocortex from patients with AD (Sethi and Lukiw 2009), and its increased expression levels have showed a significant negative correlation with decreased levels of mitochondrial voltage-dependent anion channel (VDAC1), the most abundant protein on the outer membrane of mitochondria (Kim et al. 2022). Mitochondrial malfunction is a sign of oxidative stress, inflammation, aging, and chronic degenerative diseases. VDAC1 coordinates the transmembrane transport of proteins and metabolites and regulates apoptosis. VDAC1 can also participate in the regulation of inflammation by affecting the respiratory chain and promoting the expression of cytokines (Hu et al. 2022). This supports the notion that miR-146a exerts pro-inflammatory effects when it is significantly upregulated. Several groups have analyzed the miR-146a levels in multiple biological fluids. Unfortunately, there is no consensus pattern detectable: in CSF, it has been reported to be upregulated (Alexandrov et al. 2012; Denk et al. 2015) and downregulated (Kiko et al. 2014), while expression is decreased in both serum (Dong et al. 2015) and plasma (Kiko et al. 2014). However, Maffioletti et al. (2020) found no significant differences in plasma levels of miR-146a in AD patients compared to nonaffected controls but reported a positive correlation between miR-146a levels and age (Maffioletti et al. 2020). Wang et al. (2016a) reported upregulated miR-146a expression in hippocampus from patients with AD. Their experiments in an animal model revealed that miR-146a could promote Tau protein hyperphosphorylation.

Downregulation of miR-181c in the hippocampus was observed in an AD mouse model, leading to depression of its target gene DPYSL2, a regulator of neuronal microtubule assembly, suggesting a potential role in disease pathogenesis. It has been suggested that phosphorylation of DPYSL2 is required for Aβ-induced dysfunction of cognitive memory and impairment of synaptic plasticity (Zhou et al. 2016). Furthermore, it has been shown that miR-181c could reduce apoptosis by reverting NLRP1-mediated protein effects (Wang et al. 2023). It has been reported that miR-181c is downregulated in serum (Tan et al. 2014; Abuelezz et al. 2022), while it is upregulated in plasma (Barbagallo et al. 2020).

The negative correlation between miR-107 and b-site amyloid precursor protein cleaving enzyme 1 (BACE1) in AD brains suggests a potential regulatory mechanism: decreased miR-107 levels might lead to increased BACE1 expression, thereby promoting amyloid beta accumulation and contributing to AD progression. BACE1 cleaves the β-amyloid precursor protein creating the neurotoxic β-amyloid peptide $A\beta_{1-42}$ (Wang et al. 2008). In line with this observation, it has been reported that miR-107 levels in the brain tend to decrease in correlation with the severeness of the AD pathology (Nelson and Wang 2010). Underlining its usefulness as biomarker, miR-107 has been found to be downregulated in blood samples from AD patients (Yılmaz et al. 2016).

miR-34a showed an increased expression in brain areas of patients with AD like hippocampus and medial frontal gyrus (Cogswell et al. 2008) and in temporal cortex (Sarkar et al. 2016). Interestingly, either increase or decrease in the expression of miR-34a leads to apoptosis. miR-34a expression has been found significantly altered in cortical neurons treated with neurotoxic Aβ42 and neurons from a transgenic mouse model for AD, which resulted in an aberrant increase in cyclin D1 levels, leading to cell cycle reentry and apoptosis of neurons (Modi et al. 2016). In addition, miR-34a inhibits translation of five key mitochondrial electron transport chain (ETC) proteins, preventing their replacement after protein turnover, ultimately leading to a collapse of the ETC and mitochondrial dysfunction (Russell et al. 2016). It has been shown that miR-34a is upregulated in blood mononuclear cells (Schipper et al. 2007); however, it is downregulated in plasma (Kiko et al. 2014; Cosín-Tomás et al. 2017), serum (Maldonado-Lasuncion et al. 2019), and CSF (Kiko et al. 2014) of patients with AD.

Another miRNA that has shown changes in AD is miR-125b. miR-125b may contribute to AD pathogenesis, given that elevated miR-125b levels induce tau hyperphosphorylation in primary hippocampal neurons and in the mouse dentate gyrus (Banzhaf-Strathmann et al. 2014). Moreover, miR-125b was downregulated in serum (Galimberti et al. 2014; Tan et al. 2014; Jia and Liu 2016) and upregulated in CSF (Jin et al. 2018) from patients with AD.

7.2.6 Parkinson's Disease

Parkinson's disease (PD) is a complex progressive neurodegenerative disease. The cardinal motor symptoms are tremor, rigidity, bradykinesia/akinesia, and postural instability but could also include other motor and non-motor symptoms. Non-motor symptoms are heterogenous and affect many organ systems, including gastrointestinal and genitourinary systems (Armstrong and Okun 2020; Balestrino and Schapira 2020). PD is defined by the loss of dopaminergic neurons in the substantia nigra pars compacta (SN) located in the midbrain and associated with Lewy bodies (LB), which are cytoplasmic inclusions that include insoluble α-synuclein aggregates (Simon et al. 2020). Several genes carry mutations linked to hereditary forms of PD, including SNCA, LRRK2, PARK2, PARK7, and PINK1. Mutations in SNCA, which encodes the α-synuclein protein, directly cause familial PD. These mutations trigger α-synuclein aggregation into Lewy bodies, causing severe damage in the substantia nigra pars compacta, a key brain region for motor control, and leading to persistent motor impairments in rodents (Raza et al. 2019).

miR-29 family members are associated with neuronal function, with miR-29a specifically inhibiting mitochondrial antiviral signaling protein (MAVS) expression (Yang et al. 2022b). It has been suggested that MAVS signaling is involved in the development of microglial activation-induced neurotoxicity (Cheng et al. 2017).

miR-29b emerged as a key regulator of aging and damage in the nigrostriatal dopaminergic system. Deficiency in miR-29b2/c triggered AMPK activation, concurrently suppressing NF-κB/p65 signaling in glial cells (Bai et al. 2021). miR-29c may attenuate the inflammatory and apoptotic responses in PD by directly targeting SP1 (Wang et al. 2020). There are different studies that report alterations in expression levels of miR-29 family members in different types of samples. It has been shown that serum levels of miR-29a, miR-29-b, and miR-29c are significantly downregulated in patients with PD (Bai et al. 2017); while miR-29a and miR-29c are downregulated in saliva (Jiang et al. 2021), miR-29a is downregulated in blood (Margis et al. 2011), and miR-29b and miR-29c levels are decreased in PBMCs (Martins et al. 2011). Moreover, it has been reported that miR-29c is upregulated in serum (Ozdilek and Demircan 2021).

miR-133b is expressed in midbrain (Schlaudraff et al. 2014) specifically in dopaminergic neurons (DNs) (Kim et al. 2007). miR-133b downregulation in serum has been correlated with low serum ceruloplasmin levels in patients with PD, decreased serum ceruloplasmin levels, and oxidative activity specifically exacerbate nigral iron deposition in PD patients (Zhao et al. 2014). Moreover, decreased levels of miR-133b have been reported in plasma (Zhang et al. 2017b). However, Chen et al. (2021) identified significantly elevated levels of miR-133b and miR-221-3p in the plasma of individuals with Parkinson's disease (PD), suggesting their potential as diagnostic biomarkers.

Analysis in anterior cingulate gyrus showed that upregulation of miR-221 correlates with the observed reduction of SNCA transcript levels. Upregulation of miR-221 might occur as a response to increased α-synuclein levels in pathologically affected regions of the PD brain. As a result, SNCA expression is reduced even in pathologically less affected regions (Tatura et al. 2016). In vitro studies have shown that when miR-221 is highly expressed, it promotes cell proliferation and inhibits cell apoptosis (Li et al. 2018). miR-221 levels are decreased in serum from patients with PD (Ma et al. 2016, p. 221).

Ding et al. (2016) evaluated the serum miRNA profile of PD patients through Solexa sequencing of pooled serum samples. Based on qRT-PCR assays, five differentially expressed miRNAs were identified: upregulation of miR-195 and downregulation of miR-15b, miR-181a, miR-185, and miR-221. The combination of these miRNAs could distinguish PD patients from other neurological disorders and healthy people.

Emerging evidence suggests that miR-7 acts as a potent suppressor of α-synuclein expression, effectively inhibiting α-synuclein-mediated cell death. Furthermore, its influence on BDNF expression suggests involvement in both signaling and synaptic plasticity (Junn et al. 2009; Li et al. 2019a). miR-7 is downregulated in rotenone-induced Parkinsonian rat striatum (Horst et al. 2018), and its levels are increased in peripheral blood lymphocytes (Alieva et al. 2015), CSF (Starhof et al. 2019), serum, and exosomes (Citterio et al. 2023) in patients with PD.

There are reports of other microRNAs in PD patients. Particularly, it has been reported that miR-34a expression is increased in plasmatic extracellular vesicles (Grossi et al. 2020) and downregulated in peripheral blood mononuclear cells (Yang et al. 2022a). Moreover, increased levels of miR-9 have been detected in the peripheral blood lymphocytes (Alieva et al. 2015) and downregulation of miR-433 in plasma (Zhang et al. 2017b).

7.2.7 Huntington's Disease

Huntington's disease (HD) is caused by a dominantly inherited CAG repeat expansion in exon 1 of the huntingtin gene (HTT). HTT affects many cellular functions including trafficking vesicles, coordinating cell division, regulating cell movement and transcription, mediating autophagy, and cell survival (Fields et al. 2021). The mutant HTT protein (mHTT) leads to cell death (Tabrizi et al. 2020). HD is typically characterized by a triad of symptoms: progressive involuntary choreiform movements, behavioral or cognitive symptoms, and psychiatric disturbances with slow but relentless deterioration over a period of 15–20 years (Ghosh and Tabrizi 2013).

miR-34b was the first microRNA described to be associated with HD. Gaughwin et al. (2011) found that miR-34b levels are significantly elevated in the plasma of pre-manifest HD patients. The authors of this work suggest that miR-34b regulates P53 expression, which may influence mHTT-mediated apoptosis (Gaughwin et al. 2011). Miñones-Moyano et al. (2011) observed downregulation of mir-34b and mir-34c in brain samples (amygdala, substantia nigra, frontal cortex, cerebellum) in patients with advanced PD. Their analysis showed that the decrease in either miR-34b or miR-34c in differentiated neuronal cells compromises cell viability and mitochondrial function. They found diminished metabolism activity of mitochondria accompanied by a significant loss in the mitochondrial membrane potential, increased formation of ROS, and decreased intracellular ATP levels. However, Hoss et al. (2014) found no difference in miR-34b expression in prefrontal cortex samples of patients with HD compared to control individuals.

Petry et al. (2022) analyzed a panel of microRNAs related to HD (miR-10b, miR-196b, miR-132, miR-212, miR-127, and miR-128) in brain samples (i.e., putamen and cortex). The authors found differences in the miRNA maturation steps, suggesting that alterations in autophagy might affect miRNA maturation. Furthermore, they did not find significant alterations in miRNA biogenesis in human HD cortex and blood, which suggests that miRNA maturation displays tissue-specific effects.

Other microRNAs related to HD in human samples include miR-10b overexpressed in prefrontal cortex. miR-10b may diminish translation of BDNF, a growth factor required for the survival, and differentiation of striatal neurons. Overexpression of miR-10b-5p might be harmful to neuronal cells (Hoss et al. 2015). Moreover, miR-132 is increased in plasma (Díez-Planelles et al. 2016), miR-323b is upregulated in plasma of patients with HD carrying the mutant HTT (mHTT) (Ferraldeschi et al. 2021), and miR-9 is downregulated in peripheral leukocytes, potentially controlling the transition from neural progenitor to postmitotic neurons (Chang et al. 2017).

7.2.8 Amyotrophic Lateral Sclerosis

Amyotrophic lateral sclerosis (ALS) is a progressive, fatal neuromuscular disease characterized by degeneration of the upper and lower motor neurons, ultimately resulting in dysfunction of the skeletal muscles (Feldman et al. 2022). The hallmark of ALS is progressive muscle weakness, accompanied by muscle atrophy, fasciculations, muscle cramps, and slowness of movements with muscle stiffness. The onset of muscle weakness in ALS is usually focal and typically spreads to adjacent body regions (Masrori and Van Damme 2020). The most frequent forms of familiar ALS are associated with mutations in the superoxide dismutase 1 gene (SOD1) located on chromosome 21q22 and mutations C9ORF72 on chromosome 9 (Riancho et al. 2019); however, more than 40 genes have been associated with ALS. SOD1 catalyzes the conversion of superoxide into oxygen and hydrogen peroxide, which explains its importance in the antioxidant defense of cells with aerobic metabolism, as is the case of motor neurons (De Felice et al. 2018; Riancho et al. 2019). The most commonly used mouse model for investigating SOD1 mutations is the SOD1G93A transgenic mouse model. SOD1G93A transgenic mice overexpress human mutant SOD1 (glycine to alanine at residue 93) and develop a relatively early disease onset with rapid motor function decline with transgenic mice.

miR-206 has been the most analyzed microRNA in samples of ALS patients. In silico studies have revealed that miR-206 could affect the expression of numerous mRNAs involved in crucial cellular mechanisms (Khalilian et al. 2022). It has been suggested that miR-206 delays the onset and progression of ALS by promoting the regeneration of neuromuscular synapses (Ma et al. 2015). In muscle biopsy samples of the ALS SOD1G93A mouse model, the concentration of miR-206 gradually increases with age (Nguyen et al. 2022). In humans, miR-206 serum levels are significantly upregulated at the mild stage of ALS pathology and decreases in moderate and severe stages (Dobrowolny et al. 2021), while it is upregulated in serum (Toivonen et al. 2014), blood leukocytes (Vrabec et al. 2018) and muscle samples (Pegoraro et al. 2017). However, the results in muscle tissues are unambiguous as there is a study reporting reduced levels of miR-206 in this type of sample (Jensen et al. 2016).

miR-338 is a promising ALS biomarker. Li et al. (2019b) identified a regional accumulation of glycogen in the spinal cord of ALS mice, which is caused by decreased glycogenolysis due to diminished glycogen phosphorylase (PYGB) expression. The authors of this study showed that miR-338 binds to the 3′UTR of PYGB, thereby inhibiting PYGB expression. Conversely, miR-338-3p is intensely increased in the spinal cord of ALS mice (Li et al. 2019b). miR-338 is highly expressed in ALS tissues, including neuromuscular junction characterizing ALS from normal tissues (De Felice et al. 2018). De Felice et al. (2014) found a significant increase of miR-338 in leukocytes of ALS patients compared to healthy controls (De Felice et al. 2014).

7.3 Conclusions

In this chapter, we reviewed miRNAs with strong, well-documented relationships to neurodegenerative disorders, prioritizing those with evidenced reproducibility. Reproducibility remains a significant hurdle in miRNA biomarker development. Several issues including the presence of miRNAs in different body fluids, availability of different protocols, selection of reference RNAs, and methodologies for each type of sample limit the reproducibility of their use in medical research. In neurodegenerative disorders, the diagnosis is typically based on clinical history and physical examination, which can be further confirmed by genetic testing. Despite the existence of reliable genetic markers for diagnosis and prognosis in certain scenarios, miRNAs offer valuable insights by uncovering hidden molecular events in disease development, providing a potential tool for monitoring treatment response. As indicated in Table 7.2, certain miRNAs have been discovered to be involved in multiple neurodegenerative disorders. The widespread, routine clinical use of miRNAs appears to remain distant. To bridge this gap, long-term efforts must prioritize improved reproducibility through stricter adherence to established guidelines, meticulous selection and validation of reference RNAs, and standardized assay protocols.

Table 7.2 microRNAs (miRNAS) with diagnostic/prognostic potential in more than one neurodegenerative disorder

microRNA	Disease	Sample type	Upregulated (+) Downregulated (−)	Reference
miR-25	AMD	Plasma	−	Ertekin et al. (2014)
		Serum	−	Elshelmani et al. (2020)
	SCA3	Plasma	−	Shi et al. (2014)
		Lymphoblastoid cells	+	Krauss et al. (2019)
miR-34a	AD	Hippocampus and medial frontal gyrus	+	Cogswell et al. (2008)
		Temporal cortex	+	Sarkar et al. (2016)
		Blood mononuclear cells	+	Schipper et al. (2007)
		Plasma	−	Cosín-Tomás et al. (2017)
		Serum	−	Maldonado-Lasuncion et al. (2019)
		CSF	−	Kiko et al. (2014)
	PD	Plasmatic extracellular vesicles	+	Grossi et al. (2020)
		PBMCs	−	Yang et al. (2022a)
miR-9	HD	Peripheral blood lymphocytes	−	Chang et al. (2017)
	PD	Peripheral blood lymphocytes	+	Alieva et al. (2015)

Compliance with Ethical Standards

Funding This research received no specific grant from any funding agency in the public, commercial, or not-for-profit sectors.

Conflict of Interest Authors declare that they have no conflict of interest.

Ethical Approval This chapter does not contain studies with human/animal participants performed by the authors.

References

Abeysinghe AADT, Deshapriya RDUS, Udawatte C (2020) Alzheimer's disease; a review of the pathophysiological basis and therapeutic interventions. Life Sci 256:117996. https://doi.org/10.1016/j.lfs.2020.117996

Abuelezz NZ, Nasr FE, Abdel Aal WM et al (2022) Sera miR-34a, miR-29b and miR-181c as potential novel diagnostic biomarker panel for Alzheimers in the Egyptian population. Exp Gerontol 169:111961. https://doi.org/10.1016/j.exger.2022.111961

Alexandrov PN, Dua P, Hill JM et al (2012) microRNA (miRNA) speciation in Alzheimer's disease (AD) cerebrospinal fluid (CSF) and extracellular fluid (ECF). Int J Biochem Mol Biol 3:365–373

Alieva AK, Filatova EV, Karabanov AV et al (2015) miRNA expression is highly sensitive to a drug therapy in Parkinson's disease. Parkinsonism Relat Disord 21:72–74. https://doi.org/10.1016/j.parkreldis.2014.10.018

Armstrong MJ (2021) Advances in dementia with Lewy bodies. Ther Adv Neurol Disord 14:1–14. https://doi.org/10.1177/17562864211057666

Armstrong MJ, Okun MS (2020) Diagnosis and treatment of Parkinson disease: a review. JAMA 323:548–560. https://doi.org/10.1001/jama.2019.22360

Atri A (2019) The Alzheimer's disease clinical Spectrum: diagnosis and management. Med Clin North Am 103:263–293. https://doi.org/10.1016/j.mcna.2018.10.009

Bai X, Tang Y, Yu M et al (2017) Downregulation of blood serum microRNA 29 family in patients with Parkinson's disease. Sci Rep 7:5411. https://doi.org/10.1038/s41598-017-03887-3

Bai X, Zhang X, Fang R et al (2021) Deficiency of miR-29b2/c leads to accelerated aging and neuroprotection in MPTP-induced Parkinson's disease mice. Aging 13:22390–22411. https://doi.org/10.18632/aging.203545

Balestrino R, Schapira AHV (2020) Parkinson disease. Eur J Neurol 27:27–42. https://doi.org/10.1111/ene.14108

Banzhaf-Strathmann J, Benito E, May S et al (2014) MicroRNA-125b induces tau hyperphosphorylation and cognitive deficits in Alzheimer's disease. EMBO J 33:1667–1680. https://doi.org/10.15252/embj.201387576

Barbagallo C, Mostile G, Baglieri G et al (2020) Specific signatures of serum miRNAs as potential biomarkers to discriminate clinically similar neurodegenerative and vascular-related diseases. Cell Mol Neurobiol 40:531–546. https://doi.org/10.1007/s10571-019-00751-y

Berber P, Grassmann F, Kiel C, Weber BHF (2017) An eye on age-related macular degeneration: the role of MicroRNAs in disease pathology. Mol Diagn Ther 21:31–43. https://doi.org/10.1007/s40291-016-0234-z

Boland S, Swarup S, Ambaw YA et al (2022) Deficiency of the frontotemporal dementia gene GRN results in gangliosidosis. Nat Commun 13:5924. https://doi.org/10.1038/s41467-022-33500-9

Bustin SA, Benes V, Garson JA et al (2009) The MIQE guidelines: minimum information for publication of quantitative real-time PCR experiments. Clin Chem 55:611–622. https://doi.org/10.1373/clinchem.2008.112797

Chang K-H, Wu Y-R, Chen C-M (2017) Down-regulation of miR-9* in the peripheral leukocytes of Huntington's disease patients. Orphanet J Rare Dis 12:185. https://doi.org/10.1186/s13023-017-0742-x

Chen Q, Deng N, Lu K et al (2021) Elevated plasma miR-133b and miR-221-3p as biomarkers for early Parkinson's disease. Sci Rep 11:15268. https://doi.org/10.1038/s41598-021-94734-z

Chen J, Liu C, Xu M et al (2022) Upregulation of miR-19b-3p exacerbates chronic stress-induced changes in synaptic plasticity and cognition by targeting Drebrin. Neuropharmacology 207:108951. https://doi.org/10.1016/j.neuropharm.2022.108951

Chen Q, Lin H, Li S et al (2023) Mini-αA Upregulates the miR-155-5p target gene CDK2 and plays an Antiapoptotic role in retinal pigment epithelial cells during oxidative stress. J Ophthalmol 2023:1–11. https://doi.org/10.1155/2023/6713094

Cheng J, Liao Y, Xiao L et al (2017) Autophagy regulates MAVS signaling activation in a phosphorylation-dependent manner in microglia. Cell Death Differ 24:276–287. https://doi.org/10.1038/cdd.2016.121

Chen-Plotkin AS, Unger TL, Gallagher MD et al (2012) *TMEM106B*, the risk gene for Frontotemporal dementia, is regulated by the microRNA-132/212 cluster and affects Progranulin pathways. J Neurosci 32:11213–11227. https://doi.org/10.1523/JNEUROSCI.0521-12.2012

Citterio LA, Mancuso R, Agostini S et al (2023) Serum and Exosomal miR-7-1-5p and miR-223-3p as possible biomarkers for Parkinson's disease. Biomol Ther 13:865. https://doi.org/10.3390/biom13050865

Cogswell JP, Ward J, Taylor IA et al (2008) Identification of miRNA changes in Alzheimer's disease brain and CSF yields putative biomarkers and insights into disease pathways. J Alzheimers Dis 14:27–41. https://doi.org/10.3233/JAD-2008-14103

Cosín-Tomás M, Antonell A, Lladó A et al (2017) Plasma miR-34a-5p and miR-545-3p as early biomarkers of Alzheimer's disease: potential and limitations. Mol Neurobiol 54:5550–5562. https://doi.org/10.1007/s12035-016-0088-8

De Felice B, Annunziata A, Fiorentino G et al (2014) miR-338-3p is over-expressed in blood, CFS, serum and spinal cord from sporadic amyotrophic lateral sclerosis patients. Neurogenetics 15:243–253. https://doi.org/10.1007/s10048-014-0420-2

De Felice B, Manfellotto F, Fiorentino G et al (2018) Wide-ranging analysis of MicroRNA profiles in sporadic amyotrophic lateral sclerosis using next-generation sequencing. Front Genet 9:310. https://doi.org/10.3389/fgene.2018.00310

Denk J, Boelmans K, Siegismund C et al (2015) MicroRNA profiling of CSF reveals potential biomarkers to detect Alzheimer's disease. PLoS One 10:e0126423. https://doi.org/10.1371/journal.pone.0126423

Díez-Planelles C, Sánchez-Lozano P, Crespo MC et al (2016) Circulating microRNAs in Huntington's disease: emerging mediators in metabolic impairment. Pharmacol Res 108:102–110. https://doi.org/10.1016/j.phrs.2016.05.005

Ding H, Huang Z, Chen M et al (2016) Identification of a panel of five serum miRNAs as a biomarker for Parkinson's disease. Parkinsonism Relat Disord 22:68–73. https://doi.org/10.1016/j.parkreldis.2015.11.014

Dobrowolny G, Martone J, Lepore E et al (2021) A longitudinal study defined circulating microRNAs as reliable biomarkers for disease prognosis and progression in ALS human patients. Cell Death Discov 7:4. https://doi.org/10.1038/s41420-020-00397-6

Dong H, Li J, Huang L et al (2015) Serum MicroRNA profiles serve as novel biomarkers for the diagnosis of Alzheimer's disease. Dis Markers 2015:1–11. https://doi.org/10.1155/2015/625659

ElShelmani H, Wride MA, Saad T et al (2020) Identification of novel serum MicroRNAs in age-related macular degeneration. Transl Vis Sci Technol 9:28. https://doi.org/10.1167/tvst.9.4.28

Ertekin S, Yıldırım Ö, Dinç E et al (2014) Evaluation of circulating miRNAs in wet age-related macular degeneration. Mol Vis 20:1057–1066

Feldman EL, Goutman SA, Petri S et al (2022) Amyotrophic lateral sclerosis. Lancet Lond Engl 400:1363–1380. https://doi.org/10.1016/S0140-6736(22)01272-7

Fernández-Santiago R, Iranzo A, Gaig C et al (2015) MicroRNA association with synucleinopathy conversion in rapid eye movement behavior disorder. Ann Neurol 77:895–901. https://doi.org/10.1002/ana.24384

Ferraldeschi M, Romano S, Giglio S et al (2021) Circulating hsa-miR-323b-3p in Huntington's disease: a pilot study. Front Neurol 12:657973. https://doi.org/10.3389/fneur.2021.657973

Fields E, Vaughan E, Tripu D et al (2021) Gene targeting techniques for Huntington's disease. Ageing Res Rev 70:101385. https://doi.org/10.1016/j.arr.2021.101385

Freibaum BD, Lu Y, Lopez-Gonzalez R et al (2015) GGGGCC repeat expansion in C9orf72 compromises nucleocytoplasmic transport. Nature 525:129–133. https://doi.org/10.1038/nature14974

Galimberti D, Villa C, Fenoglio C et al (2014) Circulating miRNAs as potential biomarkers in Alzheimer's disease. J Alzheimers Dis 42:1261–1267. https://doi.org/10.3233/JAD-140756

García-Morales V, González-Acedo A, Melguizo-Rodríguez L et al (2021) Current understanding of the physiopathology, diagnosis and therapeutic approach to Alzheimer's disease. Biomedicines 9:1910. https://doi.org/10.3390/biomedicines9121910

Gaughwin PM, Ciesla M, Lahiri N et al (2011) Hsa-miR-34b is a plasma-stable microRNA that is elevated in pre-manifest Huntington's disease. Hum Mol Genet 20:2225–2237. https://doi.org/10.1093/hmg/ddr111

Ghosh R, Tabrizi SJ (2013) Clinical aspects of Huntington's disease. In: Nguyen HHP, Cenci MA (eds) Behavioral neurobiology of Huntington's disease and Parkinson's disease. Springer, Berlin, pp 3–31

Grossi I, Radeghieri A, Paolini L et al (2020) MicroRNA-34a-5p expression in the plasma and in its extracellular vesicle fractions in subjects with Parkinson's disease: an exploratory study. Int J Mol Med 47:533–546. https://doi.org/10.3892/ijmm.2020.4806

He Q-Q, Xiong L-L, Liu F et al (2016) MicroRNA-127 targeting of mitoNEET inhibits neurite outgrowth, induces cell apoptosis and contributes to physiological dysfunction after spinal cord transection. Sci Rep 6:35205. https://doi.org/10.1038/srep35205

Hébert SS, Wang W-X, Zhu Q, Nelson PT (2013) A study of small RNAs from cerebral Neocortex of pathology-verified Alzheimer's disease, dementia with Lewy bodies, hippocampal sclerosis, Frontotemporal lobar dementia, and non-demented human controls. J Alzheimers Dis 35:335–348. https://doi.org/10.3233/JAD-122350

Horai T, Boku S, Okazaki S et al (2020) miR-19b is elevated in peripheral blood of schizophrenic patients and attenuates proliferation of hippocampal neural progenitor cells. J Psychiatr Res 131:102–107. https://doi.org/10.1016/j.jpsychires.2020.09.006

Horst CH, Schlemmer F, De Aguiar MN et al (2018) Signature of aberrantly expressed microRNAs in the striatum of rotenone-induced Parkinsonian rats. Neurochem Res 43:2132–2140. https://doi.org/10.1007/s11064-018-2638-0

Hoss AG, Kartha VK, Dong X et al (2014) MicroRNAs located in the Hox gene clusters are implicated in Huntington's disease pathogenesis. PLoS Genet 10:e1004188. https://doi.org/10.1371/journal.pgen.1004188

Hoss AG, Labadorf A, Latourelle JC et al (2015) miR-10b-5p expression in Huntington's disease brain relates to age of onset and the extent of striatal involvement. BMC Med Genet 8:10. https://doi.org/10.1186/s12920-015-0083-3

Hu H, Guo L, Overholser J, Wang X (2022) Mitochondrial VDAC1: a potential therapeutic target of inflammation-related diseases and clinical opportunities. Cells 11:3174. https://doi.org/10.3390/cells11193174

Huang F, Zhang L, Long Z et al (2014) miR-25 alleviates polyQ-mediated cytotoxicity by silencing ATXN3. FEBS Lett 588:4791–4798. https://doi.org/10.1016/j.febslet.2014.11.013

Jellinger KA (2018) Dementia with Lewy bodies and Parkinson's disease-dementia: current concepts and controversies. J Neural Transm 125:615–650. https://doi.org/10.1007/s00702-017-1821-9

Jensen L, Jørgensen LH, Bech RD et al (2016) Skeletal muscle Remodelling as a function of disease progression in amyotrophic lateral sclerosis. Biomed Res Int 2016:1–12. https://doi.org/10.1155/2016/5930621

Jia L-H, Liu Y-N (2016) Downregulated serum miR-223 servers as biomarker in Alzheimer's disease. Cell Biochem Funct 34:233–237. https://doi.org/10.1002/cbf.3184

Jiang C, Qin B, Liu G et al (2016) MicroRNA-184 promotes differentiation of the retinal pigment epithelium by targeting the *AKT2* /mTOR signaling pathway. Oncotarget 7:52340–52353. https://doi.org/10.18632/oncotarget.10566

Jiang Y, Chen J, Sun Y et al (2021) Profiling of differentially expressed MicroRNAs in saliva of Parkinson's disease patients. Front Neurol 12:738530. https://doi.org/10.3389/fneur.2021.738530

Jin Y, Tu Q, Liu M (2018) MicroRNA-125b regulates Alzheimer's disease through SphK1 regulation. Mol Med Rep 18:2373–2380. https://doi.org/10.3892/mmr.2018.9156

Junn E, Lee K-W, Jeong BS et al (2009) Repression of alpha-synuclein expression and toxicity by microRNA-7. Proc Natl Acad Sci USA 106:13052–13057

Juźwik CA, Drake S, Zhang Y et al (2019) microRNA dysregulation in neurodegenerative diseases: a systematic review. Prog Neurobiol 182:101664. https://doi.org/10.1016/j.pneurobio.2019.101664

Karamazovova S, Matuskova V, Ismail Z, Vyhnalek M (2023) Neuropsychiatric symptoms in spinocerebellar ataxias and Friedreich ataxia. Neurosci Biobehav Rev 150:105205. https://doi.org/10.1016/j.neubiorev.2023.105205

Kaur G, Singh N (2023) Inflammation and retinal degenerative diseases. Neural Regen Res 18:513. https://doi.org/10.4103/1673-5374.350192

Khalilian S, Hosseini Imani SZ, Ghafouri-Fard S (2022) Emerging roles and mechanisms of miR-206 in human disorders: a comprehensive review. Cancer Cell Int 22:412. https://doi.org/10.1186/s12935-022-02833-2

Khan I, Preeti K, Fernandes V et al (2022) Role of MicroRNAs, Aptamers in Neuroinflammation and neurodegenerative disorders. Cell Mol Neurobiol 42:2075–2095. https://doi.org/10.1007/s10571-021-01093-4

Kiko T, Nakagawa K, Tsuduki T et al (2014) MicroRNAs in plasma and cerebrospinal fluid as potential markers for Alzheimer's disease. J Alzheimers Dis 39:253–259. https://doi.org/10.3233/JAD-130932

Kim J, Inoue K, Ishii J et al (2007) A MicroRNA feedback circuit in midbrain dopamine neurons. Science 317:1220–1224. https://doi.org/10.1126/science.1140481

Kim SJ, Russell AE, Wang W et al (2022) miR-146a Dysregulates energy metabolism during Neuroinflammation. J Neuroimmune Pharmacol 17:228–241. https://doi.org/10.1007/s11481-021-09999-y

Krauss S, Nalavade R, Weber S et al (2019) Upregulation of miR-25 and miR-181 family members correlates with reduced expression of ATXN3 in Lymphocytes from SCA3 patients. MicroRNA 8:76–85. https://doi.org/10.2174/2211536607666180821162403

Kuo S-H (2019) Ataxia. Continuum(MinneapMinn) 25:1036–1054. https://doi.org/10.1212/CON.0000000000000753

Lee Y, Samaco RC, Gatchel JR et al (2008) miR-19, miR-101 and miR-130 co-regulate ATXN1 levels to potentially modulate SCA1 pathogenesis. Nat Neurosci 11:1137–1139. https://doi.org/10.1038/nn.2183

Lee EB, Lee VM-Y, Trojanowski JQ (2011) Gains or losses: molecular mechanisms of TDP43-mediated neurodegeneration. Nat Rev Neurosci 13:38–50. https://doi.org/10.1038/nrn3121

Li L, Xu J, Wu M, Hu JM (2018) Protective role of microRNA-221 in Parkinson's disease. Bratisl Med J 119:22–27. https://doi.org/10.4149/BLL_2018_005

Li B, Jiang Y, Xu Y et al (2019a) Identification of miRNA-7 as a regulator of brain-derived neurotrophic factor/α-synuclein axis in atrazine-induced Parkinson's disease by peripheral blood and brain microRNA profiling. Chemosphere 233:542–548. https://doi.org/10.1016/j.chemosphere.2019.05.064

Li C, Wei Q, Gu X et al (2019b) Decreased Glycogenolysis by miR-338-3p promotes regional glycogen accumulation within the spinal cord of amyotrophic lateral sclerosis mice. Front Mol Neurosci 12:114. https://doi.org/10.3389/fnmol.2019.00114

Li L, Lai K, Gong Y et al (2020) Downregulation of miR-146a-5p inhibits choroidal neovascularization via the NF-κB signaling pathway by targeting OTUD7B. Curr Eye Res 45:1514–1525. https://doi.org/10.1080/02713683.2020.1772831

Litwińska Z, Sobuś A, Łuczkowska K et al (2019) The interplay between systemic inflammatory factors and MicroRNAs in age-related macular degeneration. Front Aging Neurosci 11:286. https://doi.org/10.3389/fnagi.2019.00286

Lu TX, Rothenberg ME (2018) MicroRNA. J Allergy Clin Immunol 141:1202–1207. https://doi.org/10.1016/j.jaci.2017.08.034

Lynch DR, Schadt K, Kichula E et al (2021) Friedreich ataxia: multidisciplinary clinical care. J Multidiscip Healthc 14:1645–1658. https://doi.org/10.2147/JMDH.S292945

Ma G, Wang Y, Li Y et al (2015) MiR-206, a key modulator of skeletal muscle development and disease. Int J Biol Sci 11:345–352. https://doi.org/10.7150/ijbs.10921

Ma W, Li Y, Wang C et al (2016) Serum miR-221 serves as a biomarker for Parkinson's disease. Cell Biochem Funct 34:511–515. https://doi.org/10.1002/cbf.3224

Maffioletti E, Milanesi E, Ansari A et al (2020) miR-146a plasma levels are not altered in Alzheimer's disease but correlate with age and illness severity. Front Aging Neurosci 11:366. https://doi.org/10.3389/fnagi.2019.00366

Magen I, Yacovzada N-S, Warren JD et al (2023) microRNA-based predictor for diagnosis of frontotemporal dementia. Neuropathol Appl Neurobiol 49:e12916. https://doi.org/10.1111/nan.12916

Mahishi LH, Hart RP, Lynch DR, Ratan RR (2012) miR-886-3p levels are elevated in Friedreich ataxia. J Neurosci 32:9369–9373

Maldonado-Lasuncion I, Atienza M, Sanchez-Espinosa MP, Cantero JL (2019) Aging-related changes in cognition and cortical integrity are associated with serum expression of candidate MicroRNAs for Alzheimer disease. Cereb Cortex 29:4426–4437. https://doi.org/10.1093/cercor/bhy323

Margis R, Margis R, Rieder CRM (2011) Identification of blood microRNAs associated to Parkinson's disease. J Biotechnol 152:96–101. https://doi.org/10.1016/j.jbiotec.2011.01.023

Martins M, Rosa A, Guedes LC et al (2011) Convergence of miRNA expression profiling, α-Synuclein Interacton and GWAS in Parkinson's disease. PLoS One 6:e25443. https://doi.org/10.1371/journal.pone.0025443

Masrori P, Van Damme P (2020) Amyotrophic lateral sclerosis: a clinical review. Eur J Neurol 27:1918–1929. https://doi.org/10.1111/ene.14393

Matos CA, De Almeida LP, Nóbrega C (2019) Machado-Joseph disease/spinocerebellar ataxia type 3: lessons from disease pathogenesis and clues into therapy. J Neurochem 148:8–28. https://doi.org/10.1111/jnc.14541

Ménard C, Rezende FA, Miloudi K et al (2016) MicroRNA signatures in vitreous humour and plasma of patients with exudative AMD. Oncotarget 7:19171–19184. https://doi.org/10.18632/oncotarget.8280

Miñones-Moyano E, Porta S, Escaramís G et al (2011) MicroRNA profiling of Parkinson's disease brains identifies early downregulation of miR-34b/c which modulate mitochondrial function. Hum Mol Genet 20:3067–3078. https://doi.org/10.1093/hmg/ddr210

Modi PK, Jaiswal S, Sharma P (2016) Regulation of neuronal cell cycle and apoptosis by MicroRNA 34a. Mol Cell Biol 36:84–94. https://doi.org/10.1128/MCB.00589-15

Mody S, Joshi A (2023) Age-related macular degeneration and its association with neurodegenerative disorders. Cureus 15:e34920. https://doi.org/10.7759/cureus.34920

Murad N, Kokkinaki M, Gunawardena N et al (2014) miR-184 regulates ezrin, LAMP-1 expression, affects phagocytosis in human retinal pigment epithelium and is downregulated in age-related macular degeneration. FEBS J 281:5251–5264. https://doi.org/10.1111/febs.13066

Nelson PT, Wang W-X (2010) MiR-107 is reduced in Alzheimer's disease brain Neocortex: validation study. J Alzheimers Dis 21:75–79. https://doi.org/10.3233/JAD-2010-091603

Nguyen TPN, Kumar M, Fedele E et al (2022) MicroRNA alteration, application as biomarkers, and therapeutic approaches in neurodegenerative diseases. Int J Mol Sci 23:4718. https://doi.org/10.3390/ijms23094718

Olney NT, Spina S, Miller BL (2017) Frontotemporal Dementia. Neurol Clin 35:339–374. https://doi.org/10.1016/j.ncl.2017.01.008

Ozdilek B, Demircan B (2021) Serum microRNA expression levels in Turkish patients with Parkinson's disease. Int J Neurosci 131:1181–1189. https://doi.org/10.1080/00207454.2020.1784165

Pegoraro V, Merico A, Angelini C (2017) Micro-RNAs in ALS muscle: differences in gender, age at onset and disease duration. J Neurol Sci 380:58–63. https://doi.org/10.1016/j.jns.2017.07.008

Persengiev S, Kondova I, Otting N et al (2011) Genome-wide analysis of miRNA expression reveals a potential role for miR-144 in brain aging and spinocerebellar ataxia pathogenesis. Neurobiol Aging 32:2316.e17–2316.e27. https://doi.org/10.1016/j.neurobiolaging.2010.03.014

Petry S, Keraudren R, Nateghi B et al (2022) Widespread alterations in microRNA biogenesis in human Huntington's disease putamen. Acta Neuropathol Commun 10:106. https://doi.org/10.1186/s40478-022-01407-7

Piscopo P, Grasso M, Puopolo M et al (2018) Circulating miR-127-3p as a potential biomarker for differential diagnosis in Frontotemporal dementia. J Alzheimers Dis 65:455–464. https://doi.org/10.3233/JAD-180364

Raza C, Anjum R, Shakeel NUA (2019) Parkinson's disease: mechanisms, translational models and management strategies. Life Sci 226:77–90. https://doi.org/10.1016/j.lfs.2019.03.057

Riancho J, Gonzalo I, Ruiz-Soto M, Berciano J (2019) Why do motor neurons degenerate? Actualisation in the pathogenesis of amyotrophic lateral sclerosis. Neurologia 34:27–37

Romano GL, Platania CBM, Drago F et al (2017) Retinal and circulating miRNAs in age-related macular degeneration: an in vivo animal and human study. Front Pharmacol 8:1–10. https://doi.org/10.3389/fphar.2017.00168

Russell AE, Doll DN, Sarkar SN, Simpkins JW (2016) TNF-α and beyond: rapid mitochondrial dysfunction mediates TNF-α-induced neurotoxicity. J Clin Cell Immunol 7:467. https://doi.org/10.4172/2155-9899.1000467

Sanford AM (2018) Lewy body dementia. Clin Geriatr Med 34:603–615. https://doi.org/10.1016/j.cger.2018.06.007

Sarkar S, Jun S, Rellick S et al (2016) Expression of microRNA-34a in Alzheimer's disease brain targets genes linked to synaptic plasticity, energy metabolism, and resting state network activity. Brain Res 1646:139–151. https://doi.org/10.1016/j.brainres.2016.05.026

Schipper HM, Maes OC, Chertkow HM, Wang E (2007) MicroRNA expression in Alzheimer blood mononuclear cells. Gene Regul Syst Biol 1:263–274. https://doi.org/10.4137/GRSB.S361

Schlaudraff F, Gründemann J, Fauler M et al (2014) Orchestrated increase of dopamine and PARK mRNAs but not miR-133b in dopamine neurons in Parkinson's disease. Neurobiol Aging 35:2302–2315. https://doi.org/10.1016/j.neurobiolaging.2014.03.016

Schneider R, McKeever P, Kim T et al (2018) Downregulation of exosomal miR-204-5p and miR-632 as a biomarker for FTD: a GENFI study. J Neurol Neurosurg Psychiatry 89:851–858. https://doi.org/10.1136/jnnp-2017-317492

Sehar U, Rawat P, Reddy AP et al (2022) Amyloid Beta in aging and Alzheimer's disease. Int J Mol Sci 23:12924. https://doi.org/10.3390/ijms232112924

Sethi P, Lukiw WJ (2009) Micro-RNA abundance and stability in human brain: specific alterations in Alzheimer's disease temporal lobe neocortex. Neurosci Lett 459:100–104. https://doi.org/10.1016/j.neulet.2009.04.052

Sezgin M, Bilgic B, Tinaz S, Emre M (2019) Parkinson's disease dementia and Lewy body disease. Semin Neurol 39:274–282. https://doi.org/10.1055/s-0039-1678579

Sheinerman KS, Tsivinsky VG, Crawford F et al (2012) Plasma microRNA biomarkers for detection of mild cognitive impairment. Aging 4:590–605. https://doi.org/10.18632/aging.100486

Shi Y, Huang F, Tang B et al (2014) MicroRNA profiling in the serums of SCA3/MJD patients. Int J Neurosci 124:97–101. https://doi.org/10.3109/00207454.2013.827679

Simon DK, Tanner CM, Brundin P (2020) Parkinson disease epidemiology, pathology, genetics, and pathophysiology. Clin Geriatr Med 36:1–12. https://doi.org/10.1016/j.cger.2019.08.002

Solis E, Hascup KN, Hascup ER (2020) Alzheimer's disease: the link between amyloid-β and neurovascular dysfunction. J Alzheimers Dis 76:1179–1198. https://doi.org/10.3233/JAD-200473

Soto M, Iranzo A, Lahoz S et al (2022) Serum MicroRNAs predict isolated rapid eye movement sleep behavior disorder and Lewy body diseases. Mov Disord 37:2086–2098. https://doi.org/10.1002/mds.29171

Srinivasan SR, Shakkottai VG (2019) Moving towards therapy in SCA1: insights from molecular mechanisms, identification of novel targets, and planning for human trials. Neurotherapeutics 16:999–1008. https://doi.org/10.1007/s13311-019-00763-y

Starhof C, Hejl A, Heegaard NHH et al (2019) The biomarker potential of cell-free microRNA from cerebrospinal fluid in Parkinsonian syndromes. Mov Disord 34:246–254. https://doi.org/10.1002/mds.27542

Tabrizi SJ, Flower MD, Ross CA, Wild EJ (2020) Huntington disease: new insights into molecular pathogenesis and therapeutic opportunities. Nat Rev Neurol 16:529–546. https://doi.org/10.1038/s41582-020-0389-4

Tan L, Yu J-T, Liu Q-Y et al (2014) Circulating miR-125b as a biomarker of Alzheimer's disease. J Neurol Sci 336:52–56. https://doi.org/10.1016/j.jns.2013.10.002

Tatura R, Kraus T, Giese A et al (2016) Parkinson's disease: SNCA-, PARK2-, and LRRK2- targeting microRNAs elevated in cingulate gyrus. Parkinsonism Relat Disord 33:115–121. https://doi.org/10.1016/j.parkreldis.2016.09.028

Thomas CJ, Mirza RG, Gill MK (2021) Age-related macular degeneration. Med Clin North Am 105:473–491. https://doi.org/10.1016/j.mcna.2021.01.003

Toivonen JM, Manzano R, Oliván S et al (2014) MicroRNA-206: a potential circulating biomarker candidate for amyotrophic lateral sclerosis. PLoS One 9:e89065. https://doi.org/10.1371/journal.pone.0089065

Ulańczyk Z, Sobuś A, Łuczkowska K et al (2019) Associations of MicroRNAs, angiogenesis-regulating factors and CFH Y402H polymorphism—an attempt to search for systemic biomarkers in age-related macular degeneration. Int J Mol Sci 20:5750. https://doi.org/10.3390/ijms20225750

Van Den Berg MMJ, Krauskopf J, Ramaekers JG et al (2020) Circulating microRNAs as potential biomarkers for psychiatric and neurodegenerative disorders. Prog Neurobiol 185:101732. https://doi.org/10.1016/j.pneurobio.2019.101732

Volovikov EA, Davidenko AV, Lagarkova MA (2020) Molecular mechanisms of Spinocerebellar ataxia type 1. Russ J Genet 56:129–141. https://doi.org/10.1134/S102279542002012X

Vrabec K, Boštjančič E, Koritnik B et al (2018) Differential expression of several miRNAs and the host genes AATK and DNM2 in leukocytes of sporadic ALS patients. Front Mol Neurosci 11:106. https://doi.org/10.3389/fnmol.2018.00106

Walker Z, Costa DC, Walker RWH et al (2004) Striatal dopamine transporter in dementia with Lewy bodies and Parkinson disease: a comparison. Neurology 62:1568–1572. https://doi.org/10.1212/01.WNL.0000123248.39847.1D

Wang W-X, Rajeev BW, Stromberg AJ et al (2008) The expression of MicroRNA miR-107 decreases early in Alzheimer's disease and May accelerate disease progression through regulation of β-site amyloid precursor protein-cleaving enzyme 1. J Neurosci 28:1213–1223. https://doi.org/10.1523/JNEUROSCI.5065-07.2008

Wang G, Huang Y, Wang L-L et al (2016a) MicroRNA-146a suppresses ROCK1 allowing hyperphosphorylation of tau in Alzheimer's disease. Sci Rep 6:26697. https://doi.org/10.1038/srep26697

Wang L, Lee A, Wigg J et al (2016b) miR-126 regulation of angiogenesis in age-related macular degeneration in CNV mouse model. Int J Mol Sci 17:895. https://doi.org/10.3390/ijms17060895

Wang R, Yang Y, Wang H et al (2020) MiR-29c protects against inflammation and apoptosis in Parkinson's disease model in vivo and in vitro by targeting SP1. Clin Exp Pharmacol Physiol 47:372–382. https://doi.org/10.1111/1440-1681.13212

Wang T, Long Q, Hu Y et al (2023) miR-181c-5p suppresses neuronal pyroptosis via NLRP1 in Alzheimer's disease. Behav Brain Res 447:114387. https://doi.org/10.1016/j.bbr.2023.114387

Wojciechowska M, Krzyzosiak WJ (2011) CAG repeat RNA as an auxiliary toxic agent in polyglutamine disorders. RNA Biol 8:565–571. https://doi.org/10.4161/rna.8.4.15397

Yang P, Lin G, Wang M et al (2022a) Long non-coding RNA ANRIL interacts with microRNA-34a and microRNA-125a, and they all correlate with disease risk and severity of Parkinson's disease. J Clin Lab Anal 36:e24037. https://doi.org/10.1002/jcla.24037

Yang Y-L, Lin T-K, Huang Y-H (2022b) MiR-29a inhibits MPP + – induced cell death and inflammation in Parkinson's disease model in vitro by potential targeting of MAVS. Eur J Pharmacol 934:175302. https://doi.org/10.1016/j.ejphar.2022.175302

Yang JM, Kim SJ, Park S et al (2023) Exosomal miR-184 in the aqueous humor of patients with central serous chorioretinopathy: a potential diagnostic and prognostic biomarker. J Nanobiotechnology 21:242. https://doi.org/10.1186/s12951-023-02019-6

Yılmaz ŞG, Erdal ME, Özge AA, Sungur MA (2016) Can peripheral MicroRNA expression data serve as Epigenomic (upstream) biomarkers of Alzheimer's disease? OMICS J Integr Biol 20:456–461. https://doi.org/10.1089/omi.2016.0099

Younes K, Miller BL (2020) Frontotemporal dementia neuropathology, genetics, neuroimaging, and treatments. Psychiatr Clin North Am 43:331–344. https://doi.org/10.1016/j.psc.2020.02.006

Zhang J, Wang J, Zheng L et al (2017a) miR-25 mediates retinal degeneration via inhibiting ITGAV and PEDF in rat. Curr Mol Med 17:359–374. https://doi.org/10.2174/1566524018666171205122540

Zhang X, Yang R, Hu B-L et al (2017b) Reduced circulating levels of miR-433 and miR-133b are potential biomarkers for Parkinson's disease. Front Cell Neurosci 11:170. https://doi.org/10.3389/fncel.2017.00170

Zhao N, Jin L, Fei G et al (2014) Serum microRNA-133b is associated with low ceruloplasmin levels in Parkinson's disease. Parkinsonism Relat Disord 20:1177–1180. https://doi.org/10.1016/j.parkreldis.2014.08.016

Zhou H, Zhang R, Lu K et al (2016) Deregulation of miRNA-181c potentially contributes to the pathogenesis of AD by targeting collapsin response mediator protein 2 in mice. J Neurol Sci 367:3–10. https://doi.org/10.1016/j.jns.2016.05.038

Chapter 8
MicroRNAs and Immune-Related Disorders

Hoda Y. Abdallah, Rasha Mohamed Helmy Elkaffas, and Sarah Ahmed Aglan

Abstract Autoimmune disorders have a high prevalence worldwide with several presentations causing considerable morbidity and mortality. Most autoimmune disorders exhibit a multifactorial nature that often involves both genetic factors and environmental factors. Among the latest causes of autoimmune disorders are epigenetic mechanisms, especially those involving miRNAs. MiRNAs showed critical activities in regulating immune reactions and immune cell development. MiRNAs that are highly expressed by immune system cells—are called "immune-miRNAs." These miRNAs play pivotal roles in immune system responses and include miR-16, miR-17/92 cluster, miR-21, miR-124, miR-142, miR-146a, miR-155, miR-181, and miR-223. This chapter presents the key roles of immune-miRNAs in the regulation of the immune system, their contributions in autoimmune diseases, and their functions in major autoimmune diseases including dermatological disorders, rheumatological diseases, gastrointestinal related diseases, disorders of the nervous system, and endocrine disorders.

Keywords Immune-miRNAs · Psoriasis · Vitiligo · Lichen planus · Systemic lupus erythematosus · Rheumatoid arthritis · Celiac disease · Inflammatory bowel diseases · Endocrine autoimmune disorders

H. Y. Abdallah (✉)
Medical Genetics Unit, Histology and Cell Biology Department, Faculty of Medicine, Suez Canal University, Ismailia, Egypt

Center of Excellence in Molecular and Cellular Medicine, Faculty of Medicine, Suez Canal University, Ismailia, Egypt
e-mail: hoda_ibrahim1@med.suez.edu.eg

R. M. H. Elkaffas
Clinical and Chemical pathology department, Faculty of Medicine, Cairo University, Cairo, Egypt

S. A. Aglan
Chemical Pathology Department, Medical Research Institute, Alexandria University, Alexandria, Egypt

L. M. Vaschetto (ed.), *miRNAs, Human Health and Diseases*, Epigenetics and Human Health 13, https://doi.org/10.1007/978-3-031-64788-8_8

List of Abbreviations

AC	Adenyl cyclase
AchRs	Acetylcholine receptors
AE2	Anion exchanger 2
AHSCT	Autologous hematopoietic stem cell transplantation
AIH	Autoimmune hepatitis
AILDs	Autoimmune liver diseases
AP-1	Activator Protein 1
APAF1	Apoptotic peptidase activating factor 1
APCs	Antigen presenting cells
BACH1	BRCA1-interacting protein C-terminal helicase 1
BAFF-R	B cell-activating factor receptor
BCL2	B cell CLL/lymphoma 2
BCL6	B cell lymphoma-6 protein
BCR	B cell antigen receptor
BD	Behçet Disease
BLIMP-1	B Lymphocyte-induced maturation protein-1
C/EBPβ	CCAAT/enhancer-binding protein-β
CASP10	Caspase 10
CD	Cluster differentiation
CD	Celiac Disease
CIS	Clinically isolated syndrome
CLDN1	Claudin-1
CSF1R	Colony stimulating factor 1 receptor
CYLD	Lysine deubiquitinase
DCs	Dendritic cells
Dicer	Double-Stranded RNA-Specific Endoribonuclease
DNMT1	DNA methyltransferase 1
Drosha	Double-Stranded RNA-Specific Endoribonuclease
EBF1	Early B cell factor 1
EGFR	Epidermal growth factor receptor
ERAP1	Endoplasmic reticulum aminopepetidase1
Ets-1	V-ets avian erythroblastosis virus E26 oncogene homolog 1
FGF9	Fibroblast growth factor 9
FLSs	Synovial like fibroblasts
FOXO3	Forkhead box O3
FOXP1	Forkhead box P1
FOXP3	Forkhead box P3
GARP	Glycoprotein A repetitions predominant
GC	Germinal center
GIT	Gastrointestinal tract
HT	Hashimoto thyroiditis
IBD	Inflammatory bowel diseases

IGF2BP1	Insulin-like growth factor 2 mRNA-binding protein 1
IKKα	Inhibitor of Nuclear Factor Kappa B Kinase
IL	Interleukin
INF-γ	Interferon-gamma
IRAK-1	IL-1 receptor-associated kinase 1
IRF3	Interferon regulatory factor 3
ITP	Immune Thrombocytopenia
JAK	Janus kinase
KLF4	Krueppel-like factor 4
KLRG1	Killer cell lectin -like receptor subfamily G
LCs	Skin- resident Langerhans cells
LPS	Lipopolysaccharide
LrP4	Lipoprotein receptor related protein-4
LXR	Liver X receptors
LYN	Lck/Yes novel tyrosine kinase
MAD2L1	Mitotic arrest deficient-like 1
MAPK	Mitogen-activated protein kinase
MBP	Myelin basic protein
mDC	Myeloid dentritic cells
MDDC	Monocyte-derived dendritic cells
MEF2C	Myocyte enhancer factor 2C
MG	Myasthenia gravis
MHC	Major histocompatibility complex
microRNAs	miRNAs
MKK4	Mitogen-activated protein kinasekinase 4
MMP-1	Matrix metalloproteinase 1
MP	Macrophage polarization
MS	Multiple sclerosis
MSC	Mesenchymal stem cells
MuSK	Muscle specific kinase
MyD88	Myeloid differentiation primary response 88
MZB	Marginal zone B cells
NETs	Neutrophil extracellular traps
NF-κB	Nuclear factor kappa B
NLRP	NLR Family Pyrin Domain Containing
NOX	Neutrophils, activation of NADPH oxidase
NR4A3	Nuclear receptor subfamily 4 group A member 3
NUMB	Numb homolog
PAD4	Protein-arginine deiminase type 4
PAMPs	Pathogen associated molecular patterns
PARP-1	Poly-adenosine diphosphatase-ribose polymerase-1
PBC	Primary biliary cholangitis
pDC	Plasmacytoid dendritic cells
PDCD4	Programmed Cell Death 4
PIAS3	Protein inhibitor of activated STAT3

PMN	Polymorphonuclear cells
PPARγ	Peroxisome proliferator-activated receptor
PPMS	Primary progressive MS
PPP6C	Protein phosphatase 6 catalytic subunit
PRRs	Pattern recognition receptors
PSC	Primary sclerosing cholangitis
PTEN	Phosphatase and tensin homolog
RA	Rheumatoid arthritis
RANKL	Receptor activator of NF-κB ligand
RANTES	Regulated on activation normal T cell expressed and secreted
RBPJ	Recombination signal binding protein for immunoglobulin kappa J region
RNF114	Ring finger 114 protein
RRMS	Relapsing-remitting MS
RUNX1	Runt-related transcription factor 1
SHIP1	Src homology-2 domain-containing inositol 5-phosphatase 1
SLAM	Signaling lymphocyte activation molecule
SLE	Systemic lupus erythematosus
SMAD4	Mothers against decapentaplegic homolog 4
SMAD7	SMAD family member 7
SOCS1	Suppressor of cytokine signaling 1
SPMS	Secondary progressive MS
STAT	Signal transducer and activator of transcription
STK40	Serine/threonine kinase 40
T1DM	Type 1 Diabetes Mellitus
TAB2	Transforming growth factor β-activated kinase 1-binding protein 2
TCR	T cell receptors
Teff	T effector cells
Tfh	T follicular helper cells
Tg	Thyroglobulin
TGF-β	Transforming growth factor beta
TGFBR2	Transforming growth factor-beta receptor type 2
Th	T helper cells
TIRs	Toll-like and interleukin-1 receptors
TLR	Toll like receptor
TNF	Tumor necrosis factor
TPO	Thyroperoxidase
TPO-RA	Thrombopoietin receptor agonists
TRADD	Tumor necrosis factor receptor 1 (TNFR1)-associated death domain protein
TRAF6	TNF receptor-associated factor 6
Treg	T regulatory cells
TRIF	TIR domain-containing adaptor inducing interferon-β
USP2	Ubiquitin-specific peptidase
ZO-1	Zonula occludens

8.1 Introduction

The immune system's pathogenic state, in which it is unable to identify self-antigens and tissues and mounts an immune response against the host's own cells and tissues, is what characterizes autoimmune disease (Marshall et al. 2018). This pathogenic state occurs as a result of pathological stimulation or physiological immune reactions in immune cells, especially autoreactive T cells, or immunological components, such as autoantibodies (Nüssing et al. 2020). In the normal physiological body state, immune self-components are tolerated by a variety of regulatory mechanisms which are activation-induced cell death, anergy, clonal ignorance, and a network of peripheral regulatory cells. These regulatory mechanisms stop the activation of self-reactive lymphocytes which are the factors that cause the central and peripheral immune tolerance (Catalán et al. 2021). By means of cell-to-cell contact with autoreactive T or B cells or soluble mediators such anti-inflammatory cytokines, regulatory cells and molecules aid in the regulation of autoimmunity (Sogkas et al. 2021). There are several levels of autoimmunity: a low level that is considered "physiological" and is necessary for lymphocyte selection and immune system homeostasis; an intermediate level that is characterized by circulating autoantibodies and small tissue infiltrates that do not have a clinical impact; and a pathogenic level of autoreactivity that is linked to immune system-mediated organ damage (Chandan et al. 2020).

The immune system is a harmonious complementary group of cells and chemicals that protect our bodies from pathogens and harm. It acts through two "lines of defense," namely, innate and adaptive immunities, where the innate is the first line of defense. An eminent difference between the two lines is that the adaptive immunity is antigen-specific and has a memory for antigens, unlike innate immunity (Marshall et al. 2018). While the innate immunity is the first line of defense, it is non-specific and depends on "pattern recognition receptors (PRRs)" such as toll like receptor (TLR) (Chandan et al. 2020) for pathogens that share specific structure known as PAMPs or "pathogen associated molecular patterns," such as lipopolysaccharides or in case of viral infection, double-stranded RNA. Upon antigen recognition, an intracellular cascade of signaling pathways is activated, leading to the induction of inflammation. These pathways include Nuclear Factor Kappa B (NF-κB), interferon regulatory factor 3 (IRF3), and mitogen-activated protein kinase (MAPK) signaling (Chandan et al. 2020).

The innate immune system has two main functions: first, it produces cytokines and chemokines (signaling molecules) like tumor necrosis factor (TNF), interleukin-1 (IL-1), interleukin-6 (IL-6), and interferon-gamma (IFN-γ) to recruit immune cells. Second, it directly eliminates pathogens through phagocytosis (engulfing and destroying them) and antigen presentation (preparing them for adaptive immune response). This system comprises various cell types, including macrophages, neutrophils, dendritic cells, natural killer cells, mast cells, eosinophils, basophils, and innate lymphoid cells. Antigen presenting cells (APCs), such as dendritic cells and macrophages, exert their action through major histocompatibility complex (MHC), a protein on their surface, which presents fragments of the foreign antigen to T cells (Marshall et al. 2018).

Macrophages can adapt to their environment by polarizing into two main types: classically activated M1 macrophages and alternatively activated M2 macrophages. Lipopolysaccharide (LPS), interferon-gamma (IFN-γ), and tumor necrosis factor alpha (TNFα) activate M1 macrophages through pathways like NF-κB, Signal transducer and activator of transcription 1 (STAT1), STAT5, Interferon regulatory factor 3 (IRF3), and IRF5. M1 macrophages exhibit a pro-inflammatory and antimicrobial phenotype, meaning they promote inflammation and kill pathogens but can also damage tissues. In contrast, interleukin-4 (IL-4) and interleukin-13 (IL-13) primarily activate M2 macrophages through STAT3, STAT6, and peroxisome proliferator-activated receptor (PPARγ) pathways. M2 macrophages have an anti-inflammatory and tissue repair phenotype. M2 functions as immunoregulator, anti-inflammatory, angiogenetic, and tumor forming phenotype (Yao et al. 2019).

Unlike the innate immune system, the adaptive immune system is highly specific. It can distinguish between the body's own molecules (self) and foreign molecules (non-self), which helps prevent attacks on healthy tissues (autoreactivity). The two main cell types of the adaptive immune system are B cells and T cells. B cells differentiate into plasma cells upon activation, which produce antibodies to target specific antigens. T cells, on the other hand, become activated upon interaction with antigen-presenting cells (APCs) that display processed antigen fragments on their surface. Activated T cells differentiate into two main types: CD8+ cytotoxic T cells, which eliminate infected cells, and CD4+ helper T cells, which orchestrate the immune response by regulating the activity of other immune cells rather than directly killing pathogens. (Marshall et al. 2018). Th response involves the activity of several Th subsets: Th1, Th2, Th17, and Regulatory T cells (Tregs). Th1 produces IFN-γ activating macrophages, Th2 generates cytokines activating mast cells and promoting IgE production from B cells, and Th17 secretes IL-17, which is associated with inflammation in chronic diseases. Tregs, a specialized subset of CD4+ T lymphocytes, play a critical role in maintaining immune tolerance by suppressing immune responses that could attack healthy tissues (autoimmunity). Dysregulation in the Treg population is thought to be a major culprit in autoimmune and inflammatory diseases (Marshall et al. 2018). Tregs exert their suppressive effects through multiple mechanisms, including secreting immunosuppressive cytokines, directly suppressing other immune cells through cell contact, and even modifying or destroying antigen-presenting cells (APCs). An intricate interplay between the transcription factor FOXP3 and interleukin-2 (IL-2) is crucial for Treg development and regulation. IL-2 induces the expression of FOXP3, the master regulator of Treg function. Subsequently, FOXP3 regulates CD25 expression, which comprises the alpha chain subunit of the IL-2 receptor, thus providing a positive feedback loop for maintaining Tregs survival (Consonni et al. 2021).

Autoimmune disorders represent a significant global health burden, affecting an estimated 7–9% of the population (Angum et al. 2020). These diseases exhibit a female predominance and can manifest at any point in life, from childhood to adulthood. Furthermore, they are associated with considerable morbidity and mortality

(Angum et al. 2020). Importantly, autoimmune diseases encompass a diverse range of clinical presentations, now categorized into distinct disease phenotypes based on the specific tissues and antigens targeted by the immune system (Marshall et al. 2018). Despite this heterogeneity, the most common classification system groups them based on the affected tissue and clinical manifestations, categorizing them as organ-specific or systemic autoimmune diseases (Fig. 8.1).

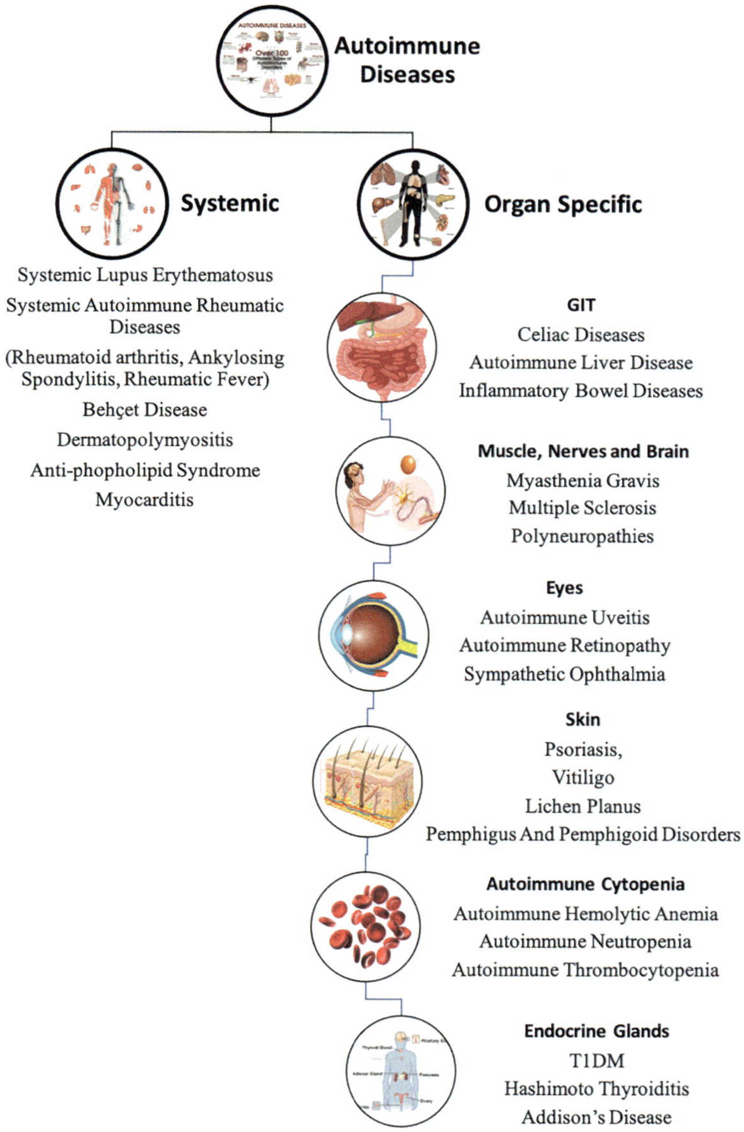

Fig. 8.1 Classification of autoimmune diseases based on the affected organ. GIT, gastrointestinal tract; T1DM, type 1 diabetes mellitus

Many factors may contribute to aberrant autoimmunity, and thus autoimmune disorders can exhibit a multifactorial nature that often involves both genetic and environmental factors (Gutierrez-Arcelus et al. 2016). In addition, a common feature of autoimmune diseases is their tendency to appear in families, suggesting genetic susceptibility (Zakharova et al. 2019). The most potent genetic influence on susceptibility to autoimmunity is MHC, which is known to affect susceptibility to a variety of autoimmune diseases. Since MHC works at the level of antigen presentation by APCs for T cell recognition, an antigen cannot become an autoantigen or elicit a response if it is not presented by the host's MHC (Jurewicz and Stern 2019; Ishina et al. 2023). In addition, lower expression levels of the proper MHC molecules may contribute to an increased threshold for recognition of antigens by immune cells and immune responses (Khan and Wang 2020).

Environmental factors have recently gained attention for their roles in triggering autoimmunity. These factors include infection, gut microbiota, and physical or environmental agents (Gutierrez-Arcelus et al. 2016; Hou et al. 2022). Current data suggest that environmental factors contribute up to 70% to the loss of tolerance (Hou et al. 2022). The gut microbiota, for example, can be influenced by the motility of the gastrointestinal tract, smoking, and intake of alcohol and pharmaceutical medications, including antibiotics and nonsteroidal anti-inflammatory drugs (Hou et al. 2022). On other hand, the gut microbiome interacts not only with the host but also with other organisms and environmental factors and acts as a barrier to enteropathogenic infections. The function of the gut microbiome may be disrupted by malnutrition or perturbations in immune system function (Mazzone et al. 2019).

Emerging research suggests epigenetics, characterized by meiotically and mitotically heritable alterations in gene expression that can be induced by environmental factors without altering the underlying DNA sequence, as a potential contributor to autoimmune disorders (Surace and Hedrich 2019). Epigenetic pathways are known to play a critical role in the development and function of the immune system. A growing body of research is exploring the potential link between epigenetic modifications and the development of autoimmune diseases (Bure et al. 2022; Wu et al. 2023). Epigenetic changes, such as DNA methylation, histone modifications, and noncoding RNA pathways, which include microRNAs (miRNAs), are involved in the pathogenesis of autoimmune diseases through regulatory gene expression mechanisms (Scherm and Daniel 2020).

Specifically, with the increased recognition that miRNAs have critical activities in regulating immune reactions and immune cell development, current investigations have found connections between miRNAs and autoimmunity (Hippen et al. 2018; Kmiołek and Paradowska-Gorycka 2022). Currently, it is well known that dysregulated miRNA expression may lead to immune tolerance breakdown and the development of autoimmune diseases. Disruption of miRNA synthesis by depleting Double-Stranded RNA-Specific Endoribonuclease (Drosha) or Double-Stranded RNA-Specific Endoribonuclease (Dicer) enzymes in Tregs has shown to trigger the complete loss of immune regulatory function of Tregs, ultimately resulting in the development of inflammatory disease and fatal autoimmunity (O'connell et al. 2010).

In this chapter, we will present the key roles of miRNAs in the regulation of the immune system, their functions in autoimmune diseases, and the role of miRNAs in each of the major autoimmune diseases.

8.2 MiRNAs in the Immune Response

The immune response can be defined as the body's ability to recognize and react towards different pathogens through its two main lines of defense. Both innate and adaptive immune responses are highly regulated, and several studies have revealed the role of miRNAs in regulating both immune responses. Supported by recent data collected in human and mice samples, several miRNAs have shown to exhibit highly specific expression patterns in organs associated with the immune system, which emphasizes the role of miRNA in regulating immune responses through its role during development, maturation, activation, proliferation, functioning, and survival of different immune cells (O'connell et al. 2010).

8.2.1 MiRNAs and Innate Immune Response

As previously mentioned, innate immune response is a rapid, non-specific inflammatory response predominantly mediated by myeloid cells such as macrophages, dendritic cells, monocytes and neutrophils. The recognition of pathogens commonly involves tissue macrophages and dendritic cells through the binding of PRRs with PAMPs, from which the immune response evolves through signals from PRRs. The best characterized PRRs are the Toll-like and interleukin-1 receptors (TIRs), which are divided into TLRs (TLR-1 to TLR-11) and the interleukin (IL)-1 receptors. Investigations using animal models and different immune cells have revealed that the activation of TIRs and the TNFα receptor may lead to the expression of several miRNAs. These miRNAs include, among others, let-7, miR-9, miR-99b, let-7e, miR-125a, miR-132, miR-146a, miR-146b, miR-155, miR-187, and miR-223 (Tsitsiou and Lindsay 2009).

8.2.1.1 Regulatory Roles of miRNAs on Macrophages and Monocytes

Macrophages and monocytes are essential cells of the innate immune response that initiate the production of cytokines when exposed to any infectious and inflammatory stimuli. The transformation of different macrophage phenotypes whether the classically activated macrophages and its pro-inflammatory activity or the alternatively activated macrophages and its anti-inflammatory reactions is reflected in various diseases or inflammatory conditions. It has been shown that miRNAs play a critical role in the regulation of macrophage polarization (MP), where several miRNAs can be induced by TLRs signaling in macrophages and target mRNAs encoding key molecules involved in MP. Furthermore, it has been highlighted that miRNAs can affect and modify all stages of the macrophage life cycle (Roy 2016).

In response to a broad range of TLRs and cytokine signals, miR-155 is strongly induced in macrophages by NF-κB and activator protein 1 (AP-1) (Mehta and Baltimore 2016). It is among the first miRNAs to be identified to play diverse roles in inflammation. Different studies had revealed that miR-155 may exert both positive and negative effects on LPS signaling. It enhances classical macrophage activation by repression of negative regulators of pro-inflammatory responses, including suppressor of cytokine signaling 1 (SOCS1), Src homology-2 domain-containing inositol 5-phosphatase 1 (SHIP1), and B cell lymphoma-6 protein (BCL6) (Roy 2016). On another note, it has been shown that miR-155 may directly target transcripts coding for several pro-apoptotic and anti-inflammatory proteins involved in LPS signaling, such as Fas-associated death domain protein, IKKε, and receptor-interacting serine-threonine kinase 1, thus acting as a negative regulator of overwhelming inflammatory responses (Wu et al. 2016). In addition to its role in inflammation, miR-155 drives myeloid cell differentiation presumptively by downregulating PU.1, BRCA1-interacting protein C-terminal helicase 1 (BACH1; known also as FANCJ), colony stimulating factor 1 receptor (CSF1R), CCAAT/enhancer-binding protein-β (C/EBPβ), and other regulators of myeloid cell differentiation (Mehta and Baltimore 2016).

Similar to miR-155, which exerts negative regulation on excessive inflammation, miR-147 inhibits inflammatory signaling triggered by TLR2, TLR3, and TLR4 through both MyD88 and TRIF pathways (Curtale et al. 2019). Additionally, miR-92a targets Mitogen-activated protein kinase kinase 4 (MAKK4), acting as another inhibitor of the TLR4-induced inflammatory response in macrophages. Furthermore, miR-24 overexpression promotes an anti-inflammatory phenotype. MiR-24 acts as a negative regulator of macrophage classical activation by LPS and enhances alternative activation under conditions of polarization and plasticity. Fordham et al. reported that miR-24 mediated inhibition of LPS-induced cytokine secretion is dependent upon macrophage activation state at the point of stimulation, which may be due to the degree to which p110δ is involved in the intracellular signaling pathway/s that transduce receptor ligation into cytokine induction (Fordham et al. 2015). Further studies evidenced various miRNAs implicated in macrophages immune response, namely miR-146, miR-146a, miR-125, miR-21, and miR-223 (Li et al. 2010; Chen et al. 2013; Curtale et al. 2019).

8.2.1.2 Regulatory Roles of miRNAs on Dendritic Cells

Dendritic cells (DCs) are the most efficient antigen-presenting cells of the immune system and they are required for induction and evolution of both types of immune response, as well as maintaining self-tolerance (Al-Ashmawy 2018). Originating from a common DC progenitor in the bone marrow, five different sets of human circulating DCs have been identified: two distinct sets of conventional DCs (cDCs); $CD11b^+$-like and $CD8\alpha^+$-like that are responsible of priming of $CD4^+$ and $CD8^+$ T cells, respectively; skin-resident Langerhans cells (LCs) that probably have immunoregulatory role, plasmacytoid dendritic cells (pDCs) that secrete type I IFN in response to viruses, and monocyte-derived DCs (MDDCs) that has a microbicidal activity and are induced in response to infections (Turner et al. 2011).

MiR-155 expression is essential for DC activation, probably through the regulation of SOCS1 protein levels. It has been evidenced that DCs lacking miR-155 are defective at inducing T cell activation, and hence, miR-155 seems to act as a positive modulator of inflammatory cytokine production in DCs. Particularly, it upregulates IL-1 signaling in MDDCs by regulating the levels of TAK1-binding protein 2 (TAB2) (Testa et al. 2017).

MiR-146a and miR-146b are naturally expressed in DCs at low levels, but during DCs differentiation from monocytes, they are markedly upregulated but do not affect DCs maturation. Further, in mature DCs, these miRNAs modulates the production of pro-inflammatory cytokines. Lastly, through targeting TNFR-associated factor 6 (TRAF6) and IL-1R-associated kinase 1(IRAK-1) in DCs, both of these miRNAs increases DC spontaneous apoptosis (Turner et al. 2011). Therefore, both miR-146a and miR-146b show a distinctive role in the regulation of dendritic cell survival and inflammatory activation (Testa et al. 2017). Other miRNAs involved in regulating DCs development are miR-21 and miRNA-34, which upon their inhibition, an inhibition of MDDC differentiation was proved. This is through their translational repression of Wnt-1 and Jagged-1 proteins. These two proteins are reported to be lowered during MDCC differentiation; thus, upon their repression, miR-21 and miRNA-34 permit MDDC differentiation (Turner et al. 2011; Chandan et al. 2020).

8.2.1.3 Regulatory Roles of miRNAs on Granulocytes

MiRNAs play an essential role in the proliferation and functioning of granulocytes, with a special mention for miR-223, miR-155, and miR-27a. MiR-155 has been implicated in the regulation of SHIP1 levels in granulocyte development and macrophage proliferation upon inflammation (Wu et al. 2016). MiR-223 regulates granulocyte differentiation through KMef2c, a transcription factor necessary for myeloid progenitor proliferation, and control of neutrophil cell function induced by LPS through MYD88/NF-κB regulation (Johnnidis et al. 2008). miR-27a plays a role in granulocytic differentiation and it has been reported that high levels of pri-miR-27a in granulocytes (Gilicze et al. 2014). Other miRNAs influencing granulocyte fate include miR-21 and miR-196b promoters, which directly binds to the zinc finger protein GFI1, a transcriptional repressor that is crucial for granulocyte differentiation, as well as miR-130a, also involved in granulocyte differentiation by direct modulation of SMAD4 (Mehta and Baltimore 2016).

8.2.2 *MiRNAs and the Adaptive Immune Response*

The two main classes of adaptive immune responses, namely the antibody responses (antibody production) and cell-mediated immune responses (cytotoxic effector cells), are accomplished by the activation and proliferation of T and B cells in response to infections. It has been emphasized that miRNAs play a critical role in modulating adaptive immunity by regulating the development, activation, survival, and clonal expansion of T and B cells (Chen et al. 2018a, b).

8.2.2.1 Regulatory Roles of miRNAs on T Cells

Mature peripheral T cells consist primarily of three distinct subpopulations: helper T cells (CD4+ T cells), cytotoxic T cells (CD8+ T cells), and regulatory T cells (Treg cells). These T cell subsets play critical roles in the immune response. Following antigen presentation by MHC+ antigen-presenting cells (APCs), a series of signaling cascades are triggered, leading to the development and activation of specific T cell populations, ultimately resulting in targeted inflammatory responses (Min 2018). Specific miRNA expression profiles have been reported in different T cell subsets and stages of development, revealing the regulatory role of miRNAs on these signaling pathways. As well as disruption in the biogenesis of miRNAs can cause conditional removal of DICER in the early developmental stage, which leads to reduced T cell count, disturbed differentiation of T-helper cells and cytokine production, along with decreased survival of αβ-expressing thymocytes and increased apoptosis are also observed in peripheral T cells (Rodríguez-Galán et al. 2018). MiR-155 plays an essential role in CD4+ T cell differentiation, and it has been observed that miR-155 overexpression is associated with T cell differentiation into Th1 cells. It is also involved in the development of effector- or central-memory cells from activated CD8+ T cells and also plays a role in Treg homeostasis (Baumjohann and Ansel 2013). MiR-142 appears to downregulate glycoprotein A repetitions, predominant (GARP) expression, which suppresses the proliferation of CD25+ helper T. Also, miR-29 is shown to downregulate the T helper type 1 activation via targeting IFN-γ (Ma et al. 2011). Other miRNAs, such as miR-150 and members of the let-7 family, have been involved in the development of effector- or central-memory cells from activated CD 8+ T cells, driven by the function of IL-2 or IL-15 (Almanza et al. 2010).

Treg cells limit immune responses and prevent autoimmunity. MicroRNAs play an essential role in their proliferation, differentiation, and functions. MiR-155 increases the differentiation of Treg cells by activating the transcription of FOXP3, a marker crucial for their suppressive function (Lv et al. 2019). Yao et al. reported that miR-155 plays a role in Treg cell differentiation by regulating Janus kinase/signal transducer and activator of transcription (JAK/STAT) via SOCS1 targeting, which is a critical negative regulator of the JAK/STAT signaling pathway (Yao et al. 2012). MiR-146a is essential for the Treg-mediated suppression of the IFNγ dependent Th1 response, through the downregulation of the Stat1 expression (Liu et al. 2020). MiR-10a plays a pivotal role in Treg cell differentiation and function. Its expression is induced by TGF-β or retinoic acid, which restricts Treg differentiation into follicular helper T cells via targeting transcriptional suppressor Bcl-6 and co-suppressor Ncor2 (Takahashi et al. 2012). MiR-142-3p modulates the development and function of Tregs through targeting adenyl cyclase (AC9) and glycoprotein A repetitions predominant (GARP) through distinct mechanisms (Liu et al. 2020). Among the natural Treg miRNA signature are microRNAs miR-21, miR-31, miR-23, miR-24, miR-29a, miR-95, miR-145, miR-210, and miR-2909, which have the potential to regulate different stages of Tregs development, proliferation, and functions (Liu et al. 2020).

The differential expression of various miRNAs, including members of the miR-181 family, the miR-17/92 clusters, miR-214, let-7, miR-29, miR-125, and miR-216, has been observed in the signaling cascade downstream of T cell activation (Luan et al. 2018). MiR-17-92 increases T cell survival during development by suppressing the expression of pro-apoptotic proteins (like BIM and BCL2L11) and PTEN (Liu et al. 2014).

8.2.2.2 Regulatory Roles of miRNAs on B Cells

B cells are responsible for the antibody-mediated response. Many studies highlight the role of various miRNAs in the development and differentiation of B cells as well as their function in regulating B cell tolerance. The expression of miRNAs in naive, germinal central (GC), and post-germinal central B cells has been reported. Moreover, the deletion of DICER or Ago2 has been shown to block the transition of pro-B to pre-B cell development and reduce B cell numbers, implying a role for miRNAs in the development and maturation of B cells (Chen et al. 2018a, b).

The conversion of pro-B to the pre-B cells is blocked by the expression of miR-34a, by inhibiting the expression of fork head box transcription factor, Foxp1. The absence of Foxp1 results in a severe blockade in the development of B cells (Danger et al. 2014). Moreover, the conversion of pre-B cells to mature B cells needs high levels of miR-150 to downregulate c-Myb expression and thus guarantee normal B cell development (Tan et al. 2009). MiR-155 blocks the expression of PU.1 and SHIP1 in B cells and its deficiency leads to an increased number of GC B cells, and decreased IgG production and maturation (Danger et al. 2014). MiR-146a suppresses the development of marginal zone B (MZB) cells from immature B cells through the inhibition of the Notch2 pathway, while miR-125b allows B cell differentiation into plasma cells by targeting the transcription factors IRF4 and B Lymphocyte-induced maturation protein-1 (BLIMP-1) (King et al. 2016). B cell tolerance is initiated in the bone marrow at the immature stage of B cell development by a number of mechanisms known as central tolerance. MiR-148 plays a critical role in the regulation of B cell tolerance, and its dysregulation is an important driver of autoimmune disease. MiR-148a protects immature B cells from B cell antigen receptor (BCR) induced apoptosis by suppressing the expression of three genes implicated in autoimmunity: *GADD45a*, *PTEN,* and *BCL2l11* (which encodes the pro-apoptotic factor Bim) (Gonzalez-Martin et al. 2016).

8.3 Differential Expression of Immune-Related miRNAs in Autoimmune Disorders

Altered miRNA expression profiles can serve as potential biomarkers for diagnostic, prognostic, and therapeutic prediction purposes in autoimmune disorders (Condrat et al. 2020). Mature miRNAs are released into the circulatory system via the excretion of extracellular vesicles, exosomes, or proteins. So, they can be easily detected in any biological fluid, such as blood, urine, and saliva, thus creating miRNA signature for each disease (Jia and Wei 2020). MiRNAs highly expressed by immune system cells are called "immune-miRNAs." These miRNAs play pivotal roles in immune system responses and additional novel immune-miRNAs are now being identified (Stelzer et al. 2016). Immune-miRNAs include, among others, miR-16, miR-17/92 cluster, miR-21, miR-124, miR-142, miR-146a, miR-155, miR-181, and miR-223 (Table 8.1).

8.3.1 MiR-16

MiR-16 gene is located on chromosome 13q14.2 (Yan et al. 2019). MiR-16 directly targets Programmed Cell Death 4 (PDCD4), affecting the mitogen-activated protein kinase (MAPK), NF-κB pathways, and downstream inflammatory cytokines, and eventually suppressing the release of inflammatory macrophages (Lai et al. 2016). MiR-16 also plays a critical role in inflammation-induced insulin sensitivity via the attenuation of pro-inflammatory responses mediated by macrophages such as IL-6, TNF-α, and IFN-β, thus, improving insulin sensitivity (Lai et al. 2016).

8.3.2 MiR-17/92 Cluster

The cluster genes for miRNA-17/92 are the host genes for a group of at least six miRNAs located on chromosome 13q31.3 (miR-17, miR-18a, miR-19a, miR-20a, miR-19b-1, and miR-92a-1). These miRNAs have been involved in various cellular processes including proliferation, cell survival, angiogenesis, and differentiation (Yan et al. 2019). The cluster plays particularly critical roles in B cell development and enforces miRNA function as oncogenes, causing B cell lymphomas (Okeke and Uzonna 2019). Other critical roles include maturation and development of T-helper subsets (Th) and TREGS, regulation of T cell activation and additional functions related to the differentiation of monocytes to macrophages and pathways for macrophage activation (Roberts et al. 2021a, b).

Table 8.1 Immune-miRNAs altered in autoimmune disorders

MiRNA	Genomic Location	Function	Gene targets	Immune- effect	References
MiR-16	Chromosome 13q14.2	- Inhibit cell proliferation and invasion, - Promote cell apoptosis and suppress cell cycle progression	"miRDB Database" 1416 "miRTargetLink 2.0 Database" 2279	- ↓ Pro-inflammatory factors (IL-6 and TNF-α) mRNA expression - ↑ Anti-inflammatory factor IL-10 mRNA expression	Yan et al. 2019; Lai et al. 206; Chen et al., 2020; Kern et al. 2021
MiR-17/92 cluster	Chromosome 13q31.3	- Involved in cell survival, proliferation, differentiation, and angiogenesis	"miRDB Database" miR-17 = 1333 miR-18a = 382 miR-19a = 756 miR-19b = 755 miR-20a = 1381 miR-92a = 322 "miRTargetLink 2.0 Database" miR-17 = 1815 miR-18 = 349 miR-19a = 131 miR-19b = 132 miR-20a = 1659 miR-92a = 101	- ↑ Response of Th1 - ↑ Proliferation and survival of T-effector cell - ↑ Production of Th2 cytokine - ↓ Differentiation of Treg - ↑ Differentiation and function of Tfh cells migration - ↑ Development and survival of B-cells - ↓ Differentiation of macrophages	Yan et al. 2019; Okeke & Uzonna, 2019; Roberts et al., 2021; Chen et al.2020; Kern et al. 2021
MiR-21	Chromosome 17q23.	- Associated with exposure to inflammatory stimuli and immune cell activation	"miRDB Database" 469 "miRTargetLink 2.0 Database" 922	- ↑ T cell immune reaction towards Th2 - ↓ Proinflammatory Th1 immune response - ↓ Production of certain inflammatory cytokines by T cells and phagocytes	Yan et al., 2019, Chi et al. 2022; Soto-Heredero et al. 2023; Qin et al., 2016; Chen et al. 2020; Kern et al. 2021
MiR-124	Chromosome 8p23.1	- Involved in cell maturation, differentiation, proliferation, metastasis, apoptosis, and autophagy	"miRDB Database" 515 "miRTargetLink 2.0 Database" 101	- ↑ IL-7 generation and maintenance of memory T cells and benefits survival - ↑ IL-2, IFN-γ, and TNF-α. - Modulates T-helper cell differentiation - ↑ Activated brain microglia - Negatively regulate TNF-α production	Yan et al. 2019; Tazi et al. 2021; Roberts et al. 2021; Chen et al., 2020; Kern et al. 2021
MiR-142	Chromosome 17q22	- Has a role in the development and function of megakaryocytes, mast cells, dendritic cells, and erythrocytes	"miRDB Database" 1133 "miRTargetLink 2.0 Database" 363	- ↓ Peripheral T cells and altered B cell development - Important for mature T cell effector responses and proliferation	Yan et al. 2019; Arif et al., 2020; Wang et al. 2019, Chen et al. 2020; Kern et al. 2021
MiR-146a	Chromosome 5q33.3	- Negative regulator of the IFN pathway and immune response, reduce inflammatory cytokine production	"miRDB Database" 488 "miRTargetLink 2.0 Database" 296	- ↓ TNFα and IL-6 immune pathways - ↓ NF-κB activity - ↓ Differentiation of Th1 cell - ↓ Inflammatory response of Th1	Yan et al. 2019; Liu et al. 2018; Chen et al. 2023; Pashangzadeh et al. 2021; Chen et al. 2020; Kern et al. 2021

(continued)

Tab. 8.1 (continued)

MiR-155	Chromosome 21q21.3	- Proinflammatory effect through direct targeting of SOCS-1 - Positively regulates IFNγ production in CD4+ and CD8+ T-cells	"miRDB Database" 701 "miRTargetLink 2.0 Database" 1149	- ↑ NF-κB activity - ↑ Accumulation and action of T effector cell - ↑ Cell function of Th17	Yan et al. 2019; Mashima et al. 2015; Bell-Hensley et al. 2023; Chen et al. 2020; Kern et al. 2021
MiR-181a	Chromosome 1q32.1	- A potent anti-inflammatory miRNA - suppression of NF-κB signaling	"miRDB Database" 1408 "miRTargetLink 2.0 Database" 686	- ↑ TCR signaling power and sensitivity - ↓ Th1 cell differentiation - ↑ Treg development - ↓ Dendritic cell response - ↑ M2 macrophage polarization	Yan et al. 2019; Kim et al. 2021;Braicu et al. 2019; Lewis et al. 2022; Chen et al., 2020; Kern et al. 2021
MiR-223	Chromosome Xq12	- Anti-inflammatory miRNA - Negative regulator of granulocyte production and inflammatory response - target IKKα is up-regulated, leading to repression of non-canonical NF-κB signaling, thus preventing hyperactivation of macrophages	"miRDB Database" 584 "miRTargetLink 2.0 Database" 551	- ↓ Granulocyte inflammatory - ↓ NFκB-, TLR-, and MAPK-signaling - ↓ Activity of NLRP inflammasome	Yan et al. 2019; Yuan et al. 2018; Roffel et al. 2020; Hirschberger et al. 2018; Chen et al. 2020; Kern et al. 2021

CD4+ Cluster differentiation 4, *CD8+* Cluster differentiation 8, *IKKα* Inhibitor of Nuclear Factor Kappa B Kinase, *IL-2* Interleukin-2, *IL-6* Interleukin-6, *IL-7* Interleukin-7, *IL-10* Interleukin-10, *IFNγ* Interferon-gamma, *MAPK* Mitogen activated protein kinase, *NFκB* Nuclear factor kappa B, *NLRP* NLR Family Pyrin Domain Containing, *q* Long arm of chromosome, *Th1* T helper cell 1, *Th2* T helper cell 2, *Th17* T helper cell 17, *TLR* Toll like receptor, *TNF-α* Tumor necrosis Factor alpha, *SOCS-1* Suppressor of Cytokine Signaling 1

8.3.3 MiR-21

MiR-21 gene is located on chromosome 17q23.1 (Yan et al. 2019). It is widely expressed in many cell types beside leukocytes. The increased expression of miR-21-5p is associated with any inflammatory stimuli and activation of immune cells. Additionally, miR-21 acts as an oncogene and its upregulation results in B cell lymphoma in experimental animals (Chi et al. 2022), thus enforcing the concept of miR-21 association with the development of many cancers (Soto-Heredero et al. 2023). In aging, miR-21 is overexpressed in T cells, decreasing memory T cells generation and ultimately leading to immune system alterations (immunosenescence) (Qin et al. 2016).

8.3.4 MiR-124

MiR-21 gene is located on chromosome 8p23.1 (Yan et al. 2019). The most abundant site for miR-124 expression is neuronal cells and it also shows increased expression in immune cells and organs, including bone marrow, thymus, lymph node, and peripheral blood mononuclear cells (Tazi et al. 2021). Inflammatory or immunological signals induce miR-124 expression, and thus it functions as a negative regulator for such signals, creating a negative feedback for maintaining homeostasis (Roberts et al. 2021a, b).

8.3.5 *MiR-142*

MiR-142 is one of the highly evolutionarily conserved miRNAs and its encoding gene is located on chromosome 17q22 (Yan et al. 2019). It is widely considered to be specific for the hematopoietic lineage (Chi et al. 2022). MiR-142 has many roles in the hematopoietic lineages development and the proper function of the immune system (Arif et al. 2020). The regulation of miR-142 isoforms lineage-enriched expression may be related to the presence of super-enhancers associated with MIR-142 promotor activity which are continuously active in immune and hematopoietic cell types (Arif et al. 2020) and epigenetic factors, such as context-specific and/or lineage changes in the DNA methylation of the MIR-142 promotor (Wang et al. 2019).

8.3.6 *MiR-146a*

MiR-146a gene is located on chromosome 5q33.3 (Yan et al. 2019). It is a potent negative regulator of both types of immune responses (Liu et al. 2018a, b). MiR-146a expression is driven by the NF-kB signaling pathway (Chen et al. 2023), and it causes immune cell activation via Toll-like receptor activation or T cell receptor signaling. Specifically, miR-146a-5p limits cellular activation by decreasing NF-kB signaling transducers IRAK1 and TRAF6 through negative feedback regulation (Pashangzadeh et al. 2021).

8.3.7 *MiR-155*

MiR-155 gene is located on chromosome 21q21.3 (Yan et al. 2019). It is a master regulator of immune system responses. MiR-155 showed to be vital for lymphoid and myeloid cell development and homoeostasis in experimental animals (Mashima 2015). Thus, it has been proposed that miR-155 has a critical role in the activation of immune cells (Bell-Hensley et al. 2023). One of the most important immunological functions of miR-155 is repressing the cytokine signaling 1 via Suppressor of Cytokine Signaling 1 gene (SOCS1), a potent inhibitor of Janus kinase (JAK)-signal transducer and activator of transcription (STAT) signaling pathways (Bell-Hensley et al. 2023).

8.3.8 MiR-181a

MiR-181 gene is located on chromosome 1q32.1 (Yan et al. 2019). It is predominantly expressed in hematopoietic progenitor cells, where plays critical roles in immune cell lineage differentiation (Kim et al. 2021). MiR-181 family members (miR-181a-1, miR-181a-2, miR-181b-1, miR-181b-2, miR-181c, and miR-181d) have a major role in the immune system by acting as a metabolic controller owing to their critical roles in lymphoid and myeloid lineage differentiation (Braicu et al. 2019). MiR-181 is considered an anti-inflammatory miRNA with a potent role in both innate and adaptive immune cells (Kim et al. 2021). The anti-inflammatory effects of miR-181 are attributed to the direct targeting of importin α3 and the lysine deubiquitinase (CYLD) via suppressing NF-κB signaling (Lewis et al. 2022). In aging and immunosenescence, the expression of miR-181 family members is reduced, ultimately causing dysregulation in the immune system function (Yuan et al. 2018).

8.3.9 MiR-223

MiR-223 gene is located on chromosome Xq12 (Yan et al. 2019). It is important regulator of innate immunity (Roffel et al. 2020). miR-223 expression is restricted to myeloid-derived cell types of the immune system, in which miR-223-3p mediates both myeloid cell differentiation and activation (Hirschberger et al. 2018). In acute lung injury, an immune response is reinforced by shuttling miR-223 by polymorphonuclear cells (PMNs) to the lung epithelial cells, thereby alleviating inflammation. This immune response occurs by direct targeting poly-adenosine diphosphatase-ribose polymerase-1 (PARP-1) (Hirschberger et al. 2018). Also, it has been shown that exosomal miR-223 suppresses inflammation in monocytes via targeting NLR Family Pyrin Domain containing (NLRP) 3′UTR and IL6 coding sequence in patients undergoing cardiac bypass surgery suppress (Roffel et al. 2020).

8.4 MiRNAs in Autoimmune Diseases

8.4.1 Autoimmune Dermatological Disorders

The skin is the body's largest organ, comprising two main layers: the epidermis and dermis. Keratinocytes are the most abundant cells in the epidermis, followed by melanocytes, Merkel cells, and Langerhans cells. The epidermis consists of four main layers, namely, stratum basale or basal cell layer (dividing cell layer), stratum spinosum, stratum granulosum, and stratum corneum (keratinized squamous cells) (Arda et al. 2014). In the epidermis, the basal layer is responsible for the proliferation and differentiation of the supra-basal layer. P63 is a protein highly expressed in the basal cells and responsible for its proliferation. Its expression is suppressed by miR-203, which is abundantly found in the supra-basal layer (Yi and Fuchs 2010). Stratum spinosum, the thickest layer of epidermis, is composed of multiple highly mitotic keratinocyte layers. However, they exhibit a significant decrease in proliferative activity as they move away from the basal layer. They are accountable for creating the skin barrier by forming intercellular junctions (Arda et al. 2014). E-cadherin, a crucial protein for cell adhesion, inhibits cell permeability. It is commonly observed to be downregulated during epithelial-mesenchymal transition. MiR-200 and miR-205 were found to inhibit ZEB1 and ZEB2, which are E-cadherin suppressors, thereby maintaining the skin's epithelial stability (Yi and Fuchs 2010).

8.4.1.1 Psoriasis

Psoriasis is an autoimmune inflammatory disorder caused by dysregulation in the inflammatory pathway, ultimately leading to NF-κB activation and consequent increased expression of cytokines. Several factors contribute to the development of the disease, including genetic, epigenetic, and environmental factors. Additionally, the imbalance between pro-inflammatory cytokines and anti-inflammatory mediators contributes to the underlying etiology of psoriasis. This disorder is characterized by increased proliferation of keratinocytes and infiltration of immune cells, resulting in erythematous scaly skin lesions. Several miRNAs are involved in the pathogenesis of the disease (Dopytalska et al. 2023).

MiR-31 has been found to be highly expressed in psoriatic lesions. Studies demonstrated its effect on the activation of the NF-κB pathway, leading to the proliferation of keratinocytes. Activation occurs by targeting serine/threonine kinase 40 (STK40), which negatively regulates NF-κB (Xu et al. 2013). Yan et al. (2015) demonstrated that miR-31 inhibits protein phosphatase 6 (PPP6) and limits the G1-S transition, increasing cells in the S phase. They also reported decreased expression of PPP6 in psoriatic epidermis. Experimentally administered anti-miR-31 in a mouse model reduced keratinocyte proliferation (Yan et al. 2015). In contrast, miR-125b is negatively expressed in psoriatic lesions and has a negative regulatory effect on NF-κB through targeting ubiquitin-specific peptidase (USP2), ultimately leading to decreased cytokine production and decreased inflammation (Wei et al. 2017). MiR-125b also targets fibroblast growth factor receptor 2 (FGFR2), a stimulator of keratinocyte proliferation (Xu et al. 2011).

MiR-21 is highly expressed in keratinocytes (Masalha et al. 2018). In addition, Degueurce et al. (2016) and Jia et al. (2019) highlighted its role in proliferation of epidermal cells and apoptosis inhibition. The authors demonstrated that miR-21 targets SMAD7, an inhibitor for TGFβ, leading to increased expression of TGFß, thereby promoting inflammation and upregulating miR-21 (Degueurce et al. 2016). MiR-21 has been found to suppress apoptosis through the downregulation of Caspase 8, a proteolytic enzyme involved in apoptosis (Jia et al. 2019).

Due to its function in the normal epidermis, miR-203 is expected to increase in psoriasis (Xiao et al. 2020). MiR-203 exerts its effect by downregulating LXR-α/PPARγ, which are inhibitors of keratinocyte proliferation (Xiao et al. 2020).

Several microRNAs, such as miR-146a, miR-155, and miR-210, are also involved in the pathogenesis of immune dysfunction in psoriasis. MiR-146a, a known suppressor of innate immunity (Liu et al. 2018a, b), was reported to target TNF receptor-associated factor 6 (TRAF6) and IL-1 receptor-associated kinase 1 (IRAK1), which regulates the balance between Treg and Th17 (Xia et al. 2012). However, there are conflicting reports regarding the correlation between its level of expression in patients' sera/plasma and disease severity (Leal et al. 2021; Shen et al. 2022). Xu et al. (2017) found that miR-155 is upregulated in skin lesions of psoriatic patients. The authors also reported that miR-155 promotes the proliferation of keratinocytes and inhibits its apoptosis by modulating the phosphatase and tensin homolog (PTEN) pathway (Xu et al. 2017). MiR-210 has been found to be upregulated in CD4+ T cells, affecting the development and function of Treg by targeting Forkhead box P3 (FOXP3) (Zhao et al. 2014).

8.4.1.2 Vitiligo

Vitiligo is an autoimmune inflammatory destruction of melanocytes caused by several interacting factors, including polygenic variants, immune dysregulation involving both innate and adaptive immunity (Spritz and Andersen 2017), as well as oxidative damage (Yi et al. 2019).

Upon exposure to triggers, reactive oxygen species (ROS) produced by melanocytes results in the destruction of cellular components and production of damage associated molecular patterns (DAMPs) such as melanocyte-specific peptides. Subsequently, dendritic cells present those DAMPs to T cells leading to INF-γ production followed by recruitment of autoreactive CD8+ T cells. Unopposed by Treg, CD8+ T cells activation eventually results in melanocytes apoptosis. (Hlača et al. 2022). Increased Teff population (Th17 and Th22) as well as decreased Treg and Th1 is observed in vitiligo. miR-21-5p downregulation in vitiligo has been found to affect the balance between Treg/Teff through targeting STAT3 (Huo et al. 2021).

A single nucleotide polymorphism (SNP) within the miR-196a2 gene (rs11614913, T > C) has emerged as a point of interest in vitiligo research. Studies by Huang et al. (2013) suggest that individuals with the CC genotype exhibit a lower susceptibility to developing vitiligo compared to those with other genotypes. Furthermore, Cui et al. (2015) reported higher tyrosinase levels in individuals with the TT and TC genotypes compared to CC. Since high tyrosinase levels are associated with an increased risk of vitiligo (Cui et al. 2015), these findings suggest a potential link between the miR-196a2 polymorphism and disease development. It is important to note that these studies were conducted on the Han Chinese population, and further investigations are needed to explore generalizability to other ethnicities.

Vitiligo is characterized by the loss of adhesion of melanocytes and keratinocytes, where the e-cadherin protein plays a key role in this process (Su et al. 2019).

MiR-9 is highly expressed in skin lesions of patients suffering from vitiligo (Su et al. 2019) as well as in their plasma (Abdallah et al. 2022). MiR-9 directly impacts melanocyte adhesion by targeting e-cadherin in cells subjected to UVB treatment to induce vitiligo re-pigmentation (Su et al. 2019).

MiR-124 plays various roles in different pathologic conditions, especially in inflammatory responses. Although it is mainly expressed in brain tissues, it was reported to be overexpressed in vitiligo patients' plasma and under-expressed in rheumatoid arthritis (RA) tissues. In addition, it acts as a tumor suppressor in melanoma (Abdallah et al. 2022).

MiR-25 exhibits increased expression in both the skin of vitiligo-affected individuals and serum, demonstrating a correlation with disease activity. Shi et al. (2016) indicated that oxidative stress can induce miR-25 production by demethylating its gene promoter. Inhibition of stem cell factor production from keratinocytes ultimately impairs their protective paracrine effects on melanocyte survival. The authors also reported its positive effect on apoptotic pathways involving caspase 3 and 9. These effects ultimately lead to the destruction of melanocytes, highlighting the crucial role of miR-25 in the impairment of pigment production by inhibiting tyrosinase (Shi et al. 2016).

8.4.1.3 Lichen Planus

Lichen planus is a chronic inflammatory disease caused by the infiltration of T lymphocytes as well as mast cells in the subepithelial layer of the skin and mucous membrane. Regulated on activation normal T cell expressed and secreted (RANTES), a special class of chemokine that recruits inflammatory cells was reported to be secreted by T cells in oral lichen planus (OLP) lesions, leading to mast cell degranulation and release of TNFα (Zhao et al. 2001). Mehdipour et al. (2018) reported decreased levels of miR-125a in the saliva of OLP patients as well as increased miR-21 expression (Mehdipour et al. 2018). MiR-125a has been discovered to target KLF13, which is a positive regulator for RANTES (Zhao et al. 2010). Danielsson et al. (2012) reported miR-125b downregulation and upregulation of both miR-203 and miR-21 in OLP lesions. MiR-125b targets MMP-2, a protein that inhibits the proliferation of keratinocytes and induces apoptosis. Wang et al. (2016) reported that miR-125b exerts its action through the Pi3K/Akt/mTOR pathway. They also reported a decreased level of miR-125b in OLP lesions (Wang et al. 2016). Moreover, El-Rifaie et al. (2017) observed a decreased expression of miR-125b in classical cutaneous lichen planus, while miR-203 expression was low, which contradicts findings in OLP (El-Rifaie et al. 2017).

8.4.1.4 Pemphigus and Pemphigoid Disorders

Pemphigus and pemphigoid are serious autoimmune disorders caused by autoantibodies against desmoglein (Dsg) proteins 3 and 1, resulting in the loss of adhesion between cells, basal cell shrinkage, and supra-basal cell acantholysis (Cipolla et al.

2016). These autoimmune disorders are characterized by the presence of immune reactants in the skin and circulating autoantibodies to skin components (Brooks et al. 1989). Several miRNAs have been implicated in the development of pemphigus and pemphigoid disorders, with some showing a correlation with disease activity, such as miR-338-3p (Lin et al. 2018) and miR-326 (He et al. 2022). Furthermore, He et al. (2022) investigated several miRNAs for their diagnostic sensitivity and specificity of pemphigus vulgaris. Their study demonstrated that the combination of miR-155 and miR-584-5p has a high diagnostic sensitivity and specificity (95% and 100%, respectively) for the disease. This finding could serve as an alternative diagnostic marker in Dsg 1 and 3 antibodies negative patients with pemphigus vulgaris (He et al. 2022).

8.4.2 Autoimmune Rheumatological Diseases

8.4.2.1 Systemic Lupus Erythematosus

Systemic lupus erythematosus (SLE) is a chronic autoimmune disease affecting several organs. It is caused by an immunoregulatory dysfunction in which genetic, epigenetic, and environmental factors play pivotal roles. Several pathways have been described, among them aberrant innate immunity, a key player in SLE pathogenesis. DCs play a critical role in the innate immune response to foreign antigens, with recognition mediated by toll-like receptors (TLRs). However, erroneous presentation of self-antigens by DCs can disrupt immune tolerance and contribute to autoimmune diseases. Two main subtypes of DCs exist: myeloid DCs and plasmacytoid DCs (pDCs). pDCs are responsible for type 1 interferon (IFN-1) production upon recognition of viral nucleic acids like ssRNA and DNA. Notably, imbalances in DC subtypes, particularly an upregulation of pDCs, are implicated in the immunopathogenesis of SLE (Chi et al. 2021). Self-nucleic acid is believed to be delivered to TLR-7 expressed in pDC of SLE patients, resulting in increased levels of INF-1 (Sakata et al. 2018). MicroRNAs (miRNAs) further regulate this pathway; for instance, miR-155 is upregulated in DCs stimulated by TLR7 ligands, leading to the inhibition of IFN-1 production (Zhao et al. 2010).

Decreased levels of miR-155 have been reported both in sera and CD4+ T cells of SLE patients (Wang et al. 2010; Rasmussen et al. 2015). Let-7c was reported by Kim et al. (2013) to be upregulated in DC promoting proinflammatory phenotype. It suppresses “B lymphocyte-induced maturation protein-1” (Blimp1), leading to transcriptional suppression in DC (Kim et al. 2013).

Impaired clearance of cellular debris and formation of neutrophil extracellular traps (NETs) are believed to be a crucial mechanism for the pathogenesis of SLE. NETs formation is a mechanism by which neutrophils enhance their antimicrobial function. Upon stimulation of neutrophils, activation of several pathways occurs. NADPH oxidase (NOX) complex and subsequent protein-arginine deiminase type 4 (PAD4) pathway activation result in induction of chromatin de-condensation and NET formation. The decondensed chromatin is then released outside the neutrophils acting as traps, immobilizing pathogens for their consequent phagocytosis. However, NET formation or "NETosis" are non-specific and if uncontrolled could lead to tissue damage. Decreased clearance of released NETs could lead to more tissue damage through exacerbation of inflammation. Moreover, it could act as a potential trigger for the generation of autoantibodies against nuclear material released (Mutua and Gershwin 2021). Some miRNAs frequently reported to be dysregulated in SLE were found to be involved in the regulation of the formation of NETs. In a mouse model, Arroyo et al. (2021) demonstrated that miR-146a deficient neutrophils, especially aged ones, are more prone to NET formation. Hawez et al. (2019) reported that PAD4 was directly induced by miR-155 in neutrophils leading to NET formation induction.

Adaptive immunity dysfunction has also a significant role in SLE pathogenesis. There is an imbalance between effector (Teff) and T regulatory (Treg) cells. This imbalance results in a decrease in Treg cells responsible for self-tolerance and an increase in Teff cells, particularly Th17 and T follicular helper (Tfh) cells. The Teff cells promote inflammation by secreting cytokines and enhance autoantibodies production by B cells (Shan et al. 2020). Figure 8.2 and Table 8.2 show the detailed role of miR-125a in promoting this imbalance.

Another interesting pathological feature in SLE patients is the hypomethylation of CD4+ T cells. Decreased expression of miR-142-3p/5p (Ding et al. 2012) as well as elevated levels of miR-126 (Zhao et al. 2011), miR-29b (Qin et al. 2013), and miR-21 (Pan et al. 2010) was associated with DNA hypomethylation in CD4+ T cells.

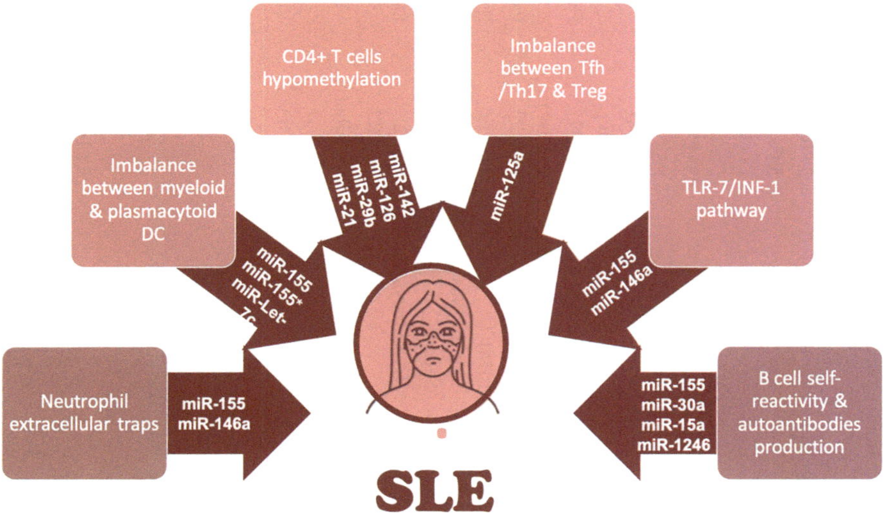

Fig. 8.2 Systemic lupus erythematosus (SLE) pathogenesis and involved miRNAs. DC: dendritic cells, NETs: neutrophil extracellular traps, TLR: toll-like receptor, INF: interferon, Tfh: T cell follicular helper, Th17: T helper 17

Table 8.2 Differentially regulated miRNAs. Samples used for detection, targets genes and/or pathways, and effects in different autoimmune diseases

MiRNA	Regulation	Sample detection	Target gene/ pathway	Disease	Effect	Reference
miR-31	↑	Keratinocytes	STK40 PPP6C	Psoriasis	Hyperplasia and pr006Fliferation of keratinocytes	Xu et al. (2013) and Yan et al. (2015)
miR-21	↑	Keratinocytes T cells	Caspase-8 SMAD7	Psoriasis	Hyperplasia of keratinocytes and decrease apoptosis	Degueurce et al. (2016) and Masalha et al. (2018)
miR-203	↑	Keratinocytes	P63 LXR-α PPARγ	Psoriasis	Keratinocytes proliferation	Yi and Fuchs (2010) and Xiao et al. (2020)
miR-155	↑	Skin lesion	PTEN pathway	Psoriasis	Treg differentiation and keratinocytes apoptosis	Xu et al. (2017)
miR-210	↑	CD4+ T cells	FOXP3	Psoriasis	Treg function	Zhao et al. (2014)
miR-146a	↑/↓	PBMCs Skin lesions Serum	TRAF6 IRAK1	Psoriasis	Suppressor of innate immunity	Xia et al. (2012), Leal et al. (2021) and Shen et al. (2022)
miR-125b	↓	Keratinocytes	USP2 FGFR2	Psoriasis	Suppress keratinocytes proliferation	Xu et al. (2011) and Wei et al. (2017)
miR-9	↑	Skin lesions Plasma	E-cadherin	Vitiligo	Decrease adhesion between melanocytes and keratinocytes	Su et al. (2019) and Abdallah et al. (2022)
miR-25	↑	Skin lesions Serum	Tyrosinase	Vitiligo	Impair pigment production Melanocyte destruction in oxidative stress Melanocyte apoptosis	Shi et al. (2016)
miR-423	↑	Plasma	–	Vitiligo	Oxidative stress	Abdallah et al. (2022)
miR-125b	↓	Lesion	MMP-2	Oral/cutaneous lichen planus	Proliferation of keratinocytes	Wang et al. (2016)
miR-125a	↓	Saliva	KLF13	Oral lichen planus	Mast cell degranulation	Zhao et al. (2010)

miR-338-3p	↑	PBMCs	TRADD RNF114	Pemphigus and pemphigoid diseases	Promote apoptosis and proliferation as well as indirectly inducing inflammation through NF-κB	Wilson et al. (2009), Liu et al. (2018a, b)
miR-584-5p	↑	Serum PBMCs	KLRG1	Pemphigus	KLRG1binds to e-cadherin which could impair its function in cell adhesion	
miR-155	↓	Serum CD4+ T cells B cells	– PU.1	SLE	MiR-155 was found to decrease IFN-1 production MiR-155 suppress PU-1 which is a regulator of B cell development	Tessmer et al. (2007), Wang et al. (2010), Zhou et al. (2010), Rasmussen et al. (2015), Cipolla et al. (2016) and Aboelenein et al. (2017)
miR-155*	↑	DC	–	SLE	MiR-155* was found to increase IFN-1 production	Zhou et al. (2010)
Let-7c	↑	DC	Blimp1	SLE	Promotes Proinflammatory phenotype in DC	Kim et al. (2013)
miR-146a	↓	Whole blood	TLR-7 TRAF6 IRAK1	SLE	Low expression leads to increased INF-1 levels	Tang et al. (2009)
miR-142-3p/5p	↓	CD4+ T cells	SLAM	SLE	Alter histone modification and DNA methylation Activation of T cell	Ding et al. (2012)
miR-126/ miR-29b miR-21	↑	CD4+ T cells	DNMT1	SLE	DNA hypomethylation	Pan et al. (2010), Zhao et al. (2011) and Qin et al. (2013)
miR-125a	↓	CD4+ T cells	STAT3 KLF13	SLE	Stat3 is a pivotal transcriptional factor required for TH17 differentiation. It is upregulated in response to low miR-125a expression KLF13 regulator of RANTES results in recruitment of CD4+ and CD8+	Tang et al. (2009) and Zhao et al. (2010)

(continued)

Table 8.2 (continued)

MiRNA	Regulation	Sample detection	Target gene/ pathway	Disease	Effect	Reference
miR-30a	↑	PBMCs	LYN	SLE	Suppress LYN which in turn increase proliferation and antibodies production by B cells	Liu et al. (2013)
miR-15a	↑	B cells in mice	Bcl-2	SLE	Downregulation of Bcl-2 leading to apoptosis of B-10 regulatory subset and increase in B-2 pathogenic with increased production of autoantibodies in mice model	Yuan et al. (2012)
miR-1246	↓	B cells	EBF1	SLE	Activation and proliferation of B cells by loss of inhibitory effect of miR-1246 on EBF1 due to its decreased expression	Luo et al. (2015)
miR-23b	↓	CD4+ T	Hes-1 notch	BD	Activation of the Notch pathway leading to Treg/Th1/ Th17 imbalance	Qi et al. (2014)
miR-19b-3p	↑	Plasma-derived exosomes	CD46	BD	Downregulation of CD40 cause Treg/ Th1/Th17 imbalance	Jiang et al. (2023)
miR-326	↑	PBMCs	Ets-1	BD	Negative regulation of Ets-1 leads to Treg/Th1/Th17 imbalance	Du et al. (2009) and Jadideslam et al. (2019)
miR-155	↑/↓	↑ CD4 + T cells ↓ DC	Ets-1 TAB2/Akt/ mTOR pathway	BD	Treg/Th1/Th17 imbalance Decreased expression of TAB2 involved in autophagy leading to increase pro-inflammatory cytokines	Na et al. (2016) and Liang et al. (2021)
miR-196a2	↓	PBMCs	Bach1/ho-1	BD	Low expression of ho-1 and subsequent increase of pro-inflammatory cytokines	Zou et al. (2016a, b)
miR-106b	↓	Synovial tissue	RANKL	RA	Its low levels promote osteoclastogenesis	Liu et al. (2022)
miR-125a	↓	CD14+ PBMCs	TRAF6	RA	Its low levels promote osteoclastogenesis	Guo et al. (2014)

miR-140-3p	↓	Chondrocytes	FGF9	RA	Chondrocytes apoptosis and cartilage destruction	Chen et al. (2018a, b)
miR-26a-5p	↑	FLSs	PTEN/PI3K/AKT pathway	RA	FLSs proliferation, invasion and decrease their apoptosis	Huang et al. (2019)
miR-203	↑	FLSs	NF-κB	RA	Increased secretion of MMP-1 and IL-6	Stanczyk et al. (2011)
miR-155	↑	CD14+ monocytes in peripheral blood and synovium	CASP10 and APAF1	RA	Suppress CASP10 and APAF1 (apoptotic genes) thus promoting survival and inflammatory activity of monocytes	Rajasekhar et al. (2017)
miR-146a	↑	Synovial tissue	Notch1	RA	Inhibits Notch1 and results in suppression of the switch to macrophage M1	Nakasa et al. (2008) and Huang et al. (2016)
miR-221–3p	↑	Synovial tissue	JAK3/STAT3	RA	Suppress JAK3/STAT3 leading to shifting of M2 to M1 macrophages phenotype	Quero et al. (2020)
miR-21	↓	PBMCs	STAT3/ Foxp3	RA	Imbalance Treg/Th17	Dong et al. (2014)
miR-301a-3p	↑	PBMCs	PIAS3	RA	Inhibit PIAS3 which increases STAT3 leading to increased Th17 population	Tang et al. (2016)
miR-449a	↑	Small intestine	NOTCH 1 KLF4	CD	Decreased differentiation of intestinal cells	Felli et al. (2022)
miR-192-5p	↓	Duodenal biopsies/ Plasma	MAD2L1	CD	MAD2L1 causes cell cycle arrest	Buoli Comani et al. (2015)
miR-31-5p	↓	Duodenal biopsies/ Plasma	FOXP3	CD	Regulate adaptive immune response	Buoli Comani et al. (2015)
miR-338-3p	↓	Duodenal biopsies	RUNX1	CD	Regulate adaptive immune response	Buoli Comani et al. (2015)
miR-122a	↑	Colon biopsy	EGFR	IBD	Increase zonulin and hence enhance intestinal permeability	Fasano (2011) and Zhang et al. (2017)

(continued)

Table 8.2 (continued)

MiRNA	Regulation	Sample detection	Target gene/ pathway	Disease	Effect	Reference
miR-675-5p	↑	Serum	ZO-1 E-cadherin	IBD	Inhibit ZO-1 and E-cadherin leading to increased intestinal permeability	Zou et al. (2016a, b) and Shaker et al. (2023)
miR-132	↑	Colon	FOXO3a	IBD	Promotes inflammation through suppression of FOXO3a with subsequent enhancement of NF-κB pathway	Kim et al. (2016)
miR-223	↑	Colon	FOXO3a	IBD		
miR-223-3p miR-21-5p	↓	Peripheral B cells	TGFBR2, MEF2C, FOXP1 and RBPJ	PBC	Regulate B cell function and differentiation	Wang et al. (2017)
miR-506	↑	Cholangiocytes	AE2	PBC	Impair biliary secretion	Banales et al. (2012)
miR-150-5p	↑	Serum	NOTCH3	MG	T cell differentiation	Ghisi et al. (2011)
miR-181a	↑	Serum	SMAD7	DM type 1	Suppress SMAD7 leading to TGF-β activation and β cell apoptosis	Nabih and Andrawes (2016)
miR-142-5p	↑	Serum Thyroid tissue	CLDN1	HT	Suppress CLDN1 resulting in impairment of thyroid epithelial barrier	Zhu et al. (2016)
miR-125a-3p	↓	PBMC	IL-23R	HT	Th17 differentiation and survival	Peng et al. (2015)
miR-557	↑	Plasma	Caspase 3 Bax	ITP	Promotes megakaryocyte apoptosis	Zufferey et al. (2017)
miR-106b-5p	↑	PBMCs	NR4A3	ITP	Suppression of NR4A3 and subsequently Foxp3 which regulates Treg differentiation, thus, leading to immune imbalance of Treg/Th17	Li et al. (2020)

miR-155-5p	↑	PBMCs/ macrophages	SOCS1	ITP	Suppression of SOCS1 leading to macrophage M2 polarization and subsequent suppression of "platelet-related factors"	Chang et al. (2020)
Let-7b-5p	↑	B cell CD19+	BAFF-R	ITP	Increase expression of BAFF-R, a receptor for BAFF which promotes B cell survival	Wang et al. (2021a, b)

AE2, Anion exchanger 2, *APAF* Apoptotic peptidase activating factor 1, *BAFF-R* B cell-activating factor receptor, *BCL2* B cell CLL/lymphoma 2, *BD* Behçet Disease, *Blimp1* B lymphocyte-induced maturation protein-1, *CASP10* Caspase 10, *CD* Celiac Disease, *CLDN1* Claudin-1, *DC* Dendritic cells, *DNMT1* DNA methyltransferase 1, *EBF1* Early B cell factor 1, *EGFR* Epidermal growth factor receptor, *Ets-1* V-ets avian erythroblastosis virus E26 oncogene homolog 1, *FGF9* Fibroblast growth factor 9, *FLSs* Synovial like fibroblasts, *FOXO3* Forkhead box O3, *FOXP1* Forkhead box P1, *FOXP3* Forkhead box P3, *HT* Hashimoto thyroiditis, *IBD* Inflammatory bowel diseases, *IL* Interleukin, *IL-6* Interleukin 6, *INF-1* Interferon-1, *IRAK-1* IL-1 receptor-associated kinase 1, *ITP* Immune Thrombocytopenia, *JAK3* Janus kinase 3, *KLF4* Krueppel-like factor 4, *KLRG1* Killer cell lectin -like receptor subfamily G, *LXR* Liver X receptors, *LYN* Lck/Yes novel tyrosine kinase, *MAD2L1* Mitotic arrest deficient-like 1, *MEF2C* Myocyte enhancer factor 2C, *MG* Myasthenia gravis, *MMP-1* Matrix metalloproteinase-1, *NF-κB* Nuclear factor kappa B, *NR4A3* Nuclear receptor subfamily 4 group A member 3, *PBC* Primary biliary cholangitis, *PIAS3* Protein inhibitor of activated STAT3, *PPARγ* Peroxisome proliferator-activated receptor, *PPP6C* Protein phosphatase 6 catalytic subunit, *PTEN* Phosphatase and tensin homolog, *RA* Rheumatoid arthritis, *RANKL* Receptor activator of NF-κB ligand, *RBPJ* Recombination signal binding protein for immunoglobulin kappa J region, *RNF114* Ring finger 114 protein, *RUNX1* Runt-related transcription factor 1, *SLAM* Signaling lymphocyte activation molecule, *SLE* Systemic lupud erythematosus, *SMAD7* SMAD family member 7, *SOCS1* Suppressor of cytokine signaling 1, *STAT* Signal transducer and activator of transcription, *STK40* Serine/threonine kinase 40, *TAB2* Transforming growth factor β-activated kinase 1-binding protein 2, *TGFBR2* Transforming growth factor-beta receptor type 2, *TGF-β* Transforming growth factor beta, *Th* T helper cell, *TLR* Toll like receptor, *TNF* Tumor necrosis factor, *TRADD* Tumor necrosis factor receptor 1 (TNFR1)-associated death domain protein, *TRAF6* TNF receptor-associated factor 6, *Treg* T regulatory cells, *USP2* Ubiquitin-specific peptidase, *ZO-1* Zonula occludens

Abnormal activation of B cells through TLR-7 and B cell antigen receptor is another pathology in SLE that result in B cell autoreactivity. TLR-7/ 9 can recognize pathogen DNA and RNA as well as self (Celhar et al. 2012). Detailed miRNAs contribution to aberrant B cell activation is depicted in Table 8.2 and Fig. 8.2.

8.4.2.2 Behçet Disease

Behçet disease (BD) is characterized by recurring oral and genital ulcers, as well as ocular and vascular pathology involvement due to the interplay between autoimmunity and autoinflammatory pathways (Salmaninejad et al. 2019) (Gül 2005). The inflammation in Behçet's disease is thought to be mediated by cytokines derived from Th1 lymphocytes (Oztas et al. 2005). HLA-B*51 susceptibility is a critical factor in the pathogenesis of BD. Cavers et al. (2022) have recently identified a high risk of developing BD in individuals carrying HLA-B*51 and the ERAP1 haplotype 10. This was ascribed to an alteration in CD8 + T cell immunogenicity and the disruption of Treg/Th1/Th17 balance (Cavers et al. 2022). It has been demonstrated that numerous miRNAs affect pathogenic pathways in BD. Figure 8.3 depicts miRNAs implicated in Treg/Th1/Th17 imbalance. MiR-155 and miR-196a2 have been involved in the regulation of pro-inflammatory cytokines production, a well-known hallmark of BD pathogenesis (Table 8.2; Liang et al. 2021, Zou et al. 2016a, b).

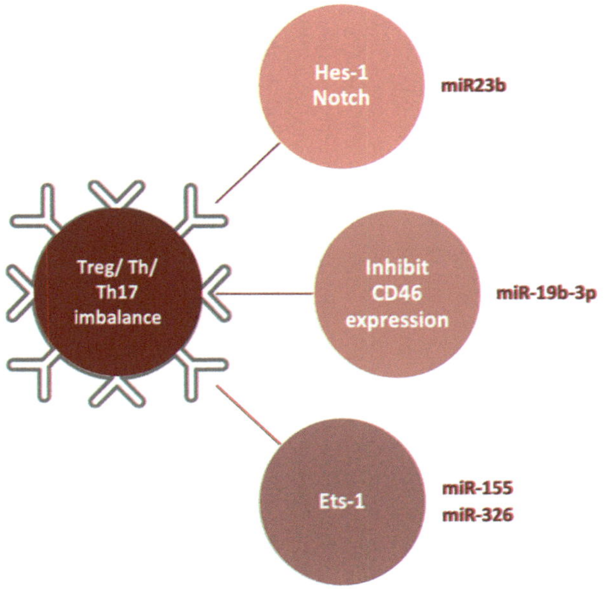

Fig. 8.3 MiRNAs involvement in different pathways in Behçet disease leading to an imbalance between Tref, Th1, and Th17. Treg: T regulatory cell, Th1: T helper cell 1, Th17: T helper cell 17, Ets-1: V-ets avian erythroblastosis virus E26 oncogene homolog 1

8.4.2.3 Rheumatoid Arthritis

Rheumatoid arthritis (RA) is a chronic autoimmune disorder that leads to inflammation and impairment of joints along with extra-articular manifestations. It is defined by impaired innate and adaptive immune responses resulting in an imbalance between Treg and Th17 as well as B cell autoreactivity triggered by the activation of TLR and the production of autoantibodies (rheumatoid factor, anti-citrullinated protein), IL-1, TNFα, and pro-inflammatory cytokines. Moreover, altered synovial-like fibroblasts (FLSs), synovial tissue macrophages, osteoblasts, osteoclasts, and chondrocytes represent hallmarks for RA pathogenesis (Peng et al. 2023). Genetic susceptibility plays a crucial role in RA pathology. Various variants have been linked to the disease risk, including SNP rs3027898 (A/C) in IL-1 receptor-associated kinase 1 (IRAK-1), a protein-target of miR-146a (Yang et al. 2017) and rs3746444 (T/C) of miR-499 gene (Liu et al. 2021).

FLS's primary role is to produce synovial fluid and extracellular matrix to protect the cartilage surface, as well as to lubricate and nourish it (Peng et al. 2023). However, in RA, they exhibit abnormal features, such as decreased apoptosis, increased proliferation, migration, and invasion (Mousavi et al. 2021). There is also an increased influx of monocytes from the bone marrow into the synovium, which become activated and transform into macrophages, leading to additional tissue damage (Huang et al. 2021). An imbalance between M1 and M2 macrophages is observed in RA, with M1 macrophages promoting inflammation and bone resorption, while M2 macrophages contribute to inflammation resolution. Monocytes in RA primarily differentiate into classical macrophage M1, which differentiate into pro-inflammatory macrophages and osteoclasts (Cutolo et al. 2022). The altered inflammatory micro-environment in joints affected by RA promotes osteoclast differentiation as well as inhibition of osteogenesis, altogether leading to bone resorption and erosion. Osteoclastic differentiation and activation occur due to the effect of copious levels of inflammatory cytokines, autoantibodies, and activation of the "receptor activator of NF-κB ligand" (RANKL) pathway. Osteoclasts reside at the interface between the periarticular surface of the bone and the synovium, causing bone erosion (Schett and Gravallese 2012). MiRNA aberrant expression in RA and their effects are illustrated in Fig. 8.4, while targeted genes and specimens used for their detection are displayed in Table 8.2.

The probability of developing RA can be predicted through miRNA biomarkers. Ouboussad et al. (2017) reported that miR-22, miR-382, and miR-486-3p were differentially expressed in the serum of individuals who progressed to the development of RA. MiRNAs are characterized by a unique profile in RA seropositive relatives (Anaparti et al. 2017). This suggests a significant genetic background of the disease and a potential predictive role in RA. MiRNAs are also investigated for the prediction of severity and treatment response in RA. MiR-125b expression in PBMCs correlates inversely with RA severity and increases in responders to treatment (Hruskova et al. 2016; Duroux-Richard et al. 2014).

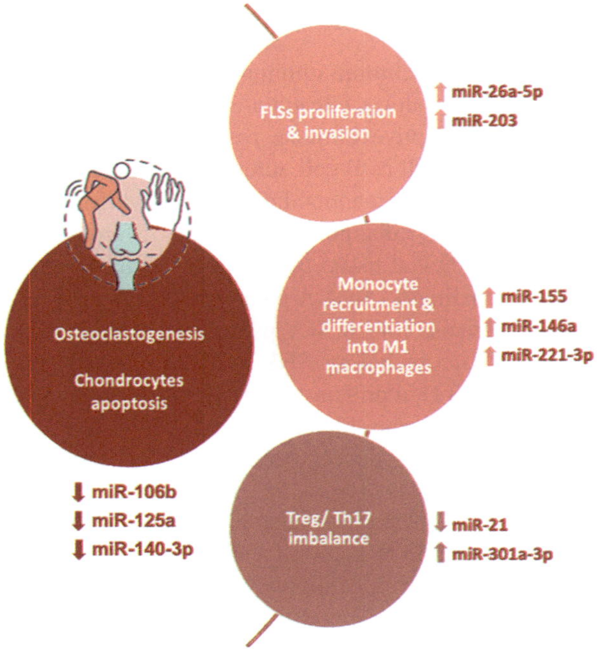

Fig. 8.4 Different pathological outcomes in rheumatoid arthritis and the effect of altered miRNA expression. FLSs: fibroblast like synoviocytes, Treg: T regulatory cells, Th17: T helper 17 cells

8.4.3 Gastrointestinal Related Autoimmune Diseases

8.4.3.1 Celiac Disease (CD)

Celiac disease (CD) is an autoimmune enteropathy involving a dysfunctional adaptive immune response to gluten containing diet in genetically susceptible individuals, particularly carriers of HLA- DQ2 and/or HLA- DQ8. However, environmental factors such as childhood infections play an important role in CD development. It has been suggested that increased intestinal mucosal permeability in susceptible individuals allows the entry of gliadin (gluten protein) to be deaminated by tissue transglutaminase (TTG), resulting in deamidated gliadin production, which is highly antigenic (Parzanese et al. 2017). It is worth mentioning that the antigenicity of the processed gluten is variable according to the gut microbiome. This data has led scientists to study the effect of probiotics on gluten processing with possibility of production of less immunogenic peptides that would serve as potential treatment of CD. However, further studies are still needed to properly identify the microbial signature in celiac disease and to elucidate the role of probiotics in CD treatment (Pecora et al. 2020). Deamidated gliadin is then presented by antigen-presenting cells carrying HLA-DQ2 /DQ8, inducing the production of copious amounts of cytokines by CD4+ T-lymphocytes in the lamina propria, resulting in expansion and differentiation of B lymphocytes producing anti-TTG and anti-gliadins (Mazzarella 2015; Parzanese et al. 2017).

Several studies have been conducted to investigate the role of miRNAs in CD, some describing as tools in diagnosis and monitoring treatment adherence. In contrast, other studies examined their role in the pathogenesis of the disease. Felli et al. (2022) reported that serum levels of miR-192-5p, miR-215-5p, and miR-125b-5p can discriminate between CD, healthy controls, and gluten-free diet in pediatric cohort. Furthermore, Bascuñán et al. (2020) also reported several miRNAs upregulated in PBMCs, monocytes, and plasma of CD patients. In pediatric patients, Capuano et al. (2011) reported increased expression of miR-449 in small intestine biopsies and further demonstrated that miR-449 exhibits an inhibitory effect on NOTCH and KLF4, affecting intestinal cell maturation and differentiation towards goblet cell lineage. Several other miRNAs have been found to be implicated in CD pathogenesis (see Table 8.2).

8.4.3.2 Inflammatory Bowel Diseases (IBD), Crohn's Disease, and Ulcerative Colitis

Inflammatory bowel diseases (IBD), Crohn's disease, and ulcerative colitis (UC) are complex autoimmune disorders affecting the gastrointestinal tract (GIT), which cause chronic inflammation that occur mainly in two forms: Crohn's and ulcerative colitis. Crohn's disease is characterized by inflammation across the GIT in a patchy manner, whereas UC causes inflammation affecting the innermost mucosa of the large intestine. The interplay between genetic susceptibility, environmental, and microbial factors is widely recognized in IBD. The GIT is exposed to enormous amounts of antigens. Therefore, maintaining a high level of self-tolerance is crucial. Microbial imbalance or dysbiosis in the GIT is considered a significant factor contributing to the pathogenesis of IBD. Disruption of the mucosal barrier is another factor contributing to IBD's pathogenesis. It results in microbial invasion, macromolecular influx, and inflammation. (Yu 2018). MiRNAs have been implicated in the pathogenesis of IBD through their effects on the epithelial barrier in the intestine by affecting crucial proteins involved in epithelial cell adhesion such as Zonula occludens (ZO)-1 (Fasano 2011) and e-cadherin (Zou et al. 2016a, b). Zonulin, a protein usually secreted to dismantle ZO-1, increases intestinal permeability to flush the bacteria away from the gut as a defense mechanism (Fasano 2011). Zonulin's increased expression is a direct effect of miR-122a targeting the epidermal growth factor receptor (EGFR), a regulator of Zonulin production. Zhang et al. (2017) reported that upregulation of miR-122a in colonic tissues of IBD patients may be involved in IBD pathology (Zhang et al. 2017). Shaker et al. (2023) demonstrated that serum levels of miR-675-5p are higher in IBD patients than in healthy individuals (Shaker et al. 2023). It was reported that miR-675-5p decreases cell adhesion through its inhibitory effect on ZO-1 and e-cadherin (Zou et al. 2016a, b). Moreover, miR-132 and miR-223 were found to be upregulated in IBD, leading to enhancement of the NF-κB pathway through targeting FOXO3a, hence promoting inflammation (Kim et al. 2016).

The diagnostic capability of miRNAs in IBD has been an intriguing area of research, with several studies reporting a correlation between miRNA levels and IBD diagnosis/activity. Lin et al. (2014) reported a panel of differentially expressed miRNAs (miR-31, miR-206, miR-424, and miR-146a) in UC and Crohn's colonic tissues. On the other hand, Schönauen et al. 2018 demonstrated the promising potential of miRNAs obtained from fecal samples, particularly miR-16, miR-155, and miR-223, in the diagnosis of IBD. These miRNAs may potentially serve as surrogate markers for fecal calprotectin and CRP as well as predicting disease activity. MiRNA prediction of activity was also investigated by Sun et al. (2017). The authors reported a negative correlation between miR-125a levels in serum with disease severity and activity. Schaefer et al. (2015) conducted a study proposing the use of miRNA profiling to differentiate between Crohn's disease and UC using a panel of miRNAs (miR-19a, miR-21, miR-31, miR-101, miR-146a, and miR-375).

8.4.3.3 Autoimmune Liver Diseases

Autoimmune liver diseases (AILDs) are disorders caused by a dysfunctional immune system, namely dysregulated T cell populations and B cell autoantibodies production in a genetically susceptible individual. AILDs include autoimmune hepatitis (AIH), primary sclerosing cholangitis (PSC), and primary biliary cholangitis (PBC) (Huang et al. 2020).

Research has been conducted to investigate the diagnostic role of miRNAs in PBC and its potential correlation with disease progression. Tan et al. (2014) demonstrated that the serum levels of miR-122-5p, miR-141–3p, and miR-26b-5p can contribute to the diagnosis of PBC with a sensitivity of 80.5% and specificity of 88.3%. Wang et al. (2017) reported decreased expression of miR-223-3p and 21-5p in peripheral B cells (PBCs) and demonstrated its correlation with disease progression. They also discovered that these miRNAs target several genes involved in B cell function and differentiation, as shown in Table 8.2 (Wang et al. 2017). MiR-506 was found to be upregulated in cholangiocytes in PBC, as reported in a study conducted by Banales et al. (2012). Researchers also demonstrated that miR-506 inhibits "Cl−/HCO 3− anion exchanger 2" (AE2), leading to decreased bile secretion.

Studies have shown aberrant expression of miRNAs in PSC, indicating that there is a downregulation of miR-200 and upregulation of miR-193b, miR-122, miR-885-5p, miR-1281, miR-126, miR-26a, and miR-30b in the sera of PSC patients (Bernuzzi et al. 2016; Voigtländer et al. 2015). Conversely, limited studies using samples of AIH patients have been conducted. Migita et al. (2015) study reported the upregulation of serum miR-21 in AIH type 1 patients and showed a positive correlation with alanine aminotransferase levels. The authors also demonstrated a decrease in levels of miR-21 and miR-122 in AIH patients with cirrhosis

(Migita et al. 2015). Numerous studies have been carried out on animal specimens, identifying potential targets for pathways involved in the pathogenesis of the disease (Huang et al. 2020).

8.4.4 *Autoimmune Disorders of the Nervous System*

8.4.4.1 Multiple Sclerosis

Multiple sclerosis (MS) is a neurodegenerative disorder caused by immune dysfunction leading to a chronic inflammatory process. Symptomatology and course of the disease are heterogeneous because of its complex pathology. Some patients experience clinically isolated syndrome (CIS), relapsing-remitting course (RRMS), or secondary progressive MS (SPMS), while others suffer from primary progressive MS (PPMS). Genetic susceptibility plays a pivotal role in MS pathogenesis. However, environmental factors influencing the penetrance of genetic factors are crucial for its development (Dendrou et al. 2015). Among the environmental factors involved in this process is the infection with EBV, which was identified as a direct culprit for MS development. It is believed that EBV infection causes B cell reprogramming in latently infected cells and possible autoreactivity, as a result of molecular mimicry, towards the chronic presentation of viral antigens (Soldan and Lieberman 2023). Myelin basic protein (MBP) and EBNA-1 have been suggested as potential examples of molecular mimicry. Several viral proteins and miRNAs have been identified to directly affect chemokines and cytokines, deregulating innate and adaptive immunity (Soldan and Lieberman 2023). For instance, EBV-miR-BHRF1–2-5p targets the MALT1 gene affecting Treg development, while EBV-miR-BHRF1–3 targets PTEN and CXCL-11, affecting Th1 and Tfh responses (Mohammed 2020). Genetic susceptibility plays a prominent role in MS development risk. Individuals carrying HLA-DRB1*1501 and HLA-DRA1*0101 have a high risk to develop MS as these HLA alleles bind to EBV peptides while cross reacting to self-protein expressed in the brain. Another interesting finding is that HLA-DR15 haplotype (HLA-DRB1*1501 and HLA-DRA1*0101) has been found to facilitate EBV B cell entry, which suggest an increased risk of developing MS (Soldan and Lieberman 2023).

The blood-brain barrier (BBB) strictly restricts leukocyte trafficking to the cerebrospinal fluid CSF, thereby limiting the involvement of the adaptive immune response within the CNS. However, this restriction is not absolute. For purposes of immune surveillance, a select population of lymphocytes can gain access to the CSF (Dendrou et al. 2015). Upon activation within the CNS, these lymphocytes can differentiate into pro-inflammatory phenotypes, including IFNγ-producing Th1 cells and IL-17-producing Th17 cells (Dendrou et al. 2015).

Vistbakka et al. (2018) identified upregulated levels of miR-191-5p and miR-24-3p in the serum of patients with RRMS. Furthermore, they observed a positive correlation between miR-24-3p levels and the disability progression index. In a separate study, Vistbakka et al. (2017) reported the upregulation of miR-191-5p, miR-128-3p, miR-24-3p, and miR-376c-3p in serum samples from patients diagnosed with PPMS.

Sanders et al. (2016) identified downregulation of several miRNAs in peripheral CD4+ T cells isolated from patients with SPMS. These miRNAs included miR-21-5p, miR-26b-5p, miR-29b-3p, miR-142-3p, and miR-155-5p. Notably, the majority of these downregulated miRNAs target genes are known to regulate Th1 and Th17 differentiation (Sanders et al. 2016; Smith et al. 2012). In a separate study investigating the potential of miRNAs as diagnostic biomarkers for differentiating between RRMS and SPMS disease courses, Gandhi et al. (2013) reported that miR-454 and miR-145 expression levels could distinguish between these two subtypes (Gandhi et al. 2013).

MiRNA expression patterns in response to MS therapy have garnered increasing research interest. Ingwersen et al. (2015) investigated miRNA expression changes following 1 year of natalizumab treatment. Their findings revealed upregulation of miR-18a, miR-20b, miR-29a, and miR-103, while miR-326 was downregulated. However, due to the limited number of non-responders in their study cohort, the authors were unable to explore the potential correlation between miRNA profiles and treatment response (Ingwersen et al. 2015). Eftekharian et al. (2019) further explored this concept, demonstrating downregulation of miR-204-5p specifically in male non-responders to Fingolimod therapy (Eftekharian et al. 2019).

Autologous hematopoietic stem cell transplantation (AHSCT) has emerged as a promising treatment strategy for MS. This approach is associated with immune response modulation, as evidenced by alterations in miRNA expression. Arruda et al. (2015) have shown that following clinical remission induced by AHSCT, FOXP3, and FOXO1, key regulators of Tregs are upregulated to normal control levels. This upregulation coincides with normalization of miR-142-3p and miR-16 expression, ultimately leading to an increase in the Treg population (Arruda et al. 2015). A comprehensive review by Pérez and Eisele (2022) highlights the broad role of miRNA expression patterns in response to various MS treatment modalities.

8.4.4.2 Myasthenia Gravis

Myasthenia gravis (MG) is characterized by the presence of autoantibodies against acetylcholine receptors (AchRs) and, to a lesser extent, against muscle-specific kinase (MuSK) and lipoprotein receptor-related protein-4 (LrP4). While antibodies play a significant role in the disease, T cell subtypes and their corresponding cytokines also play a crucial role in the pathogenesis of MG, along with genetic predisposition and, in some instances, thymoma (Dresser et al. 2021). MG can be classified into different clinical subtypes based on the presence of specific antibodies: AchR antibodies positive (including ocular MG, generalized AchR positive MG, and thymoma associated), MuSK antibodies positive, double seronegative (negative for both AchR and MuSK antibodies), LrP4 antibodies positive, and pediatric MG (Dresser et al. 2021). Multiple studies have investigated the potential diagnostic and predictive functions of miRNAs in MG. In AchR + MG, Beretta et al. (2022) reported upregulation of miR-30e-5pin MG and demonstrated its correlation with disease course. Punga et al. (2014) found that miR-150-5p was highly expressed in the sera of MG patients. Furthermore, it was significantly decreased after thymectomy (Punga et al. 2014), which is consistent with the findings of Molin et al. (2018). MiR-150 has been reported to have a role in T cell differentiation by targeting the NOTCH3 gene (Ghisi et al. 2011). The authors also detected upregulation in miR-21-5p expression as well as a downregulation of miR-27a-3p levels in MG patients' sera. Both miRNAs are involved in immune modulation, Th1 regulation, and natural killer (NK) cytotoxicity, respectively (Punga et al. 2014). MuSK+ MG showed upregulation of different miRNAs in patients' sera. These miRNAs include miR-151a-3p, let-7a-5p, let-7f-5p, and miR-423-5 (Punga et al. 2016). Furthermore, downregulation of miR-210–3p and miR-324–3p in plasma samples of MuSK+ MG patients has been reported (Sabre et al. 2018).

8.4.5 *Endocrine Autoimmune Disorders*

8.4.5.1 Diabetes Mellitus (DM) Type 1

Diabetes mellitus (DM) type 1 is an autoimmune disease that targets β cells in the pancreas, typically affecting patients at young ages. Genetic susceptibility plays a significant role in its pathogenesis. Approximately 90% of DM type 1 carry either HLA-DR3-DQ2 or HLA-DR4-DQ8 haplotypes (Paschou et al. 2018). DM type 1 can present independently or as a part of autoimmune polyendocrine syndromes (APS). It can be accompanied by Addison's disease, thyroiditis, celiac, and vitiligo (Eisenbarth and Gottlieb 2004). Along with genetic susceptibility, epigenetics, immune dysfunction, and environmental factors such as viral infection (enteroviruses and rubella) contribute to the development of DM type 1 (Paschou et al. 2018).

Autoreactive T cells induce apoptosis in pancreatic β cells by producing pro-inflammatory cytokines that activate the caspase cascade, ultimately leading to apoptosis (Paschou et al. 2018). Lakhter et al. (2018) found a significant increase in the serum level of miR-21-5p in children with type 1 DM. The authors demonstrated that it interacts with caspase3 in pancreatic β cells, regulating apoptosis (Lakhter et al. 2018). B cell CLL/lymphoma 2 (BCL2) was identified as a target for miR-21, which promotes β cell apoptosis (Sims et al. 2017). Nielsen et al. (2012) reported 12 highly expressed miRNAs in children with type 1 DM, which include miR-152, miR-30a-5p, miR-181a, miR-24, miR-148a, miR-210, miR-27a, miR-29a, miR-26a, miR-27b, miR-25, and miR-200a. Many of these miRNAs have been associated with cell apoptosis. They also observed that miR-25 is negatively correlated with the residual function of β cells (Nielsen et al. 2012).

Dysfunction immune response is manifest in DM through the production of autoantibodies such as anti-GAD65, anti-tyrosyl phosphatase (IA-2), anti-insulin (IAA), and anti-zinc transporter (ZnT8). Yang et al. (2015) reported a negative correlation between miR-146a in PBMCs and anti-GAD in patients with type 1 DM. They also showed that the decreased levels of miR-146a in PBMCs resulted in increased expression of several target genes, such as tumor necrosis factor receptor-associated factor 6 (TRAF6) and numb homolog (NUMB). TRAF6 is implicated in TLR inhibition, while NUMB inhibits NOTCH, which modulates Th cell response (Yang et al. 2015).

MiRNAs also have been shown to play a role in complications associated with Type 1 diabetes. Upon exposure to glucose, miR-21 is upregulated in mice renal mesangial cells, causing PTEN suppression and increased TOR complex 1 (TORC1) activity. This led to renal cell hypertrophy and fibronectin expression (Dey et al. 2011). Barutta et al. (2013) investigated the potential of urinary exosomal microRNAs (miRNAs) as biomarkers for early detection of diabetic nephropathy. Their study identified upregulation of miR-145 and miR-130a, and downregulation of miR-155 and miR-424 in the urine of patients with early-stage diabetic nephropathy. These findings suggest that these specific miRNAs may serve as valuable tools for early disease diagnosis (Barutta et al. 2013).

The EURODIAB Prospective complication study examined the relationship between miR-21 levels in the serum of patients with diabetes and their complications. Moreover, researchers found that miR-126 levels are reduced in diabetic patients compared to healthy controls and it is inversely linked to diabetic retinopathy (Barutta et al. 2017).

8.4.5.2 Hashimoto Thyroiditis

Hashimoto's thyroiditis (HT) represents the most common etiology of hypothyroidism in iodine-replete regions. This autoimmune disorder is characterized by progressive destruction of thyroid follicles due to T cell infiltration, ultimately leading to glandular atrophy and fibrosis. Genetic predisposition and environmental factors are believed to play a crucial role in the development of immune tolerance breakdown and subsequent disease onset. Anti-thyroperoxidase (TPO) and anti-thyroglobulin (Tg) antibodies are frequently detected in Hashimoto's thyroiditis. This condition can occur in isolation or co-exist with other autoimmune diseases. Furthermore, it is strongly associated with an increased risk of developing papillary thyroid cancer (Ragusa et al. 2019). MiRNAs serve as potential biomarkers for diagnosing Hashimoto and could be correlated with the levels of antibodies. Trummer et al. (2022) reported upregulation of several miRNAs (miR-21-5p, miR-22-3p, miR-22-5p, miR-142-3p, miR-146a-5p, miR-301–3p and miR-451) in patients' sera and demonstrated an association between miR-22-5p, miR-142-3p, and antibody levels. High expression of miR-142-5p and miR-142-3p, as well as miR-146a, in thyroid glands of HT has been reported by Zhu et al. (2016). They identified claudin-1 (CLDN1) as a target of miR-142-5p and demonstrated that miR-142-5p suppresses CLDN1, a protein crucial for cellular tight junction, resulting in impairment of the thyroid epithelial barrier. Furthermore, Zhu et al. (2016) showed a significant positive correlation between miR-142-5p expression and thyroglobulin antibody level. Peng et al. (2015) reported downregulation of miR-125a-3p in PBMCs of HT patients. Authors further demonstrated that miR-125a-3p directly targets IL-23R, leading to its suppression (Peng et al. 2015). IL-23R has been previously reported to promote Th17 cell differentiation, survival, and function (McGeachy et al. 2009).

8.4.5.3 Immune Thrombocytopenia

Immune thrombocytopenia (ITP) is an autoimmune disorder characterized by the destruction of platelets and impairment of their production. It is caused by autoreactive antibodies against mainly platelet glycoprotein (GP) IIb/IIIa and/or the GP Ib–IX–V complex to a lesser extent. Another defect in self-tolerance manifests in the presence of autoreactive CD8+ T cells, Treg dysfunction and Th reactivity to platelets, and impaired antigen presentation (Zufferey et al. 2017). Zhang et al. (2016) conducted an informative study on bone marrow mesenchymal stem cells (MSC), a subset of cells important in immune regulation of both T and B cells. Authors reported impaired immune regulation as well as enhanced apoptosis of MSCs in ITP patients (Zhang et al. 2016). Wang et al. (2020) observed elevated levels of miR-98-5p in MSC of ITP patients targeting insulin-like growth factor 2 mRNA-binding protein 1 (IGF2BP1) promoting apoptosis (Wang et al. 2020).

In 2021, Wang et al. reported that miR-557 is upregulated in the plasma of ITP patients, targeting apoptotic genes such as caspase-3 and Bax genes. They also demonstrated that miR-557 inhibition increases megakaryocytes and platelet levels and has a significant positive impact on ITP in experimental rats (Wang et al. 2021a, b). Liu et al. (2016) analyzed the expression of several miRNAs in PBMCs and found that miR-326, miR-146a, and miR-142-3p are more strongly downregulated in ITP patients than in a control group. Moreover, they reported that miR-146a and miR-142-3p correlate negatively with Th17 frequency and miR-146a correlates positively with Treg frequency (Liu et al. 2016). Table 8.2 outlines additional miRNAs that play a role in regulating T cells, B cells, and macrophages in ITP.

Corticosteroids, intravenous immunoglobulins (IVIg), and thrombopoietin receptor agonists (TPO-RA) are among the treatment options for ITP. MiRNAs are also used to evaluate responses to these therapeutic regimens and aid in unraveling their roles in disease pathogenesis. Garabet et al. (2020) showed that plasma miR-199a-5p and miR-221–3p had a potential role in the prediction of response to TPO-RA. Plasma miR-199a-5p and miR-33a-5p were found to have a high diagnostic accuracy, (AUC: 0.93) for ITP diagnosis (Garabet et al. 2020).

8.5 Conclusion

Several miRNAs exhibit highly specific expression patterns in organs associated with the immune system, which emphasizes the role of miRNA in regulating immune responses through its role during development, maturation, activation, proliferation, functioning, and survival of immune cells. Dysregulated miRNA expression leads to immune tolerance breakdown and the development of autoimmune diseases, thus serving as potential biomarkers for diagnostic, prognostic and therapeutic prediction purposes in autoimmune disorders.

In autoimmune dermatological disorders, miR-21, miR-31, miR-146a, miR-155, miR-203, and miR-210 are highly expressed in keratinocytes targeting SMAD, PPP6, TNF/IRAK, PTEN, LXR-α/PPARγ, and FOXP3 genes, respectively, affecting their expression in the skin epidermis and playing a part in psoriasis. On the other hand, miR-125 is downregulated in psoriatic lesions where it affects NF-κB through targeting USP2 and FGFR2. In vitiligo, miR-9, miR-25, and miR-124 showed increased expression affecting melanocytic function. In Lichen planus, miRNAs signature reported from different studies on this disease revealed downregulation of miR-125a and overexpression of miR-21. Finally, in pemphigus and pemphigoid disorders, miR-155, miR-326, miR-338-3p, and miR-584-5p showed different correlations with these group of dermatological disorders.

Autoimmune rheumatological diseases exhibit distinct patterns of miRNA dysregulation. In SLE, altered expression of miRNAs including let-7c, miR-15a, miR-21, miR-29b, miR-30a, miR-125a, miR-126, miR-142-3p/5p, miR-146a, miR-155, and miR-1246 contributes to various pathological processes. These dysregulated miRNAs are associated with the formation of neutrophil extracellular traps, an imbalance between myeloid and plasmacytoid DCs, CD4+ T cell hypomethylation, an imbalance between Tfh cells, Th17 cells, and Tregs, effects on the TLR-7/IFN-1 pathway, and B cell self-reactivity leading to autoantibody production. In contrast, Behçet's disease is characterized by dysregulation of a different set of miRNAs, namely miR-19b, miR-23b, miR-155, and miR-326. These miRNAs disrupt the Treg/Th/Th17 balance by inhibiting CD46 expression, Hes-1 notch signaling, and Ets-1, respectively. In Rheumatoid arthritis, altered miRNA causes different pathological outcomes. For example, decreased levels of miR-106b, miR-125a, and 140-3p cause increased osteoclastogenesis and chondrocytes apoptosis. Increased levels of miR-26a-5p and miR-203 cause FLSs proliferation and invasion. Also, overexpression of miR-155, miR-146a, and miR-221–3p lead to monocyte recruitment and differentiation into M1 macrophages. Finally, decreased miR-21 and increased miR-301a-3p expression causes Treg/ Th17 imbalance.

In gastrointestinal related autoimmune diseases, miR-125b-5p, miR-192-5p, and miR-215-5p showed to be potential diagnostic biomarkers in CD, discriminating diseased patients from normal controls. Also, increased levels of miR-449 exhibits an inhibitory effect on NOTCH and KLF4, affecting intestinal cell maturation and differentiation. In inflammatory bowel diseases, miR-122a, miR-132, and miR-223 target EGFR and NF-κB pathway, thus promoting inflammation. MiR-31, miR-206, miR-424, and miR-146a reported to be a potent tissue panel for IBD discrimination from normal controls. On the other hand, miR-16, miR-155, and miR-223 demonstrated promising potential for fecal miRNAs in IBD. Regarding autoimmune liver diseases, serum levels of miR-122-5p, miR-141–3p, and miR-26b-5p have been reported to contribute to the diagnosis of PBC. In PSC, downregulation of miR-200 and upregulation of miR-193b, miR-122, miR-885-5p, miR-1281, miR-126, miR-26a, and miR-30b have been detected. On the other hand, AIH patients with cirrhosis exhibit decreased levels of miR-21 and miR-122.

Regarding autoimmune disorders of the nervous system, it has been shown that serum miR-24-3p and miR-191-5p are upregulated in RRMS. Also, upregulation miR-24-3p, miR-128-3p, miR-191-5p, and miR-376c-3p has been reported in the sera of PPMS patients. On the other hand, miR-21-5p, miR-26b-5p, miR-29b-3p, miR-142-3p and miR-155-5p are downregulated in SPMS patients. In myasthenia gravis, miR-21 and miR-150-5p are highly expressed while miR-27a-3p is downregulated in the sera of MG patients. In MuSK+ MG, miR-151a-3p, let-7a-5p, let-7f-5p, and miR-423-5 are upregulated, while miR-210–3p and miR-324–3p are downregulated.

Concerning endocrine autoimmune disorders, miR-152, miR-30a-5p, miR-181a, miR-24, miR-148a, miR-210, miR-27a, miR-29a, miR-26a, miR-27b, miR-25, and miR-200a are upregulated in type 1 DM. In Hashimoto thyroiditis, serum miR-21-5p, miR-22-3p, miR-22-5p, miR-142-3p, miR-146a-5p, miR-301–3p, and miR-451 have been reported to be upregulated. Finally, in immune thrombocytopenia, plasma miR-199a-5p and miR-221–3p had a potential role in the prediction of response to TPO-RA.

Acknowledgement The authors acknowledge the feedback anonymous reviewers as well as the book's Editor for their valuable guidance throughout the revision process.

Compliance with Ethical Standards

Conflict of Interest The authors declare that they have no conflict of interest.

Ethical Approval This chapter contains no studies with human/animal participants performed by authors.

References

Abdallah HY, Abdelhamid NR, Mohammed EA (2022) Investigating melanogenesis-related microRNAs as disease biomarkers in vitiligo. Sci Rep 12(1):13526. https://doi.org/10.1038/s41598-022-17770-3

Aboelenein HR, Hamza MT, Marzouk H et al (2017) Reduction of CD19 autoimmunity marker on B cells of paediatric SLE patients through repressing PU.1/TNF-α/BAFF axis pathway by miR-155. Growth Factors 35(2–3):49–60. https://doi.org/10.1080/08977194.2017.1345900

Al-Ashmawy G (2018) Dendritic cell subsets, maturation and function. Dendritic Cells:11–24. https://doi.org/10.5772/intechopen.79926

Almanza G, Fernandez A, Volinia S, Cortez-Gonzalez X, Croce CM, Zanetti M (2010) Selected microRNAs define cell fate determination of murine central memory CD8 T cells. PLoS One 5:e11243. https://doi.org/10.1371/journal.pone.0011243

Anaparti V, Smolik I, Meng X et al (2017) Whole blood microRNA expression pattern differentiates patients with rheumatoid arthritis, their seropositive first-degree relatives, and healthy unrelated control subjects. Arthritis Res Ther. https://doi.org/10.1186/s13075-017-1459-x

Angum F, Khan T, Kaler J, Siddiqui L et al (2020) The prevalence of autoimmune disorders in women: a narrative review. Cureus 12(5). https://doi.org/10.7759/cureus.8094

Arda O, Göksügür N, Tüzün Y (2014) Basic histological structure and functions of facial skin. Clin Dermatol 32(1):3–13. https://doi.org/10.1016/j.clindermatol.2013.05.021

Arif KT, Elliott EK, Haupt LM et al (2020) Regulatory mechanisms of epigenetic miRNA relationships in human cancer and potential as therapeutic targets. Cancers 12(10):2922

Arroyo AB, Fernández-Pérez MP, Del Monte A et al (2021) miR-146a is a pivotal regulator of neutrophil extracellular trap formation promoting thrombosis. Haematologica 106(6):1636–1646. https://doi.org/10.3324/haematol.2019.240226

Arruda LC, Lorenzi JC, Sousa AP et al (2015) Autologous hematopoietic SCT normalizes miR-16, −155 and −142-3p expression in multiple sclerosis patients. Bone Marrow Transplant 50(3):380–389. https://doi.org/10.1038/bmt.2014.277

Banales JM, Sáez E, Uriz M et al (2012) Up-regulation of microRNA 506 leads to decreased Cl-/HCO3- anion exchanger 2 expression in biliary epithelium of patients with primary biliary cirrhosis. Hepatology 56(2):687–697. https://doi.org/10.1002/hep.25691

Barutta F, Tricarico M, Corbelli A et al (2013) Urinary exosomal microRNAs in incipient diabetic nephropathy. PLoS One. https://doi.org/10.1371/journal.pone.0073798

Barutta F, Bruno G, Matullo G et al (2017) MicroRNA-126 and micro−/macrovascular complications of type 1 diabetes in the EURODIAB prospective complications study. Acta Diabetol 54(2):133–139. https://doi.org/10.1007/s00592-016-0915-4

Bascuñán KA, Pérez-Bravo F, Gaudioso G et al (2020) A miRNA-based blood and mucosal approach for detecting and monitoring celiac disease. Dig Dis Sci 65(7):1982–1991. https://doi.org/10.1007/s10620-019-05966-z

Baumjohann D, Ansel KM (2013) MicroRNA-mediated regulation of T helper cell differentiation and plasticity. Nat Rev Immunol 1:666–678. https://doi.org/10.1038/nri3494

Bell-Hensley A, Das S, McAlinden A (2023) The miR-181 family: wide-ranging pathophysiological effects on cell fate and function. J Cell Physiol 238:698–713. https://doi.org/10.1002/jcp.30969

Beretta F, Huang YF, Punga AR (2022) Towards personalized medicine in myasthenia gravis: role of circulating microRNAs miR-30e-5p, miR-150-5p and miR-21-5p. Cells. https://doi.org/10.3390/cells11040740

Bernuzzi F, Marabita F, Lleo A, Carbone M (2016) Serum microRNAs as novel biomarkers for primary sclerosing cholangitis and cholangiocarcinoma. Clin Exp Immunol 185(1):61–71. https://doi.org/10.1111/cei.12776

Braicu C, Gulei D, Raduly L et al (2019) Altered expression of miR-181 affects cell fate and targets drug resistance-related mechanisms. Mol Asp Med 70:90–105. https://doi.org/10.1016/j.mam.2019.10.007

Brooks WS, Lee YY, Abell E et al (1989) Comparison of IgG subclasses and complement binding activity of autoantibodies from patients with bullous pemphigoid and pemphigus. J Clin Lab Anal 3(5):307–311. https://doi.org/10.1002/jcla.1860030509

Buoli Comani G, Panceri R, Dinelli M et al (2015) miRNA-regulated gene expression differs in celiac disease patients according to the age of presentation. Genes Nutr. https://doi.org/10.1007/s12263-015-0482-2

Bure IV, Nemtsova MV, Kuznetsova EB et al (2022) Histone modifications and non-coding RNAs: mutual epigenetic regulation and role in pathogenesis. Int J Mol Sci 23(10):5801. https://doi.org/10.3390/ijms23105801

Capuano M, Iaffaldano L, Tinto N et al (2011) MicroRNA-449a overexpression, reduced NOTCH1 signals and scarce goblet cells characterize the small intestine of celiac patients. PLoS One. https://doi.org/10.1371/journal.pone.0029094. Erratum in: PLoS One. doi:10.1371/annotation/7db01dfd-e044-41df-b1a8-9256ae11e4e2

Catalán D, Mansilla MA, Ferrier A et al (2021) Immunosuppressive mechanisms of regulatory B cells. Front Immunol 12:611795. https://doi.org/10.3389/fimmu.2021.611795

Cavers A, Kugler MC, Ozguler Y et al (2022) Behçet's disease risk-variant HLA-B51/ERAP1-Hap10 alters human CD8 T cell immunity. Ann Rheum Dis 81(11):1603–1611. https://doi.org/10.1136/ard-2022-222277

Celhar T, Magalhães R, Fairhurst AM (2012) TLR7 and TLR9 in SLE: when sensing self goes wrong. Immunol Res 53(1–3):58–77. https://doi.org/10.1007/s12026-012-8270-1

Chandan K, Gupta M, Sarwat M (2020) Role of host and pathogen-derived MicroRNAs in immune regulation during infectious and inflammatory diseases. Front Immunol. https://doi.org/10.3389/fimmu.2019.03081

Chang Y, Chen X, Tian Y et al (2020) Downregulation of microRNA-155-5p prevents immune thrombocytopenia by promoting macrophage M2 polarization via the SOCS1-dependent PD1/PDL1 pathway. Life Sci. https://doi.org/10.1016/j.lfs.2020.118057

Chen Y, Chen J, Wang H, Shi J, Wu K, Liu S et al (2013) HCV-induced miR-21 contributes to evasion of host immune system by targeting MyD88 and IRAK1. PLoS Pathog 9:e1003248. https://doi.org/10.1371/journal.ppat.1003248

Chen L, Deng H, Cui H, Fang J, Zuo Z, Deng J et al (2018a) Inflammatory responses and inflammation-associated diseases in organs. Oncotarget 9:7204. https://doi.org/10.18632/oncotarget.23208da

Chen YJ, Chang WA, Wu LY et al (2018b) Systematic analysis of differential expression profile in rheumatoid arthritis chondrocytes using next-generation sequencing and bioinformatics approaches. Int J Med Sci 15(11):1129–1142. https://doi.org/10.7150/ijms.27056

Chen D, Ji Q, Liu J et al (2023) MicroRNAs in the regulation of RIG-I-like receptor signaling pathway: possible strategy for viral infection and cancer. Biomol Ther 13(9):1344. https://doi.org/10.3390/biom13091344

Chi M, Ma K, Li Y et al (2021) Immunological involvement of MicroRNAs in the key events of systemic lupus Erythematosus. Front Immunol. https://doi.org/10.3389/fimmu.2021.699684

Chi LH, Cross RS, Redvers RP et al (2022) MicroRNA-21 is immunosuppressive and pro-metastatic via separate mechanisms. Oncogenesis 11(1):38. https://doi.org/10.1038/s41389-022-00413-7

Cipolla GA, Park JK, de Oliveira LA et al (2016) A 3'UTR polymorphism marks differential KLRG1 mRNA levels through disruption of a miR-584-5p binding site and associates with pemphigus foliaceus susceptibility. Biochim Biophys Acta 1859(10):1306–1313. https://doi.org/10.1016/j.bbagrm.2016.07.006

Condrat CE, Thompson DC, Barbu MG et al (2020) miRNAs as Biomarkers in Disease: Latest Findings Regarding Their Role in Diagnosis and Prognosis. Cells 9(2):276. https://doi.org/10.3390/cells9020276. PMID: 31979244; PMCID: PMC7072450

Consonni F, Favre C, Gambineri E (2021) IL-2 signaling axis defects: how many faces? Front Pediatr. https://doi.org/10.3389/fped.2021.669298

Cui TT, Yi XL, Zhang WG et al (2015) miR-196a-2 rs11614913 polymorphism is associated with vitiligo by affecting heterodimeric molecular complexes of Tyr and Tyrp1. Arch Dermatol Res 307(8):683–692. https://doi.org/10.1007/s00403-015-1563-1

Curtale G, Rubino M, Locati M (2019) MicroRNAs as molecular switches in macrophage activation. Front Immunol 10:799. https://doi.org/10.3389/fimmu.2019.00799

Cutolo M, Campitiello R, Gotelli E et al (2022) The role of M1/M2 macrophage polarization in rheumatoid arthritis Synovitis. Front Immunol. https://doi.org/10.3389/fimmu.2022.867260

Danger R, Braza F, Giral M, Soulillou JP, Brouard S (2014) MicroRNAs, major players in B cells homeostasis and function. Front Immunol 5:98. https://doi.org/10.3389/fimmu.2014.00098

Danielsson K, Wahlin YB, Gu X et al (2012) Altered expression of miR-21, miR-125b, and miR-203 indicates a role for these microRNAs in oral lichen planus. J Oral Pathol Med 41(1):90–95. https://doi.org/10.1111/j.1600-0714.2011.01084.x

Degueurce G, D'Errico I, Pich C et al (2016) Identification of a novel PPARβ/δ/miR-21-3p axis in UV-induced skin inflammation. EMBO Mol Med 8(8):919–936. https://doi.org/10.15252/emmm.201505384
Dendrou CA, Fugger L, Friese MA (2015) Immunopathology of multiple sclerosis. Nat Rev Immunol 15(9):545–558. https://doi.org/10.1038/nri3871
Dey N, Das F, Mariappan MM et al (2011) MicroRNA-21 orchestrates high glucose-induced signals to TOR complex 1, resulting in renal cell pathology in diabetes. J Biol Chem 29:25586–25603. https://doi.org/10.1074/jbc.M110.208066
Ding S, Liang Y, Zhao M et al (2012) Decreased microRNA-142-3p/5p expression causes CD4+ T cell activation and B cell hyperstimulation in systemic lupus erythematosus. Arthritis Rheum 64(9):2953–2963. https://doi.org/10.1002/art.34505
Dong L, Wang X, Tan J et al (2014) Decreased expression of microRNA-21 correlates with the imbalance of Th17 and Treg cells in patients with rheumatoid arthritis. J Cell Mol Med 18(11):2213–2224. https://doi.org/10.1111/jcmm.12353
Dopytalska K, Czaplicka A, Szymańska E et al (2023) The essential role of microRNAs in inflammatory and autoimmune skin diseases-a review. Int J Mol Sci. https://doi.org/10.3390/ijms24119130
Dresser L, Wlodarski R, Rezania K et al (2021) Myasthenia gravis: epidemiology, pathophysiology and clinical manifestations. J Clin Med. https://doi.org/10.3390/jcm10112235
Du C, Liu C, Kang J et al (2009) MicroRNA miR-326 regulates TH-17 differentiation and is associated with the pathogenesis of multiple sclerosis. Nat Immunol 10(12):1252–1259. https://doi.org/10.1038/ni.1798
Duroux-Richard I, Pers YM, Fabre S et al (2014) Circulating miRNA-125b is a potential biomarker predicting response to rituximab in rheumatoid arthritis. Mediat Inflamm. https://doi.org/10.1155/2014/342524
Eftekharian MM, Komaki A, Mazdeh M et al (2019) Expression profile of selected MicroRNAs in the peripheral blood of multiple sclerosis patients: a multivariate statistical analysis with ROC curve to find new biomarkers for Fingolimod. J Mol Neurosci 68(1):153–161. https://doi.org/10.1007/s12031-019-01294-z
Eisenbarth GS, Gottlieb PA (2004) Autoimmune polyendocrine syndromes. N Engl J Med 350(20):2068–2079. https://doi.org/10.1056/NEJMra030158
El-Rifaie AA, Rashed LA, Doss RW et al (2017) MicroRNAs in cutaneous lichen planus. Clin Exp Dermatol 42(8):898–901. https://doi.org/10.1111/ced.13235
Fasano A (2011) Zonulin and its regulation of intestinal barrier function: the biological door to inflammation, autoimmunity, and cancer. Physiol Rev 91(1):151–175. https://doi.org/10.1152/physrev.00003.2008
Felli C, Baldassarre A, Uva P et al (2022) Circulating microRNAs as novel non-invasive biomarkers of paediatric celiac disease and adherence to gluten-free diet. EBioMedicine. https://doi.org/10.1016/j.ebiom.2022.103851
Fordham JB, Naqvi AR, Nares S (2015) miR-24 regulates macrophage polarization and plasticity. J Clin Cell Immunol 6(5):362. https://doi.org/10.4172/2155-9899.1000362. Epub. PMID: 26807309
Gandhi R, Healy B, Gholipour T et al (2013) Circulating microRNAs as biomarkers for disease staging in multiple sclerosis. Ann Neurol 73(6):729–740. https://doi.org/10.1002/ana.23880
Garabet L, Ghanima W, Rangberg A et al (2020) Circulating microRNAs in patients with immune thrombocytopenia before and after treatment with thrombopoietin-receptor agonists. Platelets 31(2):198–205. https://doi.org/10.1080/09537104.2019.1585527
Ghisi M, Corradin A, Basso K et al (2011) Modulation of microRNA expression in human T-cell development: targeting of NOTCH3 by miR-150. Blood 117(26):7053–7062. https://doi.org/10.1182/blood-2010-12-32662

Gilicze AB, Wiener Z, Tóth S, Buzás E, Pállinger É, Falcone FH, Falus A (2014) Myeloid-derived microRNAs, miR-223, miR27a, and miR-652, are dominant players in myeloid regulation. Biomed Res Int 2014:870267. https://doi.org/10.1155/2014/870267

Gonzalez-Martin A, Adams BD, Lai M, Shepherd J, Salvador-Bernaldez M, Salvador JM et al (2016) The microRNA miR-148a functions as a critical regulator of B cell tolerance and autoimmunity. Nat Immunol 17(4):433–440. https://doi.org/10.1038/ni.3385

Gül A (2005) Behçet's disease as an autoinflammatory disorder. Curr Drug Targets Inflamm Allergy 4(1):81–83. https://doi.org/10.2174/1568010053622894

Guo LJ, Liao L, Yang L et al (2014) MiR-125a TNF receptor-associated factor 6 to inhibit osteoclastogenesis. Exp Cell Res 321(2):142–152. https://doi.org/10.1016/j.yexcr.2013.12.001

Gutierrez-Arcelus M, Rich SS, Raychaudhuri S et al (2016) Autoimmune diseases—connecting risk alleles with molecular traits of the immune system. Nat Rev Genet 17(3):160–174. https://doi.org/10.1038/nrg.2015.33

Hawez A, Al-Haidari A, Madhi R et al (2019) MiR-155 regulates PAD4-dependent formation of neutrophil extracellular traps. Front Immunol. https://doi.org/10.3389/fimmu.2019.02462

He W, Xing Y, Li C et al (2022) Identification of six microRNAs as potential biomarkers for pemphigus vulgaris: from diagnosis to pathogenesis. Diagnostics (Basel) 12(12):3058. https://doi.org/10.3390/diagnostics12123058

Hippen KL, Loschi M, Nicholls J et al (2018) Effects of microRNA on regulatory T cells and implications for adoptive cellular therapy to ameliorate graft-versus-host disease. Front Immunol 9:57. https://doi.org/10.3389/fimmu.2018.00057

Hirschberger S, Hinske LC, Kreth S et al (2018) MiRNAs: dynamic regulators of immune cell functions in inflammation and cancer. Cancer Lett 431:11–21. https://doi.org/10.1016/j.canlet.2018.05.020

Hlača N, Žagar T, Kaštelan M et al (2022) Current concepts of Vitiligo Immunopathogenesis. Biomedicines. https://doi.org/10.3390/biomedicines10071639

Hou K, Wu ZX, Chen XY et al (2022) Microbiota in health and diseases. Signal Transduct Target Ther 7(1):135. https://doi.org/10.1038/s41392-022-00974-4

Hruskova V, Jandova R, Vernerova L et al (2016) MicroRNA-125b: association with disease activity and the treatment response of patients with early rheumatoid arthritis. Arthritis Res Ther. https://doi.org/10.1186/s13075-016-1023-0

Huang Y, Yi X, Jian Z et al (2013) A single-nucleotide polymorphism of miR-196a-2 and vitiligo: an association study and functional analysis in a Han Chinese population. Pigment Cell Melanoma Res 26(3):338–347. https://doi.org/10.1111/pcmr.12081

Huang C, Liu XJ, QunZhou et al (2016) MiR-146a modulates macrophage polarization by inhibiting Notch1 pathway in RAW264.7 macrophages. Int Immunopharmacol 32:46–54. https://doi.org/10.1016/j.intimp.2016.01.009

Huang Z, Xing S, Liu M et al (2019) MiR-26a-5p enhances cells proliferation, invasion, and apoptosis resistance of fibroblast-like synoviocytes in rheumatoid arthritis by regulating PTEN/PI3K/AKT pathway. Biosci Rep. https://doi.org/10.1042/BSR20182192

Huang C, Xing X, Xiang X et al (2020) MicroRNAs in autoimmune liver diseases: from diagnosis to potential therapeutic targets. Biomed Pharmacother. https://doi.org/10.1016/j.biopha.2020.110558

Huang QQ, Doyle R, Chen SY et al (2021) Critical role of synovial tissue-resident macrophage niche in joint homeostasis and suppression of chronic inflammation. Sci Adv. https://doi.org/10.1126/sciadv.abd0515

Huo J, Liu T, Li F et al (2021) MicroRNA-21-5p protects melanocytes via targeting STAT3 and modulating Treg/Teff balance to alleviate vitiligo. Mol Med Rep. https://doi.org/10.3892/mmr.2020.11689

Ingwersen J, Menge T, Wingerath B et al (2015) Natalizumab restores aberrant miRNA expression profile in multiple sclerosis and reveals a critical role for miR-20b. Ann Clin Transl Neurol 2(1):43–55. https://doi.org/10.1002/acn3.152

Ishina IA, Zakharova MY, Kurbatskaia IN et al (2023) MHC class II presentation in autoimmunity. Cells 12(2):314

Jadideslam G, Ansarin K, Sakhinia E et al (2019) Expression levels of miR-21, miR-146b and miR-326 as potential biomarkers in Behcet's disease. Biomark Med 13(16):1339–1348. https://doi.org/10.2217/bmm-2019-0098

Jia Y, Wei Y (2020) Modulators of MicroRNA function in the immune system. Int J Mol Sci 21(7):2357. https://doi.org/10.3390/ijms21072357

Jia HY, Zhang K, Lu WJ et al (2019) LncRNA MEG3 influences the proliferation and apoptosis of psoriasis epidermal cells by targeting miR-21/caspase-8. BMC Mol Cell Biol. https://doi.org/10.1186/s12860-019-0229-9

Jiang Q, Wang Q, Tan S et al (2023) Effects of plasma-derived Exosomal miRNA-19b-3p on Treg/T helper 17 cell imbalance in Behçet's uveitis. Invest Ophthalmol Vis Sci. https://doi.org/10.1167/iovs.64.4.28

Johnnidis JB, Harris MH, Wheeler RT, Stehling-Sun S, Lam MH, Kirak O et al (2008) Regulation of progenitor cell proliferation and granulocyte function by microRNA-223. Nature 451:1125. https://doi.org/10.1038/nature06607

Jurewicz MM, Stern LJ (2019) Class II MHC antigen processing in immune tolerance and inflammation. Immunogenetics 71:171–187. https://doi.org/10.1007/s00251-018-1095-x

Kern F, Aparicio-Puerta E, Li Y et al (2021) miRTargetLink 2.0—interactive miRNA target gene and target pathway networks. Nucleic Acids Res 49(W1):W409–W416. https://doi.org/10.1093/nar/gkab297

Khan MF, Wang H (2020) Environmental exposures and autoimmune diseases: contribution of gut microbiome. Front Immunol 10:3094. https://doi.org/10.3389/fimmu.2019.03094

Kim SJ, Gregersen PK, Diamond B (2013) Regulation of dendritic cell activation by microRNA let-7c and BLIMP1. J Clin Invest 123(2):823–833. https://doi.org/10.1172/JCI64712

Kim HY, Kwon HY, Ha Thi HT et al (2016) MicroRNA-132 and microRNA-223 control positive feedback circuit by regulating FOXO3a in inflammatory bowel disease. J Gastroenterol Hepatol 31(10):1727–1735. https://doi.org/10.1111/jgh.13321

Kim C, Ye Z, Weyand CM et al (2021) miR-181a-regulated pathways in T-cell differentiation and aging. Immun Ageing 18(1):28. https://doi.org/10.1186/s12979-021-00240-1

King JK, Ung NM, Paing MH (2016) Regulation of marginal zone B-cell differentiation by MicroRNA-146a. Front Immunol 7:670. https://doi.org/10.3389/fimmu.2016.00670

Kmiołek T, Paradowska-Gorycka A (2022) miRNAs as biomarkers and possible therapeutic strategies in rheumatoid arthritis. Cells 11(3):452. https://doi.org/10.3390/cells11030452

Lai M, Gonzalez-Martin A, Cooper AB et al (2016) Regulation of B-cell development and tolerance by different members of the miR-17∼ 92 family microRNAs. Nat Commun 7(1):12207. https://doi.org/10.1038/ncomms12207

Lakhter AJ, Pratt RE, Moore RE et al (2018) Beta cell extracellular vesicle miR-21-5p cargo is increased in response to inflammatory cytokines and serves as a biomarker of type 1 diabetes. Diabetologia 61(5):1124–1134. https://doi.org/10.1007/s00125-018-4559-5

Leal B, Carvalho C, Ferreira AM et al (2021) Serum levels of miR-146a in patients with psoriasis. Mol Diagn Ther 25(4):475–485. https://doi.org/10.1007/s40291-021-00531-9

Lewis ED, Wu D, Meydani SN (2022) Age-associated alterations in immune function and inflammation. Prog Neuro-Psychopharmacol Biol Psychiatry 118:110576. https://doi.org/10.1016/j.pnpbp.2022.110576

Li T, Morgan MJ, Choksi S (2010) MicroRNAs modulate the noncanonical NF-κB pathway by regulating IKKα expression during macrophage differentiation. Nat Immunol 11:799. https://doi.org/10.1038/ni.1918

Li JQ, Tian JM, Fan XR et al (2020) miR-106b-5p induces immune imbalance of Treg/Th17 in immune thrombocytopenic purpura through NR4A3/Foxp3 pathway. Cell Cycle 19(11):1265–1274. https://doi.org/10.1080/15384101.2020.1746485. Erratum in: Cell Cycle (15):1994-1995

Liang L, Zhou Q, Feng L (2021) Decreased microRNA-155 in Behcet's disease leads to defective control of autophagy thereby stimulating excessive proinflammatory cytokine production. Arthritis Res Ther. https://doi.org/10.1186/s13075-021-02517-8

Lin J, Welker NC, Zhao Z et al (2014) Novel specific microRNA biomarkers in idiopathic inflammatory bowel disease unrelated to disease activity. Mod Pathol 27(4):602–608. https://doi.org/10.1038/modpathol.2013.152

Lin N, Liu Q, Wang M et al (2018) Usefulness of miRNA-338-3p in the diagnosis of pemphigus and its correlation with disease severity. PeerJ. https://doi.org/10.7717/peerj.5388

Liu Y, Dong J, Mu R et al (2013) MicroRNA-30a promotes B cell hyperactivity in patients with systemic lupus erythematosus by direct interaction with Lyn. Arthritis Rheum 65(6):1603–1611. https://doi.org/10.1002/art.37912

Liu SQ, Jiang S, Li C et al (2014) miR-17–92 cluster targets phosphatase and tensin homology and ikaros family zinc finger 4 to promote TH17-mediated inflammation. J Biol Chem 289:12446–12456. https://doi.org/10.1074/jbc.M114.550723

Liu L, Hua M, Liu C et al (2016) The aberrant expression of microRNAs and correlations with T cell subsets in patients with immune thrombocytopenia. Oncotarget 7(47):76453–76463. https://doi.org/10.18632/oncotarget.12949

Liu Q, Cui F, Wang M et al (2018a) Increased expression of microRNA-338-3p contributes to production of Dsg3 antibody in pemphigus vulgaris patients. Mol Med Rep 18(1):550–556. https://doi.org/10.3892/mmr.2018.8934

Liu W, Wu YH, Zhang L et al (2018b) MicroRNA-146a suppresses rheumatoid arthritis fibroblast-like synoviocytes proliferation and inflammatory responses by inhibiting the TLR4/NF-kB signaling. Oncotarget 9(35):23944. https://doi.org/10.18632/oncotarget.24050

Liu C, Li N, Liu G (2020) The role of MicroRNAs in regulatory T cells. J Immunol Res 2020: 3232061. https://doi.org/10.1155/2020/3232061. PMID: 32322593; PMCID: PMC7154970

Liu F, Liang Y, Zhao Y et al (2021) Meta-analysis of association of microRNAs genetic variants with susceptibility to rheumatoid arthritis and systemic lupus erythematosus. Medicine (Baltimore). https://doi.org/10.1097/MD.0000000000025689

Liu L, Chen H, Jiang T et al (2022) MicroRNA-106b overexpression suppresses synovial inflammation and alleviates synovial damage in patients with rheumatoid arthritis. Mod Rheumatol 32(6):1054–1063. https://doi.org/10.1093/mr/roab108

Luan X, Zhou X, Naqvi A et al (2018) MicroRNAs and immunity in periodontal health and disease. Int J Oral Sci 10:1–14. https://doi.org/10.1038/s41368-018-0025-y

Luo S, Liu Y, Liang G et al (2015) The role of microRNA-1246 in the regulation of B cell activation and the pathogenesis of systemic lupus erythematosus. Clin Epigenetics. https://doi.org/10.1186/s13148-015-0063-7

Lv M, Li Z, Liu J, Lin F, Zhang Q, Li Z, Wang Y, Wang K, Xu Y (2019) MicroRNA-155 inhibits the proliferation of CD8+ T cells via upregulating regulatory T cells in vitiligo. Mol Med Rep 20(4):3617–3624. https://doi.org/10.3892/mmr.2019.10607

Ma F, Xu S, Liu X (2011) The microRNA miR-29 controls innate and adaptive immune responses to intracellular bacterial infection by targeting interferon-gamma. Nat Immunol 12:861–869. https://doi.org/10.1038/ni.2073

Marshall JS, Warrington R, Watson W (2018) An introduction to immunology and immunopathology. Allergy, Asthma Clin Immunol 14(2):1. https://doi.org/10.1186/s13223-018-0278-1

Masalha M, Sidi Y, Avni D (2018) The contribution of feedback loops between miRNAs, cytokines and growth factors to the pathogenesis of psoriasis. Exp Dermatol 27(6):603–610. https://doi.org/10.1111/exd.13520

Mashima R (2015) Physiological roles of miR-155. Immunology 145(3):323–333. https://doi.org/10.1111/imm.12468

Mazzarella G (2015) Effector and suppressor T cells in celiac disease. World J Gastroenterol 21(24):7349–7356. https://doi.org/10.3748/wjg.v21.i24.7349

Mazzone R, Zwergel C, Artico M et al (2019) The emerging role of epigenetics in human autoimmune disorders. Clin Epigenetics 11(1):1–5. https://doi.org/10.1186/s13148-019-0632-2

McGeachy MJ, Chen Y, Tato CM et al (2009) The interleukin 23 receptor is essential for the terminal differentiation of interleukin 17-producing effector T helper cells in vivo. Nat Immunol 10(3):314–324. https://doi.org/10.1038/ni.1698

Mehdipour M, Shahidi M, Manifar S et al (2018) Diagnostic and prognostic relevance of salivary microRNA-21, −125a, −31 and -200a levels in patients with oral lichen planus - a short report. Cell Oncol (Dordr) 41(3):329–334. https://doi.org/10.1007/s13402-018-0372-x

Mehta A, Baltimore D (2016) MicroRNAs as regulatory elements in immune system logic. Nat Rev Immunol 16:279–294. https://doi.org/10.1038/nri.2016.40

Migita K, Komori A, Kozuru H et al (2015) Circulating microRNA profiles in patients with Type-1 autoimmune hepatitis. PLoS One. https://doi.org/10.1371/journal.pone.0136908

Min B (2018) Spontaneous T cell proliferation: a physiologic process to create and maintain homeostatic balance and diversity of the immune system. Front Immunol 9:547. https://doi.org/10.3389/fimmu.2018.00547

Mohammed EM (2020) Environmental influencers, MicroRNA, and multiple sclerosis. J Cent Nerv Syst Dis. https://doi.org/10.1177/1179573519894955

Molin CJ, Sabre L, Weis CA et al (2018) Thymectomy lowers the myasthenia gravis biomarker miR-150-5p. Neurol Neuroimmunol Neuroinflamm. https://doi.org/10.1212/NXI.0000000000000450

Mousavi MJ, Karami J, Aslani S et al (2021) Transformation of fibroblast-like synoviocytes in rheumatoid arthritis; from a friend to foe. Auto Immun Highlights. https://doi.org/10.1186/s13317-020-00145-x

Mutua V, Gershwin LJ (2021) A review of neutrophil extracellular traps (NETs) in disease: potential anti-NETs therapeutics. Clin Rev Allergy Immunol 61(2):194–211. https://doi.org/10.1007/s12016-020-08804-7

Na SY, Park MJ, Park S et al (2016) MicroRNA-155 regulates the Th17 immune response by targeting Ets-1 in Behçet's disease. Clin Exp Rheumatol 34(6 Suppl 102):S56–S63

Nabih ES, Andrawes NG (2016) The association between circulating levels of miRNA-181a and pancreatic Beta cells dysfunction via SMAD7 in type 1 diabetic children and adolescents. J Clin Lab Anal 30(5):727–731. https://doi.org/10.1002/jcla.21928

Nakasa T, Miyaki S, Okubo A et al (2008) Expression of microRNA-146 in rheumatoid arthritis synovial tissue. Arthritis Rheum 58(5):1284–1292. https://doi.org/10.1002/art.23429

Nielsen LB, Wang C, Sørensen K et al (2012) Circulating levels of microRNA from children with newly diagnosed type 1 diabetes and healthy controls: evidence that miR-25 associates to residual beta-cell function and glycaemic control during disease progression. Exp Diabetes Res. https://doi.org/10.1155/2012/896362. Erratum in: Exp Diabetes Res. 2012;2012:672865

Nüssing S, Trapani JA, Parish IA (2020) Revisiting T cell tolerance as a checkpoint target for cancer immunotherapy. Front Immunol 11:2461. https://doi.org/10.3389/fimmu.2020.589641

O'connell RM, Rao DS, Chaudhuri AA et al (2010) Physiological and pathological roles for microRNAs in the immune system. Nat Rev Immunol 10(2):111–122. https://doi.org/10.1038/nri2708

Okeke EB, Uzonna JE (2019) The pivotal role of regulatory T cells in the regulation of innate immune cells. Front Immunol 10:680. https://doi.org/10.3389/fimmu.2019.00680

Ouboussad L, Hunt L, Hensor EMA et al (2017) Profiling microRNAs in individuals at risk of progression to rheumatoid arthritis. Arthritis Res Ther. https://doi.org/10.1186/s13075-017-1492-9

Oztas MO, Onder M, Gurer MA et al (2005) Serum interleukin 18 and tumour necrosis factor-alpha levels are increased in Behcet's disease. Clin Exp Dermatol 30(1):61–63. https://doi.org/10.1111/j.1365-2230.2004.01684.x

Pan W, Zhu S, Yuan M et al (2010) MicroRNA-21 and microRNA-148a contribute to DNA hypomethylation in lupus CD4+ T cells by directly and indirectly targeting DNA methyltransferase 1. J Immunol 184(12):6773–6781. doi: https://doi.org/10.4049/jimmunol.0904060

Parzanese I, Qehajaj D, Patrinicola F et al (2017) Celiac disease: from pathophysiology to treatment. World J Gastrointest Pathophysiol 8(2):27–38. https://doi.org/10.4291/wjgp.v8.i2.27

Paschou SA, Papadopoulou-Marketou N, Chrousos GP et al (2018) On type 1 diabetes mellitus pathogenesis. Endocr Connect 7(1):R38–R46. https://doi.org/10.1530/EC-17-0347

Pashangzadeh S, Motallebnezhad M, Vafashoar F (2021) Implications the role of miR-155 in the pathogenesis of autoimmune diseases. Front Immunol 12:669382. https://doi.org/10.3389/fimmu.2021.669382

Pecora F, Persico F, Gismondi P et al (2020) Gut microbiota in celiac disease: is there any role for probiotics? Front Immunol. https://doi.org/10.3389/fimmu.2020.00957

Peng H, Liu Y, Tian J et al (2015) Decreased expression of microRNA-125a-3p upregulates interleukin-23 receptor in patients with Hashimoto's thyroiditis. Immunol Res 62(2):129–136. https://doi.org/10.1007/s12026-015-8643-3

Peng X, Wang Q, Li W et al (2023) Comprehensive overview of microRNA function in rheumatoid arthritis. Bone Res. https://doi.org/10.1038/s41413-023-00244-1

Pérez MMG, Eisele SJG (2022) MicroRNAs as a possible biomarker in the treatment of multiple sclerosis. IBRO Neurosci Rep 13:492–499. https://doi.org/10.1016/j.ibneur.2022.11.001

Punga T, Le Panse R, Andersson M et al (2014) Circulating miRNAs in myasthenia gravis: miR-150-5p as a new potential biomarker. Ann Clin Transl Neurol 1(1):49–58. https://doi.org/10.1002/acn3.24

Punga T, Bartoccioni E, Lewandowska M et al (2016) Disease specific enrichment of circulating let-7 family microRNA in MuSK+ myasthenia gravis. J Neuroimmunol 292:21–26. https://doi.org/10.1016/j.jneuroim.2016.01.003

Qi J, Yang Y, Hou S et al (2014) Increased notch pathway activation in Behçet's disease. Rheumatology (Oxford) 53(5):810–820. https://doi.org/10.1093/rheumatology/ket438

Qin H, Zhu X, Liang J et al (2013) MicroRNA-29b contributes to DNA hypomethylation of CD4+ T cells in systemic lupus erythematosus by indirectly targeting DNA methyltransferase 1. J Dermatol Sci 69(1):61–67. https://doi.org/10.1016/j.jdermsci.2012.10.011

Qin Z, Wang PY, Su DF et al (2016) miRNA-124 in immune system and immune disorders. Front Immunol 7:406. https://doi.org/10.3389/fimmu.2016.00406

Quero L, Tiaden AN, Hanser E et al (2020) miR-221-3p drives the shift of M2-macrophages to a pro-inflammatory function by suppressing JAK3/STAT3 activation. Front Immunol. https://doi.org/10.3389/fimmu.2019.03087

Ragusa F, Fallahi P, Elia G et al (2019) Hashimotos' thyroiditis: epidemiology, pathogenesis, clinic and therapy. Best Pract Res Clin Endocrinol Metab. https://doi.org/10.1016/j.beem.2019.101367

Rajasekhar M, Olsson AM, Steel KJ et al (2017) MicroRNA-155 contributes to enhanced resistance to apoptosis in monocytes from patients with rheumatoid arthritis. J Autoimmun 79:53–62. https://doi.org/10.1016/j.jaut.2017.01.002

Rasmussen TK, Andersen T, Bak RO et al (2015) Overexpression of microRNA-155 increases IL-21 mediated STAT3 signaling and IL-21 production in systemic lupus erythematosus. Arthritis Res Ther. https://doi.org/10.1186/s13075-015-0660-z

Roberts LB, Jowett GM, Read E, Zabinski T et al (2021a) MicroRNA-142 critically regulates group 2 innate lymphoid cell homeostasis and function. J Immunol 206(11):2725–2739. https://doi.org/10.4049/jimmunol.2000647

Roberts LB, Kapoor P, Howard JK et al (2021b) An update on the roles of immune system-derived microRNAs in cardiovascular diseases. Cardiovasc Res 117(12):2434–2449. https://doi.org/10.1093/cvr/cvab007

Rodríguez-Galán A, Fernández-Messina L, Sánchez-Madrid F (2018) Control of immunoregulatory molecules by miRNAs in T cell activation. Front Immunol 9:2148. https://doi.org/10.3389/fimmu.2018.02148

Roffel MP, Bracke KR, Heijink IH et al (2020) miR-223: a key regulator in the innate immune response in asthma and COPD. Front Med 7:196. https://doi.org/10.3389/fmed.2020.00196

Roy S (2016) miRNA in macrophage development and function. Antioxid Redox Signal 25:795–804. https://doi.org/10.1089/ars.2016.6728

Sabre L, Guptill JT, Russo M et al (2018) Circulating microRNA plasma profile in MuSK+ myasthenia gravis. J Neuroimmunol 325:87–91. https://doi.org/10.1016/j.jneuroim.2018.10.003

Sakata K, Nakayamada S, Miyazaki Y et al (2018) Up-regulation of TLR7-mediated IFN-α production by Plasmacytoid dendritic cells in patients with systemic lupus Erythematosus. Front Immunol. https://doi.org/10.3389/fimmu.2018.01957

Salmaninejad A, Zamani MR, Shabgah AG et al (2019) Behçet's disease: an immunogenetic perspective. J Cell Physiol 234(6):8055–8074. https://doi.org/10.1002/jcp.27576. Epub 2018 Oct 20

Sanders KA, Benton MC, Lea RA et al (2016) Next-generation sequencing reveals broad down-regulation of microRNAs in secondary progressive multiple sclerosis CD4+ T cells. Clin Epigenetics. https://doi.org/10.1186/s13148-016-0253-y

Schaefer JS, Attumi T, Opekun AR et al (2015) MicroRNA signatures differentiate Crohn's disease from ulcerative colitis. BMC Immunol. https://doi.org/10.1186/s12865-015-0069-0

Scherm MG, Daniel C (2020) miRNA-mediated immune regulation in islet autoimmunity and type 1 diabetes. Front Endocrinol 11:606322. https://doi.org/10.3389/fendo.2020.606322

Schett G, Gravallese E (2012) Bone erosion in rheumatoid arthritis: mechanisms, diagnosis and treatment. Nat Rev Rheumatol 8(11):656–664. https://doi.org/10.1038/nrrheum.2012.153

Schönauen K, Le N, von Arnim U et al (2018) Circulating and fecal microRNAs as biomarkers for inflammatory bowel diseases. Inflamm Bowel Dis 24(7):1547–1557. https://doi.org/10.1093/ibd/izy046

Shaker OG, Safa A, Khairy A et al (2023) Serum long noncoding RNA H19/micro RNA-675-5p axis as a probable diagnostic biomarker in inflammatory bowel disease. Mol Biol Rep. https://doi.org/10.1007/s11033-023-08777-8

Shan J, Jin H, Xu Y (2020) T cell metabolism: a new perspective on Th17/Treg cell imbalance in systemic lupus Erythematosus. Front Immunol. https://doi.org/10.3389/fimmu.2020.01027

Shen H, Wang D, Zhan M et al (2022) MicroRNA-146a and microRNA-146b deficiency correlates with exacerbated disease activity, and their longitude increment relates to etanercept response in psoriasis patients. J Clin Lab Anal. https://doi.org/10.1002/jcla.24198

Shi Q, Zhang W, Guo S et al (2016) Oxidative stress-induced overexpression of miR-25: the mechanism underlying the degeneration of melanocytes in vitiligo. Cell Death Differ 23(3):496–508. https://doi.org/10.1038/cdd.2015.117

Sims EK, Lakhter AJ, Anderson-Baucum E et al (2017) MicroRNA 21 targets BCL2 mRNA to increase apoptosis in rat and human beta cells. Diabetologia 60(6):1057–1065. https://doi.org/10.1007/s00125-017-4237-z

Smith KM, Guerau-de-Arellano M, Costinean S et al (2012) miR-29ab1 deficiency identifies a negative feedback loop controlling Th1 bias that is dysregulated in multiple sclerosis. J Immunol 189(4):1567–1576. https://doi.org/10.4049/jimmunol.1103171

Sogkas G, Atschekzei F, Adriawan IR et al (2021) Cellular and molecular mechanisms breaking immune tolerance in inborn errors of immunity. Cell Mol Immunol 18(5):1122–1140. https://doi.org/10.1038/s41423-020-00626-z

Soldan SS, Lieberman PM (2023) Epstein-Barr virus and multiple sclerosis. Nat Rev Microbiol 21(1):51–64. https://doi.org/10.1038/s41579-022-00770-5

Soto-Heredero G, Gómez de las Heras MM, Escrig-Larena JI et al (2023) Extremely differentiated T cell subsets contribute to tissue deterioration during aging. Annu Rev Immunol 41:181–205. https://doi.org/10.1146/annurev-immunol-101721-064501

Spritz RA, Andersen GH (2017) Genetics of Vitiligo. Dermatol Clin 35(2):245–255. https://doi.org/10.1016/j.det.2016.11.013

Stanczyk J, Ospelt C, Karouzakis E et al (2011) Altered expression of microRNA-203 in rheumatoid arthritis synovial fibroblasts and its role in fibroblast activation. Arthritis Rheum 63(2):373–381. https://doi.org/10.1002/art.30115

Stelzer G, Rosen N, Plaschkes I et al (2016) The GeneCards suite: from gene data mining to disease genome sequence analyses. Curr Protoc Bioinformatics 54(1):1–30. https://doi.org/10.1002/cpbi.5

Su M, Yi H, He X et al (2019) miR-9 regulates melanocytes adhesion and migration during vitiligo repigmentation induced by UVB treatment. Exp Cell Res 384(1):111615. https://doi.org/10.1016/j.yexcr.2019.111615

Sun CM, Wu J, Zhang H et al (2017) Circulating miR-125a but not miR-125b is decreased in active disease status and negatively correlates with disease severity as well as inflammatory

cytokines in patients with Crohn's disease. World J Gastroenterol 23(44):7888–7898. https://doi.org/10.3748/wjg.v23.i44.7888
Surace AE, Hedrich CM (2019) The role of epigenetics in autoimmune/inflammatory disease. Front Immunol 10:1525. https://doi.org/10.3389/fimmu.2019.01525
Takahashi H, Kanno T, Nakayamada S et al (2012) TGF-beta and retinoic acid induce the microRNA miR-10a, which targets Bcl-6 and constrains the plasticity of helper T cells. Nat Immunol 13:587–595. https://doi.org/10.1038/ni.2286
Tan LP, Wang M, Robertus JL et al (2009) miRNA profiling of B-cell subsets: specific miRNA profile for germinal center B cells with variation between centroblasts and centrocytes. Lab Investig 89:708. https://doi.org/10.1038/labinvest.2009.26
Tan Y, Pan T, Ye Y et al (2014) Serum microRNAs as potential biomarkers of primary biliary cirrhosis. PLoS One. https://doi.org/10.1371/journal.pone.0111424
Tang Y, Luo X, Cui H et al (2009) MicroRNA-146A contributes to abnormal activation of the type I interferon pathway in human lupus by targeting the key signaling proteins. Arthritis Rheum 60(4):1065–1075. https://doi.org/10.1002/art.24436
Tang X, Yin K, Zhu H et al (2016) Correlation between the expression of MicroRNA-301a-3p and the proportion of Th17 cells in patients with rheumatoid arthritis. Inflammation 39(2):759–767. https://doi.org/10.1007/s10753-016-0304-8
Tazi J, Begon-Pescia C, Campos N et al (2021) Specific and selective induction of miR-124 in immune cells by the quinoline ABX464: a transformative therapy for inflammatory diseases. Drug Discov Today 26(4):1030–1039. https://doi.org/10.1016/j.drudis.2020.12.019
Tessmer MS, Fugere C, Stevenaert F et al (2007) KLRG1 binds cadherins and preferentially associates with SHIP-1. Int Immunol 19(4):391–400. https://doi.org/10.1093/intimm/dxm004
Testa U, Pelosi E, Castelli G et al (2017) miR-146 and miR-155: two key modulators of immune response and tumor development. Non-coding RNA 3:22. https://doi.org/10.3390/ncrna3030022
Trummer O, Foessl I, Schweighofer N et al (2022) Expression profiles of miR-22-5p and miR-142-3p indicate Hashimoto's disease and are related to thyroid antibodies. Genes (Basel). https://doi.org/10.3390/genes13020171
Tsitsiou E, Lindsay MA (2009) microRNAs and the immune response. Curr Opin Pharmacol 9(4):514–520. https://doi.org/10.1016/j.coph.2009.05.003
Turner M, Schnorfeil F, Brocker T (2011) MicroRNAs regulate dendritic cell differenciation and function. J Immunol 187(8):3911–3917. https://doi.org/10.4049/jimmunol.1101137
Vistbakka J, Elovaara I, Lehtimäki T et al (2017) Circulating microRNAs as biomarkers in progressive multiple sclerosis. Mult Scler 23(3):403–412. https://doi.org/10.1177/1352458516651141
Vistbakka J, Sumelahti ML, Lehtimäki T et al (2018) Elovaara I, Hagman S. Evaluation of serum miR-191-5p, miR-24-3p, miR-128-3p, and miR-376c-3 in multiple sclerosis patients. Acta Neurol Scand 138(2):130–136. https://doi.org/10.1111/ane.12921
Voigtländer T, Gupta SK, Thum S et al (2015) MicroRNAs in serum and bile of patients with primary Sclerosing cholangitis and/or Cholangiocarcinoma. PLoS One. https://doi.org/10.1371/journal.pone.0139305
Wang G, Tam LS, Li EK et al (2010) Serum and urinary cell-free MiR-146a and MiR-155 in patients with systemic lupus erythematosus. J Rheumatol 37(12):2516–2522. https://doi.org/10.3899/jrheum.100308
Wang J, Luo H, Xiao Y et al (2016) miR-125b inhibits keratinocyte proliferation and promotes keratinocyte apoptosis in oral lichen planus by targeting MMP-2 expression through PI3K/Akt/mTOR pathway. Biomed Pharmacother 80:373–380. https://doi.org/10.1016/j.biopha.2016.02.043
Wang X, Wen X, Zhou J et al (2017) MicroRNA-223 and microRNA-21 in peripheral blood B cells associated with progression of primary biliary cholangitis patients. PLoS One. https://doi.org/10.1371/journal.pone.0184292

Wang H, Li X, Li T et al (2019) Multiple roles of microRNA-146a in immune responses and hepatocellular carcinoma. Oncol Lett 18(5):5033–5042. https://doi.org/10.3892/ol.2019.10862
Wang Y, Zhang J, Su Y et al (2020) miRNA-98-5p targeting IGF2BP1 induces Mesenchymal stem cell apoptosis by modulating PI3K/Akt and p53 in immune thrombocytopenia. Mol Ther Nucleic Acids 20:764–776. https://doi.org/10.1016/j.omtn.2020.04.013
Wang Y, Guo Y, Zhang X et al (2021a) The role and mechanism of miR-557 in inhibiting the differentiation and maturation of megakaryocytes in immune thrombocytopenia. RNA Biol 18(11):1953–1968. https://doi.org/10.1080/15476286.2021.1884783
Wang Y, Jia X, Zhou L et al (2021b) Increased let-7b-5p is associated with enhanced BAFF-R expression and B cell survival in immune thrombocytopenia. Int Immunopharmacol. https://doi.org/10.1016/j.intimp.2021.107393
Wei T, Folkersen L, Biskup E et al (2017) Ubiquitin-specific peptidase 2 as a potential link between microRNA-125b and psoriasis. Br J Dermatol 176(3):723–731. https://doi.org/10.1111/bjd.14916. Epub 2016 Dec 30
Wilson NS, Dixit V, Ashkenazi A (2009) Death receptor signal transducers: nodes of coordination in immune signaling networks. Nat Immunol 10(4):348–355. https://doi.org/10.1038/ni.1714
Wu XQ, Dai Y, Yang Y et al (2016) Emerging role of micro RNA s in regulating macrophage activation and polarization in immune response and inflammation. Immunology 148:237–248. https://doi.org/10.1111/imm.12608
Wu YL, Lin ZJ, Li CC et al (2023) Epigenetic regulation in metabolic diseases: mechanisms and advances in clinical study. Signal Transduct Target Ther 8(1):98. https://doi.org/10.1038/s41392-023-01333-7
Xia P, Fang X, Zhang ZH et al (2012) Dysregulation of miRNA146a versus IRAK1 induces IL-17 persistence in the psoriatic skin lesions. Immunol Lett 148(2):151–162. https://doi.org/10.1016/j.imlet.2012.09.004
Xiao Y, Wang H, Wang C et al (2020) miR-203 promotes HaCaT cell overproliferation through targeting LXR-α and PPAR-γ. Cell Cycle 19(15):1928–1940. https://doi.org/10.1080/15384101.2020.1783934
Xu N, Brodin P, Wei T et al (2011) MiR-125b, a microRNA downregulated in psoriasis, modulates keratinocyte proliferation by targeting FGFR2. J Invest Dermatol 131(7):1521–1529. https://doi.org/10.1038/jid.2011.55
Xu N, Meisgen F, Butler LM et al (2013) MicroRNA-31 is overexpressed in psoriasis and modulates inflammatory cytokine and chemokine production in keratinocytes via targeting serine/threonine kinase 40. J Immunol 190(2):678–688. https://doi.org/10.4049/jimmunol.1202695
Xu L, Leng H, Shi X et al (2017) MiR-155 promotes cell proliferation and inhibits apoptosis by PTEN signaling pathway in the psoriasis. Biomed Pharmacother 90:524–530. https://doi.org/10.1016/j.biopha.2017.03.105
Yan S, Xu Z, Lou F et al (2015) NF-κB-induced microRNA-31 promotes epidermal hyperplasia by repressing protein phosphatase 6 in psoriasis. Nat Commun 6:7652. https://doi.org/10.1038/ncomms8652
Yan L, Liang M, Hou X et al (2019) The role of microRNA-16 in the pathogenesis of autoimmune diseases: a comprehensive review. Biomed Pharmacother 112:108583. https://doi.org/10.1016/j.biopha.2019.01.044
Yang M, Ye L, Wang B et al (2015) Decreased miR-146 expression in peripheral blood mononuclear cells is correlated with ongoing islet autoimmunity in type 1 diabetes patients 1miR-146. J Diabetes 7(2):158–165. https://doi.org/10.1111/1753-0407.12163
Yang XK, Li P, Zhang C et al (2017) Association between IRAK1 rs3027898 and miRNA-499 rs3746444 polymorphisms and rheumatoid arthritis : a case control study and meta-analysis. Z Rheumatol 76(7):622–629. https://doi.org/10.1007/s00393-016-0169-0
Yao R, Ma YL, Liang W, Li HH, Ma ZJ, Yu X, Liao YH (2012) MicroRNA-155 modulates Treg and Th17 cells differentiation and Th17 cell function by targeting SOCS1. PLoS One 7(10):e46082. https://doi.org/10.1371/journal.pone.0046082
Yao Y, Xu XH, Jin L et al (2019) Macrophage polarization in physiological and pathological pregnancy. Front Immunol 10:792. https://doi.org/10.3389/fimmu.2019.00792

Yi R, Fuchs E (2010) MicroRNA-mediated control in the skin. Cell Death Differ 17(2):229–235. https://doi.org/10.1038/cdd.2009.92
Yi X, Guo W, Shi Q et al (2019) SIRT3-dependent mitochondrial dynamics remodeling contributes to oxidative stress-induced melanocyte degeneration in Vitiligo. Theranostics 9(6):1614–1633. https://doi.org/10.7150/thno.30398
Yu LC (2018) Microbiota dysbiosis and barrier dysfunction in inflammatory bowel disease and colorectal cancers: exploring a common ground hypothesis. J Biomed Sci. https://doi.org/10.1186/s12929-018-0483-8
Yuan Y, Kasar S, Underbayev C et al (2012) Role of microRNA-15a in autoantibody production in interferon-augmented murine model of lupus. Mol Immunol 52(2):61–70. https://doi.org/10.1016/j.molimm.2012.04.007
Yuan X, Berg N, Lee JW et al (2018) MicroRNA miR-223 as regulator of innate immunity. J Leukoc Biol 104(3):515–524. https://doi.org/10.1002/JLB.3MR0218-079R
Zakharova MY, Belyanina TA, Sokolov AV et al (2019) The contribution of major histocompatibility complex class II genes to an association with autoimmune diseases. Acta Nat 11(4(43)):4–12. https://doi.org/10.32607/20758251-2019-11-4-4-12
Zhang JM, Feng FE, Wang QM et al (2016) Platelet-derived growth factor-BB protects Mesenchymal stem cells (MSCs) derived from immune thrombocytopenia patients against apoptosis and senescence and maintains MSC-mediated immunosuppression. Stem Cells Transl Med 5(12):1631–1643. https://doi.org/10.5966/sctm.2015-0360
Zhang B, Tian Y, Jiang P et al (2017) MicroRNA-122a regulates Zonulin by targeting EGFR in intestinal epithelial dysfunction. Cell Physiol Biochem 42(2):848–858. https://doi.org/10.1159/000478629
Zhao ZZ, Sugerman PB, Zhou XJ et al (2001) Mast cell degranulation and the role of T cell RANTES in oral lichen planus. Oral Dis 7(4):246–251
Zhao X, Tang Y, Qu B et al (2010) MicroRNA-125a contributes to elevated inflammatory chemokine RANTES levels via targeting KLF13 in systemic lupus erythematosus. Arthritis Rheum 62(11):3425–3435. https://doi.org/10.1002/art.27632. PMID: 20589685
Zhao S, Wang Y, Liang Y et al (2011) MicroRNA-126 regulates DNA methylation in CD4+ T cells and contributes to systemic lupus erythematosus by targeting DNA methyltransferase 1. Arthritis Rheum 63(5):1376–1386. https://doi.org/10.1002/art.30196
Zhao M, Wang LT, Liang GP et al (2014) Up-regulation of microRNA-210 induces immune dysfunction via targeting FOXP3 in CD4(+) T cells of psoriasis vulgaris. Clin Immunol 150(1):22–30. https://doi.org/10.1016/j.clim.2013.10.009
Zhou H, Huang X, Cui H et al (2010) miR-155 and its star-form partner miR-155* cooperatively regulate type I interferon production by human plasmacytoid dendritic cells. Blood 26:5885–5894. https://doi.org/10.1182/blood-2010-04-280156
Zhu J, Zhang Y, Zhang W et al (2016) MicroRNA-142-5p contributes to Hashimoto's thyroiditis by targeting CLDN1. J Transl Med. https://doi.org/10.1186/s12967-016-0917-6
Zou J, Ji DN, Shen Y et al (2016a) Association of Reduced Heme Oxygenase-1 with decreased MicroRNA-196a2 expression in peripheral blood mononuclear cells of patients with intestinal Behcet's disease. Ann Clin Lab Sci 46(6):675–679
Zou T, Jaladanki SK, Liu L et al (2016b) H19 long noncoding RNA regulates intestinal epithelial barrier function via MicroRNA 675 by interacting with RNA-binding protein HuR. Mol Cell Biol 36(9):1332–1341. https://doi.org/10.1128/MCB.01030-15
Zufferey A, Kapur R, Semple JW (2017) Pathogenesis and therapeutic mechanisms in immune thrombocytopenia (ITP). J Clin Med. https://doi.org/10.3390/jcm6020016

Chapter 9
MicroRNAs and Cardiovascular Diseases

James M. Henderson, Floriana M. Farina, and Donato Santovito

Abstract MicroRNAs (miRNAs) are crucially involved in the embryonic development of the cardiovascular system, and they contribute to the maintenance of proper postnatal cardiac and vascular function. Dysregulation of miRNA expression is detected in response to pathological stimuli and cardiovascular risk factors and is involved in the pathogenesis of cardiovascular diseases, such as atherosclerosis and pathological remodeling in heart failure. Besides their role in posttranscriptional gene regulation, miRNAs in the cardiovascular system also feature non-conventional modes of action and localization, being also released to contribute to intercellular communication. Here, we will discuss the contribution of miRNAs in cardiac and vascular homeostasis, as well as their contribution to major cardiovascular diseases, such as atherosclerosis, myocardial infarction, and cardiac remodeling. We will focus on the role of miRNAs with preferential cardiovascular expression and with possible therapeutic relevance. Moreover, we will review the potential of using miRNAs as disease-specific diagnostic and/or prognostic biomarkers for cardiovascular disease. Finally, we will summarize the approaches to manipulate miRNA

James M. Henderson and Floriana M. Farina contributed equally with all other contributors.

J. M. Henderson · F. M. Farina
Institute for Cardiovascular Prevention (IPEK), Ludwig-Maximilians-Universität München, Munich, Germany

German Centre for Cardiovascular Research (DZHK), Partner Site Munich Heart Alliance, Munich, Germany

D. Santovito (✉)
Institute for Cardiovascular Prevention (IPEK), Ludwig-Maximilians-Universität München, Munich, Germany

German Centre for Cardiovascular Research (DZHK), Partner Site Munich Heart Alliance, Munich, Germany

Institute of Genetic and Biomedical Research (IRGB), Unit of Milan, National Research Council (CNR), Milan, Italy
e-mail: donato.santovito@med.uni-muenchen.de

L. M. Vaschetto (ed.), *miRNAs, Human Health and Diseases*, Epigenetics and Human Health 13, https://doi.org/10.1007/978-3-031-64788-8_9

expression and function highlighting the main candidates tested in clinical trials for future therapeutic applications.

Keywords MicroRNA · Cardiovascular disease · Atherosclerosis · Aneurysm · Heart failure · Myocardial infarction

Abbreviations

AAA	Abdominal aortic aneurysm
AAV	Adeno-associated virus
Ago2	Argonaute 2
Ang-II	Angiotensin-II
ApoE	Apolipoprotein E
ASO	Antisense oligonucleotides
bFGF	Basic fibroblast growth factor
CAD	Coronary artery disease
CMP	Human cardiomyocyte progenitor cell
DAMP	Danger-associated molecular pattern
DLK1	Delta-like noncanonical Notch ligand 1
DNMT3b	DNA methyltransferase 3b
EC	Endothelial cell
ECM	Extracellular matrix
EdU	5-ethynyl-2'deoxyuridine
EndMT	Endothelial -to-mesenchymal transition
eNOS	Endothelial nitric oxide synthase
FABP5	Fatty acid-binding protein 5
FADD	Fas-associated protein with death domain
FB	Fibroblast
FOXO4	Forkhead box 04
HAoSMC	Human aortic smooth muscle cells
HCV	Hepatitis C virus
HDL	High density lipoprotein
hESC-CMs	Human embryonic stem cell-derived cardiomyocyte
HF	Heart failure
HFD	High fat diet
HIF1a	hypoxia induced factor 1a
HUVEC	Human umbilical vein endothelial cell
IA	Intracranial aneurysms
IGF	Insulin growth factor
KLF4	Krüppel-like factor 4
KO	Knock-out
LOX	Lysyl oxidase
MALAT1	Metastasis-associated lung adenocarcinoma transcript 1
MHC	Myosin heavy chain

miRNA	MicroRNA
MMP	Matrix metalloproteinase
mRNA	Messenger ribonucleic acid
mTOR	Mammalian target of rapamycin
MYH11	Myosin heavy chain II
MYLK	Myosin light chain kinase
NF-κB	Nuclear factor-κB
OPN	Osteopontin
OR	Odds ratio
PAH	Pulmonary arterial hypertension
PAMP	Pathogen-associated molecular patterns
PCI	Percutaneous coronary intervention
PDGF-BB	Platelet-derived growth factor-BB
PDGFRβ	Platelet-derived growth factor receptor beta
PIK3R2	Phosphoinositol-3 kinase regulatory subunit 2
Pim-1	Proviral integration site for Moloney murine leukemia virus 1
PPAR-γ	Peroxisome proliferator-activated receptor gamma
PPRE	Peroxisome proliferators response element
PRRs	Pattern recognition receptors
PTEN	Phosphatase and tensin homolog
RAR	Retinoic acid receptor
RB1	Retinoblastoma protein
RIP1	Receptor interacting protein 1
RISC	RNA-induced silencing complex
RNA	Ribonucleic acid
ROS	Reactive oxygen species
SIRT1	Sirtuin 1
SM22-α	Transgelin
SMC	Smooth muscle cells
SM-MHC	Smooth muscle myosin heavy chain
SPRED1	Sprouty Related EVH1 Domain Containing 1
SRF	Serum response factor
TAA	Thoracic aortic aneurysm
TAC	Transverse aortic constriction
TGF-β	Transforming growth factor beta
THRAP	Thyroid receptor-associated protein
TNF-α	Tissue necrosis factor alpha
TSB	Target-site blockers
VCAM1	Vascular adhesion molecule
VSD	Ventricular septal defects
VSMC	Vascular smooth muscle cell
WT	Wild type
α-SMA	Alpha smooth muscle actin

9.1 Introduction

The crucial contribution of microRNAs (miRNAs) to the cardiovascular system has been evident since the pioneering experiments on transgenic animal models targeting miRNA biogenesis. Specifically, deletion in mice of the *Dicer1* gene, encoding for the RNase III endoribonuclease Dicer critical for miRNA processing, provoked embryonic death between days E12.5 and E14.5 of gestation due to severely defective vasculogenesis in embryos and yolk sacs (Yang et al. 2005). Beyond vasculature, two independent early reports involving mouse models with conditional *Dicer1* deletion revealed that, while depleting miRNAs in cardiomyocytes did not impair cardiac organogenesis, it resulted in defects in myocardial lineage differentiation with spontaneous cardiac remodeling and rapidly lethal phenotypes (Chen et al. 2008; da Costa Martins et al. 2008). Additionally, the deletion of other genes essential for miRNA biogenesis and function, such as Drosha, DGCR8, or Ago2, has demonstrated similar disruptions in cardiovascular development (Alisch et al. 2007; Cheloufi et al. 2010; Chen et al. 2019; Liu et al. 2004). Thus, miRNA biogenesis is an essential requirement for ensuring a normal heart structure and function and for the development of the vasculature.

Besides embryogenesis and development, the critical role of miRNAs in cardiovascular homeostasis in adults has been revealed by studies involving animal models with inducible deletion of *Dicer1*, for instance by tamoxifen-inducible Cre recombinases. Indeed, postnatal conditional deletion of *Dicer1* in vascular cells (i.e., endothelial and vascular smooth muscle cells) affected the susceptibility of developing vascular diseases, such as atherosclerosis (Hartmann et al. 2016; Zahedi et al. 2017). Similarly, cardiomyocyte-specific deletion of *Dicer1* in adult mice or the inhibition of miRNA function by impeding the assembly of RNA-induced silencing complexes (RISCs) led to transcriptional reprogramming toward a fetal gene profile, with severe cardiac remodeling and heart failure (da Costa Martins et al. 2008; La Rocca et al. 2021). While deletions of individual *MIRNA* genes have generally been found to be non-lethal, this observation likely arises from overlapping and functional redundancy (Bartel 2009; Fischer et al. 2015). Consistent with a possible role in the pathophysiology of cardiovascular diseases, alterations in the expression of miRNAs were also detected in humans in disease settings (such as atherosclerosis, myocardial infarction, and cardiac remodeling) both in tissue samples and in circulation (Cipollone et al. 2011; D'Alessandra et al. 2010; Raitoharju et al. 2011; Shen et al. 2022). Together, the current evidence supports the involvement of miRNAs in maintaining cardiovascular homeostasis in adults, highlighting their role in the mechanisms of cardiovascular diseases.

9.1.1 The Role of miRNAs in Cardiac Homeostasis

The heart is a highly intricate organ composed of various cell types and the first to develop and function during embryogenesis (Olson 2006). With the advent of single-cell and single-nuclei transcriptomic and the availability of comprehensive multiomics datasets (e.g., the Heart Cell Atlas), we have now appreciated that the human heart embraces 12 major cell types featuring over 70 different cell states (Kanemaru et al. 2023). The main cell types include atrial and ventricular cardiomyocytes, pacemaker cells, fibroblasts (FBs), endothelial cells (ECs), pericytes, vascular smooth muscle cells (VSMCs), myeloid and lymphoid immune cells, adipocytes, mesothelial cells, and neuronal cells. Many of the cells have specialized functions and their distribution varies between atrial and ventricular compartments. Single-nuclei sequencing data reveals that, on average, approximately 30–50% of cells are cardiomyocytes, 17–20% VSMCs and pericytes, 12–7% ECs, and 5–10% immune cells.

The role of miRNAs in cardiogenesis, homeostasis, and pathology emerged in recent years unveiling a complex and cell-specific contribution that is still under intense investigation. Pioneering studies assessed the effect of cardiomyocyte-specific *Dicer1* deletion at different time points showing distinct functional outcomes. Deletion of *Dicer1* during embryogenesis (in *MCH-Cre*$^{+}$;*Dicer1*$^{fl/fl}$ mice) produced a depletion of mature miRNAs (such as miR-1 and miR-133) as early as embryonic day E14.5 and, while these animals did not show overt phenotype at birth, they all died within 4 days due to heart failure and signs of dilative cardiomyopathy (Chen et al. 2008). On the other hand, postnatal deletion in 3-week-old animals using a tamoxifen-inducible Cre recombinase model provoked severe atrial dilatation with mild ventricular remodeling and premature death within 1 week (da Costa Martins et al. 2008). At the adult stage, the deletion of *Dicer1* in 8-week-old mice caused the development of severe dilative cardiomyopathy with cardiomyocyte hypertrophy, induction of a fetal transcriptional signature, and clinical features of heart failure without—however—significant worsening in survival at 4 weeks (da Costa Martins et al. 2008). These findings were further corroborated by a more recent study using a mouse model in which the miRNA-mediated gene repression was dampened by inducible inhibition of the RISC assembly. In this model, severe dilative cardiomyopathy with signs of congestive heart failure and higher 4–6-week mortality (La Rocca et al. 2021). Together, these observations reveal the importance of miRNAs in maintaining proper adult heart transcriptional landscape and in preserving homeostasis while preventing disease development.

Approximately 90% of cardiac miRNAs belong to 18 miRNA families and feature a contribution of the multiple cell types that compose the heart (Rao et al. 2009). Analyses of miRNA arrays of cardiomyocyte-specific *Dicer1* knock-out (KO) animals contributed to the identification of a set of miRNAs differentially regulated by deletion at different time points, including miR-1 and miR-133. These latter are members of a cluster termed "myo-miRNAs" including four miRNAs (i.e., miR-1, miR-133, miR-208, and miR-499) showing enriched expression in cardiac,

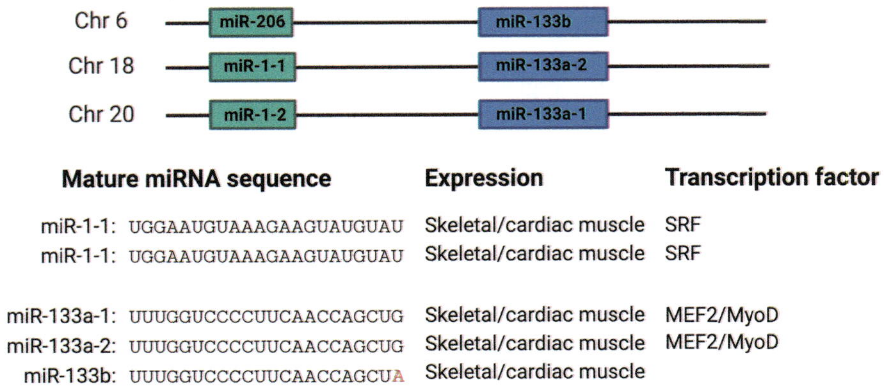

Fig. 9.1 Genomic organization of the genes encoding for miR-1, miR-133, and miR-206 in the human genome. The sequence of the mature miR-1, miR-133a, and miR-133b and the responsible transcription factors are reported below. Figure created with BioRender.com

skeletal, and smooth muscles (Beuvink et al. 2007; Chen et al. 2008; McCarthy 2008; Sempere et al. 2004). The expression of myo-miRNAs is robust at the distinct stages of heart development, with miR-1 being the most abundant in cardiomyocytes. Given these properties, the contribution of myo-miRNAs to cardiac development and function has been extensively studied (Fig. 9.1), yet additional miRNAs showed changes during heart development and upon *Dicer1* deletion, and their role should not be overlooked, especially in disease contexts. In this regard, it is noteworthy that *Dicer1* deletion in cardiomyocytes also led to the enhanced cardiac expression of miRNAs (such as miR-21 and miR-145) which are primarily expressed by other cell types (e.g., macrophages and VSMCs) rather than cardiomyocytes (da Costa Martins et al. 2008). This aspect further highlights the complexity of cell-specific mechanisms of miRNA regulation in cardiac pathophysiology.

9.1.2 *Myo-miRNAs in Cardiomyocytes and Cardiogenesis*

9.1.2.1 miR-1 and miR-133

miR-1 and miR-133 derive from processing a shared bicistronic transcript (Meder et al. 2008). The mouse genome encodes the genes for these miRNAs in chromosomes 2 and 18. In the human genomes, the genes form clusters in chromosomes 6, 18, and 20. The gene encoding for miR-133a is mapped in proximity to the *MIR1* genes in the chromosomal regions 20q13.33 and 18q11.2, as they belong to the same transcriptional unit. Conversely, the gene encoding miR-133b is in 6p12.2 and transcribed bicistrionically with miR-206 (Chen et al. 2008; Fig. 9.1). During embryonic heart development, the transcription of miR-1 and miR-133 is regulated by the transcription factors SRF, MEF2, and MyoD (Liu et al. 2007; Rao et al. 2006;

Zhao et al. 2005). In embryonic stem cells, these miRNAs collectively play a pivotal role in initiating mesoderm while suppressing endoderm differentiation (Ivey et al. 2008). With the progression of the developmental process, however, the functions of these two miRNAs diverge and while miR-1 promotes mesoderm differentiation into cardiomyocytes, miR-133 antagonizes this process (Ivey et al. 2008; Liu et al. 2008; Lu et al. 2013).

miR-1 is encoded by two highly homologous genes (i.e., *MIR1–1* and *MIR1–2*), yielding identical mature miR-1 molecules with corresponding RNA target repertoire (Safa et al. 2020). Consistently, deletion of either miR-1-1 or miR-1-2 in mice leads to a similar phenotype (Wystub et al. 2013). This is characterized by ventricular septal defects (VSDs) and severe perturbations in cardiac electrophysiology by targeting genes such as *Kcnd2*, primarily impacting potassium channel regulation (Yang et al. 2007; Zhao et al. 2007). Notably, miR-1 has been reported to modulate the electrophysiology of cardiomyocytes also through a non-conventional mechanism by biophysical modulating the function of the inward rectifier potassium channel KCNJ2 affecting the homeostasis of the adult heart (Yang et al. 2021).

Besides the role in directing electrophysiological homeostasis, miR-1 also exerts regulatory control over the expression of the transcription factors Irx5 and Hand2, with the latter playing a pivotal role in the right ventricle and outflow tract development (Zhao et al. 2007; Zhao et al. 2005), further reinforcing a relevance in the mechanisms of cardiac organogenesis. In line, miR-1 drives the cardiomyocyte commitment of human embryonic stem cell-derived cardiovascular progenitors by downregulating the WNT and FGF signaling pathways (Lu et al. 2013). As a consequence, transplantation of embryonic stem cells transfected with miR-1 mimics into infarcted mouse hearts could enhance their cardiomyocyte differentiation, reduce apoptosis, and improve heart function through activation of the PTEN/Akt pathway (Huang et al. 2013b). Finally, miR-1 regulates Insulin Growth Factor (IGF)-1 in cardiomyocytes, a key regulator of their metabolism, excitation-contraction coupling, and action potential. On the other hand, the downstream IGF-1 pathway regulates miR-1 expression via the Foxo3a transcription factor, revealing further interplay regulating the equilibrium between miR-1 and its targets (Elia et al. 2009a). Together, these findings not only reveal the crucial role of miR-1 in cardiac development, cardiomyocyte commitment, and homeostasis but also highlight a possible therapeutic relevance.

Closely analogous to miR-1, miR-133a also derives from two highly homologous genes, giving rise to two identical mature miRNAs, respectively referred to as miR-133a-1 and miR-133a-2 (Ohanian et al. 2013). A second member of the miR-133 family (i.e., miR-133b) differs from miR-133a by a single nucleotide at the 3′ end and is bicistronically transcribed with miR-206 primarily in the skeletal muscle. Notably, mice lacking either miR-133a-1 or miR-133a-2 do not display an evident phenotype, but the simultaneous deletion of both miR-133a genes in mice results in a variable fetal heart phenotype marked by VSDs, underscoring the key role of miR-133a in prompting adult cardiomyocyte commitment (Liu et al. 2008). Furthermore, miR-133a mutants surviving after birth exhibit increased cardiomyocyte proliferation and smooth muscle gene expression, eventually leading to dilated cardiomyopathy (Liu et al. 2008). The mechanism beyond this phenotype is likely to depend on the distinct role of miR-133 in modulating serum response factor (SRF) and cyclin D2 activity, both indispensable for cardiac and smooth muscle cell differentiation and growth (Liu et al. 2008). Indeed, cyclin D2 affects cardiomyocyte proliferation through its impact on the phosphorylation status of the retinoblastoma protein (RB1) during the G1 phase of the cell cycle (Busk et al. 2005; Toischer et al. 2017). In a reciprocal relationship, the activation of SRF triggers the transcription of miR-1 and miR-133, emphasizing the interplay and tiny regulatory equilibrium ruling this pathway (Zhao et al. 2005). As such, miR-1 and miR-133 have recently been shown to play an essential role in cardiac development and function, with their absence leading to ventricular septal defects, electrophysiological disruptions, and regulatory control over crucial transcription factors and signaling pathways in both mice and human cardiovascular progenitors.

9.1.2.2 miR-208 and miR-499

The other two members of the myo-miRNAs cluster also showed robust tissue expression in the heart and a significant contribution to cardiac development (van Rooij et al. 2009). The genes encoding miR-208a, miR-208b, and miR-499 are located within introns of the genes of the myosin isoforms with a prevalent cardiac expression: *Myh6*, *Myh7*, and *Myh7b*, respectively (van Rooij et al. 2009). As members of the same family, miR-208a and miR-208b share identical seed sequences reflecting a large overlap in target transcripts, and their sequences are highly conserved amongst mammals. (van Rooij et al. 2009). However, they differ in their expression at distinct developmental stages: miR-208b is expressed in the embryonic heart tissue (and in the skeletal muscle) while miR-208a expression is restricted to the postnatal heart (Callis et al. 2009; van Rooij et al. 2007). In contrast to miR-1 and miR-133, which operate in early cardiac organogenesis, miR-208a and miR-208b take center stage in the later phases of heart development, primarily regulating the expression of cardiac myosin heavy chain (MHC) isoforms (Oliveira-Carvalho et al. 2013; van Rooij et al. 2009). In particular, miR-208 is pivotal in the mechanisms of MHC isoform switching, which is crucial for the functionality of the postnatal heart. Indeed, MHC isoforms confer distinct mechanical properties to cardiomyocytes owning different ATPase activity, and their abundance is carefully modulated temporarily and spatially during embryogenesis and in pathological conditions. In rodents, β-MHC prevails during embryonic development while the expression of α-MHC begins postnatal, with *Myh7b* expression occurring at both stages (Lyons et al. 1990). In line with a role in the late stages of organogenesis, genetic deletion of miR-208a or miR-208b does not cause embryonic death. Moreover, the deletion of miR-208b decreases the expression of β-MHC expression, while the high expression is correlated with increased cardiac fibrosis (Li et al. 2022; Zhang et al. 2022).

In contrast, deletion in miR-208a leads to reduced cardiac function in 6-month-old infants due to abnormalities of the sarcomeres (van Rooij et al. 2007). Besides embryonic development, adult cardiomyocytes reactivate β-MHC synthesis during stress conditions, such as hypoxia or hypothyroidism (Bino et al. 2017; Morkin 2000). Notably, the genetic deletion of miR-208a impairs the ability of postnatal cardiomyocytes to upregulate β-MHC under stress and avoids the development of cardiac hypertrophy (Callis et al. 2009). Mechanistically, this effect is likely related to post-transcriptional repression of the thyroid receptor-associated protein (THRAP1), an activator of the fast skeletal muscle gene transcriptional program. Indeed, miR-208a inhibits THRAP1 during stress and dampens the downstream thyroid receptor signaling, which leads to higher expression of the β-MHC isoform with consequences on cardiac function (Callis et al. 2009; van Rooij et al. 2007). Further studies in gain- and loss-of-function mouse models have shown that miR-208a regulates other mediators involved in cardiac hypertrophy and conduction, such as myostatin and GATA4 (Callis et al. 2009). The discrepancy in the transcriptional outcomes of miR-208a and miR-208b, which have identical seed sequences, suggests that miR-208a could be the dominant upstream regulator of both β-MHC (*Myh7*), Myh7b, and their associated intronic miRNAs (i.e., miR-208b and miR-499) in the heart (van Rooij et al. 2009). Moreover, it may reflect a contribution of additional mechanisms such as—in the example—the ability of the primary transcript of miR-208b (pri-miR-208b) to recruit the enzyme EZH2 and epigenetically repress α-MHC expression (Santovito and Weber 2022).

Two isoforms exist for miR-499 (miR-499a and miR-499b) and play a role in the late stages of cardiac organogenesis, driving the differentiation of cardiomyocytes from myoblasts (Chistiakov et al. 2016). In humans, miR-499b (chr20; 34,990,400–34,990,472, negative strand) is an antisense form of miR-499a (chr. 20; 34,990,376–34,990,497, positive strand) contrary to the previously mentioned myo-miRNAs. Experiments in human cardiomyocyte progenitor cells (CMPCs) and human embryonic stem cell–derived cardiomyocytes (hESC-CMs) showed that miR-499 synergistically acts with miR-1 to promote differentiation into cardiomyocytes (Wilson et al. 2010). This effect is likely dependent on miR-499-dependent repression of the transcription factor SOX6, which is critical in developing and safeguarding cardiac and skeletal muscles (Li et al. 2013). Notably, deletion of miR-499 in mice did not provoke any overt phenotype. On the other hand, transgenic mice with miR-499 overexpression displayed cardiomyopathy with enlarged hearts and increased heart-to-body weight ratios, suggesting hypertrophy (Da Costa Martins and De Windt 2012; van Rooij et al. 2009). In agreement, miR-499 overexpression significantly altered the expression of genes involved in contractility such as *Myh7b* and skeletal muscle α-actin, determining hypertrophy. Moreover, it also affected the expression of ion channels such as KCNH6 and Kv11.1, possibly leading to impairments in conductivity (Shieh et al. 2011). Finally, miR-499 overexpression in cardiomyocytes in a rat model repressed pro-apoptotic molecules, such as Pdcd4, Pasc2, and Dyrk2, enhancing cell survival (Zhu et al. 2016b). Together, these findings suggest that miR-499 may be dispensable for heart development but may contribute to the development of pathological cardiac remodeling.

9.1.2.3 miR-17/-92 Family

Another crucial miRNA family in cardiovascular homeostasis is the miR-17/92 family. It encompasses six distinct miRNAs, namely miR-17, miR-18a, miR-19a, miR-20a, miR-19b-1, and miR-92a-1 (Bonauer and Dimmeler 2009; Concepcion et al. 2012). Multiple studies have demonstrated the crucial involvement in cardiac development of this miRNA family, and its abrogation leads to cardiac septal defects in mice with consequent postnatal mortality (Mendell 2008; Yue 2011).

9.1.3 The Role of miRNAs in Vessel Physiology and Development

miRNAs serve as pivotal regulators in the normal biology and developmental processes of blood vessels, overseeing a range of critical functions. Besides, miRNAs participate in the dynamic process of arterial remodeling during angiogenesis, and they contribute to vessel reactivity, ensuring that vessels are appropriately sized and functional. Notably, the miRNAs show remarkable cell-specificity in their expression in vascular cells which is likely conferred by the close localization of their encoding genes to cell-specific super-enhancers (Moreau et al. 2021). Hence, while the amount could be affected by external factors, specific sets of miRNAs (e.g., miR-126 in ECs, miR-143/-145 in VSMCs) will be primarily expressed in a single cell type to regulate their homeostasis.

9.1.3.1 Endothelial miRNAs in Vascular Physiology

ECs constitute the inner layer of a healthy blood vessel and play critical roles in maintaining vascular integrity and function. They regulate the vessel response to the blood flow through the production of vasodilatory molecules such as nitric oxide (Chiu et al. 1999), maintain a selectively permeable barrier (Chistiakov et al. 2015a), prevent the occurrence of platelet aggregation and thrombosis (Yau et al. 2015), and control inflammatory cell adhesion and transmigration (Siasos et al. 2015). Moreover, ECs not only uphold a safeguarding glycocalyx layer on their surfaces but also actively participate in the initial phases of vascular repair (Milusev et al. 2022). The thickness of the glycocalyx layer correlates with its functional integrity, and inflammatory conditions may promote its reduction thus contributing to compromised vessel function. Furthermore, ECs are capable of instigating angiogenesis by sprouting and migrating to form new vessel branches (Nicosia et al. 2011). These multifaceted functions ensure proper circulation, nutrient exchange, and immune response regulation, underscoring the critical role of ECs in sustaining vascular health and function, with implication for the homeostasis of multiple organs.

Among the miRNAs involved in the mechanisms of vessel development, miR-126 is one of the miRNAs with the highest enrichment in ECs. It is mandatory for a proper differentiation of embryonic EC subclusters by affecting the expression of genes mainly involved in hypoxic response and glucose metabolism, as shown by single-cell RNA sequencing in mice with genetic deletion of miR-126 (Guo et al. 2022). Multiple studies unveiled the ability of miR-126 to regulate key components in angiogenic signaling pathways, such as the vascular endothelial growth factor (VEGF) pathway, including the Sprouty-related protein SPRED1 and phosphoinositol-3 kinase regulatory subunit 2 (PIK3R2). Additionally, it can target the hypoxia-induced factor (HIF)-1α and mammalian target of rapamycin (mTOR) signaling pathways, influencing EC metabolism and angiogenic properties (Alique et al. 2019; Cao et al. 2020; Fish et al. 2008). In contrast to most miRNAs, miR-126 features functional roles of both mature strands (i.e., miR-126-3p and miR-126-5p) which regulates distinct targets to ensure proper differentiation and homeostasis both in embryos and adult ECs (Guo et al. 2022; Schober et al. 2014; Zhou et al. 2016). In the example, miR-126-3p may dampen EC inflammatory activation by repressing the expression of vascular cell adhesion molecule 1 (VCAM1) while miR-126-5p preserves the proliferative reserve of ECs by targeting the delta-like noncanonical Notch ligand 1 (DLK1) and reduce apoptosis by targeting caspase 3 through direct biophysical interactions (Harris et al. 2008; Santovito et al. 2020a; Schober et al. 2014). Finally, miR-126 mediates cell-to-cell communication upon its release in vesicles exerting a role in maintaining endothelial integrity (Jansen et al. 2013; Zernecke et al. 2009).

Several miRNA families importantly contribute to endothelial homeostasis. Besides the involvement in cardiac development, the miR-17/92 family plays also important roles in EC biology. Indeed, members of this family, including miR-17 and miR-20a, have been reported to dampen ECs migratory capacities, leading to impaired angiogenesis (Doebele et al. 2010; Ventura et al. 2008). The miR-212/-132 cluster also regulated vascular angiogenesis beyond its role in cardiac remodeling (Anand et al. 2010). Despite miR-212 and miR-132 share the same seed sequence, they exert distinct roles in endothelial physiology (Kumarswamy et al. 2014). Indeed, the EC expression of miR-132 is increased by angiogenic factors, such as VEGF, TGFβ, and bFGF. In this context, miR-132 acts as an angiogenic switch by repressing the expression of p120RasGA, leading to Ras activation, and contributing to the regulation of aortic stiffening by targeting KLF7 thus preventing endothelial-to-mesenchymal transition (EndMT) (Anand et al. 2010; Hulshoff et al. 2023). Moreover, it was shown to mediate EC proliferation, migration, and tube formation, thus regulating vessel stability (Zhang and Tao 2017). Conversely, miR-212 expression was independent by TGFβ treatment, and it was reported to be crucial for the mechanisms of EndMT (Kumarswamy et al. 2014). Finally, the miR-221/-222 is another miRNA family highly expressed in ECs and affect angiogenic processes by inhibiting the expression of c-Kit and endothelial nitric oxide synthase (eNOS) (Chistiakov et al. 2015b; Ihle et al. 2015; Rippe et al. 2012).

Besides these conserved miRNA families, other miRNAs are expressed in ECs to contribute to their properties. The hypoxia-responsive miR-210 is crucial in arterial angiogenesis under low oxygen conditions (Fasanaro et al. 2008). It enhances the adaption of ECs to the hypoxic environment and promotes angiogenesis (Ivan and Huang 2014; Zaccagnini et al. 2022). Similarly, miR-10 modulates the VEGF signaling, and its abrogation in vivo increases the expression of FLT1/sFLT1 leading to impaired angiogenic capability (Hassel et al. 2012; Wang et al. 2016). Another miRNA with pleiotropic functions in the maintenance of vascular endothelium integrity is miR-34. It is involved in processes ranging from senescence to inflammatory activation. Many reports indicated its contribution to the preservation of endothelial function by influencing the expression of genes encoding for cell cycle inhibitors (such as p16, p21, and the tumor suppressor p53) or the protein deacetylase SIRT1 which is crucially involved in cell senescence and vascular aging (Disayabutr et al. 2016; Hua et al. 2021; Overhoff et al. 2014; Raucci et al. 2021). Finally, other miRNAs are key players in the regulation of the EC inflammatory status.

For instance, miR-155 regulates vascular inflammation, function, and barrier integrity. In particular, it is involved in modulating inflammation by repressing pro-inflammatory genes, helping to control immune responses and endothelial activation. By targeting specifically VCAM1 and ICAM1, miR-155 can influence the expression of these adhesion molecules and the production of inflammatory cytokines in ECs, dampening leukocyte adhesion and migration (Cerutti et al. 2016; Shen et al. 2020). Moreover, miR-155 also instructs cytoskeleton rearrangements by targeting the myosin light chain kinase MYLK and consequently affecting EC shape to elicit diverse responses, particularly under specific shear stress conditions (Weber et al. 2014).

9.1.3.2 miRNAs in VSMC Physiology

VSMCs are the main component of the media layer of vessel walls, and due to their plasticity, they play a crucial role in regulating vascular tone, blood pressure, and tissue repair. In response to various pathological stimuli, such as injury or inflammation, VSMCs adopt a synthetic phenotype, contributing to vascular remodeling by proliferating and producing extracellular matrix proteins (Alexander and Owens 2012). Plenty of evidence shows that miRNAs are pivotal in homeostasis and function of VSMCs as well as in the mechanism of phenotype switch, wherein they control the transition between contractile and synthetic phenotypes (Davis-Dusenbery et al. 2011; Small and Olson 2011).

The best-established miRNA family regulating VSMC homeostasis is the miR-143/-145 family. Both miR-143 and miR-145 are highly expressed and enriched in VSMCs and are involved in maintaining the contractile phenotype by regulating smooth muscle-specific gene expression (Cordes et al. 2009; Elia et al. 2009b). Their role is crucial in the maintenance of the vascular tone and their deletion determined lower vessel reactivity and blood pressure in mice (Boettger et al. 2009). Multiple targets of these miRNAs are key regulators of the promotion of the contractile phenotype of VSMCs, such as smooth muscle alpha-actin (α-SMA) and smooth muscle myosin heavy chain (SM-MHC) (Boettger et al. 2009; Cordes et al. 2009). Besides its role in cardiomyocytes, miR-1 also plays a vital role in VSMCs.

As miR-143/−145, it primarily instructs a contractile phenotype by targeting crucial genes such as IGF-1, Krüppel-like factor 4 (KLF4), and proviral integration site for Moloney murine leukemia virus 1 (Pim-1) (Chen et al. 2011c; Liu et al. 2015; Xie et al. 2011). As such, miR-1 is important in maintaining an adequate vascular tone and blood flow (Jiang et al. 2010). Also the myo-miRNA miR-133a is involved in maintaining the contractile phenotype (Kang and Hata 2012). Its downregulation contributes to the VSMC phenotype switch by inhibiting the transcription factor Sp-1 (Torella et al. 2011). Furthermore, miR-133 hampers the expression of contractile genes specific to smooth muscle cells in the heart, by directly targeting smooth muscle-related mRNAs but also SRF (Torella et al. 2011), a key regulator of smooth muscle gene expression. Finally, miR-128 emerged as a crucial regulator of VSMC biology which fosters the maintenance of a contractile phenotype by targeting KLF4 and inducing the expression of SMMHC (Farina et al. 2020), as well as regulating proliferation, migration, and viability by targeting forkhead box O4 (FOXO4) and matrix metallopeptidase 9 (MMP9) (Qu et al. 2020).

Conversely, other miRNAs rather promote VSMC dedifferentiation and the acquisition of a synthetic phenotype affecting extracellular matrix deposition, reducing cell contractility, increasing proliferation and migration (Farina et al. 2020; Liu et al. 2009; Yan et al. 2020). This miRNA cluster is abundantly expressed in VSMCs and exerts a strong pro-proliferative effect by targeting p27 and p57 (Liu et al. 2009). In vitro experiments on human VSMCs showed that the inhibition of miR-21 reduced the proliferation rate by targeting AP-1 (Li et al. 2014). The miR-29 family (including miR-29a, miR-29b, and miR-29c) is recognized for its role in regulating the vascular tone of arterial vessels by targeting elastin expression (Sudo et al. 2015) and by affecting VSMC proliferation and migration to modulate vascular remodeling. In particular, miR-29b facilitates the migration of VSMCs, by inhibiting the DNA methyltransferase 3b (DNMT3b) and epigenetically controlling the expression of its downstream targets MMP-2 and MMP-9 genes, well known for their involvement in cell migration and extracellular matrix remodeling (Chen et al. 2011a, b, c). Finally, miR-26a is associated with VSMC phenotypic modulation, affecting proliferation, migration, and extracellular matrix production. It targets multiple signaling pathways involved in VSMC function, including the PTEN/AKT/mTOR and TGFβ pathways via Smad1 inhibition. Moreover, it affects MAPK6 and its downstream targets PCNA, Cyclin D1, MMP2, and MMP9 (Peng et al. 2018; Tan et al. 2017; Yang et al. 2017).

9.1.3.3 miRNAs in Mononuclear Cells and Their Role in the Healthy Vessel

Mononuclear cells (monocytes/macrophages) are professional phagocytic defenders with crucial roles in innate immune responses. While they are often studied in the context of vascular disease, they provide a key contribution to the immune surveillance of healthy arteries participating in mechanisms of preservation of the endothelial lining, regulation of inflammatory responses, clearance of cellular debris, and cholesterol homeostasis (Hernandez et al. 2022; Tomas et al. 2021).

Monocytes deriving from the bloodstream are recruited to the vessel wall, where they differentiate into macrophages and adopt distinct phenotypes driven by the local microenvironment (Ley et al. 2011). Murine monocytes are distinguished by specific surface markers, including CD11b, F4/80, CD115, and chemokine receptors such as CCR2 and CX3CR1, which identify two distinct monocyte populations, indicating classical and alternatively activated monocytes (Mantovani et al. 2005). Macrophages exhibit functional adaptability, that allows a prompt response to environmental changes (Mantovani et al. 2005). They play a crucial role in host defense, continuously monitoring their surroundings for signs of infection or tissue damage. Macrophages detect pathogen and danger-associated molecular patterns (PAMPs and DAMPs) through pattern recognition receptors (PRRs), which can be membrane-bound (e.g., Toll-like receptors) or cytosolic (e.g., RIG-I like receptors and AIM2-like receptors). PRR engagement activates macrophages, leading to an inflammatory response characterized by the production of proinflammatory cytokines and the recruitment of other immune cells to the site of inflammation (Li and Wu 2021). Activated macrophages also contribute to pathogen clearance through phagocytosis and the production of reactive oxygen and nitrogen species (Canton et al. 2021). After clearing the infection, macrophages help resolve inflammation and repair damaged tissue by phagocytosing debris and apoptotic cells, inducing a transition from pro-inflammatory to anti-inflammatory and reparative macrophages. These anti-inflammatory macrophages produce cytokines like IL-10 and IL-1Ra and growth factors like VEGF, TGF-β, and PDGF to support tissue repair and revascularization (Gordon 2003; Ley et al. 2011; Mantovani et al. 2004).

Several miRNAs have been identified as key regulators of monocyte biology in the maintenance of healthy arteries. Among them, miR-155 is a well-known miRNA involved in the regulation of both monocytes and macrophages. It is abundantly expressed in both cell types, where it plays a role in their activation, polarization, and immune responses. It has been reported that miR-155 is dependent on the nuclear factor-κB (NF-κB) and it can directly target and inhibit IL-1β, TNF-α, and IL-6, thus suggesting a role in fine-tuning monocyte inflammatory activation (Liu et al. 2019). Another well-known miRNA family in the regulation of the inflammatory status of monocytes and macrophages is miR-181. It encompasses a family of four members (miR-181a, miR-181b, miR-181c, and miR-181d), located in different chromosomes, which exert a plethora of biological roles. Among them, miR-181b regulates monocyte differentiation and function, playing an anti-inflammatory role by binding the 3′-untranslated regions (3’-UTR) of IL1a (Xie et al. 2013). Previous reports indicate that miR-181 has opposite expression in monocytes or macrophages, thus leading to speculation about a role in cellular differentiation via the modulation of the ERK-MAPK pathway (Lim et al. 2020). Moreover, miR-181b plays an important role in regulating NF-κB signaling and leukocyte recruitment in the vascular endothelium enlightening its role in the regulation of immune responses in the vasculature (Sun et al. 2012). Besides inflammatory status, miRNAs regulate the lipid metabolism of monocytes and macrophages. Among them, miR-27b is known to play a role in lipid metabolism and inflammation in mononuclear cells, which can impact vascular health. Among the most characterized miRNA in the regulation of cholesterol metabolism, the role of miR-33 has been investigated to highlight its contribution to the mechanisms of cholesterol efflux and reverse cholesterol transport. This miRNA participates in regulatory crosstalk linking the activity of the LXR receptors with the expression of the function of the cholesterol transporters ABCA1 and ABCG1, thus contributing to the complex mechanisms of maintenance of intracellular cholesterol content. Taken together, these observations reveal the crucial role of miRNAs in regulating monocyte homeostasis affecting both inflammatory activation and lipid homeostasis with important implication in the physiology of the vascular system.

9.2 miRNAs in Cardiovascular Pathologies

9.2.1 The Role of miRNAs in Heart Disease

9.2.1.1 Myocardial Infarction

Ischemic heart disease is one of the leading causes of death worldwide, with approximately 805,000 deaths from myocardial infarction occurring per year in the United States between 2005 and 2014 (Mozaffarian et al. 2016). Classically, myocardial ischemia and infarction occur when blood perfusion is insufficient for the metabolic needs of the myocardium. This often occurs as a manifestation of coronary artery disease, a condition in which the blood flow is reduced in the coronary arteries because of atherosclerosis. While atherosclerosis is a chronic disease evolving over decades, acute complications are mainly caused by superimposed thrombosis following atherosclerotic plaque erosion of rupture. Prolonged occlusion of the coronary arteries leads to permanent damage and necrosis in the heart tissue, which impairs cardiac function and determines severe clinical manifestations (Fox et al. 2007). The severity of cardiac ischemia usually correlates to the extent of the infarcted area which is also dependent on the location of the occlusion within the coronary tree and the duration of the blockage of the blood flow (Hasche et al. 1995). Coronary occlusion that is timely resolved may not lead to massive death of the myocardium; conversely, a persistent occlusion leads to cardiomyocyte death and prompt mechanisms of post-infarction cardiac remodeling and repairs. Classically, these mechanisms are characterized by two distinct phases: firstly, an inflammatory phase marked by inflammation and immune cell infiltration (particularly, neutrophils and monocyte/macrophages) clears damaged tissue and extracellular matrix (Hasche et al. 1995). This is followed by a reparative phase indicated by the resolution of inflammation, myofibroblast proliferation, scar formation, and neovascularization (Prabhu and Frangogiannis 2016). The cardiac remodeling may lead to heart chamber dilation and ventricular wall thinning. This is caused by apoptosis, necrosis, and fibrosis within cells and tissue (Sun 2009). To date, the use of interventions for prompt coronary revascularization such as percutaneous coronary intervention (PCI) and surgical revascularization has dramatically enhanced the clinical outcomes of acute myocardial infarction. However, patients with myocardial infarction have a high risk of developing future acute cardiovascular events and are prone to developing clinically relevant heart failure.

The loss of cardiomyocytes in the myocardium is a major determinant of the detrimental clinical manifestation of myocardial infarction and is a leading factor in determining the severity of acute and chronic heart failure. Both necrosis and apoptosis are responsible for the loss of cardiomyocytes and are a hallmark of ischemic cardiac disease. Considerable attention has been focused on elucidating mechanisms underlying myocyte cell death, aiming to mitigate damage and enhance function in the failing myocardium. Apoptosis significantly contributes to myocyte cell death in acute myocardial infarction, particularly in the peri-infarcted region (Saraste et al. 1997). Similarly, high rates of apoptosis characterize the subacute phase of myocardial infarction and correlate with the development of the progressive remodeling of the left ventricle and with the severity of clinical manifestations (Abbate et al. 2000, 2003; Baldi et al. 2002). Thus, apoptosis is a major determinant of infarct size, left ventricular remodeling, and progression of acute and chronic heart failure.

Several miRNAs play pro- or anti-apoptotic roles in cardiomyocytes and influence cardiomyocyte fate during ischemia by targeting genes and pathways, specifically affecting cell survival or death (Fig. 9.2). Among them, miR-93-5p and miR-138-5p protect cardiomyocytes against hypoxia-induced apoptosis by targeting PTEN and the MLK3/JNK/c-jun pathway, respectively (He et al. 2013; Ke et al. 2016). On the other hand, miR-320-3p targets IGF-1 as a crucial regulator of cardiomyocyte homeostasis and worsens apoptosis in the setting of cardiac ischemia. In line with therapeutic potential, inhibition of miR-320 in a rat model of myocardial ischemia/reperfusion suppressed early cardiomyocyte apoptosis and improved systolic cardiac function (Song et al. 2016). Besides, other miRNAs have been identified as important regulators of necrotic pathways in myocardial infarction. Among the pro-necrotic miRNAs, miR-103a-3p and miR-107 worsen the mechanisms of necrosis in cardiomyocytes by targeting Fas-associated protein with death domain (FADD) (Wang et al. 2015). In contrast, miR-30b-5p inhibits necrosis by targeting cyclophilin D, a well-established promoter of necrosis. Interestingly, enhanced expression of miR-30b-5p in the heart reduced necrosis and infarcted area size following ischemia/reperfusion injury (Wang et al. 2015). Likewise, overexpression of miR-155-5p in cardiomyocyte progenitor cells attenuates necrosis by targeting receptor-interacting protein 1 (RIP1) (Liu et al. 2011). Overall, evidence supports the contribution of miRNAs in the mechanisms of cell death during cardiac ischemia and identifies possible targets for possible therapeutic perspectives.

In addition to cell death, miRNAs have been also involved in the mechanisms of cardiac regeneration. Unfortunately, the adult mammalian heart has only a limited regenerative capacity compared to neonatal animals. Indeed, neonatal hearts have a remarkable cardiac repair and regeneration capability following ischemic injury (Porrello et al. 2011). In this context, miRNAs have been identified as both pro- and anti-regenerative. Several miRNAs have been shown to possibly induce cardiomyocyte proliferation. In particular, miR-548c-3p, miR-509-3p, and miR-23b-3p induce cardiomyocyte proliferation by inhibiting Meis1. On the other hand, miR-590 and

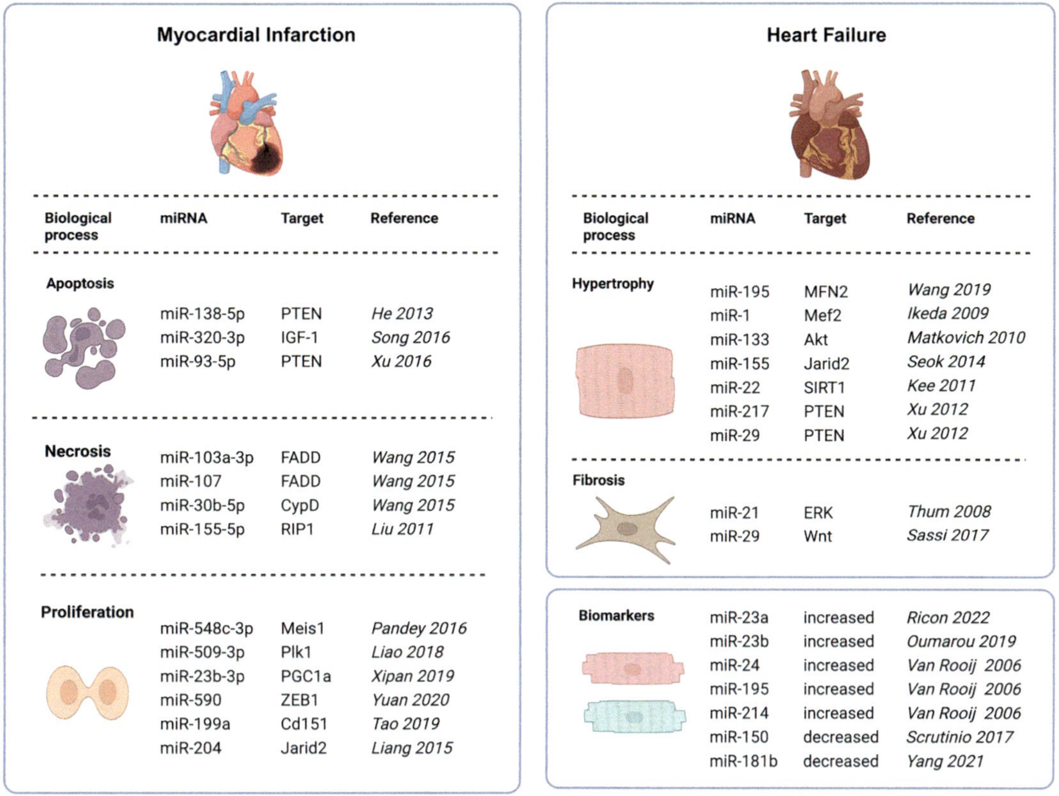

Fig. 9.2 Summary of the miRNAs involved in the regulation of disease processes in myocardial infarction and heart failure. For each miRNA, the trend of the expression and the molecular mechanism is reported. Figure created with BioRender.com

miR-199a promote the cell cycle re-entry of cardiomyocytes and their progenitors (Eulalio et al. 2012; Pandey et al. 2016). A single administration of synthetic miR-590 and miR-199a promotes cardiac repair after myocardial infarction by boosting cardiomyocyte cell cycle re-entry in neonatal mice (Lesizza et al. 2017). Similarly, miR-204 augments cardiomyocyte proliferation by targeting Jarid2 and its overexpression increases ventricular mass in mice (Liang et al. 2015). Similarly, overexpression of miRNAs of the miR-17-92 family alleviates cardiac injury after myocardial infarction in adult hearts (Chen et al. 2013). Together these observations open the intriguing perspective of leveraging the manipulation of miRNA expression after myocardial infarction as a strategy to promote cardiomyocyte regeneration and improve cardiac healing.

9.2.1.2 Heart Failure

Heart failure (HF) occurs when the heart cannot provide sufficient blood flow to the systemic circulation to meet the body's metabolic requirements because of functional or structural cardiac abnormalities impairing the ventricular systolic ejection or diastolic refilling (Leonard et al. 2012). Typically, it is the final manifestation of both cardiac injury and cardiovascular diseases. However, it can also result from prolonged arrhythmias, myocarditis, and exposure to cardiotoxic drugs (Schulte et al. 2015). In the USA, as of 2018, the nationwide mortality rate due to HF was 13.4% (Virani et al. 2020), which should be summed to the severe worsening of quality of life for patients suffering from this clinical syndrome. Mounting evidence suggests clear links between miRNAs and HF development and progression.

Pathologic remodeling in response to ischemic and mechanical injury plays an essential role in the decline of heart function during HF. This is underpinned by three major processes: hypertrophy, fibrosis, and apoptosis (discussed in the previous paragraph). Two mouse models of pressure overload-induced cardiac hypertrophy and HF (i.e., transverse aortic constriction, TAC, and expression of activated calcineurin) revealed significant changes in the cardiac expression of several miRNAs with the upregulation of mir-23a, miR-23b, miR-24, miR-195, and miR-214, while miR-150 and miR-181b were downregulated. Moreover, increased expression of miR-24, miR-125b, miR-195, miR-199a, and miR-214 (but not miR-23) was detected in cardiac specimens of patients with end-stage HF (van Rooij et al. 2006). In line with a functional role, overexpression of these miRNAs in primary cardiomyocytes led to dramatic changes in the hypertrophic response. More strikingly, overexpression of miR-195 in mice leads to increased cardiac hypertrophy, ventricular thinning, and decreased cardiac function (van Rooij et al. 2006). These findings provide support for the mechanistic role of this miRNA in the processes of cardiac remodeling. The myo-miRNAs miR-1 and miR-133 are also downregulated early during the cardiac remodeling in the TAC model in mice (Care et al. 2007; Sayed et al. 2007). In line, the overexpression of miR-1 via viral vectors in mice was shown to reverse pressure-induced cardiac hypertrophy and prevent maladaptive remodeling with improved ventricular geometry and morphology (Karakikes et al. 2013). On the other hand, inhibition of miR-1 prevented the development of hypertrophy in response to serum-enriched media in neonatal mouse cardiomyocytes (Gan et al. 2019; Rao et al. 2009; Sayed et al. 2007), possibly suggesting distinct roles of miR-1 in cardiac remodeling which are dependent on the stage of cardiac development. Similar to miR-1, adenoviral-based inhibition of miR-133 also revealed the protective role of this miRNA against the mechanisms of cardiac hypertrophy and dilatation. These data were further corroborated by the observation that overexpression of mir-133 abrogated hypertrophy induced by constitutive Akt activation or in the mouse TAC model, where this miRNA was able to reduce fibrosis and sensibly improve ventricular compliance (Matkovich et al. 2010).

Several other miRNAs have been identified to promote cardiac hypertrophy, including miR-155, miR-22, miR-217, and miR-29 (Fig. 9.2). Among them, miR-155 has been shown to promote cardiac hypertrophy by targeting prohypertrophic pathways such as inflammation and calcium signaling (Heymans et al. 2013; Seok et al. 2014; Yang et al. 2016). Similarly, miR-22 has been shown to mediate hypertrophy through the calcineurin pathway. Both cardiomyocytes transfected with miR22 inhibitors in vitro and primary cardiomyocytes isolated from miR-22-deficient mice showed higher expression of the anti-hypertrophic proteins SIRT1 and HDAC4 (Huang et al. 2013a, b; Kee and Kook 2011; Matsushima and Sadoshima 2015). Finally, miR-22, miR-217, and miR-29a have been shown to promote hypertrophy by modulating PTEN signaling, a crucial negative regulator of the Akt/mTOR pathway essential for cardiac function. Overexpression of miR-22 in hypertrophic cardiomyocytes reduces PTEN protein levels, resulting in hypertrophy, an effect that was reversed by upregulating PTEN (Tu et al. 2013; Xu et al. 2012). A similar hypertrophic effect was observed with miR-217 and miR-29a (Nie et al. 2018; Shi et al. 2019).

Notably, miRNAs have also been involved in other mechanisms relevant to cardiac remodeling. Alongside its role in cardiac hypertrophy, miR-29 has also been implicated in cardiac fibrosis by targeting numerous proteins involved in fibrosis as well as in the mechanisms of myofibroblast transdifferentiation in vitro (Zhou et al. 2012). The expression of miR-29 is downregulated in cardiac regions adjacent to the infarcted area both in mice and humans (van Rooij et al. 2008). Conversely, miR-21 in fibroblasts is selectively upregulated in mouse models of HF (Thum et al. 2008). Overexpression of miR-21 in cardiac fibroblasts decreased apoptosis and increased fibroblast growth factor (FGF) secretion through ERK activation, leading to profibrotic cardiac programming (Thum et al. 2008). Taken together, these observations reveal a key role of miRNAs in multiple mechanisms leading to cardiac remodeling and HF, possibly highlighting future targets for therapeutic purposes.

9.2.1.3 Atherosclerosis

Atherosclerosis, often considered as the silent culprit of cardiovascular disease, is a lipid-driven inflammatory disease that primarily involves the accumulation of lipids and inflammation in large arteries, potentially leading to clinical complications such as myocardial infarction and stroke. Atherosclerotic lesions are distinguished by the continuous and gradual buildup and alteration of lipids, inflammatory cells, VSMCs, and necrotic cellular remnants within the subendothelial space in the intima (Bjorkegren and Lusis 2022). The progression of atherosclerosis induces the narrowing of the arterial vessels and might lead to severe health complications, making it a significant contributor to morbidity and mortality worldwide. Within the context of atherosclerosis, miRNAs have been implicated in key pathways relevant to the pathophysiology, such as lipid metabolism (Du et al. 2021), vascular inflammation (Das and Rao 2022), and VSMC phenotypic switch, all involved in the development of atherosclerotic plaques (Farina et al. 2023). The intricate interplay between specific miRNAs and target genes has been found to contribute significantly to the initiation and progression of this chronic vascular disease (Table 9.1).

Numerous miRNAs have been identified as key players in various stages of atherosclerosis, including those impacting the very initial stages of the disease development related to endothelial dysfunction. Among them, the miR17–92 cluster is reported to regulate lipid metabolism and may play an important role in dyslipidemia and coronary artery disease (Liu et al. 2014). miRNAs belonging to this family oversee a plethora of processes, essential for atherosclerosis development. Among them, miR-20a has been described as an atheroprotective miRNA via phosphatase and tensin homolog (PTEN) inhibition, with the consequent increase in the survival and proliferation of ECs to preserve endothelial integrity (Wang et al. 2017). Furthermore, miR-20a is predicted to regulate MYF5 and PPARγ, involved in adipogenesis in both mice and humans (Guller et al. 2015). Of note, miR-20a and miR-20b can inhibit the expression of ABCA1, the cholesterol transporter to ApoA1, a critical factor for the generation of pre-β-HDL and cholesterol efflux and

Table 9.1 Summary of the main miRNAs involved in atherosclerosis

Atherosclerosis			
Cell type	miRNA	Target	Reference
Endothelial cell	miR17-92	PTEN	Wang et al. (2017)
	miR-20a	PTEN	Wang et al. (2017)
	miR-126-3p	RGS16	Zernecke et al. (2009)
	miR-126-5p	DLK1	Schober et al. (2014)
	miR-126-5p	Caspase 3	Santovito et al. (2020a, b)
	miR-34a	SIRT1	Zhang et al. (2017)
	miR-217	SIRT1	Menghini et al. (2009)
	miR-146a	SIRT1	Gong et al. (2022)
	miR-200c	ZEB1	Magenta et al. (2011)
Vascular smooth muscle cell	miR-143/miR-145	KLF4	
	miR-133	Sp-1	Torella et al. (2011)
	miR-221/miR-222	Kip1	Liu et al. (2009)
	miR-29	DNMT3b	Chen et al. (2011a, b, c)
	miR-34a	Sirt1	Badi et al. (2018)
	miR-128-3p	KLF4	Farina et al. (2020)
Myeloid cell	miR-17-5p	ABCA1	Tan et al. (2019)
	miR-212	SIRT1	Miao et al. (2018)
	miR-33a	ABCA1	Rayner et al. (2010)
	miR-33b	ABCA1	Mandolini et al. (2015)

the initiation of the reverse cholesterol transport process, essential for eliminating excess cholesterol from the cell and the body (Liang et al. 2017). Moreover, the cluster has been reported to play a crucial role in cell-to-cell communication suggesting potential regulation of cellular phenotypes and inflammation rate in a paracrine manner (Kaur et al. 2018). Another crucial miRNA in the regulation of ECs is miR-126. It is the most abundant miRNA in ECs, particularly in young cells, and the only miRNA in vertebrates whose expression is limited to EC. Several reports indicated its role in the regulation of atherosclerosis development. Of note, specifical miR-126-5p expression, but not miR-126-3p, maintains EC proliferative reserve upon hyperlipidemia by the Notch1 inhibitor delta-like 1 homolog (Dlk1), preventing atherosclerosis formation (Schober et al. 2014). Moreover, a non-canonical anti-apoptotic function of this miRNA has been reported, showing its ability to inhibit Caspase-3 within EC nuclei thus preserving endothelial integrity (Santovito et al. 2020a, b). On the other hand, miR-126-3p plays an important role in the regulation of the inflammatory EC activation by targeting VCAM1 as well as in the recruitment of endothelial progenitor cells by a paracrine mechanism involving the release via apoptotic bodies and the activation of the CXCR4-CXCL12 pathway (Harris et al. 2008; Zernecke et al. 2009). Finally, miR-200c can regulate various processes within ECs, including proliferation, migration, apoptosis, and inflammation to affect vascular remodeling. It can modulate EC function by targeting zinc finger E-box binding homeobox 1 (ZEB1) during oxidative stress, thus leading to impaired EC proliferation, death, and senescence (Magenta et al. 2011). Moreover, it is reported to impact the processes of EndMT via the SMAD7/YAP pathway during atherogenic conditions (Mao and Jiang 2021). All in all, the evidence suggests that miRNAs are crucial in regulating endothelial (dys)function already at the early stages of atherogenesis.

Atherosclerosis is strongly associated with aging and mechanisms of senescence. In this context, miR-34a expression correlates with markers of senescence in endothelial, cardiac, and splenic cells (Ito et al. 2010). Its function in senescence is mainly related to its ability to upregulate the tumor suppressor p53 by downregulating the sirtuin 1 (SIRT1) gene (Liu et al. 2017). The upregulation of miR-34a causes endothelial dysfunction by affecting the nitric oxide bioavailability, the expression of adhesion molecules, the inflammatory activation, and glucose metabolism (Raucci et al. 2021). Another miRNA related to senescence is miR-217. Similar to miR-34a, its expression is increased in ECs in late culture passages, and it is able to repress the expression of SIRT1 by directly targeting the 3'UTR of its mRNA transcript (Menghini et al. 2009). Moreover, miR-217 was reported to regulate EC inflammation, EndMT, and apoptosis via the TLR4/PI3K/Akt/NF-κB pathway (Zhang et al. 2020). Another work described how miR-217 might regulate endothelial function via downregulating VEGF and apelin receptor, activators of endothelial NO synthase (eNOS), resulting in diminished eNOS expression (de Yebenes et al. 2020). Lastly, as an aging-related miRNA, miR-146a has been linked to senescence in ECs for its regulatory role on AMPK-mediated NAD+/SIRT1 activation with consequent implications in premature aging (Gong et al. 2022).

Besides endothelium, the role of miRNAs in the modulation of VSMC phenotypic switching has emerged in the last decades. Indeed, miRNAs might independently affect the expression of mRNA markers or serve as crucial nodes in the phenotypic pathways orchestrated by other factors influencing cell phenotype. Several miRNAs, through their direct impact on VSMC phenotype, play an integral role in atherosclerotic development. Among them, the most known is the miR-143/145 cluster, given that this miRNA family holds the distinction of being the most abundant miRNA within the vascular wall. The downregulation of miR-143/145 in VSMCs plays a role in several CVD, such as hypertension and coronary artery disease (CAD). They can control VSMC phenotypes, inducing their differentiation while limiting their proliferation (Boettger et al. 2009; Cordes et al. 2009; Elia et al. 2009a, b; Zhang et al. 2016). In line with those results, miR-143 and miR-145 expression is abrogated in a plethora of in vivo CAD models and in patient samples, where proliferative and dysregulated VSMCs contribute to the lesion development (Cipollone et al. 2011; Ji et al. 2007; Zhao et al. 2015). On the contrary, miR-143 and miR-145 expression was reported to be upregulated in pulmonary arterial hypertension (PAH) induced by hypoxia in an in vivo model and in human specimens, suggesting a potential role in PAH pathophysiology, as well as in atherosclerotic plaques of hypertensive patients or in patients with clinically unstable atherosclerotic plaques (Caruso et al. 2012; Cipollone et al. 2011; Deng et al. 2015; Maitrias et al. 2015; Santovito et al. 2013).

As another miRNA strongly expressed in VSMCs, miR-133 can exert a strong role in the regulation of the phenotypic switch. Indeed, miR-133 decreases when VSMCs are primed to proliferate in vitro and following vascular injury in vivo, whereas it increases when VSMCs are induced to differentiate towards a contractile phenotype (e.g., by stimulating them with TGF-β). Notably, its role in VSMC phenotypic switching has been highlighted through the direct inhibition of the anti-proliferative and pro-apoptotic transcription factor Sp-1 (Torella et al. 2011). Moreover, miR-128-3p stands as a pivotal mediator in the phenotypic transition of VSMCs. Its protective function has been documented in maintaining the differentiation status of VSMCs through direct targeting of KLF4. This action not only contributes to the reversal of restenosis in vivo but also involves precise epigenetic modulation of Myh11. By abrogating the DNA methylation status at the promoter level of Myh11, miR-128-3p promotes its expression, further reinforcing VSMC differentiation (Farina et al. 2020). On the other side, miR-128-3p was also reported to dampen the inflammatory activation of VSMCs and to promote apoptosis via the inhibition of the FOXO4/MMP9 signaling pathway (Qu et al. 2020). In contrast, miR-221 and miR-222, aside from ECs, were found to be significantly abundant in VSMCs. Intriguingly, these miRNAs demonstrated pro-atherosclerotic properties, enhancing the proliferation and migration of VSMCs in vivo. Mechanistic experiments demonstrated that those miRNAs could impact cell proliferation via directly inhibiting p21 (Kip1) but also through the abrogation of p57(Kip2) via indirect pathways (Liu et al. 2009). Another highly enriched miRNA family in VSMCs is miR-29. Their members have been found to contribute to the regulation of extracellular matrix (ECM) proteins like collagen and elastin, whose abrogation might affect plaque development and phenotype (Liu et al. 2010). They are known to inhibit VSMC calcification by suppressing the expression of a disintegrin and metalloproteinase with thrombospondin motifs-7 (ADAMTS-7) (Du et al. 2012).

Finally, in line with its rheostatic role in senescence, miR-34a facilitates the switch of VSMC toward an osteoblastic phenotype and facilitates the development of vascular calcification, for instance in hyperphosphatemia conditions. Conversely, atherogenic and hypoxic stimuli downregulate miR-34a levels and promote VSMC proliferation and migration during atherosclerosis and PAH (Raucci et al. 2021). In conclusion, miRNAs hold the key to several pathways regulating VSMC plasticity during atherogenesis and may represent attractive therapeutic targets to modulate the progression of the disease.

Atherosclerosis progression is driven by the accumulation of lipid-laden macrophage foam cells in the inner arterial wall, which will fuel a state of chronic inflammation that promotes the formation of atherosclerotic plaques and their progression toward instability. As an example, miR-17-5p can directly target ABCA1, thus inducing cholesterol accumulation in macrophages (Tan et al. 2019). Additionally, its expression is tuned by the lncRNA metastasis-associated lung adenocarcinoma transcript 1 (MALAT1) which is able to sponge the miRNA, reducing cholesterol accumulation in these macrophages. Knocking down MALAT1 may promote cholesterol accumulation through the miR-17-5p/ABCA1 pathway (Liu et al. 2020). Moreover, miR-212 promoted lipid accumulation affecting ABCA1-dependent cholesterol efflux through the direct inhibition of SIRT1 in human macrophages (Miao et al. 2018). Lastly, miR-33 emerged as a crucial regulator of intracellular lipid metabolism by regulating cholesterol and fatty acid homeostasis. The intronic miR-33 is transcribed within the primary transcript of SREBP2 and not only reduces cellular cholesterol export via ABCA1 inhibition but is implicated in fatty acid β-oxidation by targeting CPT1A, HADHB, and CROT and reducing fatty acid degradation (Gerin et al. 2010; Marquart et al. 2010; Najafi-Shoushtari et al. 2010). Interestingly, the expression of miR-33 is increased by inflammatory clues to contribute to the inflammatory activation of macrophages, thus suggesting its role in the crosstalk between metabolic and inflammatory pathways (Ouimet et al. 2015). The inhibition of miR-33 has been proposed as an attractive therapeutic opportunity in atherosclerosis, as cholesterol efflux may clearly represent an opportunity for atherosclerosis regression. In mice, 4-week treatment with anti-miR-33 treatment increased cholesterol efflux from macrophages within the atherosclerotic plaques (Rayner et al. 2010). Unfortunately, longer treatment schedules (up to 14 weeks) did not produce significant beneficial effects on atherosclerosis while it also promoted the development of liver steatosis, likely due to the concomitant hepatic overexpression of genes regulating fatty acid synthesis (Goedeke et al. 2014; Marquart et al. 2013). However, an important caveat should be considered in translating the results of studies on miR-33 in mice to humans and other primates. Indeed, primates express two miR-33 homologs (miR-33a and miR-33b) that exhibit different transcriptional regulation and involvement in different cholesterol homeostasis. In the example, miR-33b—but not miR-33a—is upregulated in atherosclerotic plaques of patients with severe hypercholesterolemia and treatment with lipid-lowering drugs (i.e., rosuvastatin) primarily reduced the expression of miR-33b to influence the post-transcriptional regulation of ABCA1 (Mandolini et al. 2015; Santovito et al. 2020a, b). It is possible that the absence of miR-33b in mice may not allow the complete assessment of the function of the miR-33 family.

9.2.1.4 Arterial Aneurism

Aneurysms are major abnormal pathological dilations of the arteries, mainly occurring at the abdominal segment of the aorta. Aging is the primary risk factor but other factors including genetic predisposition, hypertension, cigarette smoking, male gender, and atherosclerosis might play significant roles (Bossone and Eagle 2021; Obel et al. 2021). Aneurysms are characterized by sustained inflammation, the ECM degrading enzymes activation, and a gradual thinning of the vascular wall, which might lead to vascular rupture and sudden death (Bossone and Eagle 2021).

Given their contribution to the regulation of pathways involved in inflammation and ECM homeostasis, miRNAs have been strongly implicated in the pathogenesis of arterial aneurysms. One of the most studied miRNAs in aneurysms is miR-29. As an example, miR-29 family members were reported to be upregulated in AAA tissues and further associated with the downregulation of numerous ECM components in diseased aortas (Boon et al. 2011). Moreover, miR-29b increase contributes to early aneurysm development in an in vivo murine model of Marfan syndrome (MFS), by increasing VSMC apoptosis (Merk et al. 2012). Several other miRNAs, such as miR-30c-2, miR-126, miR-146a, and miR-155, exhibit dysregulation of their expression pattern which are similar in both Thoracic Aortic Aneurysms (TAA) and Abdominal Aortic Aneurysms (AAA) (Venkatesh et al. 2017). Their possible contribution to aneurysm pathogenesis was related to the regulation of collagen and elastin degradation with consequent VSMC dysfunction. Among them, miR-126-5p expression has been shown to reduce the expansion of the abdominal aorta in mice with Ang II-induced AAA. Moreover, in vitro studies demonstrated that miR-126-5p effectively inhibits the dedifferentiation and enhances the contractility properties of aortic VSMCs induced by Ang II and PDGF-BB, by targeting VEPH1 (Shi et al. 2020). On the contrary, miR-155 has been found to be consistently upregulated in human specimens and in a murine model of Ang II-induced AAA (Zhang et al. 2018). In line with its role in regulating the inflammatory activation, mechanistic data showed that miR-155 overexpression might impact sustained inflammation in vitro due to the inhibition of IL-10 and the upregulation of IL-6, IL-1β, and TNFα, along with MMP-2, MMP-9, iNOS, and CCL2 (Zhang et al. 2018).

Among the most enriched miRNAs in VSMCs and major determinants of their contractile phenotype, miR-143 and miR-145 were found to be reduced in tissue specimens of human AAA, compared to the healthy group (Elia et al. 2009a, b). Another regulator of VSMC function, miR-21, was shown to increase upon AAA formation in multiple in vivo animal models of AAA (e.g., murine and porcine). Its expression was correlated to the reduction of its target PTEN, known to be a detrimental factor in the progression of the aneurysm via the negative regulation of the AKT pathway (Maegdefessel et al. 2012). Together, these data reveal that miRNAs crucial for the maintenance of proper VSMC homeostasis may be deregulated in the disease setting to contribute to the development of aneurysms and, possibly, their evolution toward a rupture-prone phenotype (Fig. 9.3).

miR-155 plays a significant role in regulating inflammation and macrophage function and plaque development within atherosclerosis by inhibiting CCL2

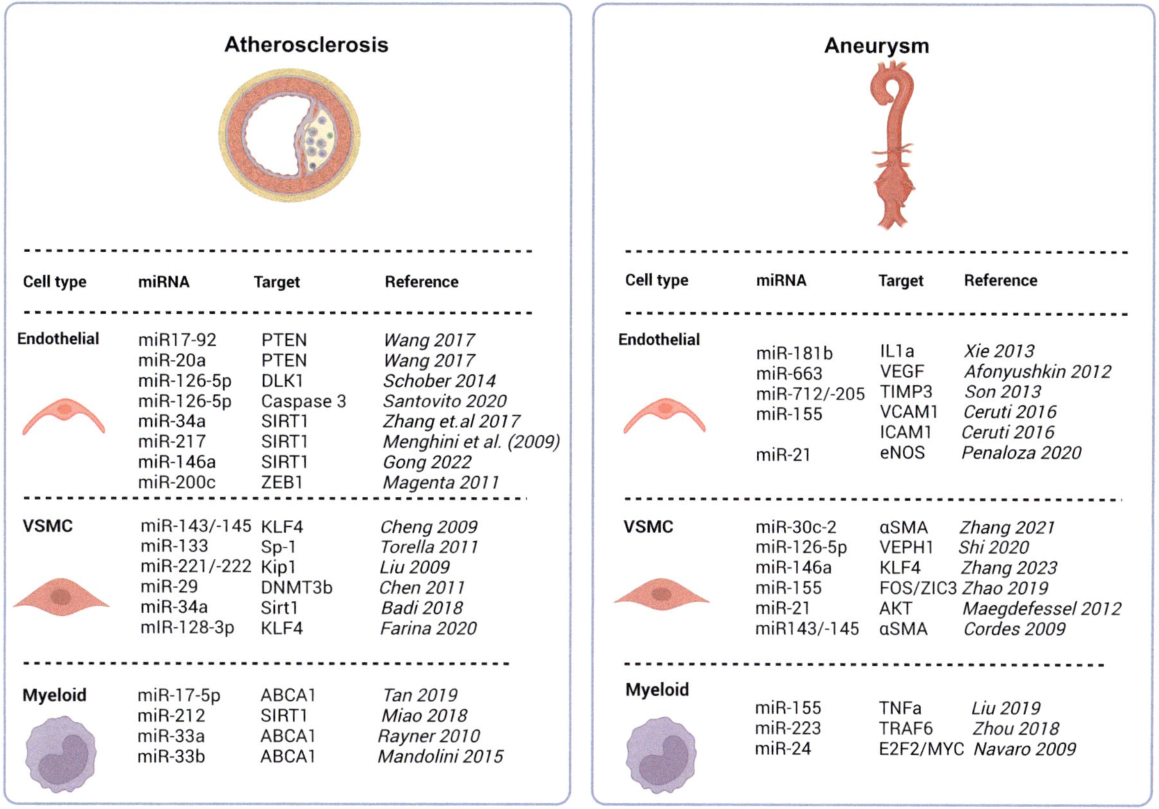

Fig. 9.3 Summary of the miRNAs involved in the pathophysiology of vascular diseases. Figure created with BioRender.com

(Nazari-Jahantigh et al. 2012). It is associated with the promotion of a pro-inflammatory phenotype in macrophages, with a strong reduction of the anti-inflammatory IL-10 and an increased release of inflammatory cytokines leading to atherosclerosis progression (Donners et al. 2012). Other studies reported that by antagomiR-155 administration atherosclerosis was reverted role via targeting the SOCS1-STAT3-PDCD4 pathway (Roselaar et al. 1996; Yang et al. 2015; Ye et al. 2016).

Among the most abundantly expressed miRNA in macrophages miR-21-5p plays a pivotal role in atherosclerotic and inflammation. Its inhibition induces apoptosis, plaque necrosis, cholesterol efflux, and vascular inflammation, by inhibiting MKK3, thus leading to the sustained activation of p38-CHOP and JNK signaling and ABCG1 degradation (Canfran-Duque et al. 2017). Moreover, miR-21 KO mice showed sustained expression of the proapoptotic XIAP-associated factor 1(XAF1) and impact on the transcription of circadian rhythms (Schober et al. 2021).

9.2.2 Circulating miRNAs As Biomarkers in Cardiovascular Disease

miRNAs have gained significant attention as potential biomarkers for cardiovascular pathologies due to their stability in bodily fluids and their association with specific cardiac conditions (Sommariva et al. 2017; D'Alessandra et al. 2010). As such, the circulating miRNA signature has been investigated in multiple cardiovascular diseases as well as in patients with cardiovascular risk factors such as diabetes and hypertension (Zampetaki et al. 2010; Li et al. 2011a, b; Santovito et al. 2014).

Elevated levels of cardiac-specific miRNAs are associated with myocardial injury, making them valuable biomarkers for acute coronary syndromes. Among abundantly expressed miRNAs in the heart, four cardiac-enriched miRNAs (miR-1, miR-133, miR-208, and miR-499) are consistently increased in the plasma of acute MI patients (Bostjancic et al. 2010; Weber et al. 2010). Additionally, circulating miR-92a and miR-181a were described as potential novel biomarkers for diagnosis of AMI patients (Zhu et al. 2016a; Zhang et al. 2017). Moreover, miR-1, miR-21, and miR-499 were shown to have a higher diagnostic value compared to troponin T (Oerlemans et al. 2012). Overall, these miRNAs can be easily detected in the blood and used to promptly assess the extent of cardiac damage and to predict outcomes.

Besides myocardial infarction, circulating miRNA profiles have been investigated also as markers of other cardiovascular diseases. Analysis of circulating miRNAs revealed a downregulation of miR-221/miR-222, coupled with the simultaneous upregulation of miR-21, miR-27b, miR-130a, and miR-210 in the serum of patients with established atherosclerosis (e.g., peripheral artery disease) (Li et al. 2011a). Of interest, their expression also reflected altered expression patterns in tissue samples obtained from the same individuals. Another study showed that patients affected by CAD displayed significantly lower circulating levels of miR-17, miR-92a, miR-126, miR-145, and miR-155 compared to healthy controls (Fichtlscherer et al. 2010). Furthermore, studies have investigated the expression of miRNAs in symptomatic vs. asymptomatic carotid atherosclerosis and identified eight deregulated miRNAs (miR-15b, miR-29c, miR-30c/d, miR-150, miR-191, miR-210, and miR-500) and, among them, miR-210 was reported to display a significant downregulation in whole plaque compared with healthy arterial tissue as well as in plasma samples. Besides its possible role as a biomarker, miR-210 located in the intima-medial layer of either control or stable plaques but completely vanished in ruptured lesions, implying a possible mechanistic role in plaque destabilization and disruption (Eken et al. 2017). Another work aimed at identifying possible correlations in miRNA expression in tissue vs. blood in baboons to better investigate their role in lesion burden and to address potential novel biomarkers in humans. Eight miRNAs (miR-17-5p, miR-144-3p, miR-148a-3p, miR-191-5p, miR-378c, miR-451a, miR-486-5p, and miR-7975) showed significant overexpression in atherosclerotic lesions vs. the healthy counterpart. Furthermore, there was a significant correlation between the expression of 14 circulating miRNAs and the severity of lesions. The research uncovered putative miRNA-targets networks, demonstrating the involvement of genes related to proliferation, immune response, and cholesterol metabolism, where miR-17-5p and miR-146a-5p were found as central nodes and miR-5001 and miR-7975 emerged as potential novel biomarkers related to early atherosclerosis (Karere et al. 2023).

Other diseases were also investigated. Initial profiling studies in patients with AAA revealed higher circulating levels of three miRNAs (miR-191-3p, miR-455-3p, and miR-1281) compared to controls, thus supporting their possible role as effective non-invasive biomarkers (Zhang et al. 2015). Moreover, in a longitudinal analysis with a 40-year follow-up, miR-126 was demonstrated to be a potential prognostic biomarker for intracranial aneurysmal ruptures. Circulating miR-126 was higher among 226 patients with ruptured intracranial aneurysms compared to 153 patients with non-ruptured aneurysms or healthy controls (Mathioudakis et al. 2023). Finally, miRNAs have been also investigated as diagnostic and prognostic markers for post-cardiac transplantation complications, such as allograft rejection. For instance, the analysis of pathological tissues revealed specific miRNA signatures for the three major mechanisms of rejection, namely acute cellular rejection (miR-27b-3p, miR-29b-3p, and miR-199a-3p), antibody-mediated rejection (miR-208a, miR-29b-3p, 135a-5p, and miR-144-3p), and mixed rejection (miR-208a, miR-126-5p, and miR-135a-5p) (Di Francesco et al. 2018). Moreover, the concomitant detection of abnormal circulating levels of miR-144-3p and miR-652-3p could enhance the diagnostic sensitivity of patients with acute cellular rejection of the transplanted graft, especially at the early stages and low pathological grades, with possible implications for early diagnosis and management of this insidious clinical condition (Perez-Carrillo et al. 2023). Altogether, miRNAs hold the promise of serving as biomarkers for early non-invasive diagnosis of many cardiovascular diseases as well as prognostic markers to tailor appropriate therapeutic and clinical management strategies.

Although the strong potential, several limitations should be considered before implementing the use of circulating miRNAs as disease biomarkers (de Gonzalo-Calvo et al. 2022). Among all, miRNA expression can significantly vary among individuals, and it can be affected by factors such as age, sex, and genetic background. Due to this aspect, it might be challenging to establish consistent baselines for diagnostic purposes (Bernardo et al. 2016; Godoy et al. 2018). In addition, due to the lack of standardized protocols for miRNA extraction, quantification, and data analysis, high variability and inconsistencies in results among different studies can occur (Wang et al. 2012; Wong et al. 2019). Moreover, the sensitivity and specificity of miRNAs as biomarkers can be a concern. Some miRNAs may be differentially expressed in multiple disease conditions, making it challenging to distinguish between different diseases with similar miRNA profiles (Chen et al. 2011a, 2015).

9.3 The Potential of miRNAs As Therapeutic Targets for Cardiovascular Diseases

The morbidity and mortality of cardiovascular disease remains a major burden for public health worldwide, despite the advances in percutaneous techniques or revascularization and the development and clinical use of pharmacological treatments such as—for instance—statins and other lipid-lowering drugs, ACE inhibitors, and beta-blockers. This underscores the need for further development of novel therapeutic approaches. Approaches for modulating the mechanisms of action of miRNAs rely on synthetic oligonucleotides that mimic or inhibit their function. Among those, miRNA mimics double-stranded sequences aimed at enhancing the expression of specific miRNA, while antisense oligonucleotides (ASOs) or target-site blockers (TSBs) are single-stranded sequences that inhibit miRNA function or its specific binding to a target mRNA, respectively. The evolving understanding of the pivotal role of miRNAs in cardiovascular diseases, as well as the development of chemical modification to increase the pharmacokinetic and pharmacodynamic of RNA-based therapeutics, has prompted efforts to develop miRNA-based therapeutics (Milani-Nejad and Janssen 2014).

While miRNA mimics have faced challenges in design and delivery, a possible avenue for enhancing miRNA function or replacing lost miRNAs is through viral vectors, such as the adeno-associated viruses (AAVs). Different AAV serotypes offer tissue-specific delivery options: AAV types 6 and 9 show skeletal muscle and heart preferences, respectively (Katwal et al. 2013; Zincarelli et al. 2010). To date, over 130 AAV-based early-phases trials are ongoing and two AAV-based therapeutics have been approved for clinical applications (i.e., voretigene-neparvovec, onasemnogene-abeparvovec), although not primarily related to miRNA-based applications in cardiovascular disease (Au et al. 2021). On the other hand, approaches to inhibit miRNA function have reached significant milestones, particularly therapeutics targeting miR-122, miR-92, and miR-132-3p. Among them, miravirsen (also known as RG-101A) is an anti-miRNA against miR-122-5p. It has long stood as a proof of principle for miRNA-based therapeutics to be applied into clinical trial for treatment of hepatitis C virus (HCV) infection and also revealed an effect in plasma lipid profile (Janssen et al. 2013). Unfortunately, the interest on further development of miravirsen have faded due to the efficacy of other drug classes and the gradual development of viral resistance (Janssen et al. 2013). More relevant to the cardiovascular system, miR-92a-3p shows high expression in ECs alongside cardiac fibroblasts and myocardial cells and is dysregulated in mouse models of vascular and myocardial injury (Bonauer et al. 2009; Hinkel et al. 2013) (Wu et al. 2011). In a pig model of ischemia/reperfusion, an anti-miRNA targeting miR-92a promoted angiogenesis and facilitated tissue repair (Hinkel et al. 2013). Based on the promising efficacy also in the large animal model, the antimiR-92a (MRG-110) has been evaluated in a phase I clinical trial administered as a single dose via i.v injection to healthy subjects (Abplanalp et al. 2020) (ClinicalTrials.gov NCT03603431). Another candidate miRNA showing potential in the clinical setting of cardiac remodeling and heart failure is miR-132. Increased expression of miR-132 leads to adverse remodeling and pathological hypertrophy in mouse and pig models of heart failure (Batkai et al. 2021; Ucar et al. 2012). To date, a synthetic lead-optimized oligonucleotide inhibitor of miR-132 (i.e., CDR132L) has completed preclinical efficacy and safety in patients with post-infarction acute heart failure in a Phase 1b study (NCT04045405). These promising results led to the progression of the clinical development of CDR132L, which is planned to be tested in phase II clinical trials. As such, CDR132L might become the first miRNA-targeting drug possibly available for the treatment of cardiovascular diseases.

9.4 Conclusion

Solid evidence supports the role of miRNAs in the cardiovascular system, from the mechanisms of embryonic development to contributions to the pathophysiology of cardiac and vascular diseases. Besides basic research aimed at investigating the mechanistic function of miRNAs in the maintenance of cardiovascular homeostasis and the development of disease, translational research has also advanced the first candidates toward the early stages of clinical application as therapeutics or disease biomarkers. Hence, future studies in the field may identify molecular targets to further improve the clinical outcomes of patients with cardiovascular diseases, which still rank as the first cause of death worldwide.

Compliance with Ethical Standards

Conflict of Interest The authors declare that they have no conflict of interest.

Ethical Approval This chapter contains no studies with human/animal participants performed by authors.

References

Abbate A, Melfi R, Patti G et al (2000) Apoptosis in recent myocardial infarction. Clin Ter 151(4):247–251. https://www.ncbi.nlm.nih.gov/pubmed/11107673

Abbate A, Biondi-Zoccai GG, Bussani R et al (2003) Increased myocardial apoptosis in patients with unfavorable left ventricular remodeling and early symptomatic post-infarction heart failure. J Am Coll Cardiol 41(5):753–760. https://doi.org/10.1016/s0735-1097(02)02959-5

Abplanalp WT, Fischer A, John D et al (2020) Efficiency and target Derepression of anti-miR-92a: results of a first in human study. Nucleic Acid Ther 30(6):335–345. https://doi.org/10.1089/nat.2020.0871

Alexander MR, Owens GK (2012) Epigenetic control of smooth muscle cell differentiation and phenotypic switching in vascular development and disease. Annu Rev Physiol 74:13–40. https://doi.org/10.1146/annurev-physiol-012110-142315

Alique M, Bodega G, Giannarelli C et al (2019) MicroRNA-126 regulates hypoxia-inducible factor-1alpha which inhibited migration, proliferation, and angiogenesis in replicative endothelial senescence. Sci Rep 9(1):7381. https://doi.org/10.1038/s41598-019-43689-3

Alisch RS, Jin P, Epstein M et al (2007) Argonaute2 is essential for mammalian gastrulation and proper mesoderm formation. PLoS Genet 3(12):e227. https://doi.org/10.1371/journal.pgen.0030227

Anand S, Majeti BK, Acevedo LM et al (2010) MicroRNA-132-mediated loss of p120RasGAP activates the endothelium to facilitate pathological angiogenesis. Nat Med 16(8):909–914. https://doi.org/10.1038/nm.2186

Au HKE, Isalan M, Mielcarek M (2021) Gene therapy advances: a meta-analysis of AAV usage in clinical settings. Front Med 8:809118. https://doi.org/10.3389/fmed.2021.809118

Badi I, Mancinelli L, Polizzotto A, Ferri D, Zeni F, Burba I, Milano G, Brambilla F, Saccu C, Bianchi ME, Pompilio G, Capogrossi MC, Raucci A (2018) miR-34a promotes vascular smooth muscle cell calcification by downregulating SIRT1 (Sirtuin 1) and Axl (AXL Receptor Tyrosine Kinase). Arterioscler Thromb Vasc Biol. https://doi.org/10.1161/ATVBAHA.118.311298

Baldi A, Abbate A, Bussani R et al (2002) Apoptosis and post-infarction left ventricular remodeling. J Mol Cell Cardiol 34(2):165–174. https://doi.org/10.1006/jmcc.2001.1498

Bartel DP (2009) MicroRNAs: target recognition and regulatory functions. Cell 136(2):215–233. https://doi.org/10.1016/j.cell.2009.01.002

Batkai S, Genschel C, Viereck J et al (2021) CDR132L improves systolic and diastolic function in a large animal model of chronic heart failure. Eur Heart J 42(2):192–201. https://doi.org/10.1093/eurheartj/ehaa791

Bernardo BC, Ooi JY, Matsumoto A et al (2016) Sex differences in response to miRNA-34a therapy in mouse models of cardiac disease: identification of sex-, disease- and treatment-regulated miRNAs. J Physiol 594(20):5959–5974. https://doi.org/10.1113/JP272512

Beuvink I, Kolb FA, Budach W et al (2007) A novel microarray approach reveals new tissue-specific signatures of known and predicted mammalian microRNAs. Nucleic Acids Res 35(7):e52. https://doi.org/10.1093/nar/gkl1118

Bino L, Prochazkova J, Radaszkiewicz KA et al (2017) Hypoxia favors myosin heavy chain beta gene expression in an Hif-1alpha-dependent manner. Oncotarget 8(48):83684–83697. https://doi.org/10.18632/oncotarget.19016

Bjorkegren JLM, Lusis AJ (2022) Atherosclerosis: recent developments. Cell 185(10):1630–1645. https://doi.org/10.1016/j.cell.2022.04.004

Boettger T, Beetz N, Kostin S et al (2009) Acquisition of the contractile phenotype by murine arterial smooth muscle cells depends on the Mir143/145 gene cluster. J Clin Invest 119(9):2634–2647. https://doi.org/10.1172/JCI38864

Bonauer A, Dimmeler S (2009) The microRNA-17-92 cluster: still a miRacle? Cell Cycle 8(23):3866–3873. https://doi.org/10.4161/cc.8.23.9994

Bonauer A, Carmona G, Iwasaki M et al (2009) MicroRNA-92a controls angiogenesis and functional recovery of ischemic tissues in mice. Science 324(5935):1710–1713. https://doi.org/10.1126/science.1174381

Boon RA, Seeger T, Heydt S et al (2011) MicroRNA-29 in aortic dilation: implications for aneurysm formation. Circ Res 109(10):1115–1119. https://doi.org/10.1161/CIRCRESAHA.111.255737

Bossone E, Eagle KA (2021) Epidemiology and management of aortic disease: aortic aneurysms and acute aortic syndromes. Nat Rev Cardiol 18(5):331–348. https://doi.org/10.1038/s41569-020-00472-6

Bostjancic E, Zidar N, Stajer D, Glavac D (2010) MicroRNAs miR-1, miR-133a, miR-133b and miR-208 are dysregulated in human myocardial infarction. Cardiology 115(3):163–169. https://doi.org/10.1159/000268088

Busk PK, Hinrichsen R, Bartkova J et al (2005) Cyclin D2 induces proliferation of cardiac myocytes and represses hypertrophy. Exp Cell Res 304(1):149–161. https://doi.org/10.1016/j.yexcr.2004.10.022

Callis TE, Pandya K, Seok HY et al (2009) MicroRNA-208a is a regulator of cardiac hypertrophy and conduction in mice. J Clin Invest 119(9):2772–2786. https://doi.org/10.1172/JCI36154

Canfran-Duque A, Rotllan N, Zhang X et al (2017) Macrophage deficiency of miR-21 promotes apoptosis, plaque necrosis, and vascular inflammation during atherogenesis. EMBO Mol Med 9(9):1244–1262. https://doi.org/10.15252/emmm.201607492

Canton M, Sanchez-Rodriguez R, Spera I et al (2021) Reactive oxygen species in macrophages: sources and targets. Front Immunol 12:734229. https://doi.org/10.3389/fimmu.2021.734229

Cao D, Mikosz AM, Ringsby AJ et al (2020) MicroRNA-126-3p inhibits Angiogenic function of human lung microvascular endothelial cells via LAT1 (L-type amino acid transporter 1)-mediated mTOR (mammalian target of Rapamycin) signaling. Arterioscler Thromb Vasc Biol 40(5):1195–1206. https://doi.org/10.1161/ATVBAHA.119.313800

Care A, Catalucci D, Felicetti F et al (2007) MicroRNA-133 controls cardiac hypertrophy. Nat Med 13(5):613–618. https://doi.org/10.1038/nm1582

Caruso P, Dempsie Y, Stevens HC et al (2012) A role for miR-145 in pulmonary arterial hypertension: evidence from mouse models and patient samples. Circ Res 111(3):290–300. https://doi.org/10.1161/CIRCRESAHA.112.267591

Cerutti C, Soblechero-Martin P, Wu D et al (2016) MicroRNA-155 contributes to shear-resistant leukocyte adhesion to human brain endothelium in vitro. Fluids Barriers CNS 13(1):8. https://doi.org/10.1186/s12987-016-0032-3

Cheloufi S, Dos Santos CO, Chong MM, Hannon GJ (2010) A dicer-independent miRNA biogenesis pathway that requires ago catalysis. Nature 465(7298):584–589. https://doi.org/10.1038/nature09092

Chen JF, Murchison EP, Tang R et al (2008) Targeted deletion of dicer in the heart leads to dilated cardiomyopathy and heart failure. Proc Natl Acad Sci USA 105(6):2111–2116. https://doi.org/10.1073/pnas.0710228105

Chen C, Tan R, Wong L et al (2011a) Quantitation of microRNAs by real-time RT-qPCR. Methods Mol Biol 687:113–134. https://doi.org/10.1007/978-1-60761-944-4_8

Chen KC, Wang YS, Hu CY et al (2011b) OxLDL up-regulates microRNA-29b, leading to epigenetic modifications of MMP-2/MMP-9 genes: a novel mechanism for cardiovascular diseases. FASEB J 25(5):1718–1728. https://doi.org/10.1096/fj.10-174904

Chen J, Yin H, Jiang Y et al (2011c) Induction of microRNA-1 by myocardin in smooth muscle cells inhibits cell proliferation. Arterioscler Thromb Vasc Biol 31(2):368–375. https://doi.org/10.1161/ATVBAHA.110.218149

Chen J, Huang ZP, Seok HY et al (2013) mir-17-92 cluster is required for and sufficient to induce cardiomyocyte proliferation in postnatal and adult hearts. Circ Res 112(12):1557–1566. https://doi.org/10.1161/CIRCRESAHA.112.300658

Chen Y, Zhao H, Tan Z et al (2015) Bottleneck limitations for microRNA-based therapeutics from bench to the bedside. Pharmazie 70(3):147–154. https://www.ncbi.nlm.nih.gov/pubmed/25980175

Chen X, Wang L, Huang R et al (2019) Dgcr8 deletion in the primitive heart uncovered novel microRNA regulating the balance of cardiac-vascular gene program. Protein Cell 10(5):327–346. https://doi.org/10.1007/s13238-018-0572-1

Chistiakov DA, Revin VV, Sobenin IA et al (2015a) Vascular endothelium: functioning in norm, changes in atherosclerosis and current dietary approaches to improve endothelial function. Mini Rev Med Chem 15(4):338–350. https://doi.org/10.2174/1389557515666150226114031

Chistiakov DA, Sobenin IA, Orekhov AN, Bobryshev YV (2015b) Human miR-221/222 in physiological and atherosclerotic vascular remodeling. Biomed Res Int 2015:354517. https://doi.org/10.1155/2015/354517

Chistiakov DA, Orekhov AN, Bobryshev YV (2016) Cardiac-specific miRNA in cardiogenesis, heart function, and cardiac pathology (with focus on myocardial infarction). J Mol Cell Cardiol 94:107–121. https://doi.org/10.1016/j.yjmcc.2016.03.015

Chiu JJ, Wung BS, Hsieh HJ et al (1999) Nitric oxide regulates shear stress-induced early growth response-1. Expression via the extracellular signal-regulated kinase pathway in endothelial cells. Circ Res 85(3):238–246. https://doi.org/10.1161/01.res.85.3.238

Cipollone F, Felicioni L, Sarzani R et al (2011) A unique microRNA signature associated with plaque instability in humans. Stroke 42(9):2556–2563. https://doi.org/10.1161/STROKEAHA.110.597575

Concepcion CP, Bonetti C, Ventura A (2012) The microRNA-17-92 family of microRNA clusters in development and disease. Cancer J 18(3):262–267. https://doi.org/10.1097/PPO.0b013e318258b60a

Cordes KR, Sheehy NT, White MP et al (2009) miR-145 and miR-143 regulate smooth muscle cell fate and plasticity. Nature 460(7256):705–710. https://doi.org/10.1038/nature08195

Da Costa Martins PA, De Windt LJ (2012) MicroRNAs in control of cardiac hypertrophy. Cardiovasc Res 93(4):563–572. https://doi.org/10.1093/cvr/cvs013

da Costa Martins PA, Bourajjaj M, Gladka M et al (2008) Conditional dicer gene deletion in the postnatal myocardium provokes spontaneous cardiac remodeling. Circulation 118(15):1567–1576. https://doi.org/10.1161/CIRCULATIONAHA.108.769984

D'Alessandra Y, Devanna P, Limana F et al (2010) Circulating microRNAs are new and sensitive biomarkers of myocardial infarction. Eur Heart J 31(22):2765–2773. https://doi.org/10.1093/eurheartj/ehq167

Das K, Rao LVM (2022) The role of microRNAs in inflammation. Int J Mol Sci 23(24). https://doi.org/10.3390/ijms232415479

Davis-Dusenbery BN, Wu C, Hata A (2011) Micromanaging vascular smooth muscle cell differentiation and phenotypic modulation. Arterioscler Thromb Vasc Biol 31(11):2370–2377. https://doi.org/10.1161/ATVBAHA.111.226670

de Gonzalo-Calvo D, Perez-Boza J, Curado J et al (2022) Challenges of microRNA-based biomarkers in clinical application for cardiovascular diseases. Clin Transl Med 12(2):e585. https://doi.org/10.1002/ctm2.585

de Yebenes VG, Briones AM, Martos-Folgado I et al (2020) Aging-associated miR-217 aggravates atherosclerosis and promotes cardiovascular dysfunction. Arterioscler Thromb Vasc Biol 40(10):2408–2424. https://doi.org/10.1161/ATVBAHA.120.314333

Deng L, Blanco FJ, Stevens H et al (2015) MicroRNA-143 activation regulates smooth muscle and endothelial cell crosstalk in pulmonary arterial hypertension. Circ Res 117(10):870–883. https://doi.org/10.1161/CIRCRESAHA.115.306806

Di Francesco A, Fedrigo M, Santovito D et al (2018) MicroRNA signatures in cardiac biopsies and detection of allograft rejection. J Heart Lung Transplant 37(11):1329–1340. https://doi.org/10.1016/j.healun.2018.06.010

Disayabutr S, Kim EK, Cha SI et al (2016) miR-34 miRNAs regulate cellular senescence in type II alveolar epithelial cells of patients with idiopathic pulmonary fibrosis. PLoS One 11(6):e0158367. https://doi.org/10.1371/journal.pone.0158367

Doebele C, Bonauer A, Fischer A et al (2010) Members of the microRNA-17-92 cluster exhibit a cell-intrinsic antiangiogenic function in endothelial cells. Blood 115(23):4944–4950. https://doi.org/10.1182/blood-2010-01-264812

Donners MM, Wolfs IM, Stoger LJ et al (2012) Hematopoietic miR155 deficiency enhances atherosclerosis and decreases plaque stability in hyperlipidemic mice. PLoS One 7(4):e35877. https://doi.org/10.1371/journal.pone.0035877

Du Y, Gao C, Liu Z et al (2012) Upregulation of a disintegrin and metalloproteinase with thrombospondin motifs-7 by miR-29 repression mediates vascular smooth muscle calcification. Arterioscler Thromb Vasc Biol 32(11):2580–2588. https://doi.org/10.1161/ATVBAHA.112.300206

Du H, Zhao Y, Li H et al (2021) Roles of MicroRNAs in glucose and lipid metabolism in the heart. Front Cardiovasc Med 8:716213. https://doi.org/10.3389/fcvm.2021.716213

Eken SM, Jin H, Chernogubova E et al (2017) MicroRNA-210 enhances fibrous cap stability in advanced atherosclerotic lesions. Circ Res 120(4):633–644. https://doi.org/10.1161/CIRCRESAHA.116.309318

Elia L, Contu R, Quintavalle M et al (2009a) Reciprocal regulation of microRNA-1 and insulin-like growth factor-1 signal transduction cascade in cardiac and skeletal muscle in physiological and pathological conditions. Circulation 120(23):2377–2385. https://doi.org/10.1161/CIRCULATIONAHA.109.879429

Elia L, Quintavalle M, Zhang J et al (2009b) The knockout of miR-143 and -145 alters smooth muscle cell maintenance and vascular homeostasis in mice: correlates with human disease. Cell Death Differ 16(12):1590–1598. https://doi.org/10.1038/cdd.2009.153

Eulalio A, Mano M, Dal Ferro M et al (2012) Functional screening identifies miRNAs inducing cardiac regeneration. Nature 492(7429):376–381. https://doi.org/10.1038/nature11739

Farina FM, Hall IF, Serio S et al (2020) miR-128-3p is a novel regulator of vascular smooth muscle cell phenotypic switch and vascular diseases. Circ Res 126(12):e120–e135. https://doi.org/10.1161/CIRCRESAHA.120.316489

Farina FM, Weber C, Santovito D (2023) The emerging landscape of non-conventional RNA functions in atherosclerosis. Atherosclerosis 374:74–86. https://doi.org/10.1016/j.atherosclerosis.2023.01.009

Fasanaro P, D'Alessandra Y, Di Stefano V et al (2008) MicroRNA-210 modulates endothelial cell response to hypoxia and inhibits the receptor tyrosine kinase ligand Ephrin-A3. J Biol Chem 283(23):15878–15883. https://doi.org/10.1074/jbc.M800731200

Fichtlscherer S, De Rosa S, Fox H et al (2010) Circulating microRNAs in patients with coronary artery disease. Circ Res 107(5):677–684. https://doi.org/10.1161/CIRCRESAHA.109.215566

Fischer S, Handrick R, Aschrafi A, Otte K (2015) Unveiling the principle of microRNA-mediated redundancy in cellular pathway regulation. RNA Biol 12(3):238–247. https://doi.org/10.1080/15476286.2015.1017238

Fish JE, Santoro MM, Morton SU et al (2008) miR-126 regulates angiogenic signaling and vascular integrity. Dev Cell 15(2):272–284. https://doi.org/10.1016/j.devcel.2008.07.008

Fox CS, Coady S, Sorlie PD et al (2007) Increasing cardiovascular disease burden due to diabetes mellitus: the Framingham heart study. Circulation 115(12):1544–1550. https://doi.org/10.1161/CIRCULATIONAHA.106.658948

Gan J, Tang FMK, Su X et al (2019) microRNA-1 inhibits cardiomyocyte proliferation in mouse neonatal hearts by repressing CCND1 expression. Ann Transl Med 7(18):455. https://doi.org/10.21037/atm.2019.08.68

Gerin I, Clerbaux LA, Haumont O et al (2010) Expression of miR-33 from an SREBP2 intron inhibits cholesterol export and fatty acid oxidation. J Biol Chem 285(44):33652–33661. https://doi.org/10.1074/jbc.M110.152090

Godoy PM, Bhakta NR, Barczak AJ et al (2018) Large differences in Small RNA composition between human biofluids. Cell Rep 25(5):1346–1358. https://doi.org/10.1016/j.celrep.2018.10.014

Goedeke L, Salerno A, Ramirez CM et al (2014) Long-term therapeutic silencing of miR-33 increases circulating triglyceride levels and hepatic lipid accumulation in mice. EMBO Mol Med 6(9):1133–1141. https://doi.org/10.15252/emmm.201404046

Gong H, Chen H, Xiao P et al (2022) miR-146a impedes the anti-aging effect of AMPK via NAMPT suppression and NAD(+)/SIRT inactivation. Signal Transduct Target Ther 7(1):66. https://doi.org/10.1038/s41392-022-00886-3

Gordon S (2003) Alternative activation of macrophages. Nat Rev Immunol 3(1):23–35. https://doi.org/10.1038/nri978

Guller I, McNaughton S, Crowley T et al (2015) Comparative analysis of microRNA expression in mouse and human brown adipose tissue. BMC Genomics 16:820. https://doi.org/10.1186/s12864-015-2045-8

Guo FH, Guan YN, Guo JJ et al (2022) Single-cell Transcriptome analysis reveals embryonic endothelial heterogeneity at spatiotemporal level and multifunctions of MicroRNA-126 in mice. Arterioscler Thromb Vasc Biol 42(3):326–342. https://doi.org/10.1161/ATVBAHA.121.317093

Harris TA, Yamakuchi M, Ferlito M et al (2008) MicroRNA-126 regulates endothelial expression of vascular cell adhesion molecule 1. Proc Natl Acad Sci USA 105(5):1516–1521. https://doi.org/10.1073/pnas.0707493105

Hartmann P, Zhou Z, Natarelli L et al (2016) Endothelial dicer promotes atherosclerosis and vascular inflammation by miRNA-103-mediated suppression of KLF4. Nat Commun 7:10521. https://doi.org/10.1038/ncomms10521

Hasche ET, Fernandes C, Freedman SB, Jeremy RW (1995) Relation between ischemia time, infarct size, and left ventricular function in humans. Circulation 92(4):710–719. https://doi.org/10.1161/01.cir.92.4.710

Hassel D, Cheng P, White MP et al (2012) MicroRNA-10 regulates the angiogenic behavior of zebrafish and human endothelial cells by promoting vascular endothelial growth factor signaling. Circ Res 111(11):1421–1433. https://doi.org/10.1161/CIRCRESAHA.112.279711

He S, Liu P, Jian Z et al (2013) miR-138 protects cardiomyocytes from hypoxia-induced apoptosis via MLK3/JNK/c-Jun pathway. Biochem Biophys Res Commun 441(4):763–769. https://doi.org/10.1016/j.bbrc.2013.10.151

Hernandez GE, Ma F, Martinez G et al (2022) Aortic intimal resident macrophages are essential for maintenance of the non-thrombogenic intravascular state. Nat Cardiovasc Res 1(1):67–84. https://doi.org/10.1038/s44161-021-00006-4

Heymans S, Corsten MF, Verhesen W et al (2013) Macrophage microRNA-155 promotes cardiac hypertrophy and failure. Circulation 128(13):1420–1432. https://doi.org/10.1161/CIRCULATIONAHA.112.001357

Hinkel R, Penzkofer D, Zuhlke S et al (2013) Inhibition of microRNA-92a protects against ischemia/reperfusion injury in a large-animal model. Circulation 128(10):1066–1075. https://doi.org/10.1161/CIRCULATIONAHA.113.001904

Hua CC, Liu XM, Liang LR et al (2021) Targeting the microRNA-34a as a novel therapeutic strategy for cardiovascular diseases. Front Cardiovasc Med 8:784044. https://doi.org/10.3389/fcvm.2021.784044

Huang ZP, Chen J, Seok HY et al (2013a) MicroRNA-22 regulates cardiac hypertrophy and remodeling in response to stress. Circ Res 112(9):1234–1243. https://doi.org/10.1161/CIRCRESAHA.112.300682

Huang F, Li ML, Fang ZF et al (2013b) Overexpression of MicroRNA-1 improves the efficacy of mesenchymal stem cell transplantation after myocardial infarction. Cardiology 125(1):18–30. https://doi.org/10.1159/000347081

Hulshoff MS, Schellinger IN, Xu X et al (2023) miR-132-3p and KLF7 as novel regulators of aortic stiffening-associated EndMT in type 2 diabetes mellitus. Diabetol Metab Syndr 15(1):11. https://doi.org/10.1186/s13098-022-00966-y

Ihle MA, Trautmann M, Kuenstlinger H et al (2015) miRNA-221 and miRNA-222 induce apoptosis via the KIT/AKT signalling pathway in gastrointestinal stromal tumours. Mol Oncol 9(7):1421–1433. https://doi.org/10.1016/j.molonc.2015.03.013

Ito T, Yagi S, Yamakuchi M (2010) MicroRNA-34a regulation of endothelial senescence. Biochem Biophys Res Commun 398(4):735–740. https://doi.org/10.1016/j.bbrc.2010.07.012
Ivan M, Huang X (2014) miR-210: fine-tuning the hypoxic response. Adv Exp Med Biol 772:205–227. https://doi.org/10.1007/978-1-4614-5915-6_10
Ivey KN, Muth A, Arnold J et al (2008) MicroRNA regulation of cell lineages in mouse and human embryonic stem cells. Cell Stem Cell 2(3):219–229. https://doi.org/10.1016/j.stem.2008.01.016
Jansen F, Yang X, Hoelscher M et al (2013) Endothelial microparticle-mediated transfer of MicroRNA-126 promotes vascular endothelial cell repair via SPRED1 and is abrogated in glucose-damaged endothelial microparticles. Circulation 128(18):2026–2038. https://doi.org/10.1161/CIRCULATIONAHA.113.001720
Janssen HL, Reesink HW, Lawitz EJ et al (2013) Treatment of HCV infection by targeting microRNA. N Engl J Med 368(18):1685–1694. https://doi.org/10.1056/NEJMoa1209026
Ji R, Cheng Y, Yue J et al (2007) MicroRNA expression signature and antisense-mediated depletion reveal an essential role of MicroRNA in vascular neointimal lesion formation. Circ Res 100(11):1579–1588. https://doi.org/10.1161/CIRCRESAHA.106.141986
Jiang Y, Yin H, Zheng XL (2010) MicroRNA-1 inhibits myocardin-induced contractility of human vascular smooth muscle cells. J Cell Physiol 225(2):506–511. https://doi.org/10.1002/jcp.22230
Kanemaru K, Cranley J, Muraro D et al (2023) Spatially resolved multiomics of human cardiac niches. Nature 619(7971):801–810. https://doi.org/10.1038/s41586-023-06311-1
Kang H, Hata A (2012) MicroRNA regulation of smooth muscle gene expression and phenotype. Curr Opin Hematol 19(3):224–231. https://doi.org/10.1097/MOH.0b013e3283523e57
Karakikes I, Chaanine AH, Kang S et al (2013) Therapeutic cardiac-targeted delivery of miR-1 reverses pressure overload-induced cardiac hypertrophy and attenuates pathological remodeling. J Am Heart Assoc 2(2):e000078. https://doi.org/10.1161/JAHA.113.000078
Karere GM, Glenn JP, Li G et al (2023) Potential miRNA biomarkers and therapeutic targets for early atherosclerotic lesions. Sci Rep 13(1):3467. https://doi.org/10.1038/s41598-023-29074-1
Katwal AB, Konkalmatt PR, Piras BA et al (2013) Adeno-associated virus serotype 9 efficiently targets ischemic skeletal muscle following systemic delivery. Gene Ther 20(9):930–938. https://doi.org/10.1038/gt.2013.16
Kaur S, Abu-Shahba AG, Paananen RO et al (2018) Small non-coding RNA landscape of extracellular vesicles from human stem cells. Sci Rep 8(1):15503. https://doi.org/10.1038/s41598-018-33899-6
Ke ZP, Xu P, Shi Y, Gao AM (2016) MicroRNA-93 inhibits ischemia-reperfusion induced cardiomyocyte apoptosis by targeting PTEN. Oncotarget 7(20):28796–28805. https://doi.org/10.18632/oncotarget.8941
Kee HJ, Kook H (2011) Roles and targets of class I and IIa histone deacetylases in cardiac hypertrophy. J Biomed Biotechnol 2011:928326. https://doi.org/10.1155/2011/928326
Kumarswamy R, Volkmann I, Beermann J et al (2014) Vascular importance of the miR-212/132 cluster. Eur Heart J 35(45):3224–3231. https://doi.org/10.1093/eurheartj/ehu344
La Rocca G, King B, Shui B et al (2021) Inducible and reversible inhibition of miRNA-mediated gene repression in vivo. elife 10. https://doi.org/10.7554/eLife.70948
Leonard BL, Smaill BH, LeGrice IJ (2012) Structural remodeling and mechanical function in heart failure. Microsc Microanal 18(1):50–67. https://doi.org/10.1017/S1431927611012438
Lesizza P, Prosdocimo G, Martinelli V et al (2017) Single-dose Intracardiac injection of pro-regenerative MicroRNAs improves cardiac function after myocardial infarction. Circ Res 120(8):1298–1304. https://doi.org/10.1161/CIRCRESAHA.116.309589
Ley K, Miller YI, Hedrick CC (2011) Monocyte and macrophage dynamics during atherogenesis. Arterioscler Thromb Vasc Biol 31(7):1506–1516. https://doi.org/10.1161/ATVBAHA.110.221127
Li D, Wu M (2021) Pattern recognition receptors in health and diseases. Signal Transduct Target Ther 6(1):291. https://doi.org/10.1038/s41392-021-00687-0

Li T, Cao H, Zhuang J et al (2011a) Identification of miR-130a, miR-27b and miR-210 as serum biomarkers for atherosclerosis obliterans. Clin Chim Acta 412(1–2):66–70. https://doi.org/10.1016/j.cca.2010.09.029

Li S, Zhu J, Zhang W et al (2011b) Signature microRNA expression profile of essential hypertension and its novel link to human cytomegalovirus infection. Circulation 124(2):175–184. https://doi.org/10.1161/CIRCULATIONAHA.110.012237

Li X, Wang J, Jia Z et al (2013) MiR-499 regulates cell proliferation and apoptosis during late-stage cardiac differentiation via Sox6 and cyclin D1. PLoS One 8(9):e74504. https://doi.org/10.1371/journal.pone.0074504

Li Y, Yan L, Zhang W et al (2014) MicroRNA-21 inhibits platelet-derived growth factor-induced human aortic vascular smooth muscle cell proliferation and migration through targeting activator protein-1. Am J Transl Res 6(5):507–516. https://www.ncbi.nlm.nih.gov/pubmed/25360215

Li X, Bi H, Xie S, Cui W (2022) MiR-208b regulates the conversion of skeletal muscle fiber types by inhibiting Mettl8 expression. Front Genet 13:820464. https://doi.org/10.3389/fgene.2022.820464

Liang D, Li J, Wu Y et al (2015) miRNA-204 drives cardiomyocyte proliferation via targeting Jarid2. Int J Cardiol 201:38–48. https://doi.org/10.1016/j.ijcard.2015.06.163

Liang B, Wang X, Song X et al (2017) MicroRNA-20a/b regulates cholesterol efflux through post-transcriptional repression of ATP-binding cassette transporter A1. Biochim Biophys Acta Mol Cell Biol Lipids 1862(9):929–938. https://doi.org/10.1016/j.bbalip.2017.06.002

Lim CX, Lee B, Geiger O et al (2020) miR-181a modulation of ERK-MAPK signaling sustains DC-SIGN expression and limits activation of monocyte-derived dendritic cells. Cell Rep 30(11):3793–3805 e3795. https://doi.org/10.1016/j.celrep.2020.02.077

Liu J, Carmell MA, Rivas FV et al (2004) Argonaute2 is the catalytic engine of mammalian RNAi. Science 305(5689):1437–1441. https://doi.org/10.1126/science.1102513

Liu N, Williams AH, Kim Y et al (2007) An intragenic MEF2-dependent enhancer directs muscle-specific expression of microRNAs 1 and 133. Proc Natl Acad Sci USA 104(52):20844–20849. https://doi.org/10.1073/pnas.0710558105

Liu N, Bezprozvannaya S, Williams AH et al (2008) microRNA-133a regulates cardiomyocyte proliferation and suppresses smooth muscle gene expression in the heart. Genes Dev 22(23):3242–3254. https://doi.org/10.1101/gad.1738708

Liu X, Cheng Y, Zhang S et al (2009) A necessary role of miR-221 and miR-222 in vascular smooth muscle cell proliferation and neointimal hyperplasia. Circ Res 104(4):476–487. https://doi.org/10.1161/CIRCRESAHA.108.185363

Liu Y, Taylor NE, Lu L et al (2010) Renal medullary microRNAs in Dahl salt-sensitive rats: miR-29b regulates several collagens and related genes. Hypertension 55(4):974–982. https://doi.org/10.1161/HYPERTENSIONAHA.109.144428

Liu J, van Mil A, Vrijsen K et al (2011) MicroRNA-155 prevents necrotic cell death in human cardiomyocyte progenitor cells via targeting RIP1. J Cell Mol Med 15(7):1474–1482. https://doi.org/10.1111/j.1582-4934.2010.01104.x

Liu F, Li R, Zhang Y et al (2014) Association of plasma MiR-17-92 with dyslipidemia in patients with coronary artery disease. Medicine (Baltimore) 93(23):e98. https://doi.org/10.1097/MD.0000000000000098

Liu K, Ying Z, Qi X et al (2015) MicroRNA-1 regulates the proliferation of vascular smooth muscle cells by targeting insulin-like growth factor 1. Int J Mol Med 36(3):817–824. https://doi.org/10.3892/ijmm.2015.2277

Liu J, Zhang C, Zhao Y, Feng Z (2017) MicroRNA control of p53. J Cell Biochem 118(1):7–14. https://doi.org/10.1002/jcb.25609

Liu ZQ, Feng J, Shi LL et al (2019) Influences of miR-155/NF-kappaB signaling pathway on inflammatory factors in ARDS in neonatal pigs. Eur Rev Med Pharmacol Sci 23(16):7042–7048. https://doi.org/10.26355/eurrev_201908_18746

Liu L, Tan L, Yao J, Yang L (2020) Long non-coding RNA MALAT1 regulates cholesterol accumulation in ox-LDL-induced macrophages via the microRNA-17-5p/ABCA1 axis. Mol Med Rep 21(4):1761–1770. https://doi.org/10.3892/mmr.2020.10987
Lu TY, Lin B, Li Y et al (2013) Overexpression of microRNA-1 promotes cardiomyocyte commitment from human cardiovascular progenitors via suppressing WNT and FGF signaling pathways. J Mol Cell Cardiol 63:146–154. https://doi.org/10.1016/j.yjmcc.2013.07.019
Lyons GE, Schiaffino S, Sassoon D et al (1990) Developmental regulation of myosin gene expression in mouse cardiac muscle. J Cell Biol 111(6 Pt 1):2427–2436. https://doi.org/10.1083/jcb.111.6.2427
Maegdefessel L, Azuma J, Toh R et al (2012) MicroRNA-21 blocks abdominal aortic aneurysm development and nicotine-augmented expansion. Sci Transl Med 4(122):122ra122. https://doi.org/10.1126/scitranslmed.3003441
Magenta A, Cencioni C, Fasanaro P et al (2011) miR-200c is upregulated by oxidative stress and induces endothelial cell apoptosis and senescence via ZEB1 inhibition. Cell Death Differ 18(10):1628–1639. https://doi.org/10.1038/cdd.2011.42
Maitrias P, Metzinger-Le Meuth V, Massy ZA et al (2015) MicroRNA deregulation in symptomatic carotid plaque. J Vasc Surg 62(5):1245–1250 e1241. https://doi.org/10.1016/j.jvs.2015.06.136
Mandolini C, Santovito D, Marcantonio P et al (2015) Identification of microRNAs 758 and 33b as potential modulators of ABCA1 expression in human atherosclerotic plaques. Nutr Metab Cardiovasc Dis 25(2):202–209. https://doi.org/10.1016/j.numecd.2014.09.005
Mantovani A, Sica A, Sozzani S et al (2004) The chemokine system in diverse forms of macrophage activation and polarization. Trends Immunol 25(12):677–686. https://doi.org/10.1016/j.it.2004.09.015
Mantovani A, Sica A, Locati M (2005) Macrophage polarization comes of age. Immunity 23(4):344–346. https://doi.org/10.1016/j.immuni.2005.10.001
Mao Y, Jiang L (2021) MiR-200c-3p promotes ox-LDL-induced endothelial to mesenchymal transition in human umbilical vein endothelial cells through SMAD7/YAP pathway. J Physiol Sci 71(1):30. https://doi.org/10.1186/s12576-021-00815-z
Marquart TJ, Allen RM, Ory DS, Baldan A (2010) miR-33 links SREBP-2 induction to repression of sterol transporters. Proc Natl Acad Sci USA 107(27):12228–12232. https://doi.org/10.1073/pnas.1005191107
Marquart TJ, Wu J, Lusis AJ, Baldan A (2013) Anti-miR-33 therapy does not alter the progression of atherosclerosis in low-density lipoprotein receptor-deficient mice. Arterioscler Thromb Vasc Biol 33(3):455–458. https://doi.org/10.1161/ATVBAHA.112.300639
Mathioudakis N, Georgakopoulou VE, Paterakis K et al (2023) Effect of circulating miR-126 levels on intracranial aneurysms and their predictive value for the rupture of aneurysms: a systematic review and meta-analysis. Exp Ther Med 26(2):411. https://doi.org/10.3892/etm.2023.12110
Matkovich SJ, Wang W, Tu Y et al (2010) MicroRNA-133a protects against myocardial fibrosis and modulates electrical repolarization without affecting hypertrophy in pressure-overloaded adult hearts. Circ Res 106(1):166–175. https://doi.org/10.1161/CIRCRESAHA.109.202176
Matsushima S, Sadoshima J (2015) The role of sirtuins in cardiac disease. Am J Physiol Heart Circ Physiol 309(9):H1375–H1389. https://doi.org/10.1152/ajpheart.00053.2015
McCarthy JJ (2008) MicroRNA-206: the skeletal muscle-specific myomiR. Biochim Biophys Acta 1779(11):682–691. https://doi.org/10.1016/j.bbagrm.2008.03.001
Meder B, Katus HA, Rottbauer W (2008) Right into the heart of microRNA-133a. Genes Dev 22(23):3227–3231. https://doi.org/10.1101/gad.1753508
Mendell JT (2008) miRiad roles for the miR-17-92 cluster in development and disease. Cell 133(2):217–222. https://doi.org/10.1016/j.cell.2008.04.001
Menghini R, Casagrande V, Cardellini M et al (2009) MicroRNA 217 modulates endothelial cell senescence via silent information regulator 1. Circulation 120(15):1524–1532. https://doi.org/10.1161/CIRCULATIONAHA.109.864629

Merk DR, Chin JT, Dake BA et al (2012) miR-29b participates in early aneurysm development in Marfan syndrome. Circ Res 110(2):312–324. https://doi.org/10.1161/CIRCRESAHA.111.253740

Miao H, Zeng H, Gong H (2018) microRNA-212 promotes lipid accumulation and attenuates cholesterol efflux in THP-1 human macrophages by targeting SIRT1. Gene 643:55–60. https://doi.org/10.1016/j.gene.2017.11.058

Milani-Nejad N, Janssen PM (2014) Small and large animal models in cardiac contraction research: advantages and disadvantages. Pharmacol Ther 141(3):235–249. https://doi.org/10.1016/j.pharmthera.2013.10.007

Milusev A, Rieben R, Sorvillo N (2022) The endothelial Glycocalyx: a possible therapeutic target in cardiovascular disorders. Front Cardiovasc Med 9:897087. https://doi.org/10.3389/fcvm.2022.897087

Moreau PR, Tomas Bosch V, Bouvy-Liivrand M et al (2021) Profiling of primary and mature miRNA expression in atherosclerosis-associated cell types. Arterioscler Thromb Vasc Biol 41(7):2149–2167. https://doi.org/10.1161/ATVBAHA.121.315579

Morkin E (2000) Control of cardiac myosin heavy chain gene expression. Microsc Res Tech 50(6):522–531. https://doi.org/10.1002/1097-0029(20000915)50:6<522::AID-JEMT9>3.0.CO;2-U

Mozaffarian D, Benjamin EJ, Go AS et al (2016) Heart disease and stroke Statistics-2016 update: a report from the American Heart Association. Circulation 133(4):e38–e360. https://doi.org/10.1161/CIR.0000000000000350

Najafi-Shoushtari SH, Kristo F, Li Y et al (2010) MicroRNA-33 and the SREBP host genes cooperate to control cholesterol homeostasis. Science 328(5985):1566–1569. https://doi.org/10.1126/science.1189123

Nazari-Jahantigh M, Wei Y, Noels H et al (2012) MicroRNA-155 promotes atherosclerosis by repressing Bcl6 in macrophages. J Clin Invest 122(11):4190–4202. https://doi.org/10.1172/JCI61716

Nicosia RF, Zorzi P, Ligresti G et al (2011) Paracrine regulation of angiogenesis by different cell types in the aorta ring model. Int J Dev Biol 55(4–5):447–453. https://doi.org/10.1387/ijdb.103222rn

Nie X, Fan J, Li H et al (2018) miR-217 promotes cardiac hypertrophy and dysfunction by targeting PTEN. Mol Ther Nucleic Acids 12:254–266. https://doi.org/10.1016/j.omtn.2018.05.013

Obel LM, Diederichsen AC, Steffensen FH et al (2021) Population-based risk factors for ascending, arch, descending, and abdominal aortic dilations for 60-74-year-old individuals. J Am Coll Cardiol 78(3):201–211. https://doi.org/10.1016/j.jacc.2021.04.094

Oerlemans MI, Mosterd A, Dekker MS et al (2012) Early assessment of acute coronary syndromes in the emergency department: the potential diagnostic value of circulating microRNAs. EMBO Mol Med 4(11):1176–1185. https://doi.org/10.1002/emmm.201201749

Ohanian M, Humphreys DT, Anderson E et al (2013) A heterozygous variant in the human cardiac miR-133 gene, MIR133A2, alters miRNA duplex processing and strand abundance. BMC Genet 14:18. https://doi.org/10.1186/1471-2156-14-18

Oliveira-Carvalho V, Carvalho VO, Bocchi EA (2013) The emerging role of miR-208a in the heart. DNA Cell Biol 32(1):8–12. https://doi.org/10.1089/dna.2012.1787

Olson EN (2006) Gene regulatory networks in the evolution and development of the heart. Science 313(5795):1922–1927. https://doi.org/10.1126/science.1132292

Ouimet M, Ediriweera HN, Gundra UM et al (2015) MicroRNA-33-dependent regulation of macrophage metabolism directs immune cell polarization in atherosclerosis. J Clin Invest 125(12):4334–4348. https://doi.org/10.1172/JCI81676

Overhoff MG, Garbe JC, Koh J et al (2014) Cellular senescence mediated by p16INK4A-coupled miRNA pathways. Nucleic Acids Res 42(3):1606–1618. https://doi.org/10.1093/nar/gkt1096

Pandey R, Yang Y, Jackson L, Ahmed RP (2016) MicroRNAs regulating meis1 expression and inducing cardiomyocyte proliferation. Cardiovasc Regen Med 3. https://www.ncbi.nlm.nih.gov/pubmed/28111633

Peng J, He X, Zhang L, Liu P (2018) MicroRNA-26a protects vascular smooth muscle cells against H2O2-induced injury through activation of the PTEN/AKT/mTOR pathway. Int J Mol Med 42(3):1367–1378. https://doi.org/10.3892/ijmm.2018.3746

Perez-Carrillo L, Sanchez-Lazaro I, Trivino JC et al (2023) Combining serum miR-144-3p and miR-652-3p as potential biomarkers for the early diagnosis and stratification of acute cellular rejection in heart transplantation patients. Transplantation 107(9):2064–2072. https://doi.org/10.1097/TP.0000000000004622

Porrello ER, Mahmoud AI, Simpson E et al (2011) Transient regenerative potential of the neonatal mouse heart. Science 331(6020):1078–1080. https://doi.org/10.1126/science.1200708

Prabhu SD, Frangogiannis NG (2016) The biological basis for cardiac repair after myocardial infarction: from inflammation to fibrosis. Circ Res 119(1):91–112. https://doi.org/10.1161/CIRCRESAHA.116.303577

Qu C, Liu X, Guo Y et al (2020) MiR-128-3p inhibits vascular smooth muscle cell proliferation and migration by repressing FOXO4/MMP9 signaling pathway. Mol Med 26(1):116. https://doi.org/10.1186/s10020-020-00242-7

Raitoharju E, Lyytikainen LP, Levula M et al (2011) miR-21, miR-210, miR-34a, and miR-146a/b are up-regulated in human atherosclerotic plaques in the Tampere vascular study. Atherosclerosis 219(1):211–217. https://doi.org/10.1016/j.atherosclerosis.2011.07.020

Rao PK, Kumar RM, Farkhondeh M et al (2006) Myogenic factors that regulate expression of muscle-specific microRNAs. Proc Natl Acad Sci USA 103(23):8721–8726. https://doi.org/10.1073/pnas.0602831103

Rao PK, Toyama Y, Chiang HR et al (2009) Loss of cardiac microRNA-mediated regulation leads to dilated cardiomyopathy and heart failure. Circ Res 105(6):585–594. https://doi.org/10.1161/CIRCRESAHA.109.200451

Raucci A, Macri F, Castiglione S et al (2021) MicroRNA-34a: the bad guy in age-related vascular diseases. Cell Mol Life Sci 78(23):7355–7378. https://doi.org/10.1007/s00018-021-03979-4

Rayner KJ, Suarez Y, Davalos A et al (2010) MiR-33 contributes to the regulation of cholesterol homeostasis. Science 328(5985):1570–1573. https://doi.org/10.1126/science.1189862

Rippe C, Blimline M, Magerko KA et al (2012) MicroRNA changes in human arterial endothelial cells with senescence: relation to apoptosis, eNOS and inflammation. Exp Gerontol 47(1):45–51. https://doi.org/10.1016/j.exger.2011.10.004

Roselaar SE, Kakkanathu PX, Daugherty A (1996) Lymphocyte populations in atherosclerotic lesions of apoE −/− and LDL receptor −/− mice. Decreasing density with disease progression. Arterioscler Thromb Vasc Biol 16(8):1013–1018. https://doi.org/10.1161/01.atv.16.8.1013

Safa A, Bahroudi Z, Shoorei H et al (2020) miR-1: a comprehensive review of its role in normal development and diverse disorders. Biomed Pharmacother 132:110903. https://doi.org/10.1016/j.biopha.2020.110903

Santovito D, Weber C (2022) Non-canonical features of microRNAs: paradigms emerging from cardiovascular disease. Nat Rev Cardiol 19(9):620–638. https://doi.org/10.1038/s41569-022-00680-2

Santovito D, Mandolini C, Marcantonio P et al (2013) Overexpression of microRNA-145 in atherosclerotic plaques from hypertensive patients. Expert Opin Ther Targets 17(3):217–223. https://doi.org/10.1517/14728222.2013.745512

Santovito D, De Nardis V, Marcantonio P et al (2014) Plasma exosome microRNA profiling unravels a new potential modulator of adiponectin pathway in diabetes: effect of glycemic control. J Clin Endocrinol Metab 99(9):E1681–E1685. https://doi.org/10.1210/jc.2013-3843

Santovito D, Egea V, Bidzhekov K et al (2020a) Noncanonical inhibition of caspase-3 by a nuclear microRNA confers endothelial protection by autophagy in atherosclerosis. Sci Transl Med 12(546). https://doi.org/10.1126/scitranslmed.aaz2294

Santovito D, Marcantonio P, Mastroiacovo D et al (2020b) High dose rosuvastatin increases ABCA1 transporter in human atherosclerotic plaques in a cholesterol-independent fashion. Int J Cardiol 299:249–253. https://doi.org/10.1016/j.ijcard.2019.07.094
Saraste A, Pulkki K, Kallajoki M et al (1997) Apoptosis in human acute myocardial infarction. Circulation 95(2):320–323. https://doi.org/10.1161/01.cir.95.2.320
Sayed D, Hong C, Chen IY et al (2007) MicroRNAs play an essential role in the development of cardiac hypertrophy. Circ Res 100(3):416–424. https://doi.org/10.1161/01.RES.0000257913.42552.23
Schober A, Nazari-Jahantigh M, Wei Y et al (2014) MicroRNA-126-5p promotes endothelial proliferation and limits atherosclerosis by suppressing Dlk1. Nat Med 20(4):368–376. https://doi.org/10.1038/nm.3487
Schober A, Blay RM, Saboor Maleki S et al (2021) MicroRNA-21 controls circadian regulation of apoptosis in atherosclerotic lesions. Circulation 144(13):1059–1073. https://doi.org/10.1161/CIRCULATIONAHA.120.051614
Schulte C, Westermann D, Blankenberg S, Zeller T (2015) Diagnostic and prognostic value of circulating microRNAs in heart failure with preserved and reduced ejection fraction. World J Cardiol 7(12):843–860. https://doi.org/10.4330/wjc.v7.i12.843
Sempere LF, Freemantle S, Pitha-Rowe I et al (2004) Expression profiling of mammalian microRNAs uncovers a subset of brain-expressed microRNAs with possible roles in murine and human neuronal differentiation. Genome Biol 5(3):R13. https://doi.org/10.1186/gb-2004-5-3-r13
Seok HY, Chen J, Kataoka M et al (2014) Loss of MicroRNA-155 protects the heart from pathological cardiac hypertrophy. Circ Res 114(10):1585–1595. https://doi.org/10.1161/CIRCRESAHA.114.303784
Shen SM, Jiang H, Zhao JN, Shi Y (2020) Down-regulation of miR-155 inhibits inflammatory response in human pulmonary microvascular endothelial cells infected with influenza a virus by targeting sphingosine-1-phosphate receptor 1. Chin Med J 133(20):2429–2436. https://doi.org/10.1097/CM9.0000000000001036
Shen NN, Wang JL, Fu YP (2022) The microRNA expression profiling in heart failure: a systematic review and meta-analysis. Front Cardiovasc Med 9:856358. https://doi.org/10.3389/fcvm.2022.856358
Shi JY, Chen C, Xu X, Lu Q (2019) miR-29a promotes pathological cardiac hypertrophy by targeting the PTEN/AKT/mTOR signalling pathway and suppressing autophagy. Acta Physiol (Oxf) 227(2):e13323. https://doi.org/10.1111/apha.13323
Shi X, Ma W, Pan Y et al (2020) MiR-126-5p promotes contractile switching of aortic smooth muscle cells by targeting VEPH1 and alleviates Ang II-induced abdominal aortic aneurysm in mice. Lab Investig 100(12):1564–1574. https://doi.org/10.1038/s41374-020-0454-z
Shieh JT, Huang Y, Gilmore J, Srivastava D (2011) Elevated miR-499 levels blunt the cardiac stress response. PLoS One 6(5):e19481. https://doi.org/10.1371/journal.pone.0019481
Siasos G, Mourouzis K, Oikonomou E et al (2015) The role of endothelial dysfunction in aortic aneurysms. Curr Pharm Des 21(28):4016–4034. https://doi.org/10.2174/1381612821666150826094156
Small EM, Olson EN (2011) Pervasive roles of microRNAs in cardiovascular biology. Nature 469(7330):336–342. https://doi.org/10.1038/nature09783
Sommariva E, D'Alessandra Y, Farina FM et al (2017) MiR-320a as a potential novel circulating biomarker of Arrhythmogenic CardioMyopathy. Sci Rep 7(1):4802. https://doi.org/10.1038/s41598-017-05001-z
Song CL, Liu B, Diao HY et al (2016) Down-regulation of microRNA-320 suppresses cardiomyocyte apoptosis and protects against myocardial ischemia and reperfusion injury by targeting IGF-1. Oncotarget 7(26):39740–39757. https://doi.org/10.18632/oncotarget.9240
Sudo R, Sato F, Azechi T, Wachi H (2015) MiR-29-mediated elastin down-regulation contributes to inorganic phosphorus-induced osteoblastic differentiation in vascular smooth muscle cells. Genes Cells 20(12):1077–1087. https://doi.org/10.1111/gtc.12311

Sun Y (2009) Myocardial repair/remodelling following infarction: roles of local factors. Cardiovasc Res 81(3):482–490. https://doi.org/10.1093/cvr/cvn333
Sun X, Icli B, Wara AK et al (2012) MicroRNA-181b regulates NF-kappaB-mediated vascular inflammation. J Clin Invest 122(6):1973–1990. https://doi.org/10.1172/JCI61495
Tan J, Yang L, Liu C, Yan Z (2017) MicroRNA-26a targets MAPK6 to inhibit smooth muscle cell proliferation and vein graft neointimal hyperplasia. Sci Rep 7:46602. https://doi.org/10.1038/srep46602
Tan L, Liu L, Jiang Z, Hao X (2019) Inhibition of microRNA-17-5p reduces the inflammation and lipid accumulation, and up-regulates ATP-binding cassette transporterA1 in atherosclerosis. J Pharmacol Sci 139(4):280–288. https://doi.org/10.1016/j.jphs.2018.11.012
Thum T, Gross C, Fiedler J et al (2008) MicroRNA-21 contributes to myocardial disease by stimulating MAP kinase signalling in fibroblasts. Nature 456(7224):980–984. https://doi.org/10.1038/nature07511
Toischer K, Zhu W, Hunlich M et al (2017) Cardiomyocyte proliferation prevents failure in pressure overload but not volume overload. J Clin Invest 127(12):4285–4296. https://doi.org/10.1172/JCI81870
Tomas L, Prica F, Schulz C (2021) Trafficking of mononuclear phagocytes in healthy arteries and atherosclerosis. Front Immunol 12:718432. https://doi.org/10.3389/fimmu.2021.718432
Torella D, Iaconetti C, Catalucci D et al (2011) MicroRNA-133 controls vascular smooth muscle cell phenotypic switch in vitro and vascular remodeling in vivo. Circ Res 109(8):880–893. https://doi.org/10.1161/CIRCRESAHA.111.240150
Tu Y, Wan L, Bu L et al (2013) MicroRNA-22 downregulation by atorvastatin in a mouse model of cardiac hypertrophy: a new mechanism for antihypertrophic intervention. Cell Physiol Biochem 31(6):997–1008. https://doi.org/10.1159/000350117
Ucar A, Gupta SK, Fiedler J et al (2012) The miRNA-212/132 family regulates both cardiac hypertrophy and cardiomyocyte autophagy. Nat Commun 3:1078. https://doi.org/10.1038/ncomms2090
van Rooij E, Sutherland LB, Liu N et al (2006) A signature pattern of stress-responsive microRNAs that can evoke cardiac hypertrophy and heart failure. Proc Natl Acad Sci USA 103(48):18255–18260. https://doi.org/10.1073/pnas.0608791103
van Rooij E, Sutherland LB, Qi X et al (2007) Control of stress-dependent cardiac growth and gene expression by a microRNA. Science 316(5824):575–579. https://doi.org/10.1126/science.1139089
van Rooij E, Sutherland LB, Thatcher JE et al (2008) Dysregulation of microRNAs after myocardial infarction reveals a role of miR-29 in cardiac fibrosis. Proc Natl Acad Sci USA 105(35):13027–13032. https://doi.org/10.1073/pnas.0805038105
van Rooij E, Quiat D, Johnson BA et al (2009) A family of microRNAs encoded by myosin genes governs myosin expression and muscle performance. Dev Cell 17(5):662–673. https://doi.org/10.1016/j.devcel.2009.10.013
Venkatesh P, Phillippi J, Chukkapalli S et al (2017) Aneurysm-specific miR-221 and miR-146a participates in human thoracic and abdominal aortic aneurysms. Int J Mol Sci 18(4). https://doi.org/10.3390/ijms18040875
Ventura A, Young AG, Winslow MM et al (2008) Targeted deletion reveals essential and overlapping functions of the miR-17 through 92 family of miRNA clusters. Cell 132(5):875–886. https://doi.org/10.1016/j.cell.2008.02.019
Virani SS, Alonso A, Benjamin EJ et al (2020) Heart disease and stroke Statistics-2020 update: a report from the American Heart Association. Circulation 141(9):e139–e596. https://doi.org/10.1161/CIR.0000000000000757
Wang K, Yuan Y, Cho JH et al (2012) Comparing the MicroRNA spectrum between serum and plasma. PLoS One 7(7):e41561. https://doi.org/10.1371/journal.pone.0041561
Wang K, An T, Zhou LY et al (2015) E2F1-regulated miR-30b suppresses Cyclophilin D and protects heart from ischemia/reperfusion injury and necrotic cell death. Cell Death Differ 22(5):743–754. https://doi.org/10.1038/cdd.2014.165

Wang X, Ling CC, Li L et al (2016) MicroRNA-10a/10b represses a novel target gene mib1 to regulate angiogenesis. Cardiovasc Res 110(1):140–150. https://doi.org/10.1093/cvr/cvw023
Wang D, Wang Y, Ma J et al (2017) MicroRNA-20a participates in the aerobic exercise-based prevention of coronary artery disease by targeting PTEN. Biomed Pharmacother 95:756–763. https://doi.org/10.1016/j.biopha.2017.08.086
Weber JA, Baxter DH, Zhang S et al (2010) The microRNA spectrum in 12 body fluids. Clin Chem 56(11):1733–1741. https://doi.org/10.1373/clinchem.2010.147405
Weber M, Kim S, Patterson N et al (2014) MiRNA-155 targets myosin light chain kinase and modulates actin cytoskeleton organization in endothelial cells. Am J Physiol Heart Circ Physiol 306(8):H1192–H1203. https://doi.org/10.1152/ajpheart.00521.2013
Wilson KD, Hu S, Venkatasubrahmanyam S et al (2010) Dynamic microRNA expression programs during cardiac differentiation of human embryonic stem cells: role for miR-499. Circ Cardiovasc Genet 3(5):426–435. https://doi.org/10.1161/CIRCGENETICS.109.934281
Wong RKY, MacMahon M, Woodside JV, Simpson DA (2019) A comparison of RNA extraction and sequencing protocols for detection of small RNAs in plasma. BMC Genomics 20(1):446. https://doi.org/10.1186/s12864-019-5826-7
Wu W, Xiao H, Laguna-Fernandez A et al (2011) Flow-dependent regulation of Kruppel-like factor 2 is mediated by MicroRNA-92a. Circulation 124(5):633–641. https://doi.org/10.1161/CIRCULATIONAHA.110.005108
Wystub K, Besser J, Bachmann A et al (2013) miR-1/133a clusters cooperatively specify the cardiomyogenic lineage by adjustment of myocardin levels during embryonic heart development. PLoS Genet 9(9):e1003793. https://doi.org/10.1371/journal.pgen.1003793
Xie C, Huang H, Sun X et al (2011) MicroRNA-1 regulates smooth muscle cell differentiation by repressing Kruppel-like factor 4. Stem Cells Dev 20(2):205–210. https://doi.org/10.1089/scd.2010.0283
Xie W, Li M, Xu N et al (2013) MiR-181a regulates inflammation responses in monocytes and macrophages. PLoS One 8(3):e58639. https://doi.org/10.1371/journal.pone.0058639
Xu XD, Song XW, Li Q et al (2012) Attenuation of microRNA-22 derepressed PTEN to effectively protect rat cardiomyocytes from hypertrophy. J Cell Physiol 227(4):1391–1398. https://doi.org/10.1002/jcp.22852
Yan Y, Xu Y, Ni G et al (2020) MicroRNA-221 promotes proliferation and migration of pulmonary arterial smooth muscle cells (PASMCs) by targeting tissue inhibitor of metalloproteinases-3 (TIMP3). Cardiovasc Diagn Ther 10(4):646–657. https://doi.org/10.21037/cdt-20-328
Yang WJ, Yang DD, Na S et al (2005) Dicer is required for embryonic angiogenesis during mouse development. J Biol Chem 280(10):9330–9335. https://doi.org/10.1074/jbc.M413394200
Yang B, Lin H, Xiao J et al (2007) The muscle-specific microRNA miR-1 regulates cardiac arrhythmogenic potential by targeting GJA1 and KCNJ2. Nat Med 13(4):486–491. https://doi.org/10.1038/nm1569
Yang Y, Yang L, Liang X, Zhu G (2015) MicroRNA-155 promotes atherosclerosis inflammation via targeting SOCS1. Cell Physiol Biochem 36(4):1371–1381. https://doi.org/10.1159/000430303
Yang Y, Zhou Y, Cao Z et al (2016) miR-155 functions downstream of angiotensin II receptor subtype 1 and calcineurin to regulate cardiac hypertrophy. Exp Ther Med 12(3):1556–1562. https://doi.org/10.3892/etm.2016.3506
Yang X, Dong M, Wen H et al (2017) MiR-26a contributes to the PDGF-BB-induced phenotypic switch of vascular smooth muscle cells by suppressing Smad1. Oncotarget 8(44):75844–75853. https://doi.org/10.18632/oncotarget.17998
Yang D, Wan X, Dennis AT et al (2021) MicroRNA biophysically modulates cardiac action potential by direct binding to Ion Channel. Circulation 143(16):1597–1613. https://doi.org/10.1161/CIRCULATIONAHA.120.050098
Yau JW, Teoh H, Verma S (2015) Endothelial cell control of thrombosis. BMC Cardiovasc Disord 15:130. https://doi.org/10.1186/s12872-015-0124-z
Ye J, Guo R, Shi Y et al (2016) miR-155 regulated inflammation response by the SOCS1-STAT3-PDCD4 Axis in Atherogenesis. Mediat Inflamm 2016:8060182. https://doi.org/10.1155/2016/8060182

Yue J (2011) miRNA and vascular cell movement. Adv Drug Deliv Rev 63(8):616–622. https://doi.org/10.1016/j.addr.2011.01.001

Zaccagnini G, Greco S, Voellenkle C et al (2022) miR-210 hypoxamiR in angiogenesis and diabetes. Antioxid Redox Signal 36(10–12):685–706. https://doi.org/10.1089/ars.2021.0200

Zahedi F, Nazari-Jahantigh M, Zhou Z et al (2017) Dicer generates a regulatory microRNA network in smooth muscle cells that limits neointima formation during vascular repair. Cell Mol Life Sci 74(2):359–372. https://doi.org/10.1007/s00018-016-2349-0

Zampetaki A, Kiechl S, Drozdov I et al (2010) Plasma microRNA profiling reveals loss of endothelial miR-126 and other microRNAs in type 2 diabetes. Circ Res 107(6):810–817. https://doi.org/10.1161/CIRCRESAHA.110.226357

Zernecke A, Bidzhekov K, Noels H et al (2009) Delivery of microRNA-126 by apoptotic bodies induces CXCL12-dependent vascular protection. Sci Signal 2(100):ra81. https://doi.org/10.1126/scisignal.2000610

Zhang L, Tao L (2017) miR-132 promotes retinal neovascularization under anoxia and reoxygenation conditions through up-regulating Egr1, ERK2, MMP2, VEGFA and VEGFC expression. Int J Clin Exp Pathol 10(8):8845–8857. https://www.ncbi.nlm.nih.gov/pubmed/31966751

Zhang W, Shang T, Huang C et al (2015) Plasma microRNAs serve as potential biomarkers for abdominal aortic aneurysm. Clin Biochem 48(15):988–992. https://doi.org/10.1016/j.clinbiochem.2015.04.016

Zhang HP, Wang YH, Cao CJ et al (2016) A regulatory circuit involving miR-143 and DNMT3a mediates vascular smooth muscle cell proliferation induced by homocysteine. Mol Med Rep 13(1):483–490. https://doi.org/10.3892/mmr.2015.4558

Zhang Y, Cheng J, Chen F et al (2017) Circulating endothelial microparticles and miR-92a in acute myocardial infarction. Biosci Rep 37(2). https://doi.org/10.1042/BSR20170047

Zhang Z, Liang K, Zou G et al (2018) Inhibition of miR-155 attenuates abdominal aortic aneurysm in mice by regulating macrophage-mediated inflammation. Biosci Rep 38(3). https://doi.org/10.1042/BSR20171432

Zhang B, Zhang YF, Li R et al (2020) MiR-217 inhibits apoptosis of atherosclerotic endothelial cells via the TLR4/PI3K/Akt/NF-kappaB pathway. Eur Rev Med Pharmacol Sci 24(24):12867–12877. https://doi.org/10.26355/eurrev_202012_24190

Zhang Y, Yuan B, Xu Y et al (2022) MiR-208b/miR-21 promotes the progression of cardiac fibrosis through the activation of the TGF-beta1/Smad-3 signaling pathway: An in vitro and in vivo study. Front Cardiovasc Med 9:924629. https://doi.org/10.3389/fcvm.2022.924629

Zhao Y, Samal E, Srivastava D (2005) Serum response factor regulates a muscle-specific microRNA that targets Hand2 during cardiogenesis. Nature 436(7048):214–220. https://doi.org/10.1038/nature03817

Zhao Y, Ransom JF, Li A et al (2007) Dysregulation of cardiogenesis, cardiac conduction, and cell cycle in mice lacking miRNA-1-2. Cell 129(2):303–317. https://doi.org/10.1016/j.cell.2007.03.030

Zhao W, Zhao SP, Zhao YH (2015) MicroRNA-143/−145 in cardiovascular diseases. Biomed Res Int 2015:531740. https://doi.org/10.1155/2015/531740

Zhou L, Wang L, Lu L et al (2012) Inhibition of miR-29 by TGF-beta-Smad3 signaling through dual mechanisms promotes transdifferentiation of mouse myoblasts into myofibroblasts. PLoS One 7(3):e33766. https://doi.org/10.1371/journal.pone.0033766

Zhou Q, Anderson C, Hanus J et al (2016) Strand and cell type-specific function of microRNA-126 in angiogenesis. Mol Ther 24(10):1823–1835. https://doi.org/10.1038/mt.2016.108

Zhu J, Yao K, Wang Q et al (2016a) Circulating miR-181a as a potential novel biomarker for diagnosis of acute myocardial infarction. Cell Physiol Biochem 40(6):1591–1602. https://doi.org/10.1159/000453209

Zhu J, Yao K, Wang Q et al (2016b) Ischemic Postconditioning-regulated miR-499 protects the rat heart against ischemia/reperfusion injury by inhibiting apoptosis through PDCD4. Cell Physiol Biochem 39(6):2364–2380. https://doi.org/10.1159/000452506

Zincarelli C, Soltys S, Rengo G et al (2010) Comparative cardiac gene delivery of adeno-associated virus serotypes 1-9 reveals that AAV6 mediates the most efficient transduction in mouse heart. Clin Transl Sci 3(3):81–89. https://doi.org/10.1111/j.1752-8062.2010.00190.x

Chapter 10
miRNAs in Aging

Hai Duc Nguyen and Min-Sun Kim

Abstract MicroRNAs (miRNAs) have been recognized as significant contributors to the process of aging, offering valuable insights into its underlying mechanisms and potential therapeutic strategies aimed at improving aging and enhancing our overall well-being. The present chapter explores the characteristics of miRNAs and their roles in the modulation of gene expression during the course of aging. The primary objective of this study was to investigate the modifications in miRNA expression patterns associated with the aging process. Additionally, we sought to explore the potential functional implications of these alterations. Moreover, we explored the influence of environmental factors on the control of miRNAs and the possible applicability of these sequences as diagnostic biomarkers and targets for therapeutic intervention in age-related diseases.

Keywords miRNAs · Aging · Biomarkers · Lifespan · Therapeutics

List of Abbreviations

3′UTR	3′ untranslated region
AMPK	AMP-activated protein kinase
DNA	Deoxyribonucleic acid
GEO	Gene expression omnibus
IL	Interleukin

H. D. Nguyen (✉)
Division of Microbiology, Tulane National Primate Research Center, Covington, LA, USA

Department of Pharmacy, College of Pharmacy and Research Institute of Life and Pharmaceutical Sciences, Sunchon National University, Suncheon, Republic of Korea
e-mail: dnguyen14@tulane.edu

M.-S. Kim
Department of Pharmacy, College of Pharmacy and Research Institute of Life and Pharmaceutical Sciences, Sunchon National University, Suncheon, Republic of Korea

L. M. Vaschetto (ed.), *miRNAs, Human Health and Diseases*, Epigenetics and Human Health 13, https://doi.org/10.1007/978-3-031-64788-8_10

IRAK1	Interleukin-1 receptor-associated kinase 1
miRNAs	MicroRNAs
mRNA	Messenger RNA
NAD^+	Nicotinamide adenine dinucleotide
NAMPT	Nicotinamide phosphoribosyltransferase
pre-miRNAs	Precursor miRNA
RISC	RNA-induced silencing complex
ROS	Reactive oxygen species
SIRT	Sirtuin
TRAF6	Tumor necrosis factor receptor-associated factor 6

10.1 Introduction

MicroRNAs (miRNAs) represent a category of small, non-coding RNA molecules that play a pivotal role in the post-transcriptional regulation of gene expression (Bushati and Cohen 2007). These small regulatory elements, typically comprising 21–25 nucleotides, are transcribed from specific genes and modulate gene expression by the targeting of messenger RNA (mRNA) molecules. miRNAs selectively bind to the 3′ untranslated region (3′UTR) of the target mRNA. This interaction subsequently triggers either mRNA degradation or inhibition of translation, delicately adjusting the protein production within the cell.

The process of aging is a complex biological phenomenon that impacts all living organisms (da Costa et al. 2016). Aging, being a natural occurrence, has garnered significant scientific attention over the course of several decades owing to its profound effects on human health, quality of life, and broader societal implications. The substantial increase in global life expectancy over the past century, resulting from healthcare advancements and improved living conditions, has contributed to the expansion of the elderly demographic (Crimmins 2015). Therefore, the investigation of the critical processes involved in aging has emerged as a prominent area of interest within the field of biomedical research. The aging process is characterized by a gradual deterioration of physiological functions, leading to increased susceptibility to age-associated diseases. This decline is attributed to a complex interplay of factors, such as genetic predisposition, DNA damage triggering cellular senescence and inflammaging, telomere shortening, and disruptions in cellular signaling pathways. Despite a significant amount of research, a complete understanding of the precise mechanisms contributing to aging is lacking.

Given the intricate nature of the aging process and its profound implications for human health, elucidating the regulatory network governing this phenomenon is imperative. miRNAs have been recognized as significant post-transcriptional regulators, playing crucial roles in molecular pathways associated with aging. The ability to finely regulate gene expression and impact diverse cellular mechanisms establishes them as potential factors in the manifestation of the aging phenotype and age-related diseases. Examination of miRNAs in relation to aging presents compelling justifications. First, miRNAs undergo dysregulation as a consequence of the aging process, resulting in modified gene expression patterns in various tissues and organs. These alterations have the potential to disturb the intricate equilibrium of gene expression networks, thereby playing a role in the occurrence of cellular dysfunction and age-related pathologies (Williams et al. 2017). Second, miRNAs demonstrate distinct expression patterns in different tissues and have the ability to simultaneously target multiple mRNAs, influencing complex signaling cascades and biological pathways (Dimmeler and Nicotera 2013). Gaining a comprehensive understanding of miRNA-mRNA interactions can yield valuable insights into the molecular mechanisms that regulate the process of aging and the development of age-related diseases. Third, miRNAs have been observed not only in tissues but also in biofluids, including blood, saliva, and urine (Layne et al. 2019). This characteristic makes them easily obtainable and holds potential for their utilization as non-invasive biomarkers in the assessment of aging and the detection of age-related diseases. The potential of biomarkers lies in their ability to serve as early diagnostic tools, enabling timely interventions, and ultimately contributing to the promotion of healthy aging. This chapter explores the complex realm of miRNAs and their involvement in the regulation of gene expression throughout the aging process. To gain insights into the aging process, we aim to (1) examine age-related changes in miRNA expression and their functional consequences, and (2) assess the impact of environmental factors on miRNA regulation, while simultaneously exploring their potential as diagnostic biomarkers and therapeutic targets for age-related diseases.

10.2 miRNAs Biogenesis and Function

miRNAs are synthesized through the transcription of specific genes within the cellular nucleus by RNA polymerase II (Lee et al. 2004). This process yields long primary miRNA transcripts, also known as primary miRNA (pri-miRNAs). Pri-miRNAs generally consist of one or multiple hairpin structures. The maturation process of primary miRNAs (pri-miRNAs) into mature miRNAs comprises a complex and highly regulated series of steps that relies on the involvement of multiple essential proteins. Within the nucleus, the microprocessor complex, consisting of the RNase III enzyme known as Drosha and the double-stranded RNA-binding protein DGCR8, has the ability to identify and cleave the pri-miRNA (Johanson et al. 2013). This enzymatic activity results in the generation of precursor miRNAs (pre-miRNAs), typically about 70 nucleotides in length (Shukla et al. 2011). The transcription of pre-miRNAs from the nucleus to the cytoplasm is facilitated by Exportin-5, which functions as a nuclear export receptor. Dicer, a RNase III enzyme, facilitates the subsequent processing of pre-miRNAs within the cytoplasm, leading to the generation of miRNA duplexes. Every duplex consists of a guide strand, the mature miRNA, and a passenger strand, also known as miRNA-3p. The guide strand is selectively integrated into the RNA-induced silencing complex (RISC), while the passenger strand is commonly subjected to degradation (O'Brien et al. 2018).

Once incorporated into RISC, miRNAs act as molecular guides to selectively target particular mRNAs. Gene regulation is achieved through two primary mechanisms when there is a complementary base pairing between the miRNA and the 3′URT of the target mRNA (O'Brien et al. 2018). When the binding between miRNA and mRNA is highly accurate, it initiates the cleavage and subsequent degradation of the target mRNA by Argonaute proteins within RISC (Sheu-Gruttadauria and MacRae 2017). The degradation process hinders the conversion of mRNA molecules into functional proteins. Translational inhibition occurs when there is imperfect binding between the miRNA and its target mRNA, resulting in a hindered process of translation. The RISC complex impedes the translational apparatus from effectively converting mRNA into proteins, leading to a decrease in protein production. By utilizing these mechanisms, miRNAs precisely adjust the expression of numerous genes simultaneously, enabling them to exert a broad regulatory impact on diverse cellular processes.

As schematized in Fig. 10.1, microRNAs (miRNAs) have a profound influence on cellular homeostasis by regulating gene expression patterns across diverse processes, including development, stress response, cell cycle control, apoptosis, immunity, and metabolism (Bhaskaran and Mohan 2014; Bueno and Malumbres 2011; Leung and Sharp 2010; Rottiers and Näär 2012; Sadri Nahand et al. 2021; Tsitsiou and Lindsay 2009). Disruption of miRNA regulation can perturb this equilibrium and lead to numerous diseases, especially aging-related diseases. A comprehensive understanding of the processes involved in miRNA biogenesis, their functional mechanisms, and their significance in cellular homeostasis is crucial to fully reveal their impact on the aging process and age-related diseases. This understanding can also provide potential avenues for therapeutic interventions and the promotion of healthy aging.

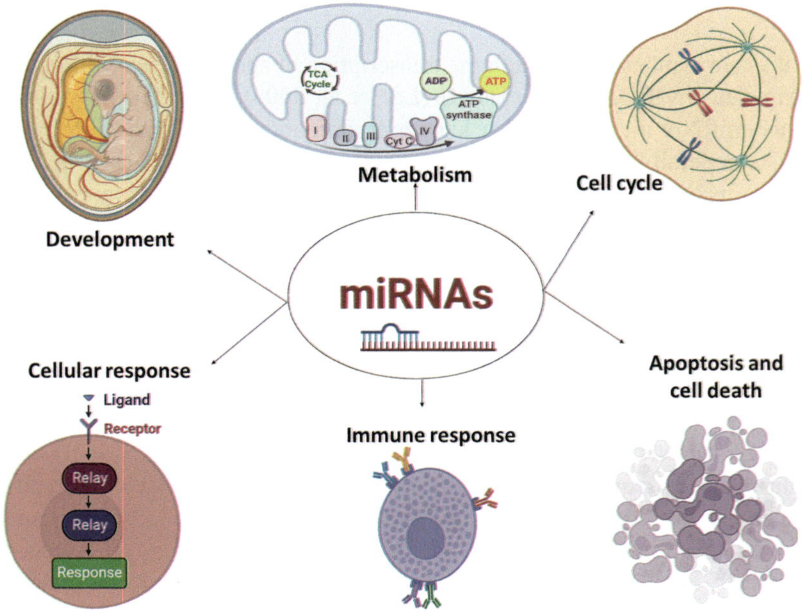

Fig. 10.1 Importance of miRNAs in maintaining cellular homeostasis

10.3 miRNAs Dysregulation in Aging

10.3.1 Changes in miRNA Expression Profiles During Aging

The phenomenon of aging is characterized by substantial alterations in miRNA expression patterns across diverse tissues and organs (Table 10.1). Changes in miRNA expression cause dysregulation of gene networks, representing a major factor in age-related changes in phenotype and an increased risk of age-related diseases. These changes encompass global miRNA dysregulation, tissue-specific miRNA expression, and dynamic changes with age (Lucanic et al. 2013). Global miRNA dysregulation is associated with both upregulation and downregulation of specific miRNAs (Kinser and Pincus 2020). Certain miRNAs consistently undergo alterations across various tissues, whereas others exhibit specific changes in particular tissues (Kinser and Pincus 2020). A defining characteristic of tissue-specific miRNA expression is the observation that different tissues display distinct miRNA expression profiles, which evolve with age (Kinser and Pincus 2020; Smith-Vikos and Slack 2012). This suggests that miRNAs play various roles in preserving tissue equilibrium throughout the aging trajectory. Moreover, the expression levels of specific miRNAs suffer changes during different stages of the aging process (Kim et al. 2017). This observation suggests that the dysregulation of miRNAs during aging is a multifaceted and continuously evolving process. Thus, changes in miRNA expression levels, which occur across different stages of aging, are not static but rather dynamic, reflecting the intricate interplay between various molecular mechanisms and environmental influences throughout the aging trajectory.

10.3.2 Impact of Altered miRNA Expression Profiles on Cellular Functions

The dysregulation of miRNAs during the aging process can significantly impact on various cellular functions and molecular pathways. These effects have the potential to contribute to the deterioration of cellular integrity and the subsequent impairment of tissue function observed during aging, including cellular senescence, deoxyribonucleic acid (DNA) damage response, oxidative stress, autophagy, and inflammation (Table 10.1 and Fig. 10.2) (Lan et al. 2021; Majidinia et al. 2020; Olivieri et al. 2021; Reddy et al. 2017; Roy 2021).

Table 10.1 Associations between miRNA expression profiles and target genes in the aging processes of the heart, brain, bone, and skeletal muscle. Up arrows (↑) indicate upregulation/overexpression/activation, while the down arrows (↓) indicate downregulation/suppression/inhibition

miRNAs	Target genes	Models	Organs	Effects on aging	Ref
↓ miR-18a, miR-19, and miR-17	*Ctgf, Tsp1, Par4, Timp1, Timp2*	Mouse, human	Heart	↓ cardiac fibrosis ↓ senescence and apoptosis ↓ cardiac remodeling	Du et al. (2015); Li et al. (2013); van Almen et al. (2011)
↑ miR-21, miR-22, miR-34a	*Jagged1, Smad7, Ogn,* Pparα, *Sirt1, Pnuts, Bcl2, Ccnd1*	Mouse, human	Heart	↑ cardiac fibrosis ↑ cellular senescence ↑ hypertrophy ↑ heart failure ↑ diastolic dysfunction ↓ myocardial autophagy	Boon et al. (2013); Gupta et al. (2016); Gurha et al. (2013); Jazbutyte et al. (2013); Yang et al. (2015b); Yuan et al. (2017); Zhou et al. (2018)
↓ miR-155-5p, miR-15a-5p, and miR-16-5p ↑ miR-106a-5p, miR-9-5p, let-7a-5p, miR-203a-3p, and hsa-miR-26a-5p	*APP, IL1B, IL10, NFKB1, IL6, GSK3B, TNF, NOS2, Trt, Scn5a, Drd2*	Mouse, human, rat	Brain	↓ Amyloid beta production ↓ neuroinflammation ↑ cognitive function	Duc Nguyen et al. (2022); Nguyen et al. (2022a, b)
↑ miR-34a, miR-181-a-1, and miR-30-e	*Sirt1 or Bcl2*	Mouse	Brain	↑ apoptosis	Khanna et al. (2011); Li et al. (2011b)
↑ miR-34c and miR-29	*Sirt1, Igf1, Cx3cl1, or Ireb2*	Killifish, mouse, human	Brain	↑ cognitive decline ↑ microglial activation ↓ loading of intracellular iron	Fenn et al. (2013); Ripa et al. (2017); Zovoilis et al. (2011)
↑ miR-34 ↓ miR-1000 and miR-186	*Eip74EF, Pcl,Su(z)12, VGlut, or Bace1*	Fly, mouse	Brain	↓ amyloid beta production ↓ neurodegeneration ↓ glutamate excitotoxicity	Kennerdell et al. (2018); Kim et al. (2016b); Liu et al. (2012); Verma et al. (2015)

(continued)

Table 10.1 (continued)

miRNAs	Target genes	Models	Organs	Effects on aging	Ref
↑ miR-218, miR-214, miR-31, miRNA-183	*Rictor, Atf4, Pten, Ho1*	Mouse, human	Bone	↑ osteoclast activity & bone loss ↓ osteoblast adhesion & survival ↓ osteoblast activity & bone formation ↓ osteogenesis ↓ bone marrow mesenchymal stem cell	Davis et al. (2017a); Ke et al. (2015); Lai et al. (2016); Li et al. (2016a); Wang et al. (2013); Weilner et al. (2016); Zhao et al. (2015a)
↑ miR-188 ↓ miR-21	*Rictor, Hdac9, Pten, Spry1, Reck, Pdcd4*	Mouse, human	Bone	↑ osteocyte apoptosis ↑ osteogenic differentiation ↑ bone formation ↓TNF-α-induced apoptosis ↑ bone resorption	Davis et al. (2017b); Hu et al. (2017); Li et al. (2015); Yang et al. (2013); Zhao et al. (2015b)
↓ miR-181a, miR-434-3p, miR-431	*Sirt1, Eif5a1, Smad4*	Mouse, monkey	Skeletal muscle	↓ regulates myotube size ↓ caspase activation ↑ myogenic differentiation ↓ myotube apoptosis	Lee et al. (2015); Pardo et al. (2017); Soriano-Arroquia et al. (2016)
↑ miR-29 ↓ miR-126	*Igf1, P85α, Bmyb, Ppm1d*	Mouse, human	Skeletal muscle	↑ cellular senescence ↑ DNA damage ↓ proliferation ↓ growth signaling related to IGF-1	Hu et al. (2014); Rivas et al. (2014); Ugalde et al. (2011)

10.3.3 How miRNA Dysregulation Contributes to Age-Related Diseases

miRNAs have been identified as crucial regulators in disease pathways, and their dysregulation has been found to exacerbate pathological processes such as neurodegenerative disease, cardiovascular disease, cancer, metabolic disorders, osteoporosis, and musculoskeletal disease (Bravo Vázquez et al. 2021; Brzeszczyńska et al. 2020; Laggerbauer and Engelhardt 2022; Li et al. 2023; Nguyen and Kim 2022b; Rottiers and Näär 2012; Smolarz et al. 2022). Table 10.2 illustrates the dysregulation of specific miRNAs in a range of neurodegenerative diseases, including Alzheimer's disease, Parkinson's disease, Huntington's disease, multiple sclerosis, and age-related macular degeneration. This dysregulation has the potential to impact on the aggregation of toxic protein aggregates (amyloid-beta, α-Synuclein, etc.) and the reduction of neuron survival and synapse plasticity (Deng et al. 2021). For

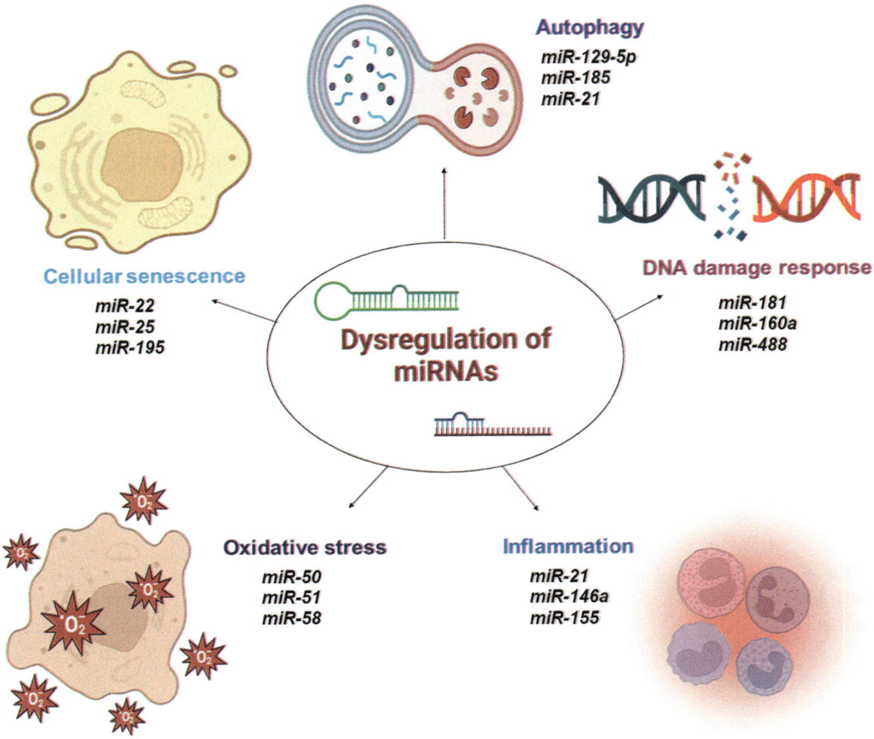

Fig. 10.2 Impact of altered miRNA expression profiles on cellular functions

further information regarding the interplay between miRNAs and neurodegenerative diseases, please refer to Chap. 7.

10.4 The Role of Specific miRNAs in Aging

microRNA-34a (miR-34a) is one of the most extensively investigated miRNAs that exhibit increased expression levels in the context of aging (Raucci and Vinci 2020). Increased expression of miR-34a has been found to play a role in the decline of cellular function and tissue aging (Kukreti and Amuthavalli 2020; Raucci et al. 2021). Thus, miR-34a plays a crucial role in regulating cellular senescence and apoptosis by targeting genes associated with the cell cycle, DNA repair, and anti-apoptotic pathways (Chang et al. 2007; Modi et al. 2016). In vivo, an age-related increase in muscle-derived miR-34a within circulating extracellular vesicles was observed in both male and female C57BL6 mice at the ages of 6 and 24 months, causing the senescence of bone marrow stem cells (Fulzele et al. 2019).

The upregulation of microRNA-146a (miR-146a) has been associated with inflammaging, a term referring to the chronic low-grade inflammation commonly

Table 10.2 Association between miRNAs, their targets, and neurodegenerative diseases. Up arrows (↑) indicate upregulation/overexpression/activation, while the down arrows (↓) indicate downregulation/suppression/inhibition

miRNAs	Targets	Models	Key outcomes	Ref
Parkinson's disease				
↑ miR-7/ miR-153, miR-205, miR-599, miR-205, miR-599, miR-29c, miR-410, miR-626	α-syn, LRRK2, PTEN, PI3K, AKT, Nrf2, Keap 1, SIRT2, p53	Primary neuron cells, G2019S (Y1699C mutation cells, SH-SY5Y cells, G2019S (Y1699C mutation cells), PC12 cells, retinal pigment epithelium cells, dopaminergic neurons cells	↓ LRRK2, PTEN, Keap 1, Nrf2 ↑ α-syn ↑ Parkinson's disease	Cho et al. (2013); Doxakis (2010); Ge et al. (2019); Li et al. (2016b, 2019a, b); Wu et al. (2019); Zhu et al. (2018a); Zou et al. (2015)
↑ miR-21, miR-126, miR-34a, miR-153, miR-27a, miR-27b, miR-505, miR-15b-5p	α-syn, Nrf2, Bax, caspase-3, HO-1, PINK1	SH-SY5Y cells, neuron cells, Hela and M17 cells	↑ α-syn, Nrf2, Bax, caspase-3 ↓ PI3K, AKT, Nrf2-HO-1, PINK1 ↓ Parkinson's disease	Kim et al. (2008, 2022); Su et al. (2016); Zhu et al. (2018a, b, 2021a, b)
↑ miR-221, miR-185, miR-212-5p	PI3K, AKT, SIRT2, p53	Rats	↑ PI3K/AKT ↓ SIRT2, p53 ↑ Parkinson's disease	Qin et al. (2021); Salama et al. (2020); Sun et al. (2020)
↑ miR-183	Bcl-2	Rats	↑ Bcl-2 ↓ Parkinson's disease	Gao et al. (2019a, b)
↓ miR-433	FGF20	Human sample	↑ FGF20 ↑ α-syn ↑ Parkinson's disease	Wang et al. (2008)
Alzheimer's disease				

(continued)

Table 10.2 (continued)

miRNAs	Targets	Models	Key outcomes	Ref
↑ miR-15b, miR-101a, miR-106b, miR-142-5p, miR-30b, miR-142-5p, miR-21, miR-104-5p, miR-15b-5p, miR-21, miR-29a ↓ miR-6845-3p, miR-124	BACE1, NF-κB, MAPK, Fyn, PDCD4, PI3K, AKT, GSK-3β, sGC, sGMP, Aβ42, Aβ25-35, ADAM10, APP, PI3K, AKT, RFX1, Wnt1, CREB	SH-SY5Y cell, CHP212 cells, HEK293 cell, human neural cells, N2a/APP695swe cells, BV2 microglia cell, peripheral blood mononuclear cells	↓ APP, Aβ, protein LC3, beclin-1, tau phosphorylation, PSD-95, ApoE, Caveolin-1-PI3K/Akt/GSK3β ↓ inflammatory cytokines ↓ Alzheimer's disease	Feng et al. (2017, 2018); Hu et al. (2018); Kang et al. (2017); Li and Wang (2018); Liu et al. (2016, 2019); Sedighi et al. (2019); Song and Kim (2017); Wang et al. (2015a, b); Xu et al. (2019a, b); Zhou et al. (2015)
↑ miR-34a, miR-144, miR-146a, miR-933, miR-138 ↓ miR212	Caspase-2, NRF2, GSH, TRAF6, NF-κB, IRAK-1, 27-OHC, RARA, NOS1	SH-SY5Y cell, HT-22 cells/C8-B4 cells/bEnd.3 cells, HAG cells, THP-1/U937/HL-60/WEHI-3 cell, human lung microvascular endothelial cells, HEK293/tau cells, human neural cells	↓ Caspase-2, NRF2, GSH ↑ p62/Beclin1 ↓ IRAK-1 ↑ IL-1/TNF ↑ inflammatory cytokines ↑ RARA/GSK-3β ↑ SNO-GAPDH/SNO-Drp1/SNO-Cdk5 ↑ tau Hyperphosphorylation ↑ Alzheimer's disease	Akhter et al. (2018); Cui et al. (2010); Dias et al. (2018); Fang et al. (2014); Kim et al. (2016a); Liu et al. (2016, 2020b); Taganov et al. (2006); Wang and Zhang (2020)

(continued)

Table 10.2 (continued)

miRNAs	Targets	Models	Key outcomes	Ref
↑ miR-31, miR-200a-3p, miR-101a, miR-101b, miR-21, miR-200b/c, miR-124, miR-107, miR-34c, miR-937, miR-107 ↓ miR-338-5p	APP/BACE1, BACE1, PRKACB, MAPK, HDAC2/HNF-4A/AMPK, PI3K/AKT/GSK-3β, delta in notch signaling pathway, PDCD10, S6K1, SYT1, GluN2A, BDNF, PDCD10	SAMP8 mice, 3xTg-AD mice, 5XFAD transgenic mice, APPswe/PS1ΔE9 transgenic mice, swAPP695, MCAO mice, Alzheimer's disease mice and flies, the mouse model insulted by 6-OHDA	↓ Aβ, tau phosphorylation, protein LC3 and beclin-1, IRS-1pSer, delta, PDCD10, GluN2A ↑ ROS-JNK-p53, Brn-4, PDCD10 ↓ Caveolin-1-PI3K/Akt/GSK3β ↓ Alzheimer's disease	Barros-Viegas et al. (2020); Higaki et al. (2018); Kang et al. (2017); Kong et al. (2015); Liu et al. (2015b, 2021); Liu and Wang (2012); Qian et al. (2019); Shi et al. (2020); Sun et al. (2018); Tang et al. (2019); Wang et al. (2015b)
↓ miR-29c, miR-101, miR-26b, miR-10a, miR-206-3p ↑ miR-125b, miR-922	BACE1, APP/RanBP9/Rab5, DUSP6/PPP1CA/Bcl-W, UCHL1, Rb1, BDNF-TrkB, BDNF	SAMP8 mice, pLSyn-miR-101 sponge mice, E18 Sprague Dawley rat, Alzheimer's disease mice, APP/PS1 mice	↑ PKA/CREB, Cdk5 ↓ p27/Kip1, BDNF, TrkB ↑ tau phosphorylation ↑ AMPK Hyperphosphorylation ↑ Alzheimer's disease	Absalon et al. (2013); Banzhaf-Strathmann et al. (2014); Barbato et al. (2020); Yang et al. (2015a); Zhao et al. (2014)
Huntington disease				
↑ miR-132	BBB, C1q	MCAO mice, APP/PS1 transgenic mice	↑ VE-cadherin, β-Catenin ↓ C1q ↓ Alzheimer's disease/Huntington's disease	Zuo et al. (2019)
Multiple sclerosis				
↑ miR-142	SOCS1, TGFBR1	Experimental autoimmune encephalomyelitis mice	↓ SOCS1, TGFBR1 ↑ Multiple sclerosis	Talebi et al. (2017)
↓ miR-155	IL-6, IL-1, TNF	SOD1 mice	↓ L-6, IL-1, TNF ↓ Multiple sclerosis / experimental autoimmune encephalomyelitis/amyotrophic lateral sclerosis	Butovsky et al. (2015)
Age-related macular degeneration				
↑ miR-34a	TREM2	C8B4 microglial cells	↓ TREM2 ↑ Age-related macular degeneration	Bhattacharjee et al. (2016)

observed during the aging process (Olivieri et al. 2021). Furthermore, miR-146a has emerged as a critical modulator of the immune response, regulating the expression of inflammatory mediators including interleukin (IL)-1α, IL-6, IL-8, interleukin-1 receptor-associated kinase 1 (IRAK1), and tumor necrosis factor receptor-associated factor 6 (TRAF6) (Bhaumik et al. 2009; Jiang et al. 2014; Yan et al. 2021a). Recently, studies have confirmed the essential role of NAD+ in SIRT-mediated anti-aging effects. This pathway is further amplified by AMPK, which promotes NAD+ and SIRT activity through NAMPT. By suppressing NAMPT, miR-164a hinders NAD+/SIRT activity, potentially mitigating the anti-aging effects of AMPK (Gong et al. 2022).

microRNA-21 (miR-21) has also emerged as a novel circulating biomarker associated with inflammaging (Olivieri et al. 2012). Studies have shown an association between miR-21 upregulation and age-related cardiac dysfunction (de Lucia et al. 2017). This miRNA is thought to exert its effects by targeting genes involved in the remodeling of the extracellular matrix and signaling pathways implicated in cardiac fibrosis and hypertrophy (de Lucia et al. 2017; Surina et al. 2021). Olivieri et al. (2012) conducted a cohort study examining miR-21 expression across diverse groups: individuals with cardiovascular disease, centenarians, and healthy controls. The study revealed a contrasting pattern, with reduced miR-21 expression levels in centenarians compared to its upregulation in cardiovascular disease patients. Additionally, a notable association was found between miR-21 and the inflammatory markers C-reactive protein and fibrinogen.

The miroRNA-29 (miR-29) family members exhibit a decrease in expression levels as the organism undergoes the aging process (Cron et al. 2020; Horita et al. 2021). These miRNAs are of significant importance in preserving tissue homeostasis and regulating genes implicated in extracellular matrix production (Dalgaard et al. 2022). A decrease in their expression is believed to play a role in the development of age-related tissue fibrosis in the liver, heart, and lungs (Kriegel et al. 2012). For example, downregulation of miR-29a, miR-29b-1, miR-29b-2, and miR-29c has been associated with the pathogenesis of osteoporosis, osteoarthritis, immune disease, and cardiorenal disease (Horita et al. 2021). Additionally, the expression of microRNA-181 (miR-181) decreases with advancing age, correlating with compromised immune functionality (Kim et al. 2021). The downregulation of this particular factor contributes to immune dysfunction and increased vulnerability to infections in old individuals, targeting genes associated with immune cell activation and functioning (Li et al. 2012). The miR-181 family members (miR-181a, miR-181b, miR-181c, and miR-181d) were also found to be implicated in the pathogenesis of vascular biology and immunity (Sun et al. 2014).

The let-7 family of microRNAs exhibits high expression levels in various tissues (e.g., brain, muscle, and retina) (Nweke and Brand 2020). They play a crucial role in regulating the differentiation process of multi-tissue stem cells. This regulation is achieved by targeting specific genes involved in aging-related pathways. It has been reported that the let-7 family members also exhibit a decrease in their expression levels as an organism undergoes the aging process (Wang et al. 2022b). These miRNAs play a crucial role in the regulation of stem cell homeostasis and the process of cellular differentiation. The decline in let-7 levels could potentially play a role in the decreased ability of aging individuals to regenerate tissues and repair damage (Nweke and Brand 2020; Wang et al. 2022b). Additionally, a study found that these miRNAs have the potential to target various constituents of the insulin receptor/IGF pathway, including both receptors and binding proteins (Nweke and Brand 2020).

10.5 Influence of Environmental Factors on miRNA Regulation in Aging

10.5.1 The Interplay Between miRNAs and Environmental Stressors

The regulation of miRNA can be influenced by a diverse range of environmental stressors, including physical, chemical, and biological factors (Balasubramanian et al. 2020; Duc Nguyen 2023; Duc Nguyen et al. 2022; Nguyen et al. 2022a; Nguyen and Kim 2022b, c). These stressors have the capacity to modify the levels and activity of miRNAs, inducing alterations in cellular functions and ultimately contributing to the aging process. First, oxidative stress involves the elevated generation of reactive oxygen species (ROS), which can affect the processes of miRNA biogenesis and stability (Reddy et al. 2017; Smith-Vikos and Slack 2012). As shown in Table 10.3, certain miRNAs (e.g., miRN-93) respond to oxidative stress by regulating the expression of antioxidant genes, while others are subjected to degradation, impacting the cellular redox balance (Bu et al. 2017). Second, exposure to ultraviolet (UV) radiation emitted by sunlight has the potential to alter the expression of miRNAs (e.g., miR-23a-5p) in the skin and other directly exposed tissues, potentially playing a role in the processes related to skin aging and photocarcinogenesis (Gerasymchuk et al. 2020). Third, dietary factors, including micronutrients and bioactive compounds, can influence the expression of miRNAs. Several dietary factors, such as resveratrol, B vitamins, and curcumin, have demonstrated the ability to regulate miRNAs (e.g., miR-128-3p) implicated in processes associated with cognitive function (Duc Nguyen 2023; Duc Nguyen et al. 2022; Nguyen et al. 2022a; Nguyen and Kim 2022c). Furthermore, disruption of miRNA expression/function can occur due to exposure to environmental pollutants and toxins. Specific miRNAs (e.g., miR-1-3p) have the potential to react to environmental stressors by regulating detoxification pathways or genes associated with cognitive impairment, apoptosis, cell death, tumorigenesis, and oxidative stress (Balasubramanian et al. 2020; Nguyen and Kim 2022b).

Table 10.3 Influence of environmental and other factors on miRNA regulation in aging. Up arrows (↑) indicate upregulation/overexpression/activation, while the down arrows (↓) indicate downregulation/suppression/inhibition

miRNAs	Target genes	Models	Organs	Effects on aging	Ref
Environmental factors					
↑ miRN-93, miR-214, miR-669c	MGST, GSTZ1, GSTM1, and GSTT1	Mouse, rats	Livers	↓ oxidative defense mechanisms	Li et al. (2011a); Maes et al. (2008)
↑ miRNA-34a, miR-335, miR-217, miR-92a, miR-21	SIRT1, SOD2, TNXRD2, SREBP2	Rats, human angiogenic progenitor cells, mesangial cells	Endothelial cells, kidney, liver	↓ oxidative defense mechanisms ↑ endothelial dysfunction ↑ reactive oxidative stress generation ↑senescence	Bai et al. (2011); Chen et al. (2015); Fleissner et al. 2010; Li et al. (2011a); Menghini et al. (2009)
↑ miR-494, miR-144, and miR-146a	c-MYC, SIRT, DJ-1, miR-144,	Pancreatic cancer cells tissues, and human diploid fibroblasts, rats	Dopaminergic neurons, pancreas, liver	↑ reactive oxidative stress generation ↑senescence ↑DNA damage ↓NRF2 expression	Csiszar et al. (2014); Faraonio et al. (2012); Liu et al. (2015a); Smith et al. (2015); Xiong et al. (2014)
↑ miR-23a-5p, miR-34c-5p, miR-30b, miR-30c, miR-148a, miR-199a-5p, miR-365, miR-1246, miR-3613-p ↓miR-155, miR-146a, miR-146b-5p, miR-181c, miR-218, miR-4281	AMBRA, MAPK, c-Jun, IL1, IL6, IL8	Human fibroblasts	Skin and hair	↑ Skin aging and photocarcinogenesis	Gerasymchuk et al. (2020); Si et al. (2019); Zhang et al. (2016)

(continued)

Table 10.3 (continued)

miRNAs	Target genes	Models	Organs	Effects on aging	Ref
↓ miR-128-3p, 155-5p, miR-15a-5p, miR-16-5p ↑ miR-9-5p, miR-106a-5p, let-7a-5p, miR-203a-3p, miR-26a-5p, and miR-204-5p	GSK3B, IL6, NFKB1, APP, etc.	Neuroblastoma SH-SY5Y cell line	Brain	↑ Cognitive function	Duc Nguyen (2023); Duc Nguyen et al. (2022); Nguyen et al. (2022a); Nguyen and Kim (2022c)

(continued)

Table 10.3 (continued)

miRNAs	Target genes	Models	Organs	Effects on aging	Ref
miR-1-3p, let-7a-5p, miR-9-5p, miR-16-5p, miR-17-5p, miR-20a-5p, miR-26a-5p, miR-26b-5p, miR-34a-5p, miR-101-3p, miR-106a-5p, miR-128-3p, miR-144-3p, miR-199a-3p, miR-204-5p, and miR-335-5p	BAX, CASP3, BCL2, TNF, and IL-1B, etc.	Mouse, rats, cells	Brain	Cognitive impairment Apoptosis Tumorigenesis Cell death Oxidative stress	Krammer et al. (2023); Maffioletti et al. (2021)
Other factors					
↑miR-98-3p, miR-181a-1, miR-30e, miR-181b, miR-669c, miR-709, miR-34a, miR-451, and miR-144	Bcl2	Aged rats, mice	Cerebral cortex, cerebrum, liver	↑ Bcl2 ↓ apoptosis ↑ detoxification activity	Zhai et al. (2022)

(continued)

Table 10.3 (continued)

miRNAs	Target genes	Models	Organs	Effects on aging	Ref
miR-1, miR-21, miR-126, miR-133a, miR-133b, miR-181a, and miR-206	HDAC4, Pax3/Pax7, Cyclin D, RAAS, VCAM-1, SRF, PTBP2, SP1, etc.	Mice	Skeletal muscle	Skeletal muscle regeneration	Ultimo et al. (2018)
miR-132, miR-291-1, miR-192/194 cluster, miR-142-3p, mir-132, let-7b, and miR-125a	CREB, BMAL1, ADCY6, etc.	Mice	Brain	Circadian timing	Gaine et al. (2018)
miR-10a, miR-15a, miR-146a, miR-15a, miR-29a, miR-126, miR-195, let-7a, let-7g, and let-7f	CRP, BDNF, SERT, SLC6A4, etc.	Human	Blood	Stressful life event Acute or chronic psychological stress Mental health	Krammer et al. (2023); Maffioletti et al. (2021)

10.5.2 How Lifestyle Factors Impact miRNA Expression in Aging

The expression of miRNAs can be influenced by dietary patterns, specifically those related to calorie intake and nutrient composition. Caloric restriction has been demonstrated to affect miRNAs linked to both longevity and age-related diseases (Zhai et al. 2022). Furthermore, engagement in regular physical activity has been found to be associated with miRNA altered expression patterns in different tissues, influencing cellular metabolic processes, inflammatory responses, and tissue maintenance (Ultimo et al. 2018). Thus, miRNA expression resulting from exercise has shown positive effects on health. Moreover, disrupted sleep patterns and chronic sleep deprivation also have the potential to induce changes in miRNA expression profiles within the brain and other bodily tissues, consequently impacting neuronal function and overall well-being (Gaine et al. 2018). The impact of psychological stress and metabolic disorders on miRNA expression in both the brain and peripheral tissues has been observed. Additionally, dysregulated miRNAs have the potential to play a role in the development of stress-related pathologies and age-related cognitive decline (Krammer et al. 2023; Maffioletti et al. 2021) (Table 10.3).

10.5.3 Epigenetic Modifications and Their Influence on miRNA Regulation

Epigenetic modifications, including DNA methylation and histone modifications, exert a pivotal influence on the regulation of miRNA expression throughout the aging process (Wang et al. 2022a). These modifications have the potential to influence the accessibility of promoter regions associated with miRNAs, resulting in modifications in the expression levels of these miRNAs. DNA methylation can be influenced by age-related disease, subsequently impacting on miRNA expression profiles. For example, a DNA-methylation study of miRNA genes found an association between miR34b, hypermethylation, vascular aging, and arteriosclerosis in 414 high-cardiovascular-risk individuals aged 55–75 years (Carrión et al. 2022). Using the Gene Expression Omnibus (GEO) database, several miRNAs were found to be related to the pathogenesis of osteoarthritis, hypomethylation, and target genes (e.g., FOXN3, TGIF2, and PAX5) (Jin et al. 2023). On the other hand, histone modifications, especially histone acetylation and methylation, can regulate the transcription of miRNAs by targeting the regulatory regions associated with miRNA genes (Wang et al. 2022a). Histone deacetylase inhibitors may modulate the expression of miRNAs, potentially mitigating age-related phenotypes. A study of 353 patients with acute lymphoblastic leukemia found that the CpG surrounding 13 microRNAs exhibit elevated levels of K9H3me2 and/or diminished levels of K4H3me3 (Roman-Gomez et al. 2009). This particular pattern of histone modifications is indicative of a compact chromatin structure associated with the suppression of gene expression. Abnormal methylation of these miRNAs is a prevalent phenomenon in acute lymphoblastic leukemia, significantly impacting on the clinical prognosis of affected individuals (Roman-Gomez et al. 2009). Aging is accompanied by alterations in chromatin remodeling, which in turn impacts the expression of miRNAs involved in aging-related pathways. In vivo studies have revealed that miR-132 specifically targets genes enriched in chromatin remodeling (Mecp2, Ep300, Jarid1a) and translation regulation (Btg2 and Paip2a) pathways (Alvarez-Saavedra et al. 2011). It plays a crucial role in coordinating chromatin remodeling and protein translation processes within the suprachiasmatic nucleus clock of mice, contributing to the precise adjustment of clock entrainment in this particular brain region (Alvarez-Saavedra et al. 2011).

Epigenetic modifications bridge the gap between environmental exposures and miRNA regulation, emerging as potential targets for interventions aimed at promoting healthy aging and alleviating age-related diseases. Integrating information on environmental and lifestyle factors with miRNA regulation holds promise for the development of personalized interventions, potentially extending health span and improving overall well-being in aging populations.

10.6 miRNAs as Biomarkers of Aging

10.6.1 Potential of miRNAs as Aging Biomarkers

miRNAs have gained considerable attention as potential indicators of aging due to their inherent stability, tissue-specific expression patterns, and their detectability in a wide range of bodily fluids (Guo et al. 2014; Li et al. 2021; Teoh and Das 2022). The alteration in miRNA expression patterns during aging and in response to various environmental stimuli can provide significant insights into an individual's physiological age and overall well-being. First, miRNAs serve as a non-invasive means of detection. They can be identified in readily obtainable biofluids, including blood, saliva, urine, and cerebrospinal fluid (Teoh and Das 2022). The non-invasive nature of miRNA samples make them highly appealing for longitudinal studies and clinical applications. Second, miRNAs exhibit tissue-specificity (Guo et al. 2014), showing distinct expression patterns in different tissues and therefore allowing for the evaluation of age-related alterations in particular organs or systems (Table 10.1). Third, miRNAs exhibit dynamic alterations. The expression profiles of miRNAs undergo changes throughout various stages of the aging process, rendering them promising indicators of both biological age and the progression of diseases (Kinser and Pincus 2020; Smith-Vikos and Slack 2012). Overall, miRNAs exhibit a high degree of stability (Li et al. 2021). They also exhibit remarkable stability in biofluids and can endure diverse storage conditions, facilitating convenient sample preservation and transportation (Teoh and Das 2022).

10.6.2 Challenges and Opportunities for the Use of miRNAs in Age-Related Diagnostics

While miRNAs exhibit significant potential as biomarkers for aging, several challenges must be overcome to effectively apply them in clinical settings. Standardization poses a significant challenge in miRNA detection and quantification, as the absence of standardized protocols limits the comparability of findings across distinct studies (Lee et al. 2017). The establishment of standardized methods is crucial to ensuring the reliability and reproducibility of miRNA biomarker assessments. The analysis of miRNA expression profiles and their interactions with various biological pathways necessitates the utilization of advanced data interpretation methods. To obtain a comprehensive understanding of age-related alterations in miRNA, it may be imperative to integrate additional omics data (Wu et al. 2021). To ascertain the efficacy of miRNAs as biomarkers for aging, it is imperative to conduct extensive longitudinal studies on a large cohort (Huan et al. 2018). These studies would serve to evaluate the predictive capacity, sensitivity, and specificity of miRNAs in this context. The diversity of populations plays a significant role in the age-related changes observed in miRNA expression (Condrat et al. 2020). Consequently, it is imperative to conduct studies encompassing a wide range of ethnic and geographic groups to ensure the global applicability of miRNA biomarkers. Age-related diseases are frequently linked to the dysregulation of miRNA, posing a challenge in distinguishing whether alterations in miRNA are solely attributable to the aging process or influenced by disease-related factors.

10.6.3 miRNAs as Indicators of Healthspan and Lifespan

miRNAs have the potential to serve as indicators of an individual's healthspan, defined as the phase of life characterized by robust health and functional autonomy before the emergence of age-related diseases (Ibáñez-Ventoso and Driscoll 2009). Several miRNAs specifically linked to the process of healthy aging have shown the potential to serve as reliable indicators of healthspan in *Caenorhabditis elegans*, bats, and vegetarians (Huang et al. 2019; Liu et al. 2020a; Vora et al. 2013). While miRNAs do not directly indicate an organism's lifespan, their expression patterns have the potential to impact various processes associated with aging and the likelihood of developing certain diseases. The investigation of miRNAs linked to exceptional longevity provides an opportunity to gain valuable knowledge about the determinants contributing to a prolonged lifespan. For example, overexpression of miR-228 and alg-2 (a miRNA-specific Argonaute protein) reduces the lifespan in *Caenorhabditis elegans* (Aalto et al. 2018; Smith-Vikos et al. 2014), while overexpression of miR-34, let-7, miR-1000, and miR-17 was found to be related to increased lifespan in *Drosophila* and mice (Chawla et al. 2016; Du et al. 2014; Gendron and Pletcher 2017; Liu et al. 2012; Verma et al. 2015).

On the other hand, miRNA-based age prediction models, alternatively referred to as epigenetic clocks, have demonstrated considerable potential for the estimation of an individual's chronological age (Huan et al. 2018). A comprehensive examination of whole-blood microRNA expression was conducted on a sample of 5221 adults, aged between 46 and 66 years, who were part of the Framingham Heart Study cohort (Huan et al. 2018). The results of this analysis revealed that participants with a predicted miRNA age exceeding their chronological age demonstrated elevated levels of blood pressure and glucose. Moreover, this group exhibited a higher prevalence of coronary heart disease and hypertension, along with an augmented risk of mortality from various causes (Huan et al. 2018). This study observed and predicted the co-expression of miR-181a-3p with three immune response genes: RAB27A, CXCL16, and SPON2. Interestingly, RAB27A also participates in the T cell activation pathway, suggesting a potential regulatory role for miR-181a-3p in immune function. Several microRNAs (miRNAs) have been implicated in regulating immune function. For instance, miR-193b-3p targets genes such as FAIM3, CTSW, ICOS, TGFBR3, DPP4, ETS1, and KIF13B, while miR-31-5p exhibits targeting capabilities towards INPP5D, CD27, TCF7, TGFBR3, FASLG, and DPP4 (Huan et al. 2018). These results suggest that age-related microRNAs and their corresponding targets hold promise for identifying accelerated aging and forecasting susceptibility to age-related diseases. The potential for enhanced accuracy in age prediction can be achieved by integrating miRNA-based epigenetic clocks with other biomarkers. Despite the presence of various challenges, continuous research and advancements in miRNA profiling technologies exhibit significant potential for their incorporation into age-related diagnostics and personalized strategies aimed at enhancing healthy aging and prolonging lifespans.

10.7 Therapeutic Implications

The discovery of innovative therapies may benefit from focusing on dysregulated miRNAs linked to age-related illnesses. These treatments may be able to delay the progression of disease and raise the standard of living for elderly people. Sponges and antisense oligonucleotides, commonly referred to as anti-miRNAs or miRNA inhibitors, have demonstrated potential for selectively modulating the expression of miRNAs implicated in age-related diseases (Ebert and Sharp 2010; Nguyen and Kim 2022a, b, 2023). By inhibiting these miRNAs, it is possible to restore the expression of the genes they target, potentially improving phenotypes associated with diseases. For example, inhibition of miR-31 has been observed to effectively suppress the aging of hair follicle stem cells in mice (Zhang and Yi 2021).On the other hand, reintroducing miRNAs into target tissues holds promise for promoting tissue repair and rejuvenation by enhancing their regulatory functions. In this regard, miR-199-3 has demonstrated the ability to augment myogenic differentiation and facilitate muscle regeneration. Studies have shown that injecting miR-199 mimics, delivering miR-199-3p, into aged mice resulted in the development of muscle fiber hypertrophy and delayed decline in muscle strength. Further substantiating the therapeutic potential of miRNAs, systemic administration of miR-199 mimics significantly enhanced muscle strength in Duchene muscular dystrophy mice (Fukuoka et al. 2021). While miRNA-based therapeutics offer promising avenues for treating age-related diseases, integrating them with established therapies or other targeted approaches holds the potential to unlock synergistic effects and further improve patient outcomes (Nguyen and Kim 2023). For example, triflusal, andrographolide, and miRNA sponges for miR-26a-5p, miR-9–5p, miR-124–3p, and miR-155–5p are promising compounds for treating heavy metal-induced stroke (Nguyen and Kim 2023). These findings suggest that the modulation of miRNAs linked to aging has the potential to offer interventions aimed at promoting healthy aging and prolonging life span. Potential interventions could encompass lifestyle modifications, pharmaceutical agents, or genetic approaches. The integration of miRNA-based interventions with additional anti-aging approaches, such as caloric restriction, exercise, or senolytic therapies, has the potential to yield comprehensive and synergistic advantages in the promotion of healthy aging (Chen et al. 2021; van Deursen 2019; Yan et al. 2021b).

The development of miRNA-based therapeutics necessitates a rigorous evaluation of safety profiles, including potential off-target effects, to ensure patient well-being. Additionally, implementing these interventions may require personalized strategies, accounting for individual variations in miRNA expression and treatment responses. Given the gradual and complex nature of aging, long-term investigations are essential to definitively assess the efficacy and safety of miRNA-based interventions over extended periods.

The development of miRNA-based interventions raises ethical concerns regarding accessibility, affordability, and equitable distribution of these potential anti-aging therapies (Winkle et al. 2021). To ensure the effectiveness and safety of these novel therapeutics, rigorous adherence to regulatory guidelines will be paramount throughout the development and approval process. Future research into miRNA holds immense promise for unraveling the mechanisms underlying aging and uncovering novel therapeutic avenues (Winkle et al. 2021). While challenges remain, the potential application of miRNA-based therapeutics in clinical settings offers a transformative outlook for the field of aging research and the management of age-related diseases. This advancement holds promise to ultimately improve the well-being and quality of life for aging populations, guiding the responsible development and implementation of these novel approaches. Continued research, collaborative efforts, and meticulous evaluation of ethical considerations are crucial to achieve healthy aging, potentially facilitated by miRNA-based therapies.

10.8 Conclusion

10.8.1 Recapitulation of the Key Findings

Throughout this chapter, we conducted a comprehensive examination of the role of miRNAs in the aging process, shedding light on their significance as potential biomarkers and therapeutic targets. The key findings underscore the dynamic nature of miRNA involvement in aging:

1. The miRNA landscape reveals a profound dysregulation in miRNA expression profiles during aging. This phenomenon manifests as a complex mosaic of molecular alterations, characterized by both upregulation and downregulation of distinct miRNAs.
2. Changes in miRNA expression patterns exert a profound influence on various cellular processes, ranging from cellular senescence and inflammation to the oxidative stress response and tissue fibrosis. Consequently, these miRNA-mediated alterations significantly contribute to the development of age-related phenotypes and diseases.

3. Environmental factors and lifestyle choices also affect miRNA regulation, highlighting the dynamic interplay between genetics and the environment in shaping the aging process. These findings suggest that environmental stressors and lifestyle factors intricately modulate the expression of miRNAs, adding a layer of complexity to our understanding of aging.
4. The potential of miRNAs as non-invasive biomarkers for healthspan, biological age, and disease susceptibility is gaining significant interest. Their stability and presence in numerous bodily fluids make them promising candidates for a wide range of applications in comprehensive health assessments.
5. miRNA-based interventions using miRNA mimics and inhibitors open the doors to the management of age-related diseases, promoting tissue repair, and facilitating regeneration. These multifaceted approaches showcase the potential of miRNAs as therapeutic targets for promoting healthier aging.

10.8.2 Importance of miRNAs in Understanding Aging Processes

The investigation of miRNAs in the context of aging has yielded significant findings concerning the molecular processes that drive age-related alterations and the emergence of age-associated diseases. Their dysregulation during the aging process is a significant factor in the development of cellular dysfunction and the subsequent decline in tissue integrity commonly observed in older individuals. Therefore, elucidating the regulatory functions of miRNAs in the aging process provide novel opportunities to explore aging. Furthermore, miRNAs possess the capacity to serve as a link connecting genetic predisposition and environmental influences, thereby providing insights into the complex interaction between genetics and lifestyle in determining the unique patterns of aging experienced by individuals.

10.8.3 Potential for miRNA-Based Interventions to Promote Healthy Aging

The utilization of miRNA-based interventions presents promising opportunities for the advancement of healthy aging and the extension of healthspan. Identifying and therapeutically targeting dysregulated miRNAs in age-related diseases present promising avenues for developing novel treatment approaches to effectively manage these conditions and enhance the overall well-being of aging individuals. Additionally, it is worth nothing that the utilization of miRNA-based strategies exhibits potential in the fields of regenerative medicine, tissue regeneration, and neuroprotection. These approaches aim to mitigate the effects of age-related tissue deterioration and promote cellular well-being, contributing to the promotion of healthy aging. The potential implementation of miRNA-based interventions, coupled with individualized strategies and ethical deliberations, holds promise for the advancing precision medicine that caters to the specific requirements of aging populations.

Compliance with Ethical Standards

Funding: none.

Conflict of Interest: The authors report no declarations of interest.

Ethical Approval: This chapter is a review of previously published accounts, as such, no animal or human studies were performed.

Author Contribution: Hai Duc Nguyen: Conceptualization, Methodology, Formal analysis, Investigation, Writing—original draft, Methodology, Investigation, Writing—review & editing, Visualization. Min-Sun Kim: Visualization.

References

Aalto AP et al (2018) Opposing roles of microRNA Argonautes during Caenorhabditis elegans aging. PLoS Genet 14:e1007379. https://doi.org/10.1371/journal.pgen.1007379
Absalon S et al (2013) MiR-26b, upregulated in Alzheimer's disease, activates cell cycle entry, tau-phosphorylation, and apoptosis in postmitotic neurons. J Neurosci 33:14645–14659. https://doi.org/10.1523/jneurosci.1327-13.2013
Akhter R et al (2018) Regulation of ADAM10 by miR-140-5p and potential relevance for Alzheimer's disease. Neurobiol Aging 63:110–119. https://doi.org/10.1016/j.neurobiolaging.2017.11.007
Alvarez-Saavedra M et al (2011) miRNA-132 orchestrates chromatin remodeling and translational control of the circadian clock. Hum Mol Genet 20:731–751. https://doi.org/10.1093/hmg/ddq519
Bai X-Y et al (2011) miR-335 and miR-34a promote renal senescence by suppressing mitochondrial antioxidative enzymes. J Am Soc Nephrol 22
Balasubramanian S et al (2020) MicroRNAs and xenobiotic toxicity: an overview. Toxicol Rep 7:583–595. https://doi.org/10.1016/j.toxrep.2020.04.010
Banzhaf-Strathmann J et al (2014) MicroRNA-125b induces tau hyperphosphorylation and cognitive deficits in Alzheimer's disease. EMBO J 33:1667–1680. https://doi.org/10.15252/embj.201387576
Barbato C et al (2020) Cognitive decline and modulation of Alzheimer's disease-related genes after inhibition of microRNA-101 in mouse hippocampal neurons. Mol Neurobiol 57:3183–3194. https://doi.org/10.1007/s12035-020-01957-8
Barros-Viegas AT et al (2020) miRNA-31 improves cognition and abolishes amyloid-β pathology by targeting APP and BACE1 in an animal model of Alzheimer's disease. Mol Ther Nucleic Acids 19:1219–1236. https://doi.org/10.1016/j.omtn.2020.01.010
Bhaskaran M, Mohan M (2014) MicroRNAs: history, biogenesis, and their evolving role in animal development and disease. Vet Pathol 51:759–774. https://doi.org/10.1177/0300985813502820
Bhattacharjee S et al (2016) MicroRNA-34a-mediated down-regulation of the microglial-enriched triggering receptor and phagocytosis-sensor TREM2 in age-related macular degeneration. PLoS One 11:e0150211. https://doi.org/10.1371/journal.pone.0150211

Bhaumik D et al (2009) MicroRNAs miR-146a/b negatively modulate the senescence-associated inflammatory mediators IL-6 and IL-8. Aging (Albany NY) 1:402–411. https://doi.org/10.18632/aging.100042
Boon RA et al (2013) MicroRNA-34a regulates cardiac ageing and function. Nature 495:107–110. https://doi.org/10.1038/nature11919
Bravo Vázquez LA et al (2021) The emerging role of microRNAs in bone diseases and their therapeutic potential. Molecules 27. https://doi.org/10.3390/molecules27010211
Brzeszczyńska J et al (2020) Role of microRNA in muscle regeneration and diseases related to muscle dysfunction in atrophy, cachexia, osteoporosis, and osteoarthritis. Bone Joint Res 9:798–807. https://doi.org/10.1302/2046-3758.911.Bjr-2020-0178.R1
Bu H et al (2017) MicroRNA regulation of oxidative stress-induced cellular senescence. Oxidative Med Cell Longev 2017:2398696. https://doi.org/10.1155/2017/2398696
Bueno MJ, Malumbres M (2011) MicroRNAs and the cell cycle. Biochim Biophys Acta 1812:592–601. https://doi.org/10.1016/j.bbadis.2011.02.002
Bushati N, Cohen SM (2007) MicroRNA functions. Annu Rev Cell Dev Biol 23:175–205. https://doi.org/10.1146/annurev.cellbio.23.090506.123406
Butovsky O et al (2015) Targeting miR-155 restores abnormal microglia and attenuates disease in SOD1 mice. Ann Neurol 77:75–99. https://doi.org/10.1002/ana.24304
Carrión RF et al (2022) Dna-methylation signature of microrna coding genes associated with chronological age in a high-cardiovascular risk mediterranean population. Sex-specific analysis. Am Heart J 254:256. https://doi.org/10.1016/j.ahj.2022.10.059
Chang TC et al (2007) Transactivation of miR-34a by p53 broadly influences gene expression and promotes apoptosis. Mol Cell 26:745–752. https://doi.org/10.1016/j.molcel.2007.05.010
Chawla G et al (2016) A let-7-to-miR-125 microRNA switch regulates neuronal integrity and lifespan in Drosophila. PLoS Genet 12:e1006247. https://doi.org/10.1371/journal.pgen.1006247
Chen Z et al (2015) Oxidative stress activates endothelial innate immunity via sterol regulatory element binding protein 2 (SREBP2) transactivation of microRNA-92a. Circulation 131:805–814. https://doi.org/10.1161/CIRCULATIONAHA.114.013675
Chen X-K et al (2021) Is exercise a senolytic medicine? A systematic review. Aging Cell 20:e13294. https://doi.org/10.1111/acel.13294
Cho HJ et al (2013) MicroRNA-205 regulates the expression of Parkinson's disease-related leucine-rich repeat kinase 2 protein. Hum Mol Genet 22:608–620. https://doi.org/10.1093/hmg/dds470
Condrat CE et al (2020) miRNAs as biomarkers in disease: latest findings regarding their role in diagnosis and prognosis. Cells 9. https://doi.org/10.3390/cells9020276
Crimmins EM (2015) Lifespan and healthspan: past, present, and promise. Gerontologist 55:901–911. https://doi.org/10.1093/geront/gnv130
Cron MA et al (2020) Decreased expression of miR-29 family associated with autoimmune myasthenia gravis. J Neuroinflamm 17:294. https://doi.org/10.1186/s12974-020-01958-3
Csiszar A et al (2014) Caloric restriction confers persistent anti-oxidative, pro-angiogenic, and anti-inflammatory effects and promotes anti-aging miRNA expression profile in cerebromicrovascular endothelial cells of aged rats. Am J Phys Heart Circ Phys 307:H292–H306. https://doi.org/10.1152/ajpheart.00307.2014
Cui JG et al (2010) Differential regulation of interleukin-1 receptor-associated kinase-1 (IRAK-1) and IRAK-2 by microRNA-146a and NF-kappaB in stressed human astroglial cells and in Alzheimer disease. J Biol Chem 285:38951–38960. https://doi.org/10.1074/jbc.M110.178848
da Costa JP et al (2016) A synopsis on aging-Theories, mechanisms and future prospects. Ageing Res Rev 29:90–112. https://doi.org/10.1016/j.arr.2016.06.005
Dalgaard LT et al (2022) The microRNA-29 family: role in metabolism and metabolic disease. Am J Physiol Cell Physiol 323:C367–c377. https://doi.org/10.1152/ajpcell.00051.2022
Davis C et al (2017a) MicroRNA-183-5p increases with age in bone-derived extracellular vesicles, suppresses bone marrow stromal (stem) cell proliferation, and induces stem cell senescence. Tissue Eng Part A 23:1231–1240. https://doi.org/10.1089/ten.TEA.2016.0525

Davis HM et al (2017b) Disruption of the Cx43/miR21 pathway leads to osteocyte apoptosis and increased osteoclastogenesis with aging. Aging Cell 16:551–563. https://doi.org/10.1111/acel.12586
de Lucia C et al (2017) MicroRNA in cardiovascular aging and age-related cardiovascular diseases. Front Med (Lausanne) 4:74. https://doi.org/10.3389/fmed.2017.00074
Deng B, Zheng Z, Zheng J, Yang W, Huang Y, Luo Y, Jin D, Shen L, Jin K, Wang Q (2021) FTD-PSP is an unusual clinical phenotype in a frontotemporal dementia patient with a novel progranulin mutation. Aging Dis 12(7):1741–1752. https://doi.org/10.14336/ad.2021.0309
Dias IHK et al (2018) miRNA 933 expression by endothelial cells is increased by 27-hydroxycholesterol and is more prevalent in plasma from dementia patients. J Alzheimers Dis 64:1009–1017. https://doi.org/10.3233/jad-180201
Dimmeler S, Nicotera P (2013) MicroRNAs in age-related diseases. EMBO Mol Med 5:180–190. https://doi.org/10.1002/emmm.201201986
Doxakis E (2010) Post-transcriptional regulation of alpha-synuclein expression by mir-7 and mir-153. J Biol Chem 285:12726–12734. https://doi.org/10.1074/jbc.M109.086827
Du WW et al (2014) miR-17 extends mouse lifespan by inhibiting senescence signaling mediated by MKP7. Cell Death Dis 5:e1355. https://doi.org/10.1038/cddis.2014.305
Du WW et al (2015) The microRNA miR-17-3p inhibits mouse cardiac fibroblast senescence by targeting Par4. J Cell Sci 128:293–304. https://doi.org/10.1242/jcs.158360
Duc Nguyen H (2023) Neurotherapeutic effects of quercetin and its metabolite compounds on cognitive impairment and Parkinson's disease: an in silico study. Eur J Drug Metab Pharmacokinet. https://doi.org/10.1007/s13318-023-00816-w
Duc Nguyen H et al (2022) Anti-inflammatory effects of B vitamins protect against tau hyperphosphorylation and cognitive impairment induced by 1,2 diacetyl benzene: An in vitro and in silico study. Int Immunopharmacol 108:108736. https://doi.org/10.1016/j.intimp.2022.108736
Ebert MS, Sharp PA (2010) MicroRNA sponges: progress and possibilities. RNA (New York, NY) 16:2043–2050. https://doi.org/10.1261/rna.2414110
Fang J et al (2014) Myeloid malignancies with chromosome 5q deletions acquire a dependency on an intrachromosomal NF-κB gene network. Cell Rep 8:1328–1338. https://doi.org/10.1016/j.celrep.2014.07.062
Faraonio R et al (2012) A set of miRNAs participates in the cellular senescence program in human diploid fibroblasts. Cell Death Differ 19:713–721. https://doi.org/10.1038/cdd.2011.143
Feng CZ et al (2017) Regulatory factor X1 depresses ApoE-dependent Aβ uptake by miRNA-124 in microglial response to oxidative stress. Neuroscience 344:217–228. https://doi.org/10.1016/j.neuroscience.2016.12.017
Feng MG et al (2018) MiR-21 attenuates apoptosis-triggered by amyloid-β via modulating PDCD4/PI3K/AKT/GSK-3β pathway in SH-SY5Y cells. Biomed Pharmacother 101:1003–1007. https://doi.org/10.1016/j.biopha.2018.02.043
Fenn AM et al (2013) Increased micro-RNA 29b in the aged brain correlates with the reduction of insulin-like growth factor-1 and fractalkine ligand. Neurobiol Aging 34:2748–2758. https://doi.org/10.1016/j.neurobiolaging.2013.06.007
Fleissner F et al (2010) Short communication: asymmetric dimethylarginine impairs angiogenic progenitor cell function in patients with coronary artery disease through a microRNA-21-dependent mechanism. Circ Res 107:138–143. https://doi.org/10.1161/CIRCRESAHA.110.216770
Fukuoka M et al (2021) MiR-199-3p enhances muscle regeneration and ameliorates aged muscle and muscular dystrophy. Commun Biol 4:427. https://doi.org/10.1038/s42003-021-01952-2
Fulzele S et al (2019) Muscle-derived miR-34a increases with age in circulating extracellular vesicles and induces senescence of bone marrow stem cells. Aging (Albany NY) 11:1791–1803. https://doi.org/10.18632/aging.101874
Gaine ME et al (2018) Sleep deprivation and the epigenome. Front Neural Circuits 12:14. https://doi.org/10.3389/fncir.2018.00014

Gao JX et al (2019a) Overexpression of microRNA-183 promotes apoptosis of substantia nigra neurons via the inhibition of OSMR in a mouse model of Parkinson's disease. Int J Mol Med 43:209–220. https://doi.org/10.3892/ijmm.2018.3982

Gao Y et al (2019b) miR-29 promotes the proliferation of cultured rat neural stem/progenitor cells via the PTEN/AKT signaling pathway. Mol Med Rep 20:2111–2118. https://doi.org/10.3892/mmr.2019.10445

Ge H et al (2019) MiR-410 exerts neuroprotective effects in a cellular model of Parkinson's disease induced by 6-hydroxydopamine via inhibiting the PTEN/AKT/mTOR signaling pathway. Exp Mol Pathol 109:16–24. https://doi.org/10.1016/j.yexmp.2019.05.002

Gendron CM, Pletcher SD (2017) MicroRNAs mir-184 and let-7 alter Drosophila metabolism and longevity. Aging Cell 16:1434–1438. https://doi.org/10.1111/acel.12673

Gerasymchuk M et al (2020) The role of microRNAs in organismal and skin aging. Int J Mol Sci 21(15):5281

Gong H et al (2022) miR-146a impedes the anti-aging effect of AMPK via NAMPT suppression and NAD+/SIRT inactivation. Signal Transduct Targeted Ther 7:66. https://doi.org/10.1038/s41392-022-00886-3

Guo Z et al (2014) Genome-wide survey of tissue-specific microRNA and transcription factor regulatory networks in 12 tissues. Sci Rep 4:5150. https://doi.org/10.1038/srep05150

Gupta SK et al (2016) Preclinical development of a microRNA-based therapy for elderly patients with myocardial infarction. J Am Coll Cardiol 68:1557–1571. https://doi.org/10.1016/j.jacc.2016.07.739

Gurha P et al (2013) microRNA-22 promotes heart failure through coordinate suppression of PPAR/ERR-nuclear hormone receptor transcription. PLoS One 8:e75882. https://doi.org/10.1371/journal.pone.0075882

Higaki S et al (2018) Defensive effect of microRNA-200b/c against amyloid-beta peptide-induced toxicity in Alzheimer's disease models. PLoS One 13:e0196929. https://doi.org/10.1371/journal.pone.0196929

Horita M et al (2021) The role of miR-29 family in disease. J Cell Biochem 122:696–715. https://doi.org/10.1002/jcb.29896

Hu Z et al (2014) MicroRNA-29 induces cellular senescence in aging muscle through multiple signaling pathways. Aging (Albany NY) 6:160–175. https://doi.org/10.18632/aging.100643

Hu CH et al (2017) miR-21 deficiency inhibits osteoclast function and prevents bone loss in mice. Sci Rep 7:43191. https://doi.org/10.1038/srep43191

Hu L et al (2018) The emerging role of microRNA-4487/6845-3p in Alzheimer's disease pathologies is induced by Aβ25-35 triggered in SH-SY5Y cell. BMC Syst Biol 12:119. https://doi.org/10.1186/s12918-018-0633-3

Huan T et al (2018) Age-associated microRNA expression in human peripheral blood is associated with all-cause mortality and age-related traits. Aging Cell 17:e12687. https://doi.org/10.1111/acel.12687

Huang Z et al (2019) Longitudinal comparative transcriptomics reveals unique mechanisms underlying extended healthspan in bats. Nat Ecol Evol 3:1110–1120. https://doi.org/10.1038/s41559-019-0913-3

Ibáñez-Ventoso C, Driscoll M (2009) MicroRNAs in C. elegans aging: molecular insurance for robustness? Curr Genomics 10:144–153. https://doi.org/10.2174/138920209788185243

Jazbutyte V et al (2013) MicroRNA-22 increases senescence and activates cardiac fibroblasts in the aging heart. Age (Dordr) 35:747–762. https://doi.org/10.1007/s11357-012-9407-9

Jiang W et al (2014) miR-146a ameliorates liver ischemia/reperfusion injury by suppressing IRAK1 and TRAF6. PLoS One 9:e101530. https://doi.org/10.1371/journal.pone.0101530

Jin L et al (2023) Osteoarthritis related epigenetic variations in miRNA expression and DNA methylation. BMC Med Genet 16:163. https://doi.org/10.1186/s12920-023-01597-6

Johanson TM et al (2013) MicroRNA-independent roles of the RNase III enzymes Drosha and Dicer. Open Biol 3:130144. https://doi.org/10.1098/rsob.130144

Kang Q et al (2017) MiR-124-3p attenuates hyperphosphorylation of Tau protein-induced apoptosis via caveolin-1-PI3K/Akt/GSK3β pathway in N2a/APP695swe cells. Oncotarget 8:24314–24326. https://doi.org/10.18632/oncotarget.15149

Ke K et al (2015) MicroRNA-183 increases osteoclastogenesis by repressing heme oxygenase-1. Bone 81:237–246. https://doi.org/10.1016/j.bone.2015.07.006

Kennerdell JR et al (2018) MiR-34 inhibits polycomb repressive complex 2 to modulate chaperone expression and promote healthy brain aging. Nat Commun 9:4188. https://doi.org/10.1038/s41467-018-06592-5

Khanna A et al (2011) Gain of survival signaling by down-regulation of three key miRNAs in brain of calorie-restricted mice. Aging (Albany NY) 3:223–236. https://doi.org/10.18632/aging.100276

Kim DH et al (2008) MicroRNA-directed transcriptional gene silencing in mammalian cells. Proc Natl Acad Sci USA 105:16230–16235. https://doi.org/10.1073/pnas.0808830105

Kim J et al (2016a) miR-27a and miR-27b regulate autophagic clearance of damaged mitochondria by targeting PTEN-induced putative kinase 1 (PINK1). Mol Neurodegener 11:55. https://doi.org/10.1186/s13024-016-0121-4

Kim J et al (2016b) miR-186 is decreased in aged brain and suppresses BACE1 expression. J Neurochem 137:436–445. https://doi.org/10.1111/jnc.13507

Kim J-H et al (2017) Reverse expression of aging-associated molecules through transfection of mirnas to aged mice. Mol Ther Nucleic Acids 6:106–115. https://doi.org/10.1016/j.omtn.2016.11.005

Kim C et al (2021) miR-181a-regulated pathways in T-cell differentiation and aging. Immun Ageing 18:28. https://doi.org/10.1186/s12979-021-00240-1

Kim SJ et al (2022) miR-146a dysregulates energy metabolism during neuroinflammation. J NeuroImmune Pharmacol 17:228–241. https://doi.org/10.1007/s11481-021-09999-y

Kinser HE, Pincus Z (2020) MicroRNAs as modulators of longevity and the aging process. Hum Genet 139:291–308. https://doi.org/10.1007/s00439-019-02046-0

Kong Y et al (2015) The role of miR-124 in Drosophila Alzheimer's disease model by targeting delta in notch signaling pathway. Curr Mol Med 15:980–989. https://doi.org/10.2174/1566524016666151123114608

Krammer UDB et al (2023) MiR-10a, miR-15a, let-7a, and let-7g expression as stress-relevant biomarkers to assess acute or chronic psychological stress and mental health in human capillary blood. Mol Biol Rep 50:5647–5654. https://doi.org/10.1007/s11033-023-08467-5

Kriegel AJ et al (2012) The miR-29 family: genomics, cell biology, and relevance to renal and cardiovascular injury. Physiol Genomics 44:237–244. https://doi.org/10.1152/physiolgenomics.00141.2011

Kukreti H, Amuthavalli K (2020) MicroRNA-34a causes ceramide accumulation and effects insulin signaling pathway by targeting ceramide kinase (CERK) in aging skeletal muscle. J Cell Biochem 121:3070–3089. https://doi.org/10.1002/jcb.29312

Laggerbauer B, Engelhardt S (2022) MicroRNAs as therapeutic targets in cardiovascular disease. J Clin Invest 132. https://doi.org/10.1172/JCI159179

Lai P et al (2016) Loss of Rictor with aging in osteoblasts promotes age-related bone loss. Cell Death Dis 7:e2408. https://doi.org/10.1038/cddis.2016.249

Lan T et al (2021) New insights into the interplay between miRNAs and autophagy in the aging of intervertebral discs. Ageing Res Rev 65:101227. https://doi.org/10.1016/j.arr.2020.101227

Layne TR et al (2019) MicroRNA detection in blood, urine, semen, and saliva stains after compromising treatments. J Forensic Sci 64:1831–1837. https://doi.org/10.1111/1556-4029.14113

Lee Y et al (2004) MicroRNA genes are transcribed by RNA polymerase II. EMBO J 23:4051–4060. https://doi.org/10.1038/sj.emboj.7600385

Lee KP et al (2015) miR-431 promotes differentiation and regeneration of old skeletal muscle by targeting Smad4. Genes Dev 29:1605–1617. https://doi.org/10.1101/gad.263574.115

Lee I et al (2017) The importance of standardization on analyzing circulating RNA. Mol Diagn Ther 21:259–268. https://doi.org/10.1007/s40291-016-0251-y

Leung AK, Sharp PA (2010) MicroRNA functions in stress responses. Mol Cell 40:205–215. https://doi.org/10.1016/j.molcel.2010.09.027
Li J, Wang H (2018) miR-15b reduces amyloid-β accumulation in SH-SY5Y cell line through targetting NF-κB signaling and BACE1. Biosci Rep 38. https://doi.org/10.1042/bsr20180051
Li N et al (2011a) Increased expression of miR-34a and miR-93 in rat liver during aging, and their impact on the expression of Mgst1 and Sirt1. Mech Ageing Dev 132:75–85. https://doi.org/10.1016/j.mad.2010.12.004
Li X et al (2011b) Circulatory miR34a as an RNAbased, noninvasive biomarker for brain aging. Aging (Albany NY) 3:985–1002. https://doi.org/10.18632/aging.100371
Li G et al (2012) Decline in miR-181a expression with age impairs T cell receptor sensitivity by increasing DUSP6 activity. Nat Med 18:1518–1524. https://doi.org/10.1038/nm.2963
Li SH et al (2013) miR-17 targets tissue inhibitor of metalloproteinase 1 and 2 to modulate cardiac matrix remodeling. FASEB J 27:4254–4265. https://doi.org/10.1096/fj.13-231688
Li CJ et al (2015) MicroRNA-188 regulates age-related switch between osteoblast and adipocyte differentiation. J Clin Invest 125:1509–1522. https://doi.org/10.1172/jci77716
Li D et al (2016a) Osteoclast-derived exosomal miR-214-3p inhibits osteoblastic bone formation. Nat Commun 7:10872. https://doi.org/10.1038/ncomms10872
Li S et al (2016b) MicroRNA-7 inhibits neuronal apoptosis in a cellular Parkinson's disease model by targeting Bax and Sirt2. Am J Transl Res 8:993–1004
Li M et al (2019a) Plant MIR167e-5p inhibits enterocyte proliferation by targeting β-catenin. Cells 8. https://doi.org/10.3390/cells8111385
Li Q et al (2019b) MicroRNA-101a regulates autophagy phenomenon via the MAPK pathway to modulate Alzheimer's-associated pathogenesis. Cell Transplant 28:1076–1084. https://doi.org/10.1177/0963689719857085
Li Z et al (2021) mRNA and microRNA stability validation of blood samples under different environmental conditions. Forensic Sci Int Genet 55:102567. https://doi.org/10.1016/j.fsigen.2021.102567
Li S et al (2023) The role of microRNAs in neurodegenerative diseases: a review. Cell Biol Toxicol 39:53–83. https://doi.org/10.1007/s10565-022-09761-x
Liu LF, Wang Y (2012) [Cellular function of microRNA-15 family]. Sheng Li Xue Bao 64:101–106
Liu N et al (2012) The microRNA miR-34 modulates ageing and neurodegeneration in Drosophila. Nature 482:519–523. https://doi.org/10.1038/nature10810
Liu Y et al (2015a) Ectopic expression of miR-494 inhibited the proliferation, invasion and chemoresistance of pancreatic cancer by regulating SIRT1 and c-Myc. Gene Ther 22:729–738. https://doi.org/10.1038/gt.2015.39
Liu Z et al (2015b) Therapeutic effects of transplantation of As-MiR-937-expressing mesenchymal stem cells in murine model of Alzheimer's disease. Cell Physiol Biochem 37:321–330. https://doi.org/10.1159/000430356
Liu W et al (2016) miR-106b inhibits tau phosphorylation at Tyr18 by targeting Fyn in a model of Alzheimer's disease. Biochem Biophys Res Commun 478:852–857. https://doi.org/10.1016/j.bbrc.2016.08.037
Liu Y et al (2019) The emerging role of the piRNA/piwi complex in cancer. Mol Cancer 18:123. https://doi.org/10.1186/s12943-019-1052-9
Liu T et al (2020a) Vegetarian diets, circulating miRNA expression and healthspan in subjects living in the Blue Zone. Precis Clin Med 3:245–259. https://doi.org/10.1093/pcmedi/pbaa037
Liu W et al (2020b) MiR-211 protects cerebral ischemia/reperfusion injury by inhibiting cell apoptosis. Bioengineered 11:189–200. https://doi.org/10.1080/21655979.2020.1729322
Liu C et al (2021) Honeysuckle-derived microRNA2911 inhibits tumor growth by targeting TGF-β1. Chin Med 16:49. https://doi.org/10.1186/s13020-021-00453-y
Lucanic M et al (2013) Age-related micro-RNA abundance in individual C. elegans. Aging (Albany NY) 5:394–411. https://doi.org/10.18632/aging.100564
Maes OC et al (2008) Murine microRNAs implicated in liver functions and aging process. Mech Ageing Dev 129:534–541. https://doi.org/10.1016/j.mad.2008.05.004

Maffioletti E et al (2021) Inflammation-related microRNAs are involved in stressful life events exposure and in trauma-focused psychotherapy in treatment-resistant depressed patients. Eur J Psychotraumatol 12:1987655. https://doi.org/10.1080/20008198.2021.1987655

Majidinia M et al (2020) MicroRNAs, DNA damage response and ageing. Biogerontology 21:275–291. https://doi.org/10.1007/s10522-020-09862-2

Menghini R et al (2009) MicroRNA 217 modulates endothelial cell senescence via silent information regulator 1. Circulation 120:1524–1532. https://doi.org/10.1161/CIRCULATIONAHA.109.864629

Modi PK et al (2016) Regulation of neuronal cell cycle and apoptosis by microRNA 34a. Mol Cell Biol 36:84–94. https://doi.org/10.1128/mcb.00589-15

Nguyen HD, Kim M-S (2022a) The effects of a mixture of cadmium, lead, and mercury on metabolic syndrome and its components, as well as cognitive impairment: genes, microRNAs, transcription factors, and sponge relationships. Biol Trace Elem Res. https://doi.org/10.1007/s12011-022-03343-y

Nguyen HD, Kim M-S (2022b) Exposure to a mixture of heavy metals induces cognitive impairment: genes and microRNAs involved. Toxicology. https://doi.org/10.1016/j.tox.2022.153164

Nguyen HD, Kim M-S (2022c) Roles of curcumin on cognitive impairment induced by a mixture of heavy metals. Neurotox Res. https://doi.org/10.1007/s12640-022-00583-7

Nguyen HD, Kim M-S (2023) In silico identification of molecular mechanisms for stroke risk caused by heavy metals and their mixtures: Sponges and drugs involved. NeuroToxicology. 96:222–239. https://doi.org/10.1016/j.neuro.2023.04.011

Nguyen HD et al (2022a) Curcumin-attenuated TREM-1/DAP12/NLRP3/caspase-1/IL1B, TLR4/NF-κB pathways, and tau hyperphosphorylation induced by 1,2-diacetyl benzene: an in vitro and in silico study. Neurotox Res. https://doi.org/10.1007/s12640-022-00535-1

Nguyen HD et al (2022b) In silico identification of the potential molecular mechanisms involved in protective effects of prolactin on motor and memory deficits induced by 1,2-Diacetylbenzene in young and old rats. NeuroToxicology 93:45–59. https://doi.org/10.1016/j.neuro.2022.09.002

Nweke EE, Brand M (2020) Downregulation of the let-7 family of microRNAs may promote insulin receptor/insulin-like growth factor signalling pathways in pancreatic ductal adenocarcinoma. Oncol Lett 20:2613–2620. https://doi.org/10.3892/ol.2020.11854

O'Brien J et al (2018) Overview of microRNA biogenesis, mechanisms of actions, and circulation. Front Endocrinol (Lausanne) 9:402. https://doi.org/10.3389/fendo.2018.00402

Olivieri F et al (2012) Age-related differences in the expression of circulating microRNAs: miR-21 as a new circulating marker of inflammaging. Mech Ageing Dev 133:675–685. https://doi.org/10.1016/j.mad.2012.09.004

Olivieri F et al (2021) miR-21 and miR-146a: The microRNAs of inflammaging and age-related diseases. Ageing Res Rev 70:101374. https://doi.org/10.1016/j.arr.2021.101374

Pardo PS et al (2017) MicroRNA-434-3p regulates age-related apoptosis through eIF5A1 in the skeletal muscle. Aging (Albany NY) 9:1012–1029. https://doi.org/10.18632/aging.101207

Qian Q et al (2019) Down-regulated expression of microRNA-338-5p contributes to neuropathology in Alzheimer's disease. FASEB J 33:4404–4417. https://doi.org/10.1096/fj.201801846R

Qin X et al (2021) MicroRNA-185 activates PI3K/AKT signalling pathway to alleviate dopaminergic neuron damage via targeting IGF1 in Parkinson's disease. J Drug Target 29:875–883. https://doi.org/10.1080/1061186x.2021.1886300

Raucci A, Vinci MC (2020) miR-34a: a promising target for inflammaging and age-related diseases. Int J Mol Sci 21. https://doi.org/10.3390/ijms21218293

Raucci A et al (2021) MicroRNA-34a: the bad guy in age-related vascular diseases. Cell Mol Life Sci 78:7355–7378. https://doi.org/10.1007/s00018-021-03979-4

Reddy PH et al (2017) MicroRNAs, aging, cellular senescence, and Alzheimer's disease. Prog Mol Biol Transl Sci 146:127–171. https://doi.org/10.1016/bs.pmbts.2016.12.009

Ripa R et al (2017) MicroRNA miR-29 controls a compensatory response to limit neuronal iron accumulation during adult life and aging. BMC Biol 15:9. https://doi.org/10.1186/s12915-017-0354-x

Rivas DA et al (2014) Diminished skeletal muscle microRNA expression with aging is associated with attenuated muscle plasticity and inhibition of IGF-1 signaling. FASEB J 28:4133–4147. https://doi.org/10.1096/fj.14-254490
Roman-Gomez J et al (2009) Epigenetic regulation of microRNAs in acute lymphoblastic leukemia. J Clin Oncol 27:1316–1322. https://doi.org/10.1200/JCO.2008.19.3441
Rottiers V, Näär AM (2012) MicroRNAs in metabolism and metabolic disorders. Nat Rev Mol Cell Biol 13:239–250. https://doi.org/10.1038/nrm3313
Roy SG (2021) Regulation of autophagy by miRNAs in human diseases. Nucleus (Calcutta) 64:317–329. https://doi.org/10.1007/s13237-021-00378-9
Sadri Nahand J et al (2021) Cell death pathways and viruses: Role of microRNAs. Mol Ther Nucleic Acids 24:487–511. https://doi.org/10.1016/j.omtn.2021.03.011
Salama RM et al (2020) Neuroprotective effect of crocin against rotenone-induced Parkinson's disease in rats: Interplay between PI3K/Akt/mTOR signaling pathway and enhanced expression of miRNA-7 and miRNA-221. Neuropharmacology 164:107900. https://doi.org/10.1016/j.neuropharm.2019.107900
Sedighi M et al (2019) Klotho ameliorates cellular inflammation via suppression of cytokine release and upregulation of miR-29a in the PBMCs of diagnosed Alzheimer's disease patients. J Mol Neurosci 69:157–165. https://doi.org/10.1007/s12031-019-01345-5
Sheu-Gruttadauria J, MacRae IJ (2017) Structural foundations of RNA silencing by argonaute. J Mol Biol 429:2619–2639. https://doi.org/10.1016/j.jmb.2017.07.018
Shi Z et al (2020) Increased miR-34c mediates synaptic deficits by targeting synaptotagmin 1 through ROS-JNK-p53 pathway in Alzheimer's disease. Aging Cell 19:e13125. https://doi.org/10.1111/acel.13125
Shukla GC et al (2011) MicroRNAs: processing, maturation, target recognition and regulatory functions. Mol Cell Pharmacol 3:83–92
Si C et al (2019) Circular RNA expression profile in human fibroblast premature senescence after repeated ultraviolet B irradiations revealed by microarray. J Cell Physiol 234:18156–18168. https://doi.org/10.1002/jcp.28449
Smith EJ et al (2015) Age-related loss of hepatic Nrf2 protein homeostasis: Potential role for heightened expression of miR-146a. Free Radic Biol Med 89:1184–1191. https://doi.org/10.1016/j.freeradbiomed.2015.11.003
Smith-Vikos T, Slack FJ (2012) MicroRNAs and their roles in aging. J Cell Sci 125:7–17. https://doi.org/10.1242/jcs.099200
Smith-Vikos T et al (2014) MicroRNAs mediate dietary-restriction-induced longevity through PHA-4/FOXA and SKN-1/Nrf transcription factors. Curr Biol 24:2238–2246. https://doi.org/10.1016/j.cub.2014.08.013
Smolarz B et al (2022) miRNAs in cancer (review of literature). Int J Mol Sci 23. https://doi.org/10.3390/ijms23052805
Song J, Kim YK (2017) Identification of the role of miR-142-5p in Alzheimer's disease by comparative bioinformatics and cellular analysis. Front Mol Neurosci 10:227. https://doi.org/10.3389/fnmol.2017.00227
Soriano-Arroquia A et al (2016) The functional consequences of age-related changes in microRNA expression in skeletal muscle. Biogerontology 17:641–654. https://doi.org/10.1007/s10522-016-9638-8
Su C et al (2016) Geniposide reduces α-synuclein by blocking microRNA-21/lysosome-associated membrane protein 2A interaction in Parkinson disease models. Brain Res 1644:98–106. https://doi.org/10.1016/j.brainres.2016.05.011
Sun X et al (2014) Role of miR-181 family in regulating vascular inflammation and immunity. Trends Cardiovasc Med 24:105–112. https://doi.org/10.1016/j.tcm.2013.09.002
Sun S et al (2018) MicroRNA-212-5p prevents dopaminergic neuron death by inhibiting SIRT2 in MPTP-induced mouse model of Parkinson's disease. Front Mol Neurosci 11:381. https://doi.org/10.3389/fnmol.2018.00381

Sun L et al (2020) MiR-107 overexpression attenuates neurotoxicity induced by 6-hydroxydopamine both in vitro and in vivo. Chem Biol Interact 315:108908. https://doi.org/10.1016/j.cbi.2019.108908

Surina S et al (2021) miR-21 in human cardiomyopathies. Front Cardiovasc Med 8:767064. https://doi.org/10.3389/fcvm.2021.767064

Taganov KD et al (2006) NF-kappaB-dependent induction of microRNA miR-146, an inhibitor targeted to signaling proteins of innate immune responses. Proc Natl Acad Sci USA 103:12481–12486. https://doi.org/10.1073/pnas.0605298103

Talebi F et al (2017) MicroRNA-142 regulates inflammation and T cell differentiation in an animal model of multiple sclerosis. J Neuroinflamm 14:55. https://doi.org/10.1186/s12974-017-0832-7

Tang H et al (2019) Activation of MT2 receptor ameliorates dendritic abnormalities in Alzheimer's disease via C/EBPα/miR-125b pathway. Aging Cell 18:e12902. https://doi.org/10.1111/acel.12902

Teoh SL, Das S (2022) MicroRNAs in various body fluids and their importance in forensic medicine. Mini Rev Med Chem 22:2332–2343. https://doi.org/10.2174/1389557522666220303141558

Tsitsiou E, Lindsay MA (2009) microRNAs and the immune response. Curr Opin Pharmacol 9:514–520. https://doi.org/10.1016/j.coph.2009.05.003

Ugalde AP et al (2011) Aging and chronic DNA damage response activate a regulatory pathway involving miR-29 and p53. EMBO J 30:2219–2232. https://doi.org/10.1038/emboj.2011.124

Ultimo S et al (2018) Influence of physical exercise on microRNAs in skeletal muscle regeneration, aging and diseases. Oncotarget 9:17220–17237. https://doi.org/10.18632/oncotarget.24991

van Almen GC et al (2011) MicroRNA-18 and microRNA-19 regulate CTGF and TSP-1 expression in age-related heart failure. Aging Cell 10:769–779. https://doi.org/10.1111/j.1474-9726.2011.00714.x

van Deursen JM (2019) Senolytic therapies for healthy longevity. Science 364:636–637. https://doi.org/10.1126/science.aaw1299

Verma P et al (2015) A neuroprotective role for microRNA miR-1000 mediated by limiting glutamate excitotoxicity. Nat Neurosci 18:379–385. https://doi.org/10.1038/nn.3935

Vora M et al (2013) Deletion of microRNA-80 activates dietary restriction to extend C. elegans healthspan and lifespan. PLoS Genet 9:e1003737. https://doi.org/10.1371/journal.pgen.1003737

Wang L, Zhang L (2020) Circulating exosomal miRNA as diagnostic biomarkers of neurodegenerative diseases. Front Mol Neurosci 13:53. https://doi.org/10.3389/fnmol.2020.00053

Wang G et al (2008) Variation in the miRNA-433 binding site of FGF20 confers risk for Parkinson disease by overexpression of alpha-synuclein. Am J Hum Genet 82:283–289. https://doi.org/10.1016/j.ajhg.2007.09.021

Wang X et al (2013) miR-214 targets ATF4 to inhibit bone formation. Nat Med 19:93–100. https://doi.org/10.1038/nm.3026

Wang J et al (2015a) Genome-wide circulating microRNA expression profiling indicates biomarkers for epilepsy. Sci Rep 5:9522. https://doi.org/10.1038/srep09522

Wang X et al (2015b) MicroRNA-138 promotes tau phosphorylation by targeting retinoic acid receptor alpha. FEBS Lett 589:726–729. https://doi.org/10.1016/j.febslet.2015.02.001

Wang K et al (2022a) Epigenetic regulation of aging: implications for interventions of aging and diseases. Signal Transduct Targeted Ther 7:374. https://doi.org/10.1038/s41392-022-01211-8

Wang Y et al (2022b) Let-7 as a promising target in aging and aging-related diseases: a promise or a pledge. Biomolecules 12. https://doi.org/10.3390/biom12081070

Weilner S et al (2016) Secreted microvesicular miR-31 inhibits osteogenic differentiation of mesenchymal stem cells. Aging Cell 15:744–754. https://doi.org/10.1111/acel.12484

Williams J et al (2017) Are microRNAs true sensors of ageing and cellular senescence? Ageing Res Rev 35:350–363. https://doi.org/10.1016/j.arr.2016.11.008

Winkle M et al (2021) Noncoding RNA therapeutics—challenges and potential solutions. Nat Rev Drug Discov 20:629–651. https://doi.org/10.1038/s41573-021-00219-z

Wu Q et al (2019) MicroRNA-599 regulates the development of Parkinson's disease through mediating LRRK2 expression. Eur Rev Med Pharmacol Sci 23:724–731. https://doi.org/10.26355/eurrev_201901_16886

Wu L et al (2021) Integrated multi-omics for novel aging biomarkers and antiaging targets. Biomolecules 12. https://doi.org/10.3390/biom12010039

Xiong R et al (2014) MicroRNA-494 reduces DJ-1 expression and exacerbates neurodegeneration. Neurobiol Aging 35:705–714. https://doi.org/10.1016/j.neurobiolaging.2013.09.027

Xu H et al (2019a) Post-transcriptional regulation of soluble guanylate cyclase that governs neuropathic pain in Alzheimer's disease. J Alzheimers Dis 71:1331–1338. https://doi.org/10.3233/jad-190743

Xu N et al (2019b) miR-132 regulates the expression of synaptic proteins in APP/PS1 transgenic mice through C1q. Eur J Histochem 63. https://doi.org/10.4081/ejh.2019.3008

Yan M et al (2021a) Expression and clinical significance of miR-146a and tumor necrosis factor receptor-associated factor 6 (TRAF6) in myasthenia gravis patient serum. Biomed Res Int 2021:5573469. https://doi.org/10.1155/2021/5573469

Yan X et al (2021b) The effect of caloric restriction on the increase in senescence-associated T cells and metabolic disorders in aged mice. PLoS One 16:e0252547. https://doi.org/10.1371/journal.pone.0252547

Yang N et al (2013) Tumor necrosis factor α suppresses the mesenchymal stem cell osteogenesis promoter miR-21 in estrogen deficiency-induced osteoporosis. J Bone Miner Res 28:559–573. https://doi.org/10.1002/jbmr.1798

Yang G et al (2015a) MicroRNA-29c targets β-site amyloid precursor protein-cleaving enzyme 1 and has a neuroprotective role in vitro and in vivo. Mol Med Rep 12:3081–3088. https://doi.org/10.3892/mmr.2015.3728

Yang Y et al (2015b) MicroRNA-34a plays a key role in cardiac repair and regeneration following myocardial infarction. Circ Res 117:450–459. https://doi.org/10.1161/circresaha.117.305962

Yuan J et al (2017) Mir-21 promotes cardiac fibrosis after myocardial infarction via targeting Smad7. Cell Physiol Biochem 42:2207–2219. https://doi.org/10.1159/000479995

Zhai J et al (2022) Caloric restriction induced epigenetic effects on aging. Front Cell Dev Biol 10:1079920. https://doi.org/10.3389/fcell.2022.1079920

Zhang C, Yi R (2021) Inhibition of microRNA turns back the CLOCK of hair follicle aging. Nat Aging 1:753–754. https://doi.org/10.1038/s43587-021-00101-y

Zhang JA et al (2016) MiR-23a-depressed autophagy is a participant in PUVA- and UVB-induced premature senescence. Oncotarget 7:37420–37435. https://doi.org/10.18632/oncotarget.9357

Zhao ZB et al (2014) MicroRNA-922 promotes tau phosphorylation by downregulating ubiquitin carboxy-terminal hydrolase L1 (UCHL1) expression in the pathogenesis of Alzheimer's disease. Neuroscience 275:232–237. https://doi.org/10.1016/j.neuroscience.2014.06.013

Zhao C et al (2015a) miR-214 promotes osteoclastogenesis by targeting Pten/PI3k/Akt pathway. RNA Biol 12:343–353. https://doi.org/10.1080/15476286.2015.1017205

Zhao W et al (2015b) MiR-21 overexpression improves osteoporosis by targeting RECK. Mol Cell Biochem 405:125–133. https://doi.org/10.1007/s11010-015-2404-4

Zhou Z et al (2015) Honeysuckle-encoded atypical microRNA2911 directly targets influenza A viruses. Cell Res 25:39–49. https://doi.org/10.1038/cr.2014.130

Zhou XL et al (2018) miR-21 promotes cardiac fibroblast-to-myofibroblast transformation and myocardial fibrosis by targeting Jagged1. J Cell Mol Med 22:3816–3824. https://doi.org/10.1111/jcmm.13654

Zhu J et al (2018a) Inhibition of microRNA-505 suppressed MPP+-induced cytotoxicity of SHSY5Y cells in an in vitro Parkinson's disease model. Eur J Pharmacol 835:11–18. https://doi.org/10.1016/j.ejphar.2018.07.023

Zhu J et al (2018b) Overexpression of miR-153 promotes oxidative stress in MPP(+)-induced PD model by negatively regulating the Nrf2/HO-1 signaling pathway. Int J Clin Exp Pathol 11:4179–4187

Zhu J et al (2021a) Downregulation of microRNA-15b-5p targeting the Akt3-mediated GSK-3β/β-catenin signaling pathway inhibits cell apoptosis in Parkinson's disease. Biomed Res Int 2021:8814862. https://doi.org/10.1155/2021/8814862

Zhu M et al (2021b) Inhibition of miR-96-5p may reduce Aβ42/Aβ40 ratio via regulating ATP-binding cassette transporter A1. J Alzheimers Dis 83:367–377. https://doi.org/10.3233/jad-210411

Zou H et al (2015) MicroRNA-29c/PTEN pathway is involved in mice brain development and modulates neurite outgrowth in PC12 cells. Cell Mol Neurobiol 35:313–322. https://doi.org/10.1007/s10571-014-0126-x

Zovoilis A et al (2011) microRNA-34c is a novel target to treat dementias. EMBO J 30:4299–4308. https://doi.org/10.1038/emboj.2011.327

Zuo X et al (2019) MicroRNA-132 attenuates cerebral injury by protecting blood-brain-barrier in MCAO mice. Exp Neurol 316:12–19. https://doi.org/10.1016/j.expneurol.2019.03.017

Chapter 11
MicroRNAs in Viral Infection

Shima Poorghobadi, Maryam Ghanbari, and Kazem Baesi

Abstract MiRNAs can regulate the cytotoxicity of viruses by controlling viral protein expression involved in innate and acquired immune pathways. They directly target viral transcripts and can either inhibit or induce viral replication. In addition, viruses have their miRNAs. Viruses translate miRNAs in their genomes to enhance viral replication, control viral latency, and reduce factors that increase inflammatory responses and leukocyte recruitment. Similar to cellular miRNAs, viral miRNAs (v-miRNAs) interact with the 3′UTR of the target mRNA.

Keywords Cellular miRNAs · Viral miRNAs · Viral replication · Viral disease · Viruses

List of Abbreviations

ACE2	Angiotensin converting enzyme 2
AGO	Argonaute
AIDS	Acquired immunodeficiency syndrome
AKT	Ak strain transforming
ALV	Avian leukosis virus
ATP6V0C	ATPase H+ transporting V0 subunit C
BAX	Bcl2-associated X protein
BDV	Borna disease virus
BLV	Bovine leukosis virus

S. Poorghobadi · K. Baesi (✉)
Hepatitis and AIDS Department, Pasteur Institute of Iran, Tehran, Iran
e-mail: k_baesi@pasteur.ac.ir

M. Ghanbari
Department of Microbial Biotechnology, Faculty of Biological Science, Tehran North Branch, Islamic Azad University, Tehran, Iran

L. M. Vaschetto (ed.), *miRNAs, Human Health and Diseases*, Epigenetics and Human Health 13, https://doi.org/10.1007/978-3-031-64788-8_11

C/EBP	CCAAT/enhancer-binding protein beta
CLDN-1	Claudin 1
CMV	Cytomegalovirus
COVID-19	Coronavirus disease-2019
COX2	Cyclooxygenase
CXCL	Chemokine (C-X-C motif) ligand
CXCL12	C-X-C motif chemokine 12
DAG1	Dystroglycan 1
DENV	Dengue virus
DNMT 3a and 3b	DNA methyl transferase 3a, 3b
EBNA2	EBV nuclear antigen 2
EBV	Epstein-Barr virus
eIF4A1	Eukaryotic translation initiation factor 4A1
FADD	Fas associated via death domain
FGFR2	Fibroblast growth factor receptor 2
HCMV	Human cytomegalovirus
HDFs	HIV-1 dependent factors
HIV	Human immunodeficiency virus
HPV	Human papillomavirus
HSCs	Hematopoietic stem cells
HSV	Herpes simplex virus
HUVECs	Human umbilical vein endothelial cells
ICP0	Infected cell protein 0
IFN	Interferon
JARID1B	Lysine-specific demethylase 1B
JEV	Japanese encephalitis virus
KDM5B	Lysine demethylase 5B
KSHV	Kaposi's sarcoma-associated herpesvirus
LCR	Locus control regions
LMP1	Latent membrane protein 1
MAPK	Mitogen-activated protein kinase
MeCP2	Methyl CpG-binding protein 2
MyD88	Myeloid differentiation primary response 88
NLRP3	Nucleotide-binding domain, leucine-rich-containing family, pyrin domain-containing-3
PGF-2	Prostaglandin
PI3K	Phosphoinositide 3-kinases
PRRSV	Porcine reproductive and respiratory syndrome virus
RIPK2	Receptor interacting serine/threonine kinase 2
RTA	Replication and transcription activator
SARS-CoV-2	Severe acute respiratory syndrome coronavirus 2
SIRT1	Sirtuin
TFPI	Tissue factor pathway inhibitor
TGF-β	Transforming growth factor β

TGFBR1	Transforming growth factor beta receptor I
TMPRSS2	Transmembrane serine protease 2
v-miRNAs	Viral miRNAs
WNV	West Nile virus
XRN1	5′–3′ exoribonuclease 1

11.1 Introduction

Viruses are considered one of the most severe human pathogens, contributing to the deaths of approximately one million people each year (Nahand et al. 2021). In the early stages of viral infection, host cell sensors detect viral products and activate signaling pathways to trigger an antiviral response. MiRNAs are part of the immune response and viruses can manipulate host miRNA levels to evade antiviral or pro-viral responses (Bruscella et al. 2017). They directly target viral transcripts and can either inhibit or induces viral replication (Barbu et al. 2020). The identification of viral miRNA targets and their function on cells is important and several miRNAs have been detected in viruses during infection in host cells (Saliminejad et al. 2019). Viruses have their own miRNAs. These viral miRNAs (v-miRNAs) alter the expression of host genes as well as the viral genes that affect its replication, latency, and immune system evasion (Mishra et al. 2020). Thus, miRNAs mediate viral–host interactions. Changes in miRNA expression can lead to immune system dysfunction (Naqvi et al. 2018). The first viral miRNAs were discovered in EBV (Epstein-Barr virus), a DNA virus, belonging to the herpesvirus family which has been shown to play a role in viral pathogenesis (Wei et al. 2013). In general, DNA viruses tend to encode functional miRNAs and RNA viruses such as influenza viruses, which replicate in the nucleus, are less able to encode v-miRNAs (Tycowski et al. 2015). However, HIV (human immunodeficiency virus), a retrovirus, can utilize host miRNAs for its replication (Sadri Nahand et al. 2020).

11.2 Role of miRNAs in Human Immunodeficiency Virus (HIV) Infection

HIV-1 virus infects humans and significantly reduces the number of CD4 T cells, making infected people very susceptible to AIDS (acquired immunodeficiency syndrome) (Poorghobadi et al. 2023). Some host cell miRNAs directly target v-mRNAs and inhibit translation while others indirectly influence host cell factors that regulate HIV-1 integration and transcription (Sadri Nahand et al. 2020). Certain miRNAs, such as miR-28, miR-125b, miR-150, and miR-223, modulate HIV-1 infection in resting CD4+ T cells (Fig. 11.1). MiR-382 is also capable of binding to the 3′UTR of v-mRNA (Huang et al. 2007). Activation of resting CD4+ T cells leads to

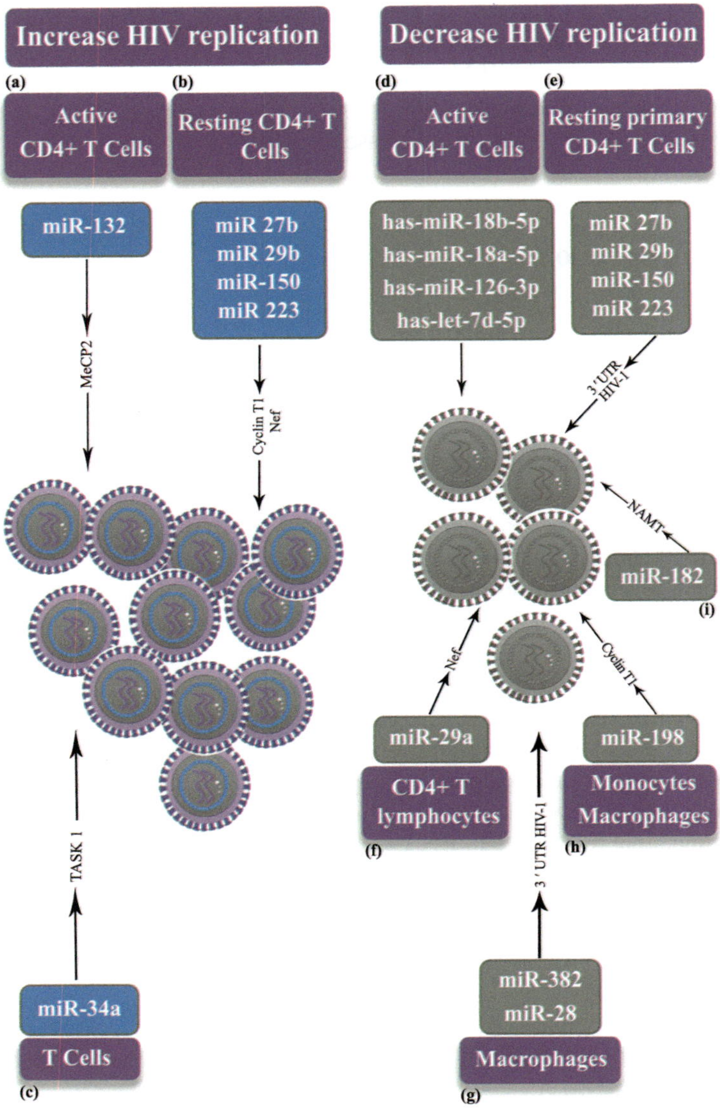

Fig. 11.1 Role of miRNAs in human immunodeficiency virus (HIV) infection. (**a**) Expression of miR-132 in active CD+ T cells and targeting of methyl CpG binding protein 2 (MeCP2) could be associated with enhanced HIV-1 replication. (**b**) In resting CD4+ T cells, miR 27b, miR 29b, miR-150, and miR 223 inhibit cyclin T1 and nef which leads to a high level of HIV-1 replication. (**c**) Expression of miR-34a in T cells via inhibiting TASK1 promotes HIV-1 replication and virus spread. (**d**) Increased expression of hsa-miRNA-18b-5p, hsa-miRNA-126-3p, hsa-let-7d-3p, and hsa-miRNA-18a-5p is associated to decreased HIV-1 replication and disease progression. (**e**) In resting primary CD4+ T cells, miR 27b, miR 29b, miR-150, and miR 223 by targeting the HIV-1 3′UTR decrease replication of virus. (**f**) In CD4+ T cell lymphocytes, miR-29a inhibits HIV-1 replication by targeting the 3′UTR of the nef gene. (**g**) Expression of miR-382 and miR-28 in monocytes is associated with promoted HIV-1 replication while in macrophages, the expression of these two miRNAs leads to a decrease in virus replication. (**h**) Expression of miR-198 in monocytes and macrophages which targets cyclin T1 inhibits HIV-1 replication. (**i**) mir-182 via inhibiting NAMT could decrease HIV-1 replication

the downregulation of miRNAs, which increases susceptibility to HIV infection. It has been suggested that these newly identified miRNAs may have anti-HIV-1 properties and could be used to control HIV-1 latency and clear viral reservoirs (Huang et al. 2007; Sadri Nahand et al. 2020). The HIV-1 virus requires HDFs (HIV-1 dependent factors) to progress through different stages in the viral life cycle. Therefore, the host cell miRNAs target and repress HDFs to inhibit HIV-1 replication (Piedade and Azevedo-Pereira 2016). Some miRNAs, such as miR-132, have been reported to enhance HIV-1 infection. MiR-132 expression increases significantly in activated CD4+ T cells than in resting cells. Furthermore, overexpression of miR-132 in Jurkat T cells also increases the prevalence of HIV-1 infection. This enhancement is attributed to the downregulation of MeCP2 (methyl CpG-binding protein 2), which increases HIV replication (Chiang et al. 2013). One miRNA that plays an important role in HIV replication is miR-198, which targets cyclin T1 mRNA in monocytes and inhibits HIV replication (Kaushik and Kumar 2022). MiR-198 can inhibit the expression of Cyclin T1, which influences the amount of HIV-1 produced in infected cells. The expression level of miR-198 is significantly reduced in macrophages, and this may be related to cyclin T1 translation and expression. It can be concluded that this miRNA might limit the spread of HIV-1 in monocytes and reduce HIV-1 replication by inhibiting other critical cellular pathways (Frattari et al. 2017; Sung and Rice 2009). In resting CD4 T cells, miR 27b, miR 29b, miR-150, and miR 223 inhibit cyclin T1. Activation of resting CD4 T cells reduces the levels of these four miRNAs, thereby disinhibiting cyclin T1 and leading to a high level of HIV replication (Chiang et al. 2012). Mir-29a inhibits HIV replication by targeting the 3′UTR of the nef gene. Nef is one of the major proteins of HIV, which is encoded by the nef gene and plays an important role in viral replication and the development of AIDS (Frattari et al. 2017). Moreover, miR-182 positively alters HIV replication by targeting NAMPT, leading to changes in the expression of SIRT1(sirtuin) (Chen et al. 2013). A total of 16 miRNAs whose expression patterns differ between healthy individuals and those with HIV have been identified. Some of these miRNAs play a role in HIV infection or immune system function. The level of several of these miRNAs decreases in the plasma of infected individuals compared to healthy individuals, such as hsa-miRNA-29b-3p, hsa-miRNA-33a-5p, hsa-miRNA-146a-5p, and hsa-miRNA-223 (Reynoso et al. 2014). Increased expression of miRNA-223, miRNA-382, miRNA-125b, and miRNA-28 leads to a decrease in HIV replication. Decreased expression of miRNAs miRNA-31, miRNA-572, miRNA-146b-5p, miRNA-16, miRNA-191, miRNA-150, and miRNA-223 was also confirmed in HIV infected individuals. The increased expression of miRNA-122 and miRNA-21, which are associated with liver cirrhosis and pulmonary arterial hypertension in AIDS patients, is likely caused by interference in the TGF-β (transforming growth factor β) signaling pathway. Additionally, miR-122, miR-22, and miR-34a are associated with liver damage in HIV patients. The increased expression of miRNA-200a, miRNA-34a, miRNA-12, and Let-7 has been shown to induce inflammatory responses. Moreover, increased plasma levels of hsa-miRNA-18b-5p, hsa-miRNA-126-3p, hsa-let-7d-3p, hsa-miRNA-18a-5p, and decreased levels of hsa-miR-424-5p and hsa-miR-34a-5p are associated to the decrease in T CD4+ cells and disease progression (Ojha et al. 2019).

11.3 The Role of miRNAs in Hepatitis C Virus (HCV) Infection

Hepatitis C virus (HCV) is a single-stranded RNA virus of the genus *Hepacivirus* with six major genotypes. HCV transmission occurs through blood, such as intravenous drug use, blood transfusions, risky sexual behavior, and mother-to-child transmission. These are the main causes of HCV infection (Pascut et al. 2021). Although HCV infection is associated with intracellular miRNA changes, the HCV genome does not encode any identified miRNA (Piedade and Azevedo-Pereira 2016). The exact mechanism by which HCV infection causes liver disease is unknown. However, miRNAs have been shown to play key roles in many processes during HCV infection. MiR-122 is one of the liver-specific miRNAs that exert important effects on the life cycle and pathogenesis of HCV (Pascut et al. 2021). A mutation in the miR-122 binding site reduces HCV RNA stability and viral replication. This miRNA stabilizes the viral genome by binding to the non-coding region of the 5′UTR of the viral genome, acting as a cap in collaboration with AGO (Argonaute) proteins, and creating a sponge effect at the site of HCV infection, ultimately preventing genome destruction by the exoribonuclease 1 (XRN1). As a result, the accumulation of viral RNA gradually leads to hepatocellular carcinoma (Hartling et al. 2016). It also plays an important role in hepatocyte homeostasis regulation, fatty acid, and cholesterol metabolism (Hartling et al. 2016). The interaction between HCV and miR-122 prevents excessive viral production and chronic infections (Hartling et al. 2016). Other miRNAs, such as miR-21-5p and miR-448, influence HCV replication. MiR-130a and miR-141 induce viral replication by inhibiting the IFN (Interferon) signaling pathway, while miR-196, miR-199a, miR-221, miR-448, and Let-7b reduce viral replication in the early stages of infection (Cheng et al. 2013; Hou et al. 2010). In later stages, miR-21, miR-155, and miR-200c contribute to infection and liver fibrosis (Hartling et al. 2016). Conversely, miR-449a, miR-107 and miR-29 have antifibrotic effects. Exosomes derived from HCV-infected hepatocytes contain high levels of miR-19a and miR-192. These exosomes can be transferred to HSCs (hematopoietic stem cells) and cause the upregulation of fibrogenic markers such as smooth muscle actin and TGF-1 (transforming growth factor), leading to liver fibrosis. Therefore, exosome-derived miR-192 and miR-19a could be potential therapeutic targets for HCV-induced liver fibrosis (Devhare et al. 2017; Kim et al. 2019). MiR-155 and miR-182 inhibit HCV entry into host cells by targeting CLDN-1 (Claudin 1) mRNA (Bala et al. 2012; Riad et al. 2018). Furthermore, the downregulation of miR-149, miR-194, miR-373, miR-638, miR-888, miR-940, miR-1181, and miR-1234 contributes to HCV pathogenicity (Joshi et al. 2022). MiR-199 and Let-7b can inhibit HCV virus replication by targeting specific sites. MiR-199 targets the ribosomal entry site, while Let-7b targets 5′UTR of NS5B in HCV (Cheng et al. 2012). MiR-181c reduces HCV infectivity by altering C/EBP (CCAAT/Enhancer-Binding Protein Beta) (Mukherjee et al. 2014). These miRNAs belong to the category of microRNAs that interfere with the viral genome and influence viral pathogenesis. For example, miR-451, Let-7, miR-135, and miR-99 bind to the factors CXCL12 (C-X-C motif chemokine 12), MyD88 (Myeloid differentiation primary response 88), TGFBR1 (Transforming growth factor beta receptor I), and RIPK2 (receptor interacting serine/threonine kinase 2) which are involved in the translation and cleavage of viral proteins (Joshi et al. 2022). Excessive levels of miR-155 have been detected in HCV-infected patients. This could activate the Wnt signaling pathway and contribute to the development of HCV-induced liver cancer. However, virus infection and replication do not influence the expression of miR-155 (Joshi et al. 2022; Vigorito et al. 2013). MiR-155 is an important regulator of the immune system and plays a role in various immune processes, including lymphocyte homeostasis, chronic inflammation, and immunological memory (Vigorito et al. 2013). The HCV virus can downregulate Let-7b in the Huh-7.5 cell line and the liver of individuals. This suggests that HCV can counteract the antiviral effects of this specific miRNA (Cheng et al. 2012). The expression levels of miRNA-638 and miRNA-149 increase following HCV infection. The reduced expression of these miRNAs may affect various signaling pathways, including PI3K-AKT (Phosphoinositide 3-kinases-Ak strain transforming), MAPK (Mitogen-activated protein kinase), TGF-β, Wnt, T cell receptor signaling, chemokines, and virus-induced carcinogenesis. Altered signaling pathways lead to immune system dysfunction and various cell differentiation disorders (Ojha et al. 2019).

11.4 The Role of miRNAs in Influenza

Influenza is a global health problem caused by the influenza virus, which belongs to the Orthomyxoviridae family (Naqvi et al. 2018). The expression profile of microRNAs in individuals infected with the influenza virus differs significantly from that in healthy individuals. Several miRNAs (e.g., miR-323, miR-491, and miR-654) can inhibit the BAX (Bcl2-associated X protein) gene, thereby preventing the spread of H1N1 influenza. Conversely, miR-155 is upregulated during influenza infection and activates type 1 interferon signaling, ultimately triggering the host innate immune response in macrophages. One study found that miR-29 levels increase during early stages of influenza A, inhibiting DNMT3a and DNMT3b (DNA methyl transferases 3a and 3b) and simultaneously inducing the expression of COX2 (Cyclooxygenase) and PGF-2 (Prostaglandin). This pathway ultimately leads to the production of IFN-1 as an important factor in activating the immune response (Fang et al. 2012). MiR-7, miR-132, miR-146a, miR-187, miR-200c, and miR-1275 are shown to reduce the expression of IRAK1 and MAPK antiviral proteins, which in turn increases virus replication (Naqvi et al. 2018). In 2013, researchers identified 14 miRNAs with significant altered expression changes in the blood of infected people. Of these, five miRNAs (miR-1260, miR-335, miR-664 miR-26a, and miR-576-3p) showed overexpression patterns (Tambyah et al. 2013).

11.5 The Role of miRNAs in Coronavirus Disease-2019 (COVID 19)

Coronaviruses are a type of single-stranded RNA virus. Symptoms of the new coronavirus include fever, cough, and severe pneumonia (Fani et al. 2021). Certain miRNAs expressed in airway cells have been found to either induce or inhibit respiratory coronavirus infections. These miRNAs play roles in viral replication, pathogenesis, and suppression of antiviral responses. This suggests that cellular miRNAs could be potential candidates for the development of miRNA-based therapies for coronavirus diseases (Fani et al. 2021). Excessive expression of a particular microRNA in the lung epithelium can protect against infection and the spread of viruses, while low expression can make individuals more susceptible to infection. Researchers have identified 128 human microRNAs that can potentially target the SARS-CoV-2 (Severe acute respiratory syndrome coronavirus 2) genome. Of these, six microRNAs showed differential expression during in vitro infection. Four microRNAs showed decreased expression (let-7a-3p, miR-135b-5p, miR-16-2-3p, and miR1275), while two exhibited increased expression (miR-155-3p and miR-139-5p) (Kaur and Kaur 2023). Li et al. examined 73 miRNAs and discovered that 35 miRNAs showed overexpression in SARS-CoV-2 infected individuals, while 38 miRNAs exhibited lower expression compared to healthy individuals (Li et al. 2020). ACE2 (Angiotensin Converting Enzyme 2) has been proposed as a cellular receptor of SARS CoV-2. The virus uses the ACE2 receptor to enter cells and the viral S protein is then broken down using the host proteases. Then, TMPRSS2 (transmembrane serine protease 2) penetrates the ACE2-virus spike complex (Davidson et al. 2020). Therefore, inhibiting ACE2 and TMPRSS2 can prevent virus entry into the cell, ultimately rendering useful outcomes to identify microRNAs capable of inhibiting/blocking these two factors (Lin et al. 2023). MiR-200c and mir-125b reduce the risk of infection by decreasing ACE2 expression (Lin et al. 2023). The JARID1B) Lysine-specific demethylase 1B (gene encodes KDM5B (lysine demethylase 5B) and may indirectly influence ACE2/TMPRSS2 expression by repressing the transcription of hsa-let-7e/hsa-mir-125a and hsa-mir-141. Some miRNAs influence ACE2 and TMPRSS2 expression, while others bind directly to the genome sequence of the virus and affect its virulence (Chow and Salmena 2020; Nersisyan et al. 2020; Widiasta et al. 2020). Through bioinformatics analysis, some host miRNAs such as miR-17-5p, miR-20b-5p and miR-323a-5p and miR-21-3p were found to be attached to different RNA transcripts of SARS-CoV-2 (M, N, ORF3a, ORF1ab, and the S gene) and have a significant impact on reducing virus replication (Yang et al. 2022). SARS-CoV-2 can alter the expression patterns of the host cell's miRNAs, and mutations in the viral genome help the virus evade the immune system and increase virus replication (Barbu et al. 2020). Some miRNAs like miR-155, which are overexpressed in infected patients, play a role in the cytokine storm that elicits immune and inflammatory responses and increases viral pathogenicity. To prove the role of this miRNA, anti-mir-155 was used in mouse models and reduction in the expression level of mir-155 resulted in mouse survival by enhancing antiviral responses and reducing cytokine storm (Soni et al. 2020).

11.6 The Role of miRNAs in Human Papillomavirus (HPV) Infection

Human papillomavirus (HPV) is the main cause of cervical cancer and one of the most common sexually transmitted infections in men and women worldwide. This virus infects the keratinocytes of the mucous membranes or skin. Most types of HPV do not cause symptoms and malignant lesions, as they are usually eliminated by the immune system, although prolonged and chronic infections are caused by different high-risk HPV types like types 16, 18, 31, 33, and 35 that lead to disease and malignancies, especially cervical cancer. It has been shown that the expression of miR-27b, miR-24-1, miR-3654, miR-647, and miR-1914 is decreased in the genital warts, which is attributed to low-risk HPV types like types 6, 11, 42, 43, and 61 (Varesano et al. 2023). Viruses develop clever mechanisms to successfully maintain the viral life cycle. Even in cervical cancer, miRNAs play a role in manipulating cellular and viral gene expression that drives cancer development, evading the host's innate or adaptive immune system (Al-Eitan et al. 2019; Sammarco et al. 2020). For example, miR-125b can significantly reduce HPV DNA synthesis in NIH 3T3 cells susceptible to HPV infection. Studies have shown that miR-125b is downregulated in a variety of cancers and has putative targets that include several oncogenes such as FGFR2 (fibroblast growth factor receptor 2) and several members of the mitogen-activated signal transduction pathway (Nuovo et al. 2010). MiR-34a is downregulated in productive premalignant HPV infections and cervical cancer cells. The high-risk HPV E6 oncoprotein mediates the degradation of p53, one of the transcriptional regulatory factors of miR-34a. The viral oncoprotein destabilizes p53 and reduces the expression of mir-34a, a tumor suppressor, and increases expression of cell cycle regulators and cell growth (Chen and Zhao 2015). In total, five putative papillomavirus miRNAs were encoded by HPV 16, HPV 38, HPV 68, HPV 45, and HPV 6. None of the HPV miRNAs resemble to known human miRNA sequences. Among them, HPV16-miR-H1-1 is located in the E1 region of the coding strand and HPV16-miR-H2-1 is located in the negative strand corresponding to the LCR (locus control regions) region, playing an important role in viral infection (Satapathy et al. 2017).

11.7 The Role of miRNAs in Herpes Simplex Virus (HSV) Infection

The herpes simplex virus (HSV) is a type of double-stranded DNA virus that encodes at least 70 proteins and 37 miRNAs (Duan et al. 2023). HSV-1 uses host miRNAs to increase its latency, supporting the virus's ability to spread in the population (Cokarić Brdovčak et al. 2018). The host cell miR-138 has a target site in the 3′UTR of the HSV-1 ICP0 (Infected Cell Protein 0) mRNA. Therefore, it is believed that miR-138 could limit the expression of ICP0 as a lytic factor in neurons and facilitate the latency period (Sha et al. 2017). Host miRNAs can influence HSV-1 replication indirectly through various mechanisms, which can be divided into three categories: (a) miRNAs that alter the expression of important factors for viral replication, (b) miRNAs that prolong the survival of infected cells, and (c) miRNAs that target host immune factors. These miRNAs can have either pro- or anti-viral effects. HSV-1 infects cells and induces massive changes in the host cell's miRNAs (Cokarić Brdovčak et al. 2018).

MiR-373 is involved in cell growth, apoptosis, and immune response, being upregulated in infected HeLa cells. Transient overexpression of miR-373 slightly increased virus yield, while inhibition of miR-373 decreased virus load. Therefore, miR-373 is an exciting new potential target against HSV-1 (Xie et al. 2018). Finally, miR-146a is a host cell miRNA upregulated during HSV-1 infection. MiR-146a regulates immunity and inflammatory host responses by targeting many genes, including TNF receptor-related factors (Hill et al. 2009).

11.8 The Role of miRNAs in Cytomegalovirus Infection

Human cytomegalovirus is a double-stranded linear DNA virus that belongs to the Herpesviridae family. This virus encodes microRNAs that alter the responses of the host cell's microenvironment and defense system. HCMV (human cytomegalovirus) infection remains latent for a long time, thereby prevention of HCMV infection is important in immunocompromised patients, including tissue transplanted recipients, patients with HIV or leukemia, and in the first trimester of pregnancy (Abdalla et al. 2020). The HCMV genome encodes miRNAs that inhibit viral DNA replication. MiR-UL112-3p is the most studied HCMV miRNA, it is expressed in fibroblast cells at early stages of acute infection, and its levels increase during prolonged infection. MiR-UL112-3p reduces the expression of IE72, which regulates virus replication (Lau et al. 2016). MiR-US25-1-5p and miR-US25-2-5p are expressed during the early stage of infection and they are accumulated with prolonged infection. These two miRNAs alter the expression of host genes that play fundamental roles in cellular activities. For example, miR-US25-2-5p inhibits the expression of IE72, eIF4A1 (eukaryotic translation initiation factor 4A1) and miR-US25-1-5p inhibits the ATP6V0C (ATPase H+ transporting V0 subunit C) and lead to a significant reduction in viral replication (Jiang et al. 2015). The level of miR-US33-5p increases in patients with HMCV compared to healthy individuals, and its expression decreases DNA synthesis and viral replication (Dong et al. 2017). The expression of miR-UL148D increases during the late stages of HCMV infection and promotes its infectivity by targeting the 3′UTR of the IRE-5 gene (Pan et al. 2016). Finally, HCMV-encoded miRNAs protect the virus from immune system recognition by inducing inhibition of cell apoptosis (Abdalla et al. 2020).

11.9 The Role of miRNAs in Epstein-Barr Virus (EBV) Infection

EBV is a double-stranded DNA virus that can cause infectious mononucleosis in adolescents. It is also implicated in several malignancies, including Burkitt lymphoma and Hodgkin lymphoma (Iizasa et al. 2020). Moreover, viral miRNAs repress target genes in EBV and the host genome, allowing the virus to evade host immune surveillance and promote the growth of infected cells (Iizasa et al. 2020). In this regard, it has been shown that miR-BART2 inhibits the transition of EBV virus from the latent to the lysis phase by downregulating viral DNA polymerase (Edwards et al. 2008). MiR-BART20-5p is a microRNA that inhibits ZTA and RTA (replication and transcription activator) genes, which are responsible for inducing the expression of lytic phase genes and viral replication, respectively. This suggests that EBV utilizes its microRNAs to maintain latent infection in host cells by suppressing the viral lytic cycle (Wang et al. 2018). LMP1 (latent membrane protein 1) is a protein that plays a crucial role in cell proliferation and transformation. It has been shown that LMP1 activates several cellular pathways, including PI3K, MAPK, IRF7, NF-κB, JAK/STAT, and AP-1 (Cai et al. 2006; Edwards et al. 2008). Several EBV microRNAs, including miR-BART-1p, miR-BART16, BART17-5p, miR-BART5-5p, and miR-BART19-5p, have been found to inhibit the translation of LMP-1 mRNA, which is associated with the viral latency phase (Cai et al. 2006, Edwards et al. 2008). EBV miRNAs inhibit the expression of LMP2A to stimulate specific T cell responses. Furthermore, miR-BART6-5p directly targets Dicer to regulate host miRNA production, and also suppresses EBNA2 (Epstein-Barr Virus Nuclear Antigen 2), highlighting its role in EBV latency and carcinogenesis (Wang et al. 2018). MiR-BART15 suppresses inflammation by targeting the inflammatory protein NLRP3 (nucleotide-binding domain leucine-rich-containing family, pyrin domain-containing-3). This suppression allows for the EBV infection (Wang et al. 2018). Also, miR-BHRF1-3 targets the chemokine CXCL-11(chemokine (C-X-C motif) ligand 11), which induces interferon and interacts with the chemokine receptor CXCR3 on T cells. This suggests that viral miRNAs may manipulate the host's cytokine networks to evade the immune system (Wang et al. 2018). In conclusion, EBV is a successful human pathogen that can persist in humans without causing disease (Kuzembayeva et al. 2014).

11.10 The Role of miRNAs in EBOLA Infection

The Ebola virus is a highly dangerous pathogen for humans, with the Zaire strain having a mortality rate over 50% (Bettini et al. 2023). Ebola virus infection may cause immunosuppression, inflammation, vascular disruption, blood coagulation, and multi-organ failure, resembling septic shock (Feldmann and Geisbert 2011).

EBOV (Ebola virus) may produce miRNAs that can be used as biomarkers for diagnosing and predicting the outcome of Ebola virus infection. These miRNAs could also be targeted for therapeutic purposes (Liang et al. 2014). The viral envelope glycoprotein (GP) causes inflammation and cell damage, particularly at the protein level. Overexpression of GP leads to detachment of endothelial cells in infected vessels and cytotoxicity (Feldmann and Geisbert 2011). Upon expression of Ebola virus glycoprotein (EBOV GP), hsa-miR-1246, hsa-miR-320a, and hsa-miR-196b-5p were found to be induced in HUVECs (human umbilical vein endothelial cells). The targets of these microRNAs, including TFPI (Tissue Factor Pathway Inhibitor), DAG1 (Dystroglycan1), and Caspase 8 and FADD (Fas Associated via Death Domain) are significantly reduced in cells expressing the EBOV virus-GFP. These results suggest that such microRNA inhibitors could be potential therapies to treat Ebola virus infections (Sheng et al. 2014). MiR-MAY-251 and miR-MAK-403 are EBV microRNAs with a large number of human host genes. These microRNAs can regulate various genes involved in the hemorrhagic phenotype, viral replication, and immune system modulation. It has been suggested that miR-MAK-403 may also be involved in coagulation disorders and fibrinolysis, which are important factors in the pathogenesis of Ebola virus (Diallo et al. 2022). Zebov-miR-1-5p generated only by EBOV reduces the expression of importin-a5 at the post-transcriptional level. This reduction leads to disruption in the interferon signaling pathway and higher mortality rates in affected patients (Liu et al. 2016).

11.11 MiRNAs in Other Viruses

In addition to the mentioned viruses, there are relation between other viruses and miRNAs associated with KSHV (Kaposi's sarcoma-associated herpesvirus), ALV (avian leukosis virus), BLV (bovine leukosis virus), WNV (West Nile virus), JEV (Japanese encephalitis virus), DENV (Dengue virus), PRRSV (porcine reproductive and respiratory syndrome virus), Poliovirus, and BDV (Borna disease virus) (Table 11.1).

Table 11.1 Cellular miRNAs in other viruses

Virus	miRNA	Target	References
KSHV	miR-K5,K9,K10	BCALF1	Bellare and Ganem (2009), Ziegelbauer et al. (2009)
	miR-K9-K1,K11,K3-3p,K6-3p	THBS1	Samols et al. (2007)
Avian leukosis virus	miR-23b	IRF1	Li et al. (2015)
BLV	BLV-miR-B4	HBP1, PXDN	Rosewick et al. (2013)
WNV	KUN-miR-1	GATA4	Hussain et al. (2012)
	miR-6124	CTCF, ECOP	Slonchak et al. (2016)
	miR-532-5p	SESTD1, TAB3	Slonchak et al. (2016)
DENV	miR-30e	IκB/IFN-β, OAS1, MxA, IFITM1	Diosa-Toro et al. (2017)
	miR-3614-5p	ADAR1	Diosa-Toro et al. (2017)
	miR-146a	TRAF6	Diosa-Toro et al. (2017)
JEV	miR-301	IRF1, SOCS5, IFN-β	Hazra et al. (2017)
	miR-33a	EEF1A	Chen et al. (2016)
	miR-146a	TRAF6	Sharma et al. (2015)
	miR-432	SOCS5	Sharma et al. (2016)
	miR-155	SHIP1	Thounaojam et al. (2014a, b)
	miR-15b	RNF125	Zhu et al. (2015)
	miR-29b	TNFAIP3	Thounaojam et al. (2014a, b)
PRRSV	mir-181	CD163	Gao et al. (2013)
Poliovirus	mir-555	hnRNP C	Shim et al. (2016)
Borna disease virus	miR-155	SOCS3	Zhai et al. (2013)

BCLAF1, BCL2 associated transcription factor 1; THBS1, thrombospondin 1; IRF1, IFN regulatory factor 1; HBP1, HMG-box transcription factor 1; PXDN, peroxidasin; GATA4, GATA binding protein 4; CTCF, CCCTC-binding factor; ECOP, EGFR-coamplified and overexpressed protein; SESTD1, SEC14 and spectrin domain containing 1; TAB3, TGF-β activated kinase 1 (MAP3K7) binding protein 3; OAS1, 2′–5′-oligoadenylate synthetase 1; MxA, myxovirus-resistance protein 1; IFITM1, interferon induced transmembrane protein 1; ADAR1, adenosine deaminase 1; TRAF6, tumor necrosis factor receptor associated factor 6; SOCS5, suppressor of cytokine signaling 5; SHIP1, SH-2 containing inositol 5′ polyphosphatase 1; RNF125, ring finger protein 125; TNFAIP3, TNF alpha induced protein 3

11.12 Conclusion

In summary, human and viral miRNAs are interrelated and play critical roles during viral infections, especially during viral entry, replication, pathogenicity, and cell death/survival. Crosstalk between these can be a major factor that determines the severity of the disease in the host.

Acknowledgements We would like to thank our colleagues in the Hepatitis and AIDS Department, Pasteur Institute of Iran.

Compliance with Ethical Standards

Conflicts of Interest: The authors declare no conflict of interests.

Ethical Approval: This chapter does not contain any studies with human participants performed by any of the authors.

References

Abdalla AE, Mahjoob MO, Abosalif KOA et al (2020) Human cytomegalovirus-encoded microRNAs: A master regulator of latent infection. Infect Genet Evol 78:104119

Al-Eitan LN, Alghamdi MA, Tarkhan AH et al (2019) Gene expression profiling of microRNAs in HPV-induced warts and normal skin. Biomolecules 9(12):757

Bala S, Tilahun Y, Taha O et al (2012) Increased microRNA-155 expression in the serum and peripheral monocytes in chronic HCV infection. J Transl Med 10:1–10

Barbu MG, Condrat CE, Thompson DC et al (2020) MicroRNA involvement in signaling pathways during viral infection. Front Cell Dev Biol 8:143

Bellare P, Ganem D (2009) Regulation of KSHV lytic switch protein expression by a virus-encoded microRNA: an evolutionary adaptation that fine-tunes lytic reactivation. Cell Host Microbe 6(6):570–575

Bettini A, Lapa D, Garbuglia AR (2023) Diagnostics of Ebola virus. Front Public Health 11:1123024

Bruscella P, Bottini S, Baudesson C et al (2017) Viruses and miRNAs: more friends than foes. Front Microbiol 8:824

Cai X, Schäfer A, Lu S et al (2006) Epstein–Barr virus microRNAs are evolutionarily conserved and differentially expressed. PLoS Pathog 2(3):e23

Chen J, Zhao K-N (2015) HPV-p53-miR-34a axis in HPV-associated cancers. Ann Transl Med 3:21

Chen X-Y, Zhang H-S, Wu T-C et al (2013) Down-regulation of NAMPT expression by miR-182 is involved in Tat-induced HIV-1 long terminal repeat (LTR) transactivation. Int J Biochem Cell Biol 45(2):292–298

Chen Z, Ye J, Ashraf U et al (2016) MicroRNA-33a-5p modulates Japanese encephalitis virus replication by targeting eukaryotic translation elongation factor 1A1. J Virol 90(7):3722–3734

Cheng J-C, Yeh Y-J, Tseng C-P et al (2012) Let-7b is a novel regulator of hepatitis C virus replication. Cell Mol Life Sci 69:2621–2633

Cheng M, Si Y, Niu Y et al (2013) High-throughput profiling of alpha interferon-and interleukin-28B-regulated microRNAs and identification of let-7s with anti-hepatitis C virus activity by targeting IGF2BP1. J Virol 87(17):9707–9718

Chiang K, Sung T-L, Rice AP (2012) Regulation of cyclin T1 and HIV-1 replication by microRNAs in resting CD4+ T lymphocytes. J Virol 86(6):3244–3252

Chiang K, Liu H, Rice AP (2013) miR-132 enhances HIV-1 replication. Virology 438(1):1–4

Chow J, Salmena L (2020) Prediction and analysis of SARS-CoV-2-targeting MicroRNA in human lung epithelium. Genes 11(9):1002

Cokarić Brdovčak M, Zubković A, Jurak I (2018) Herpes simplex virus 1 deregulation of host microRNAs. Noncoding RNA 4(4):36

Davidson AM, Wysocki J, Batlle D (2020) Interaction of SARS-CoV-2 and other coronavirus with ACE (angiotensin-converting enzyme)-2 as their main receptor: therapeutic implications. Hypertension 76(5):1339–1349

Devhare PB, Sasaki R, Shrivastava S et al (2017) Exosome-mediated intercellular communication between hepatitis C virus-infected hepatocytes and hepatic stellate cells. J Virol 91(6). https://doi.org/10.1128/jvi.02225-02216

Diallo I, Husseini Z, Guellal S et al (2022) Ebola virus encodes two microRNAs in Huh7-infected cells. Int J Mol Sci 23(9):5228
Diosa-Toro M, Echavarría-Consuegra L, Flipse J et al (2017) MicroRNA profiling of human primary macrophages exposed to dengue virus identifies miRNA-3614-5p as antiviral and regulator of ADAR1 expression. PLoS Negl Trop Dis 11(10):e0005981
Dong J, Bao J, Feng R et al (2017) Circulating microRNAs: a novel potential biomarker for diagnosing acute aortic dissection. Sci Rep 7(1):12784
Duan Y, Sun L, Li Q (2023) Herpes simplex virus 1 microRNAs: an update. Intervirology 66(1):97–110
Edwards RH, Marquitz AR, Raab-Traub N (2008) Epstein-Barr virus BART microRNAs are produced from a large intron prior to splicing. J Virol 82(18):9094–9106
Fang J, Hao Q, Liu L et al (2012) Epigenetic changes mediated by microRNA miR29 activate cyclooxygenase 2 and lambda-1 interferon production during viral infection. J Virol 86(2):1010–1020
Fani M, Zandi M, Ebrahimi S et al (2021) The role of miRNAs in COVID-19 disease. Future Virol 16(4):301–306
Feldmann H, Geisbert TW (2011) Ebola haemorrhagic fever. Lancet 377(9768):849–862
Frattari G, Aagaard L, Denton PW (2017) The role of miR-29a in HIV-1 replication and latency. J Virus Erad 3(4):185–191
Gao L, Guo X-k, Wang L et al (2013) MicroRNA 181 suppresses porcine reproductive and respiratory syndrome virus (PRRSV) infection by targeting PRRSV receptor CD163. J Virol 87(15):8808–8812
Hartling HJ, Ballegaard VC, Nielsen NS et al (2016) Immune regulation in chronic hepatitis C virus infection. Scand J Gastroenterol 51(11):1387–1397
Hazra B, Kumawat KL, Basu A (2017) The host microRNA miR-301a blocks the IRF1-mediated neuronal innate immune response to Japanese encephalitis virus infection. Sci Signal 10(466):eaaf5185
Hill JM, Zhao Y, Clement C et al (2009) HSV-1 infection of human brain cells induces miRNA-146a and Alzheimer-type inflammatory signaling. Neuroreport 20(16):1500
Hou W, Tian Q, Zheng J et al (2010) MicroRNA-196 represses Bach1 protein and hepatitis C virus gene expression in human hepatoma cells expressing hepatitis C viral proteins. Hepatology 51(5):1494–1504
Huang J, Wang F, Argyris E et al (2007) Cellular microRNAs contribute to HIV-1 latency in resting primary CD4+ T lymphocytes. Nat Med 13(10):1241–1247
Hussain M, Torres S, Schnettler E et al (2012) West Nile virus encodes a microRNA-like small RNA in the 3′ untranslated region which up-regulates GATA4 mRNA and facilitates virus replication in mosquito cells. Nucleic Acids Res 40(5):2210–2223
Iizasa H, Kim H, Kartika AV et al (2020) Role of viral and host microRNAs in immune regulation of Epstein-Barr virus-associated diseases. Front Immunol 11:367
Jiang S, Qi Y, He R et al (2015) Human cytomegalovirus microRNA miR-US25-1-5p inhibits viral replication by targeting multiple cellular genes during infection. Gene 570(1):108–114
Joshi N, Chandane Tak M, Mukherjee A (2022) The involvement of microRNAs in HCV and HIV infection. Ther Adv Vaccines Immunother 10. https://doi.org/10.1177/25151355221106104
Kaur B, Kaur J (2023) miRNAomics in COVID-19. In: Barh D, De Car Azevedo VA (eds) Omics approaches and technologies in COVID-19. Elsevier, Amsterdam, pp 145–160
Kaushik P, Kumar A (2022) Emerging role and function of miR-198 in human health and diseases. Pathol Res Pract 229:153741
Kim JH, Lee CH, Lee S-W (2019) Exosomal transmission of microRNA from HCV replicating cells stimulates transdifferentiation in hepatic stellate cells. Mol Ther Nucleic Acids 14:483–497
Kuzembayeva M, Hayes M, Sugden B (2014) Multiple functions are mediated by the miRNAs of Epstein-Barr virus. Curr Opin Virol 7:61–65
Lau B, Poole E, Van Damme E et al (2016) Human cytomegalovirus miR-UL112-1 promotes the down-regulation of viral immediate early-gene expression during latency to prevent T-cell recognition of latently infected cells. J Gen Virol 97(9):2387

Li Z, Chen B, Feng M et al (2015) MicroRNA-23b promotes avian leukosis virus subgroup J (ALV-J) replication by targeting IRF1. Sci Rep 5(1):10294
Li J, Li P, Zhang G et al (2020) CircRNA TADA2A relieves idiopathic pulmonary fibrosis by inhibiting proliferation and activation of fibroblasts. Cell Death Dis 11(7):553
Liang H, Zhou Z, Zhang S et al (2014) Identification of Ebola virus microRNAs and their putative pathological function. Sci China Life Sci 57:973–981
Lin Y, Sun Q, Zhang B et al (2023) The regulation of lncRNAs and miRNAs in SARS-CoV-2 infection. Front Cell Dev Biol 11
Liu Y, Sun J, Zhang H et al (2016) Ebola virus encodes a miR-155 analog to regulate importin-α5 expression. Cell Mol Life Sci 73:3733–3744
Mishra R, Kumar A, Ingle H et al (2020) The interplay between viral-derived miRNAs and host immunity during infection. Front Immunol 10:3079
Mukherjee A, Shrivastava S, Bhanja Chowdhury J et al (2014) Transcriptional suppression of miR-181c by hepatitis C virus enhances homeobox A1 expression. J Virol 88(14):7929–7940
Nahand JS, Shojaie L, Akhlagh SA et al (2021) Cell death pathways and viruses: role of microRNAs. Mol Ther Nucleic Acids 24:487–511
Naqvi AR, Shango J, Seal A et al (2018) Herpesviruses and MicroRNAs: new pathogenesis factors in oral infection and disease? Front Immunol 9:2099
Nersisyan S, Engibaryan N, Gorbonos A et al (2020) Potential role of cellular miRNAs in coronavirus-host interplay. PeerJ 8:e9994
Nuovo GJ, Wu X, Volinia S et al (2010) Strong inverse correlation between microRNA-125b and human papillomavirus DNA in productive infection. Diagn Mol Pathol 19(3):135
Ojha R, Nandani R, Pandey RK et al (2019) Emerging role of circulating microRNA in the diagnosis of human infectious diseases. J Cell Physiol 234(2):1030–1043
Pan C, Zhu D, Wang Y et al (2016) Human cytomegalovirus miR-UL148D facilitates latent viral infection by targeting host cell immediate early response gene 5. PLoS Pathog 12(11):e1006007
Pascut D, Hoang M, Nguyen NN et al (2021) HCV proteins modulate the host cell miRNA expression contributing to hepatitis C pathogenesis and hepatocellular carcinoma development. Cancers 13(10):2485
Piedade D, Azevedo-Pereira JM (2016) MicroRNAs, HIV and HCV: a complex relation towards pathology. Rev Med Virol 26(3):197–215
Poorghobadi S, Agharezaei M, Ghanbari M et al (2023) Discordant immune response among treatment experienced patients infected with HIV-1: Crosstalk between MiRNAs expression and CD4+ T cells count. Int Immunopharmacol 114:109533
Reynoso R, Laufer N, Hackl M et al (2014) MicroRNAs differentially present in the plasma of HIV elite controllers reduce HIV infection in vitro. Sci Rep 4(1):5915
Riad SE, Elhelw DS, Shawer H et al (2018) Disruption of Claudin-1 expression by miRNA-182 alters the susceptibility to viral infectivity in HCV cell models. Front Genet 9:93
Rosewick N, Momont M, Durkin K et al (2013) Deep sequencing reveals abundant noncanonical retroviral microRNAs in B-cell leukemia/lymphoma. Proc Natl Acad Sci USA 110(6):2306–2311
Sadri Nahand J, Bokharaei-Salim F, Karimzadeh M et al (2020) MicroRNAs and exosomes: key players in HIV pathogenesis. HIV Med 21(4):246–278
Saliminejad K, Khorram Khorshid HR, Soleymani Fard S et al (2019) An overview of microRNAs: Biology, functions, therapeutics, and analysis methods. J Cell Physiol 234(5):5451–5465
Sammarco ML, Tamburro M, Pulliero A et al (2020) Human papillomavirus infections, cervical cancer and microRNAs: an overview and implications for public health. MicroRNA 9(3):174–186
Samols MA, Skalsky RL, Maldonado AM et al (2007) Identification of cellular genes targeted by KSHV-encoded microRNAs. PLoS Pathog 3(5):e65
Satapathy S, Batra J, Jeet V et al (2017) MicroRNAs in HPV associated cancers: small players with big consequences. Expert Rev Mol Diagn 17(7):711–722
Sha H-H, Wang D-D, Chen D et al (2017) MiR-138: A promising therapeutic target for cancer. Tumor Biol 39(4). https://doi.org/10.1177/1010428317697575

Sharma N, Verma R, Kumawat KL et al (2015) miR-146a suppresses cellular immune response during Japanese encephalitis virus JaOArS982 strain infection in human microglial cells. J Neuroinflamm 12(1):1–16
Sharma N, Kumawat KL, Rastogi M et al (2016) Japanese encephalitis virus exploits the microRNA-432 to regulate the expression of suppressor of cytokine signaling (SOCS) 5. Sci Rep 6(1):27685
Sheng M, Zhong Y, Chen Y et al (2014) Hsa-miR-1246, hsa-miR-320a and hsa-miR-196b-5p inhibitors can reduce the cytotoxicity of Ebola virus glycoprotein in vitro. Sci China Life Sci 57:959–972
Shim B-S, Wu W, Kyriakis CS et al (2016) MicroRNA-555 has potent antiviral properties against poliovirus. J Gen Virol 97(3):659–668
Slonchak A, Shannon RP, Pali G et al (2016) Human microRNA miR-532-5p exhibits antiviral activity against West Nile virus via suppression of host genes SESTD1 and TAB3 required for virus replication. J Virol 90(5):2388–2402
Soni DK, Cabrera-Luque J, Kar S et al (2020) Suppression of miR-155 attenuates lung cytokine storm induced by SARS-CoV-2 infection in human ACE2-transgenic mice. bioRxiv. https://doi.org/10.1101/2020.12.17.423130
Sung T-L, Rice AP (2009) miR-198 inhibits HIV-1 gene expression and replication in monocytes and its mechanism of action appears to involve repression of cyclin T1. PLoS Pathog 5(1):e1000263
Tambyah PA, Sepramaniam S, Mohamed Ali J et al (2013) microRNAs in circulation are altered in response to influenza A virus infection in humans. PLoS One 8(10):e76811
Thounaojam MC, Kaushik DK, Kundu K et al (2014a) Micro RNA-29b modulates Japanese encephalitis virus-induced microglia activation by targeting tumor necrosis factor alpha-induced protein 3. J Neurochem 129(1):143–154
Thounaojam MC, Kundu K, Kaushik DK et al (2014b) MicroRNA 155 regulates Japanese encephalitis virus-induced inflammatory response by targeting Src homology 2-containing inositol phosphatase 1. J Virol 88(9):4798–4810
Tycowski KT, Guo YE, Lee N et al (2015) Viral noncoding RNAs: more surprises. Genes Dev 29(6):567–584
Varesano S, Pulliero A, Martorana E et al (2023) Screening of precancerous lesions in women with human papillomavirus (HPV) infection by molecular typing and microRNA analysis. J Pers Med 13(3):531
Vigorito E, Kohlhaas S, Lu D et al (2013) miR-155: an ancient regulator of the immune system. Immunol Rev 253(1):146–157
Wang M, Yu F, Wu W et al (2018) Epstein-Barr virus-encoded microRNAs as regulators in host immune responses. Int J Biol Sci 14(5):565
Wei Y-F, Cui G-Y, Ye P et al (2013) MicroRNAs may solve the mystery of chronic hepatitis B virus infection. World J Gastroenterol WJG 19(30):4867
Widiasta A, Sribudiani Y, Nugrahapraja H et al (2020) Potential role of ACE2-related microRNAs in COVID-19-associated nephropathy. Noncoding RNA Res 5(4):153–166
Xie Y, He S, Wang J (2018) MicroRNA-373 facilitates HSV-1 replication through suppression of type I IFN response by targeting IRF1. Biomed Pharmacother 97:1409–1416
Yang C-Y, Chen Y-H, Liu P-J et al (2022) The emerging role of miRNAs in the pathogenesis of COVID-19: Protective effects of nutraceutical polyphenolic compounds against SARS-CoV-2 infection. Int J Med Sci 19(8):1340
Zhai A, Qian J, Kao W et al (2013) Borna disease virus encoded phosphoprotein inhibits host innate immunity by regulating miR-155. Antiviral Res 98(1):66–75
Zhu B, Ye J, Nie Y et al (2015) MicroRNA-15b modulates Japanese encephalitis virus-mediated inflammation via targeting RNF125. J Immunol 195(5):2251–2262
Ziegelbauer JM, Sullivan CS, Ganem D (2009) Tandem array-based expression screens identify host mRNA targets of virus-encoded microRNAs. Nat Genet 41(1):130–134

Chapter 12
Inter-kingdom Signalling: Exploring the Relationships Among Microorganisms and MicroRNAs in Human Health

Iván Fernández-Rengel, Laura C. Terrón-Camero, and Eduardo Andrés-León

Abstract Since their discovery in the 1990s, microRNAs have been linked to numerous relevant biological processes such as development, proliferation, and immune response. Any minimal alteration in the expression profiles of these non-coding RNAs had been associated with the onset and development of complex diseases such as cancer or infectious diseases. miRNAs have been unveiled in nearly all domains of life, spanning from viruses to protozoa. Investigating how microRNAs from microorganisms have the capacity to influence host genes opens new avenues for advancing our comprehension of human health. This chapter delves into the signalling frontier between kingdoms, uncovering the multifaceted relationships between microorganisms and miRNAs in the regulation of pathways related to human health. We explore how virus, bacteria, fungi, and protozoa can modulate host messenger RNA expression profiles, allowing evading the immune responses, activating metabolic functions, and even disease susceptibility. Understanding these complex signalling networks offers new insights into a variety of diseases, including infectious diseases and cancer, as well as the potential for innovative therapeutic interventions. This manuscript integrates the most recent findings to provide an overview of the exciting and evolving field of cross-realm signalling, shedding light on connections between microorganisms' miRNAs in the maintenance of human health.

I. Fernández-Rengel
Bioinformatics Unit, Institute of Parasitology and Biomedicine López-Neyra (IPBLN), CSIC, Granada, Spain

GENYO, Centre for Genomics and Oncological Research: Pfizer - University of Granada - Andalusian Regional Government, Granada, Spain

L. C. Terrón-Camero · E. Andrés-León (✉)
Bioinformatics Unit, Institute of Parasitology and Biomedicine López-Neyra (IPBLN), CSIC, Granada, Spain
e-mail: eduardo.andres@csic.es

L. M. Vaschetto (ed.), *miRNAs, Human Health and Diseases*, Epigenetics and Human Health 13, https://doi.org/10.1007/978-3-031-64788-8_12

Keywords miRNA · milRNA · vmiRNA · Cancer · Infectious diseases · Virus · Bacteria · Fungi · Protozoa · Host regulation

List of Abbreviations

3′UTR	3′ untranslated region
AGO	Argonaute
BART	Bam HI-A region rightward transcript
BHRF1	Bam HI fragment H rightward open reading frame I
BL	Burkitt's lymphoma
BRCC3	BRCA1/BRCA2 complex subunit 3
CDS	Coding sequence
CRC	Colorectal cancer
DCL	Dicer-like
DGCR8	DiGeorge syndrome critical region gene 8
EBOV	Ebola virus
EBV	Epstein-Barr virus
EVs	Extracellular vesicles
GBM	Glioblastoma multiforme
GC	Gastric carcinoma
HAdV	Human adenovirus
HCC	Hepatocellular carcinoma
HCMV	Human cytomegalovirus
HHV-8	Human herpesvirus 8
HL	Hodgkin's lymphoma
HPV	Human papillomavirus
HSV-1	Herpes simplex virus 1
HSV-2	Herpes simplex virus 2
IA	Invasive aspergillosis
IFN-β	Interferon-beta
IGRs	Intergenic regions
JCPyV or JCV	John Cunningham polyomavirus
KSHV	Kaposi's sarcoma-associated herpesvirus
lncRNAs	Long non-coding RNAs
LPS	Lipopolysaccharide
MCPyV	Merkel cell polyomavirus
MICB	MHC class I-related B chain
milRNAs	miRNA-like small RNAs
ncRNAs	Non-coding RNAs
NPC	Nasopharyngeal carcinomas
OMVs	Outer membrane vesicles
ORF	Open reading frame
PEL	Primary effusion lymphoma
piRNAs	PIWI-interacting RNAs

PPAR	Peroxisome proliferator-activated receptor
pri-miRNAs	miRNAs
PTEN	Phosphatase and tensin homolog receptor
RISC	RNA-induced silencing complex
RNAi	RNA-interference
SASP	Senescence-associated secretory phenotype
siRNAs	Small interfering
sncRNAs	Small non-coding RNAs
snoRNAs	Small nucleolar RNAs
sRNAs	Small RNA
TIMP3	Targets tissue inhibitors of metalloprotease 3
TNF-α	Tumor necrosis factor-alpha
TNRC6A	Trinucleotide repeat-containing gene 6A
tRFs	tRNA-derived fragments
TUSC3	Tumor suppressor factor candidate 3
vmiRNAs	Viral miRNAs
vsRNAs	Very small RNAs

12.1 Introduction

Traditionally, the fundamental principle of molecular biology asserted that genetic information is contained within DNA, then transcribed into RNA, and finally translated into proteins. This idea of unidirectionality in the flow of genetic information suggests that most genes code for proteins and these explain the phenotypic characteristics of the organism (Mattick 2003). Nowadays, it has been established that most human RNA transcripts do not carry instructions for protein synthesis. Instead, non-coding RNAs (ncRNAs) play a crucial role in governing cell physiology and influencing various cellular functions. Among these ncRNAs, a subset is engaged in controlling genes at multiple levels, encompassing epigenetic gene suppression and post-transcriptional management of mRNA stability (Panni et al. 2020). Due to their relevant role, they are also implicated in the development of numerous diseases (Lopez-Jimenez and Andres-Leon 2021).

ncRNA are classified based on their size in two groups. On one hand, long non-coding RNAs (lncRNAs) are those molecules of at least 200 nucleotides with diverse functions in the cell. On the other hand, small non-coding RNAs (sncRNAs) are short molecules involved in the post-translational regulation of gene expression, and include, among others, small nucleolar RNAs (snoRNAs), small nuclear RNAs (snRNAs), small interfering (siRNAs), PIWI-interacting RNAs (piRNAs), and microRNAs (miRNAs). miRNAs have been extensively studied because of their great relevance in many cellular processes and for its role in numerous diseases, including different types of cancer, cardiovascular diseases such as hypertrophy and ischemia, and mental disorders, including schizophrenia and depression (Hombach and Kretz 2016).

Nonetheless, the involvement of miRNAs from pathogens like viruses, bacteria, fungi, or protozoans in the modulation of host genes has received considerably less attention. Consequently, the objective of this chapter is to gather the latest insights on the roles of miRNAs in mediating interactions between pathogens and human cells in the context of human diseases.

12.1.1 miRNAs: Small Non-coding Regulatory RNA Molecules

miRNAs are a group of highly conserved sncRNA molecules with an average length of 22 nucleotides whose role is to inhibit gene expression at the post-transcriptional level (Bartel 2004; Bhaskaran and Mohan 2014), either by binding to their target mRNA and inhibiting translation of the transcript, or by recruiting decay factors, inducing their destabilization and degradation (Bhaskaran and Mohan 2014). They were discovered in 1993, when the research groups led by Ambros and Ruvkun identified the first known miRNA, lin-4, in *Caenorhabditis elegans* (Lee et al. 1993). In addition, these researchers found that lin-4 downregulated lin-14 gene expression by binding to its 3′ UTR region by base complementarity (Wightman et al. 1993). Since then, miRNAs have been found in all major domains of life and some of them were shown to be highly conserved across species (Grimson et al. 2007). In fact, tens of thousands of publications have detailed their properties such as their molecular structure, their biogenesis, and the functions they perform in the cell (O'Brien et al. 2018). Furthermore, they have been demonstrated to have a crucial function, as they are indispensable in development and participate in a multitude of biological processes (Fu et al. 2013). Moreover, alterations in gene regulation mediated by miRNAs are associated with a multitude of human diseases (O'Brien et al. 2018).

In the realm of microorganisms, it is increasingly evident that miRNAs represent a noteworthy component of gene regulation. These sncRNA molecules have been discovered in various microorganism species, underscoring their prevalence and potential significance. While miRNAs in microorganisms are recognized to influence gene expression, their precise roles and mechanisms are areas of ongoing investigation.

In the case of viruses, both DNA and RNA viruses contain genes that encode miRNAs known as viral miRNAs (vmiRNAs). To date, more than 250 vmiRNAs are known (Pan et al. 2019). Virus-derived miRNAs target several mRNAs from host cells and thus regulate the function of numerous genes (Skalsky and Cullen 2010). Furthermore, vmiRNAs confer some advantages in regulating cellular activity while minimizing the stimulation of the immune response (Dass et al. 2023).

Regarding bacteria, it's worth noting that small RNA (sRNAs) molecules, which possess the capacity to modulate gene expression, have also been discovered. *Escherichia coli* was the first species in which they were discovered (Argaman et al. 2001). Some of these sncRNAs have been reported to function as a regular miRNA. However, they could range in size from 50 to 400 nucleotides and could be encoded in genomic regions either proximal (*cis*-encoded) or distal (*trans*-encoded) to the genes they influence. Moreover, RNAs with a size of less than 16 nucleotides belonging to the category of vsRNAs (very small RNAs) were recently discovered in bacteria (Diallo et al. 2022a). Among them we can highlight tRNA-derived fragments (tRFs) that are believed to repress translation in a manner similar to miRNAs, being able to regulate multiple mRNA targets (Kumar et al. 2014).

In 2010, Lee et al. discovered the first miRNAs in fungi within *Neurospora crassa*. They named these novel molecules as miRNA-like small RNAs (milRNAs) due to the striking similarities in the biogenesis process shared with conventional miRNAs. Some indications point towards the regulatory role of milRNAs in *Neurospora* resembling that of animal and plant miRNAs. However, it's important to note that milRNAs exhibit remarkable flexibility in both their sizes and the structures of their pre-miRNAs, suggesting a different biogenesis mechanism (Lee et al. 2010). As of now, a plethora of milRNAs has been identified in numerous fungal species.

Furthermore, miRNAs exhibit ubiquitous expression across a wide spectrum of eukaryotic parasites, although the available information remains limited. These molecules may have a role in orchestrating post-transcriptional gene regulation across various facets, including inflammatory and immune responses, cell cycle control, apoptosis, autophagy, and cytoskeleton reorganization (Acuña et al. 2020; Riahi Rad et al. 2021).

12.1.2 miRNA Biogenesis

Most of the miRNAs known to date, whether from animals or plants, are transcribed by RNA polymerase II or III (O'Brien et al. 2018). This is followed by a series of molecular processes that will lead to the maturation of the immature miRNA. Most genes encoding miRNAs are transcribed into RNAs with a specific molecular structure called primary miRNAs (pri-miRNAs) which then become precursor miRNAs (pre-miRNAs) that will mature into miRNAs per se. However, there have been identified variations in the synthesis pathways which result in different isoforms known as "IsomiRs" derived from the same miRNA gene (Bartel 2018). MiRNA isoforms are heterogeneous and vary in length, sequence or both as a consequence of the alternative cleavage by Drosha and Dicer, 5′ or 3′ at the respective ends.

In this section, we will discuss the main miRNA synthesis pathways. Depending on the identity of the miRNA, its biogenesis will follow a set of transformations and processes grouped into two types of pathways, classified as canonical or non-canonical.

12.1.2.1 Canonical Biogenesis Pathway

Among the different miRNA synthesis pathways, canonical biogenesis is the major pathway by which miRNAs are processed (see Fig. 12.1). Most miRNA coding transcripts are initially transcribed by RNA polymerase II as long pri-miRNA with a 5′ guanosine cap and a 3′ polyadenylated tail. This structure is processed by a multiprotein complex called Microprocessor resulting in a pre-miRNA with an approximate size of 70–120 bp and a stem-loop structure (Bhaskaran and Mohan 2014). The typical structure of the Microprocessor complex consists of at least two proteins. On the one hand, it contains Drosha, a type III RNAase enzyme highly conserved in animals with a molecular weight of approximately 160 kDa, which dimerizes with the other indispensable component of the complex, the

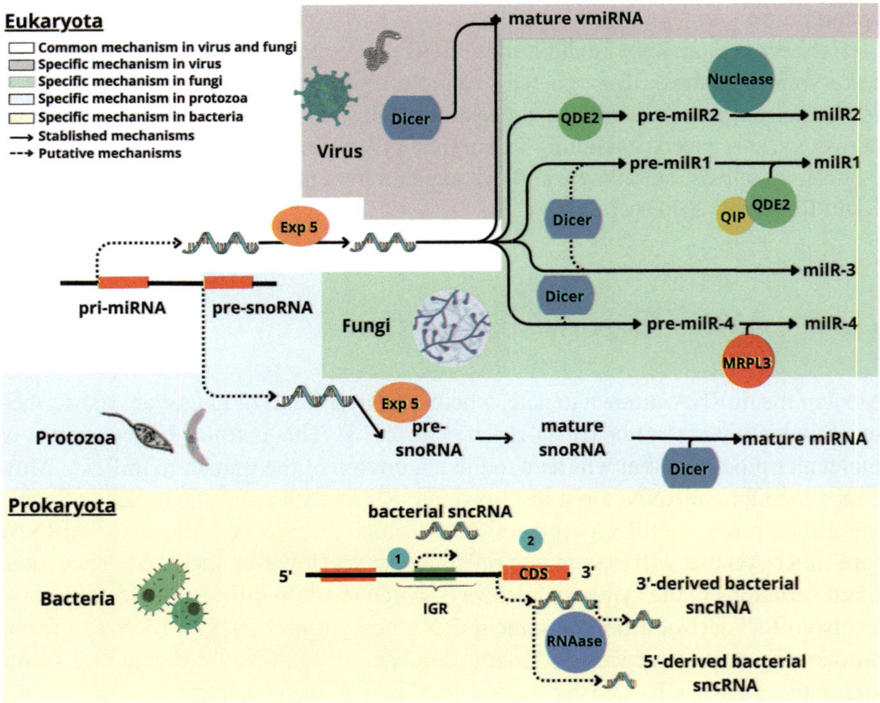

Fig. 12.1 Biogenesis pathways of miRNAs. Each group of organisms has particularities in their sncRNA biogenesis process. Viral miRNAs are transcribed in the cell nucleus, exported to the cytoplasm, and cleaved by Dicer to give rise to mature vmiRNAs. In fungi, after transcription and export to the nucleus, different proteins act in the maturation of pri-milRNAs. In milR1, in addition, recruitment of QIP and QDE2 is necessary for the final maturation of the transcript. Catalytic activity of QDE2 is indispensable for the formation of pre-milR2. In the case of pre-milR4, after cleavage by Dicer, MRPL3 leads its maturation. In protozoa, mature snoRNAs serve as substrate for Dicer. In bacteria, most sncRNAs are transcribed from non-coding genes located in intergenic regions (IGRs). Alternatively, they can be synthesized from coding genes, by premature termination in the 5′ region, or by mRNA processing, giving rise to sncRNAs derived from the 3′ or 5′ ends

double-stranded RNA-binding protein DiGeorge syndrome critical region gene 8 (DGCR8) (Bhaskaran and Mohan 2014). DGCR8 recognizes N6-methyladenylated GGAC and other motifs in the apical region of the pre-miRNA hairpin through two double-stranded RNA-binding domains located toward its C-terminal end, exposing the duplex region for subsequent Drosha action (Leitão and Enguita 2022). This new pre-miRNA has a typical structure with a 5′ phosphate and two 3′ overhang nucleotides. It is subsequently transported to the nucleus by exportin 5 in a Ran protein-dependent process (Bhaskaran and Mohan 2014). In the cytoplasm occurs the maturation of pre-miRNA into 22-nucleotide-long duplexes by Dicer-1, a type III RNase, together with accessory proteins, resulting in a mature miRNA duplex (Zhang et al. 2004). The two miRNA strands are subsequently separated, influenced by diverse factors such as the thermodynamic imbalance within the duplex and the stability of base pairing at the 5′ end. One of these strands, designated as the guide strand, teams up with the previously mentioned RNA binding proteins, which encompass trinucleotide repeat-containing gene 6A (TNRC6A). This partnership established in coordination with catalytic Argonaute (AGO) proteins, leading to the formation of a micro ribonuclear protein complex known as the RNA-induced silencing complex (RISC) (Bhaskaran and Mohan 2014).

12.1.2.2 Non-canonical Biogenesis Pathway

Besides the conventional pathway, several alternative pathways have been identified and thoroughly characterized. These alternative routes involve different combinations of essential proteins typically associated with the canonical pathway, notably Drosha, Dicer, exportin 5, and AGO-2. Two primary non-canonical miRNA biogenesis pathways are identified, namely, Drosha/DGCR8-independent and Dicer-independent (O'Brien et al. 2018). An example of the first is constituted by molecules that play a dual function as introns and as miRNAs called "mirtrons" (see Fig. 12.1). These are debranched introns that are incorporated into the biogenesis pathway as pre-miRNA hairpins after processing by the spliceosome. Following splicing, pre-miRNAs originating from miRtrons are transported to the cytoplasm with the assistance of XPO5 and are subsequently subject to Dicer-mediated processing, ultimately giving rise to mature miRNAs (Goymer 2007). miRtrons can be differentiated from conventional miRNAs by identifying specific characteristics such as structural irregularities in the stem region, guanine composition, free energy of the hairpin, and the length of the hairpin (Rorbach et al. 2018).

In the second case, miRNAs that do not depend on Dicer are generated from endogenous shRNA transcripts, generating molecules with insufficiently long hairpins to be processed by Dicer (see Fig. 12.1). Instead, it is processed by AGO-2, which directs the cleavage of the 3′-arm of the hairpin. Following that, the 3′–5′ trimming of the 5p strand generated the mature miRNA. Through this pathway, miR-451 is processed, which is the non-canonical miRNA whose function has been most clearly elucidated in vertebrate mammals (O'Brien et al. 2018; Bartel 2018).

12.1.2.3 miRNA Biogenesis in Viruses, Bacteria, Fungi, and Protozoa

In the case of viruses, the majority of vmiRNA biogenesis processes rely on Dicer activity (see Fig. 12.1). One of the most extensively investigated virus-associated RNAs, the human adenovirus (HAdV), utilizes the host's miRNA synthesis machinery to generate its vmiRNAs (Nanbo et al. 2021). Dicer plays a highly specific role in cleaving the terminal stem of vaRNAI and vaRNAII (Punga et al. 2020). The precise location of Dicer's cleavage is particularly significant because variations in this site lead to vmiRNAs with distinct seed sequences. Consequently, these vmiRNAs possess the potential to target diverse mRNA molecules (Kamel et al. 2014).

When comparing sncRNAs in eukaryotes and bacteria, eukaryotic miRNAs display differences from bacterial sRNAs in terms of their shared biogenesis pathways and their reliance on specific proteins such as AGO and protein complexes like RISC for their functionality (Wagner and Romby 2015). The conventional sRNA biogenesis pathways entail the production of sRNAs through the transcription of distinct noncoding genes, typically located in proximity to protein-coding genes (see Fig. 12.1). Alternatively, sRNAs can be generated through various mechanisms, including premature transcription termination in the 5′ region, initiating transcription within the coding sequence (CDS) while utilizing the same terminator, or via the processing of mRNA molecules (Ponath et al. 2022). The majority of these sRNAs are transcribed from standalone sncRNA genes situated in intergenic regions (IGRs) between annotated protein-coding genes (Chao et al. 2012).

Animal and plant miRNAs typically adhere to well-established canonical biogenesis pathways. However, when delving into the genetic and molecular aspects of fungi, particularly in the context of *N. crassa* milRNAs, researchers have unveiled an astonishing diversity in biogenesis mechanisms. These pathways engage a variety of RNAi components, such as Dicer, the argonaute QDE-2, QIP (a potential RNase III domain-containing protein), MRPL3, and other unknown nucleases, all contributing to the production of diverse milRNAs (see Fig. 12.1). Out of the four most prevalent milRNAs, only milR-3 aligns with the canonical plant miRNA pathway, which primarily relies on Dicer for pre-milRNA processing. In contrast, the biogenesis of milR-4 is partially dependent on Dicer and necessitates MRPL3. The investigation into milR-1 and milR-2 milRNA maturation has shed light on the distinct and essential roles played by the Argonaute QDE-2 in two innovative sRNA biogenesis pathways. For milR-1, the pri-milR-1, which is characterized by its hairpin structure, undergoes an initial processing step by Dicer, yielding a double-stranded pre-milR-1. Subsequently, QDE-2 binds to pre-milR-1 and recruits the exonuclease QIP for further processing, ultimately leading to the maturation of milR-1. In this pathway, QDE-2 functions as a scaffold in milRNA maturation rather than a slicer. Instead, milR-2 is generated independently of Dicer. In contrast, the catalytic activity of QDE-2 is indispensable for producing mature milR-2. This suggests that, after binding to pre-milR-2, QDE-2 cleaves the milRNA* strand, enabling additional milRNA processing by another unidentified nuclease (Dang et al. 2011).

One of the earliest protozoa to be extensively studied to understand the biogenesis of its miRNAs and their role in its life cycle was *Giardia lamblia*. Shortly before this, a homologous argonaute and a functional Dicer were identified in this ancestral eukaryote, despite its apparent inability to engage in RNA interference (RNAi). The Argonaute protein, a key element in growth, exhibits a specific binding affinity to the m7G cap, suggesting potential involvement in miRNA-mediated translation repression (see Fig. 12.1). The snoRNAs identified in *Giardia* typically range from 60 to 160 nucleotides in length, a size conducive to folding into hairpin structures suitable for Dicer-mediated cleavage. Consequently, mature snoRNAs could serve as direct substrates for Dicer without requiring further processing. Given the minimalist nature of *Giardia*, it is plausible to propose that this organism employs a single RNA processing mechanism for both snoRNAs and miRNA precursors exhibiting snoRNA-like characteristics (Saraiya and Wang 2008).

12.1.3 Function and Molecular Mechanisms of miRNAs

The function of miRNAs is related to post-transcriptional regulation of gene expression through the RNA-interference (RNAi) pathway (Pillai et al. 2004). In the majority of cases, miRNAs engage in interactions with the 3′ untranslated region (UTR) of target mRNAs to repress their expression (Ha and Kim 2014). The specific sequence pairing is determined by nucleotides in position 2–7 in the 5′ region of mature microRNAs, known as the "seed" sequence of the miRNA (see Fig. 12.2). In animals, this sequence complementarity is typically partial, and miRNAs generally inhibit mRNA translation or target mRNA degradation by removing the 5′ cap and the 3′ poly-A tail (Eulalio et al. 2007; O'Brien et al. 2018).

While significant advancements have been achieved in understanding the origin and roles of miRNAs, the mechanisms through which miRNAs regulate gene expression remain a subject of scientific discussion. Notably, various published studies have demonstrated that miRNAs in animal cells can exert repression on gene expression through as many as nine distinct mechanisms. On one hand, inhibitory processes have been observed due to the binding of miRNA to the 40S ribosomal complex for the formation of the initiation complex and its binding to the 60S ribosomal complex. In addition, the inhibition of translation elongation and premature termination of translation has been observed through ribosome disassembly or mRNA sequestration in P-bodies. On the other hand, miRNAs are also capable of destabilizing and/or degrading mRNA or protein during translation. Finally, transcription inhibition has also been reported through chromatin reorganization mediated by miRNAs (Morozova et al. 2012). In addition, miRNAs have been reported to be transported between various subcellular compartments to regulate translation and transcription rates (Makarova et al. 2016) and even affect neighboring cells, distant cells, and even cells of different species. This characteristic is crucial for enabling host-pathogen interactions mediated by miRNAs (He and Pan 2022).

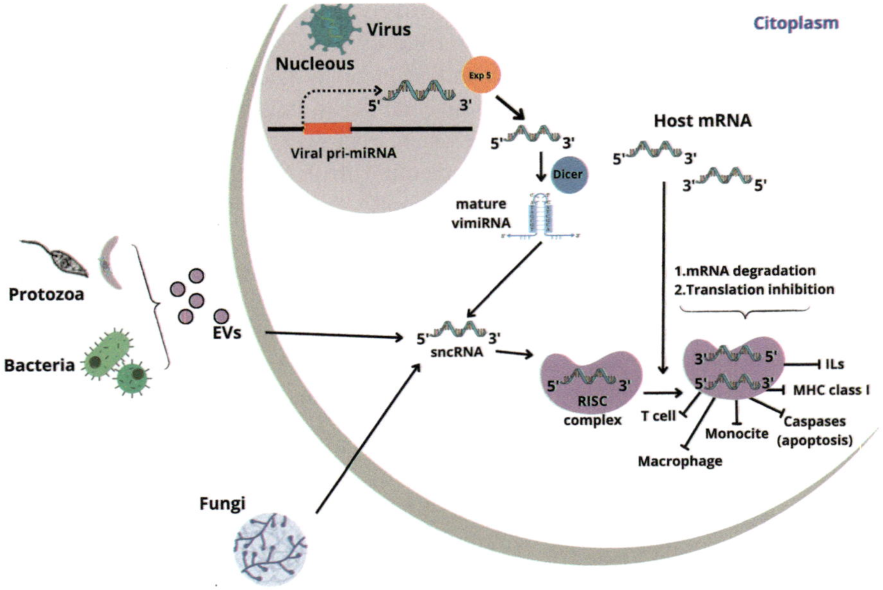

Fig. 12.2 Gene regulation by sncRNAs promote immune system evasion. Viral sncRNAs are transcribed in the cell nucleus and subsequently exported to the cytoplasm via exportin 5, whereas protozoa, bacteria, and fungi release sncRNAs into the cell. Once inside the cytoplasm, sncRNAs are loaded into the RISC complex that binds to the host target mRNA, resulting in its degradation or inhibition of its translation. As a result, different elements of the immune system are mainly inhibited, leading to immune evasion

12.1.3.1 Dynamics of Viral miRNAs: Mechanisms of Host-Virus Gene Control

Viral miRNAs regulate host genes for their own purpose. Associations of vmiRNAs with AGO proteins have been reported, leading to the formation of the miRISC complex, which can regulate both host and self-genes (Punga et al. 2020). In addition, it has also been reported that viruses and host cell miRNAs employ diverse mechanisms to interfere with each other (see Fig. 12.2). Viruses can disrupt or modify the host cell miRNA pathway through interactions with key proteins, create and regulate their own miRNAs (Grundhoff and Sullivan 2011; Harwig et al. 2014), or exploit host cell miRNAs to support their replication cycle (Luna et al. 2015). Conversely, host cells can counter viral miRNAs with their own endogenous miRNAs (Fruci et al. 2017).

12.1.3.2 Small RNAs (sRNAs) in Bacterial Post-transcriptional Regulation: Insights, Mechanisms, and Gram-Specific Variations

sRNAs in bacteria perform a role akin to that of miRNAs in animals by acting as post-transcriptional regulators of significant cellular processes (see Fig. 12.2). It seems that fundamental aspects of sRNA regulation are preserved across all bacterial species. It has been described that the sRNA-mRNA interaction represses gene expression mainly by two processes: by hindering the ribosome binding site or by decreasing mRNA stability and consequently leading to transcript degradation (King et al. 2022). For example, McaS is an sRNA identified in *E. coli*, which interacts with multiple mRNAs as well as RNA-binding proteins. This molecule exerts negative regulation on the expression of csgD. Moreover, it has also been reported that McaS also has a positive regulatory effect on flagellum synthesis when bacteria move into the stationary phase of growth, which occurs after base pairing with the 5′-UTR of the flhDC operon (Boehm and Vogel 2012).

It is noteworthy that differences exist in the molecular process by which gene regulation is carried out between gram-positive and gram-negative bacteria. In this process, the importance of RNA chaperones in Gram-negative species in mediating the RNA-mRNA interaction has been reported. In contrast, Hfq RNA chaperones are absent or not required in Gram-positive bacteria, where dependence on C- or CU-rich motifs for base pairing has been reported (Jørgensen et al. 2020).

12.1.3.3 Molecular Functional Characterization of milRNAs and Their Impact on Fungal Pathogenicity

The functional characterization of miRNA-like small RNAs (milRNAs) in certain fungi remains an ongoing process (see Fig. 12.2). Among the earliest organisms in which several milRNAs have been functionally assessed is *Verticillium dahliae*. The precursor gene VdMILR1 is transcribed by RNA polymerase II and can produce mature VdmilR1 independently of the Dicer-like (DCL) and Argonaute (AGO) proteins of *V. dahliae*. Notably, VdR3, a protein featuring an RNase III domain, appears to be a player in VdmilR1 biogenesis. VdmilR1 targets a putative protein-coding gene, VdHy1, within its 3′UTR, resulting in transcriptional repression through an enhancement of histone H3K9 methylation at the VdHy1 *locus*. A comprehensive examination of pathogenicity implies that VdHy1 is a critical component for the virulence of the fungus (Jin et al. 2019).

12.1.3.4 Molecular Functional Characterization of milRNAs in Protozoa

Protozoan miRNAs possess the capability to modulate gene expression post-transcriptionally by binding to both the 3′ and 5′ untranslated regions (3′UTRs) and the coding regions of target mRNAs inducing translation inhibition or mRNA destabilization and degradation (see Fig. 12.2). However, the molecular mechanisms of miRNA regulation in protozoa are not fully understood. Research has shown that miRNA pathways in protozoan parasites exhibit different characteristics from those in other metazoan organisms, possibly shedding light on the evolution of miRNA-induced silencing networks (Zheng et al. 2013). In addition, it has been reported that the RISC complex is more heterogeneous in sequence compared to other animals. Current knowledge of miRNAs in protozoa is limited, and more research is needed to elucidate the molecular machinery related to miRNA regulation in protozoa (Rojas-Pirela et al. 2022).

12.2 Inter-kingdom miRNAs on Human Diseases: Infectious Diseases

The discovery and characterization of miRNAs have opened new perspectives in our understanding of gene regulation and its implications for human health. miRNAs have predominantly been studied in the context of the human genome, where they have a key role in the post-transcriptional modulation of gene expression. However, new research has revealed a remarkable and hitherto underappreciated dimension of miRNA biology: the existence of communication between miRNAs from different kingdoms. This phenomenon encompasses the transfer of miRNAs between species, transcending kingdom boundaries in the biological hierarchy, and has profound implications for human disease (Cai et al. 2019). Moreover, these interactions play a role in human diseases (see Figs. 12.2 and 12.3). In this section, we compile the information obtained so far on interactions via miRNA, milRNAs, vmiRNAs, and bacterial sRNA, occurring between species belonging to different kingdoms in infectious diseases and cancer.

Infectious diseases arise from the colonization and replication of pathogenic microorganisms into the host organism where the interaction between miRNAs from infectious agents and host exert a significant influence in the development of pathology (Aguilar et al. 2019).

The initial studies aiming on the involvement of miRNAs in infections were focused on viruses (Bruscella et al. 2017). These pioneering studies unveiled the expression of miRNAs by DNA viruses. In the last few years, miRNAs have come to be recognized as influential elements in the interplay between bacterial pathogens and their host. Similarly, bacteria have exhibited complex mechanisms that facilitate the establishment of an immunotolerant environment and/or the manipulation of host physiological processes for their benefit, thus contributing to pathogen survival, replication, and induction of latency/persistence through sncRNA mediation (see Fig. 12.2). Moreover, emerging research suggests that miRNAs play a fundamental role in modulating the host's response to microorganisms, influencing infection progression, immune responses, and disease outcomes. The sncRNAs produced by bacteria and viruses related to infectious diseases collected in this review are

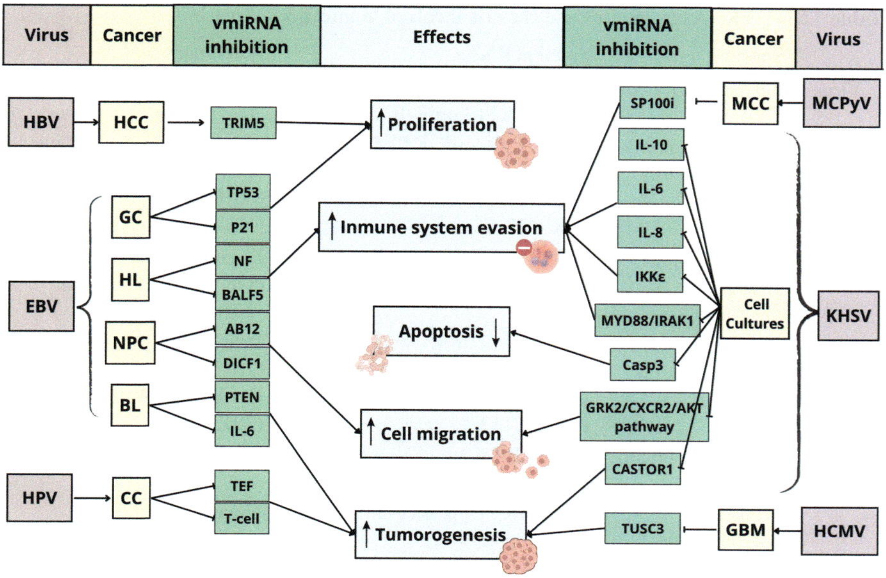

Fig. 12.3 Viral miRNAs involved in different types of cancer. The main vmiRNA-producing oncoviruses, tumor processes they regulate, and associated cancer types are shown. HCC, hepatocellular carcinoma; GC, Gastric cancer; HL, Hodgkin lymphoma; NPC, nasopharyngeal cancer; BL, Burkitt's lymphoma; CC, cervical cancer; MCC, Merkel cell carcinoma; GBM, glioblastoma multiforme

shown in Table 12.1. Similarly, other infectious agents like fungi and parasites have also been documented (Aguilar et al. 2019).

12.2.1 Role of Viral miRNAs in Human Genome During Disease Infections

The dynamic interaction between viruses and their hosts involves complex mechanisms that facilitate viral survival and replication in the host environment. One such mechanism, which has attracted increasing interest, is the use of viral miRNAs to control host gene expression. This phenomenon highlights the sophisticated strategies that viruses employ to manipulate host cellular processes to their advantage (Skalsky and Cullen 2010).

In this context, viral miRNAs have surfaced as potent controllers of host gene expression, effectively coordinating numerous mRNA targets (see Table 12.1). Unlike viral proteins, these miRNAs often evade detection by the host immune system, enhancing their efficacy in shaping the course of infection. This essay delves into the multiple functions of viral miRNAs, shedding light on their impact on viral persistence, host immune responses, and intercellular communication. Through an

Table 12.1 sncRNAs from virus and bacteria involved in infectious diseases

	Species	Small ncRNAs	References
Virus	Epstein-Barr virus	miR-BART2-3p, miR-BART2-5p, miR-BART13, miR-BART15	De Re et al. (2020); Kawano et al. (2013)
	Kaposi's sarcoma-associated virus	miR-K12-9, miR-K12-5, miR-K12-1, miR-K12-11, miR-K12-3, miR-K12-4-3p, miR-K12-6	Abend et al. (2012); Gallo et al. (2020); Qin et al. (2010, 2012); Suffert et al. (2011); Hansen et al. (2010)
	Herpex simplex virus 1	miR-H6, miR-H-27, miR-H2-3p, miR-H8	Duan et al. (2012, 2019); Wu et al. (2013); Enk et al. (2016)
	Human cytomegalovirus	miR-UL112, miR-UL112-3p, miR-US25-2-3p, miR-US4-1, miR-UL148D, miR-US25-1, miR-UL112-1	Dass et al. (2023); Kim et al. (2012); Esteso et al. (2014); Hook et al. (2014); Qi et al. (2013)
	John Cunningham polyomavirus	JCV-miR-J1-3p, JCV-miR-J1-5p	Lagatie et al. (2013)
	Merkel cell polyomavirus	mcp-miR-J1-3p, mcp-miR-J1-5p	Lagatie et al. (2013)
	SARS-CoV-2	miRNA-147-3p, miRNA-359-5p, -miRNA-N-28612, miRNA-N-29094, miRNA-N-29443	Khan et al. (2020); Walsh et al. (2013); Meng et al. (2021)
	Ebola virus	miR-MAY-251, miR-MAK-403	Diallo et al. (2022b)
Bacteria	*Legionella pneumophila*	RsmY	Sahr et al. (2022)
	Pseudomonas aeruginosa	sRNA-52320	Wang and Fu (2019)
	Listeria monocytogenes	sRNA rli32	Frantz et al. (2019); Abdullah et al. (2012)

exploration of recent research, we uncover the strategic relevance of viral miRNAs in the context of host-virus interactions, elucidating their pivotal role in the establishment and progression of infections (Auvinen 2017). Numerous studies provide insights into the diverse landscape of viral miRNAs in various infections, highlighting their roles in manipulating host processes and their contribution to pathogenic outcomes. The best-characterized examples are reviewed below.

12.2.1.1 Herpesviridae

Viral miRNAs are predominantly found in the Herpesviridae family, and it has been established that specific miRNAs encoded by herpesviruses are linked to viral latency and its associated pathogenic consequences (Piedade and Azevedo-Pereira 2016). *Kaposi's sarcoma-associated herpesvirus* (KSHV), also known as Human herpesvirus 8 (HHV-8), is a linear double-stranded DNA oncovirus belonging to the Gammaherpesvirinae family. KSHV encodes a total of 25 miRNAs, originating from 12 viral pre-miRNAs. The Kaposi promoter orchestrates the regulation of KSHV miRNA gene clusters during viral latency. Numerous KSHV pre-miRNA genes are apparently intergenic, situated between the Kaposi gene sequence and open reading frame (ORF) 71 within the KSHV genome (Qin et al. 2017). This virus is able to express miRNAs to regulate host cellular mRNAs in order to evade the host immune response. For example, TLR/IL-1R signalling is mediated by two KSHV-encoded miRNAs: miR-K12-9 which regulates IRAK1 and by miR-K12-5 which regulates MYD88 (Abend et al. 2012). Another example is miR-K12-1 which has been reported to regulate the MHC class I-related B chain (MICB) mRNA to prevent NK cell recognition, or miR-K12-11 that is able to regulate SMAD5, a TGF-beta pathway intermediate promoting cell survival and virus-induced oncogenesis (Gallo et al. 2020). Moreover, miR-K12-1, miR-K12-3, and miR-K12-4-3p, in a cooperative approach suppresses caspase-3 and inhibit caspase-3-induced apoptosis (Suffert et al. 2011). Some KSHV-encoded miRNAs can increase the susceptibility of cells to be infected. This is the case for miRNA miR-K12-1, miR-K12-9, and miR-K12-11, which have also been described to increase the expression of a viral fusion entry receptor xCT, making macrophages and endothelial cells more susceptible to infection (Qin et al. 2010). Specifically, miR-K12-11 together with miR-K12-6 are able to reduce the expression of the transition factor v-MAF which is homologous to the MAF oncogene, which in turn down-regulates xCT (Hansen et al. 2010; Qin et al. 2012).

An additional illustration within the Gammaherpesvirinae subfamily pertains to research on the Epstein-Barr virus (EBV), a double-stranded DNA virus classified under the Gammaherpesvirinae subfamily. Its genome encodes 45 mature microRNAs derived from 25 precursors, which are distributed across two regions of the genome: BHRF1 (Bam HI fragment H rightward open reading frame I) and BART (Bam HI-A region rightward transcript). The BART region is further divided into subclusters 1 and 2, with the miRNAs ebv-miR-BART2-5p and ebv-miR-BART2-3p located downstream of these two clusters (De Re et al. 2020). Notably, EBV miR-BART2-5p, along with 13 and 15 other miRNAs, have been implicated in active chronic EBV infections characterized by elevated morbidity and mortality rates (Kawano et al. 2013).

In the subfamily Alphaherpesvirinae, herpes simplex virus 1 (HSV-1) and 2 (HSV-2) encode 27 and 24 functional vmiRNAs, respectively, with several of these sequences being shared between the two serotypes. The vmiRNAs encoded in this group of viruses are mainly involved in the regulation of viral proliferation and modulation of the immune system. For example, miR-H6 appears to play a crucial role in HSV-1 latency (Duan et al. 2012) and is usually detected in HSV-1 infection targeting ICP4, inhibiting replication in human corneal epithelial cells. In contrast, miR-H-27 decreases the transcriptional output of immediate-early and early viral genes by targeting and blocking KLHL24, facilitating cell proliferation and thus viral population growth (Wu et al. 2013). Moreover, miR-H2-3p vmiRNA targets the DDX41 gene, reducing the host immune response (Duan et al. 2019), and MiR-H8 stops viral replication and modulates the immune system indirectly by targeting the GPI gene (Enk et al. 2016).

Within the Betaherpesvirinae subfamily, there has been a particular focus on the study of human *cytomegalovirus* (HCMV). A total of 21 miRNAs, originating from HCMV, have been identified, and these viral miRNAs target multiple host genes associated with the immune system. For example, HCMV-miR-UL112 is involved in the down-regulation of NKG2D immune recognition by targeting MICB. Moreover, it acts in tandem with cellular miR-376a to suppress MICB expression (Dass et al. 2023). In turn, HCMV-miR-UL112-3p targets TLR2 to inhibit signalling through NF-kB (Kim et al. 2012). Therefore, miR-US25-2-3p, targets tissue inhibitors of metalloprotease 3 (TIMP3), reducing recognition by NK cells (Esteso et al. 2014). miR-US4-1 also contributes to immune evasion by silencing ERAP1, which is involved in processing and presentation by MHC class I to cytotoxic T cells. HCMV miR-UL148D targets the chemokine ligand 5 (C-C motif) gene (CCL5) (Kim et al. 2012). This sequence (miR-UL148D), along with miAR-UL112 and miR-US25-2-3p, plays a role in HCMV NK evasion. These examples respond to the need of the virus to avoid the cellular immune response by gene silencing of different mRNAs (Dass et al. 2023). On the other hand, it has also been described that HCMV-miR-US25-1 hinders cell cycle progression by post-transcriptional silencing of more than 20 mRNAs, including CD147, CCNE2 and BRCA1/BRCA2 complex subunit 3 (BRCC3) (Hook et al. 2014). In turn, miR-US25-2-3p regulates eIF4AI expression levels, possibly enhancing latency and cell phase regulation (Qi et al. 2013).

12.2.1.2 Polyomaviridae

As we have seen above, most of the vmiRNAs have been studied in the family Herpesviridae. However, miRNAs regulating host genes have also been described in other viral families. Merkel cell polyomavirus (MCPyV) and JC polyomavirus (JCPyV or JCV), belonging to the family Polyomaviridae, encode 5p and 3p miRNAs, respectively, which are cleaved from a common precursor transcript and show perfect homology with their target sequences (Lagatie et al. 2013). The role appears to be specialized in downregulating host antiviral immune responses through two mechanisms: *in cis*, which involves regulation of the accumulation of viral gene products that might otherwise trigger enhanced host immune responses, and *in trans*, which involves downregulation of host transcripts associated with immune responses (Zou and Imperiale 2021). MCPyV polyomavirus miRNAs appear to be crucial in establishing persistence; however, the genes it regulates in the host are unknown (Theiss et al. 2015).

12.2.1.3 Coronaviridae

Due to the worldwide pandemic, the coronaviridae family has been extensively studied in recent years. In respiratory disease resulting from coronavirus infection, the miRNA molecules of the virus play an active role, being involved in the progression of the virus life cycle, as well as in the regulation of the inflammatory response and modulation of the immune system (Panda et al. 2022). An in silico study identified 170 mature miRNAs from SARS-CoV-2 that are likely to regulate 3377 human genes (Khan et al. 2020). This interaction between viral miRNAs and human genes can modulate multiple signalling pathways involved in various biological processes, including insulin regulation. These encompass T cell-mediated immunity, autophagy, FGF receptor binding, type I interferon signalling, TNF-alpha, PI3K/Akt, NF-κB, VEGF, TGF-beta, ErbB, mTOR, MAPK signalling pathways, as well as the Notch signalling pathway. These pathways have the potential to impact the development of both the heart and brain (Khan et al. 2020).

Some studies suggest that SARS-CoV-2 employs viral miRNAs that interfere with human miRNAs responsible for regulating processes related to fatty acids, insulin resistance, renal and myocardial functions, and peroxisome proliferator-activated receptor (PPAR)-activated signalling (Khan et al. 2020). Furthermore, miRNA-147-3p has been studied for its role in priming TMPRSS2 gene expression in human gastrointestinal tissue. This effect facilitates the entry of the virus into the tissue and can protect its mRNA from degradation by the host (Walsh et al. 2013). Another vmiRNA, miRNA-359-5p, upregulates MYH9 expression, promoting adhesion, invasion, migration, and release into the host cell (Panda et al. 2022). Additionally, Meng et al. identified three significant viral miRNAs (v-miRNA-N-28612, v-miRNA-N-29094, and v-miRNA-N-29443) that could potentially regulate host genes SNCA, ACO1, BNIP3L, DMBX1, CLDN10, and BCAS1 transcripts. Finally, SARS-CoV-2 viral miRNAs have also been shown to regulate host inflammasome components such as caspase 1, IL-1, and NLRP3 (Meng et al. 2021).

These findings highlight the intricate relationship between viral miRNAs and host genes, shedding light on their potential roles in the pathogenesis and progression of COVID-19. Further research is essential to fully understand the mechanisms and implications of these interactions.

12.2.1.4 Filoviridae

The infectious agent responsible for ebola disease consists of a group of viruses categorized within the *Ebola virus* (EBOV) genus of the Filoviridae family. EBOV is a negative-strand RNA virus that replicates in the cytoplasm and can lead to a highly lethal hemorrhagic fever. EBOV can encode its own vmiRNAs as a means to manipulate host immune responses. Several research groups have made predictions about the possible existence of miRNAs encoded by EBOV, until recent research has been able to experimentally validate the existence of miR-MAY-251 and miR-MAK-403, which are capable of regulating host mRNA, and this modulation could be related to hemorrhagic phenotyping, regulation of virus replication, and regulation of host immune responses (Diallo et al. 2022b). Other studies have detected vmiRNA in several models of infection, including humans, and involving different EBOV variants; however, their functions are unknown (Duy et al. 2018).

12.2.2 Bacteria

In the field of study of "miRNAs like" synthesized by bacteria, most of the knowledge we currently have comes from studies carried out in Gram-negative bacteria, specifically in *E. coli* and species of the genus *Salmonella* (see Table 12.1). It is known that most of these sncRNAs are not constitutively expressed, but are usually expressed as a response mechanism to perturbations in the environment or to different phases of the organism's life cycle. However, understanding the involvement of sncRNAs in host gene regulation during the infectious process remains limited (Wagner and Romby 2015).

Given the relevance that this regulation of gene expression via sncRNA may have in host-pathogen interaction, it is interesting to study these molecules, especially in the case of human diseases (Choi et al. 2017). Recently, several studies have suggested that bacteria secrete sRNAs and tRNAs through extracellular vesicles (EVs) in Gram-positive bacteria or in outer membrane vesicles (OMVs) in Gram-negative bacteria, which can enter the interior of the host cell (Ahmadi Badi et al. 2017; Choi et al. 2017; Ahmadi Badi et al. 2020; Stanton 2021). Nevertheless, only a limited number of investigations have documented the stable transport of bacterial RNAs in OMVs or their association with proteins and biomolecules as a complexes (Ghosal 2018; Quendera et al. 2020). Alternatively, these RNAs have been detected in a free, unbound state circulating in the plasma of healthy human donors (Cui et al. 2019). Here, we review the most relevant findings, as well as the best characterized interactions between bacterial sRNAs and host mRNA in the context of bacterial infection.

12.2.2.1 Gram-Negative Bacteria

In gram-negative bacteria, the increase of OMVs is enabled by the influence of sRNAs through a mechanism conserved throughout evolution (Argaman et al. 2001; Song et al. 2008). OMVs typically exhibiting a diameter ranging from approximately 20–200 nm that harbor proteins, lipids, nucleic acids, and other molecules (Reza Aghasadeghi et al. 2011; Sedaghat et al. 2019). In the context of OMVs, the enclosed sRNAs have been posited as potential agents of interspecies communication. This is substantiated by their demonstrated capacity to modulate gene expression across various cell types and species, as referenced in (Koeppen et al. 2016; Choi et al. 2017). Furthermore, these sRNAs have been associated with the pathogenesis of a wide range of infectious diseases and chronic inflammatory conditions, including *Helicobacter pylori* infection (Keenan et al. 2000), Crohn's disease (Rolhion et al. 2005), atherosclerosis, and Alzheimer's disease (Chen et al. 2023). Taken together, within the context of host-pathogen interaction, pathogens may actively modify host cellular processes to their advantage by modulating the host mRNA expression profile (Sarshar et al. 2022).

Legionella pneumophila is a gram-negative bacterium parasitic on aquatic protozoa capable of causing severe respiratory disease in humans (Christie 2019). Sahr et al. shown that *L. pneumophila* releases EVs containing the sncRNA RsmY and tRNA-Phe which can regulate genes such as IRAK1 and RIG-I, modulating the immune system in human macrophages and monocytes (Sahr et al. 2022).

Pseudomonas aeruginosa is also a pathogenic microorganism of humans that lives in the mucus layer present on the surface of lung epithelial cells. This bacteria is capable of releasing OMVs loaded with sRNAs that diffuse through the mucus layer and internalize into lung epithelial cells. Specifically, a 23 nt tRNA fragment called sRNA-52320 is transported to target cells by this pathway. This sRNA has been shown to target different genes of the lipopolysaccharide-stimulated MAPK signalling pathway, attenuating IL-8 release and neutrophil migration into the lungs in murine models. Thus, sRNA-52320 may enhance *P. aeruginosa's* capacity to initiate a chronic lung infection in immunocompromised subjects, potentially influencing and limiting the host immune response (Wang and Fu 2019). Other gram-negative pathogenic bacteria in humans also use OMVs to introduce sRNAs into host cells. *Aggregatibacter actinomycetemcomitans* is a periodontal pathogen that has the ability to cross the blood-brain barrier, leading to severe symptoms. This pathogen secretes sRNAs within OMVs, and these sRNAs stimulate the production of tumor necrosis factor-alpha (TNF-α) by activating the NF-κB signalling pathways in macrophages. This, in turn, promotes the secretion of pro-inflammatory cytokines in the brain (Choi et al. 2017; Han et al. 2019; Ha et al. 2020). In addition, other periodontal pathogens, such as *Treponema denticola* and *Porphyromonas gingivalis*, also release sRNAs within OMVs. These sRNAs have been found to suppress the production of IL5, IL13, IL15 in T cells. This modulation of cytokine production could influence the host's immune response and potentially play a role in the development of periodontal diseases. *E. coli* also releases sRNAs within OMVs. These sRNAs are transferred into bladder epithelial cells and have the ability to suppress lipopolysaccharide (LPS)-induced IL-1α secretion (Dauros-Singorenko et al. 2020). *Listeria monocytogenes* secretes OMVs that contain the sRNA rli32. This sRNA enhances the synthesis of interferon-beta (IFN-β) in bone marrow-derived macrophages. Consequently, this process promotes the intracellular growth of the pathogen (Frantz et al. 2019; Abdullah et al. 2012).

12.2.2.2 Gram-Positive Bacteria

Some studies have shown the role of these sncRNAs in the regulation of gene expression in other intracellular pathogens. Gram-positive bacteria usually fulfil this purpose by EVs including sRNA of between 10 and 40 nt. While recent research has demonstrated the capacity of EV sRNAs to impact host cell biology, no studies have as yet pinpointed the specific sRNAs responsible for mediating the host response to EVs.

Staphylococcus aureus secretes EVs that carry immunostimulatory components in addition to sRNAs, contain other nucleic acids and peptidoglycan. These components have the ability to activate innate immune receptors and trigger autophagy in host A549 epithelial cells. This illustrates the complex interaction between bacterial pathogens such as *S. aureus* and host cells, in which EVs serve as a vehicle for transporting immunostimulatory molecules that regulate the host response (Bitto et al. 2021).

Another study elucidates the infection process of human THP-1 macrophage cells by *Mycobacterium marinum*, wherein the bacterium releases sRNA that binds to the host's complex RISC, thus disrupting gene expression. This finding suggests a potential mechanism for host regulation (Furuse et al. 2014).

12.2.3 Fungi

Currently, miRNA and milRNA-related research on eukaryotic organisms focuses mainly on animals and plants, leaving miRNAs from the fungal kingdom in the background. The fungus *Beauveria bassiana* has been documented to produce miRNA-bba-milR1, which plays a role in modulating and weakening the immune system of the mosquito *Anopheles stephensi* (Cui et al. 2019). Another study revealed that the pathogen *N. bombycis* produces Nb-milR8, increasing its ability to cause disease in its host, the silkworm *Bombyx mori* (Dong et al. 2021).

In the human context, it has been shown that host cells can release miRNA in response to fungal infections, sometimes through extracellular vesicles (Halder et al. 2022; Gupta et al. 2014). In addition, in silico studies performed in the fungus *Agaricus bisporus* have identified the ability to produce numerous miRNA-like sRNAs. This study revealed that approximately one-third of these miRNAs could regulate genes involved in various signalling pathways associated with human diseases, including 52 related to cancer, 28 to infectious processes and neurodegenerative diseases (Marin et al. 2022). Additionally, it has been proposed that the pathogenic fungus *Botrytis cinerea* could release miRNAs that have the ability to bind to RISC complex and consequently inhibit the expression of host immune-related genes, preventing the host immune response (Weiberg et al. 2013). In summary, we can conclude that the interaction between fungi and humans, and its possible implication in disease, remains a largely unexplored area of research.

12.2.4 Protozoa

Protozoan parasites, such as *Toxoplasma*, *Leishmania*, *Trypanosoma,* and *Plasmodium*, possess the capability to manipulate host cell's cellular machinery and signalling pathways to their advantage, ultimately fostering pathogenicity within the host cell. Delving into the role of miRNAs as potent regulators in orchestrating these mechanisms can greatly advance our comprehension of the biology and pathogenicity of these parasites (Acuña et al. 2020; Barrow et al. 2020).

Notably, *Toxoplasma* harbors its own miRNA processing machinery and exhibits the ability to discharge exosomes containing them (Menard et al. 2019). Additionally, distinct expression patterns have been documented across various strains of *Toxoplasma* (Wang et al. 2012). While there is a possibility that *T. gondii* can export miRNAs to its hosts for direct regulatory purposes, the specific targets and pathways modulated by these miRNAs remain undisclosed (Saçar et al. 2014). Analysis has unveiled that certain parasite, such as *Plasmodium*, *Cryptosporidium*, as well as various *Leishmania* and *Trypanosoma* species, lack a complete apparatus for sRNA biogenesis. Remarkably, *P. falciparum* demonstrates the capacity to exploit human cellular machinery to modulate the expression of its own genes. Inside the parasite, hAgo2 forms a complex with particular human miRNAs, such as let-7a and miR15a, enabling them to target genes within *P. falciparum*, as illustrated, for instance, in the targeting of the *P. falciparum* Rad54 gene (Lodde et al. 2022).

A set of microRNAs have also been described in *Trypanosoma brucei* targeting mRNAs corresponding to surface glycoprotein. Furthermore, clusters of multiple identical copies of miRNA forks have been discovered, suggesting their potential role as genetic switches that modulate host-parasite interactions (Mallick et al. 2008). Furthermore, certain computational tools have indicated that the putative target genes of miRNA-like elements expressed in *Leishmania major* exhibit associations with multidrug-resistant proteins, including ATP-binding cassette transporters and ribosomal proteins, as well as exonucleases, hydrolases, and RNA-binding proteins (Chandra Sahoo et al. 2013).

12.3 Inter-kingdom miRNA Regulation on Cancer

Cancer biology is a remarkably complex and profound field of study, in which a multitude of molecular and genetic processes underlying the development of malignant tumors are involved. Understanding this complexity is essential to advance in the prevention and treatment of cancer, one of the leading causes of mortality in the world. The majority of published research related with sncRNAs centers on the characterization of cellular miRNAs, understanding of how vmiRNAs, sRNAs, and milRNAs produced by organisms across various biological kingdoms can impact tumor development is not yet as comprehensive. Below, we present the most relevant and up-to-date information in this field. The sncRNAs related with cancer reviewed here is shown in Table 12.2.

12.3.1 The Role of Oncoviral miRNAs in the Human Genome

The molecular regulation of the development of cancer is a multifactorial process that is complex to address. Moreover, the complexity is intensified by the addition of a new factor, as it is roughly calculated that around 15–20% of human cancers are

Table 12.2 sncRNAs from virus and bacteria involved in cancer

	Species	Small ncRNAs	Cancer type	References
Virus	Human papillomavirus	miR-H1, miR-H3, miR-H5, miR-H6, mir-H1-1, miR-H2-1	Cervical cancer	Virtanen et al. (2016); Qian et al. (2013); Martinez-Zapien et al. (2016)
	Hepatitis B virus	HBV-miR-2, HBV-miR-3	Hepatocellular carcinoma	Yang et al. (2017); Yao et al. (2019)
	Merkel cell polyomavirus	mvc-miR-M1-5p, mvc-miR-M1-3p	Merkel cell carcinoma	Lee et al. (2011)
	Epstein-Barr virus	miR-BART-19-3p, miR-BART-13-3p, miR-BART2-5p	Hodgkin lymphoma	De Re et al. (2020)
		miR-BART-6-3p	Burkitt's lymphoma	Ambrosio et al. (2014)
		miR-BART20-5p, miR-BART17-5p, miR-BART5-5p, miR-BART7-3p,	Gastric cancer	Nachmani et al. (2009); De Re et al. (2020); Shinozaki-Ushiku et al. (2015)
		miR-BART13-3p, miR-BART3, miR-BART1-5p, miR-BART22	Nasopharyngeal cancer	De Re et al. (2020); Lei et al. (2013); Lyu et al. (2018); Lung et al. (2009)
	KSHV	miR-K12-1, miR-K12-3, miR-K12-4-p, miR-K4-5p, miR-K10a, miR-K12-10, miR-K12-11, miR-K3, miR-K12-5, miR-K12-9, miR-K12-7	Cell cultures	Chen et al. (2016); Hu et al. (2015); Li et al. (2019); Abend et al. (2010); Lei et al. (2012); Gottwein et al. (2007); Abend et al. (2012); Liang et al. (2011)
	Human cytomegalovirus	miR-UL112-3p, miR-UL70-3p	Glioblastoma multiforme	Liang et al. (2017); Ulasov et al. (2017)
Bacteria	*P. gingivalis*	sRNA-23392	Squamous cell carcinoma	Liu et al. (2021)
	H. pylori	sRNA-2509025, sRNA-989262	Gastric adenocarcinoma	Zhang et al. (2020)

directly or indirectly induced by viral infections. Some of these oncogenic viruses encode vmiRNAs in their genetic material and it has been reported that they can regulate degradative target mRNAs participate in the control of cellular proliferation leading the cell to an abnormal proliferative state (Krump and You 2018; Hull et al. 2022).

Oncovirus-related cancers have higher incidence rates among immunosuppressed patients (Grulich et al. 2007). In addition, reports suggest the existence of effector proteins and shared pathways between tumor suppressor signalling and innate immune signalling, exemplified by factors such as p53 and p21 (see Fig. 12.3). In addition, vmiRNAs play a critical role in viral propagation and persistence. Therefore, the main functions of these vmiRNAs can be synthesized in regulating the cellular splicing machinery (Boudreault et al. 2019), influencing cellular pathways implicated in the process of tumorigenesis and transitioning from the lytic phase to the latent phase to evade the immune system. In the process of tumorigenesis, miRNAs are able to influence various stages such as cell growth and proliferation, DNA damage response, apoptosis, and cell cycle regulation (Gallo et al. 2020). Again, we note in oncogenic viruses that the use of vmiRNAs seems to carry the advantage of not triggering the immune response as might the use of viral proteins (Krump and You 2018). We will review in more detail the targets and role of these miRNAs in the best characterized oncoviruses (see Table 12.2).

12.3.1.1 Human Papillomavirus

Among the oncoviruses, human papillomavirus (HPV) is notably prevalent and is a primary causative agent in the majority of cervical cancer cases, as well as other malignancies in areas like the vagina, vulva, penis, anus, mouth, and throat (Francies and Dlamini 2021). They have been reported several vmiRNAs in different HPV species, including HPV 16, 17, 37, and 41 (Chirayil et al. 2018), with disparities observed among the singular viruses studied (Weng et al. 2018). Within cervical cells infected by HPV-16, various viral vmiRNAs, including miR-H1, miR-H3, miR-H5, and miR-H6, were identified. These vmiRNAs are directed toward TEF-1, a transcription factor that plays a role in regulating cell proliferation and migration. Consequently, the modulation of TEF-1 activity has downstream effects, activating early promoters E6 and E7 (Virtanen et al. 2016; Qi et al. 2013; Ishiji et al. 1992). This activation involves E6-mediated p53 degradation and E7-mediated cell cycle-interrupting retinoblastoma degradation, ultimately promoting cell survival and leading to a tumorigenic process (Martinez-Zapien et al. 2016). It has been determined that miR-H1-1 has the capability to inhibit T cell activation and immune system development by targeting TCEA1, RAG1, ITGAM, CHD7, and BCL11A genes (Martinez-Zapien et al. 2016). In a similar manner, miR-H2-1 targets FOXP1, JAK2 , PKNOX1, SP3, XRCC4 proteins, which are essential for both the activation of T cells and the maturation of the immune system (Qian et al. 2013).

12.3.1.2 Hepatitis B and C Viruses

Based on the most recent statistics, an estimated 50–80% of Hepatocellular carcinoma (HCC) cases worldwide can be attributed to HBV infection (Venook et al. 2010). Notably, a substantial body of recent research has demonstrated the significant involvement of miRNAs in HBV replication and the pathogenesis of HBV-associated HCC (Jiang et al. 2021). While the existence of miRNAs encoded in the HCV genome has not been reported, numerous studies have focused on identifying vmiRNAs involved in the oncogenic process. Recently, it has been found that a vmiRNA called HBV-miR-2 may serve a dual function as a tumor suppressor and promoter of the oncogenic process. HBV-miR-2 inhibits TRIM35, a protein known to induce apoptosis and prevent cell proliferation in hematopoietic cells, thus acting as a tumor suppressor. In addition, HBV-miR-2 plays a role in the stabilization of RAN, a protein closely related to cell cycle regulation, nucleocytoplasmic transport, and cell transformation, thus favoring cancer development (Yao et al. 2019). In addition, it has been suggested that RAN may be involved with epithelial-mesenchymal transition in HCC cells (Song et al. 2017) and with the proliferative pathways of HCC and ovarian cancer (Wang et al. 2017).

Another miRNA, HBV-miR-3, has been linked to tumorigenesis in persistent HBV infection. HBV-miR-3 inhibits the expression of HBsAg, HBeAg, and HBc viral proteins and viral production, with the aim of avoiding excessive damage to host cells, which allows the infection to become chronic. In the long term, this infection may cause mutations that may trigger an oncogenic effect. Thus, HBV-miR-3 could indirectly influence the occurrence of HCC (Yang et al. 2017).

12.3.1.3 Merkel Cell Polyomavirus

Merkel cell polyomavirus (MCPyV) is a dsDNA polyomavirus capable of causing Merkel cell carcinoma in immunocompromised patients (Buck et al. 2016; Clarke et al. 2015). The vmiRNA family known as mvc-miR-M1, consisting of mcv-miR-M1-5p and mcv-M1-3p, both derived from a common pre-miRNA, demonstrates the capacity to regulate gene expression across phases of the virus life cycle and plays a significant role in evading the host immune response (Gallo et al. 2020; Lee et al. 2011). Akhbari et al. demonstrated that MCV-miR-M1 downregulates sp100i gene expression by decreasing SP100 protein levels, which in turn promoted decreased CXCL8 secretion, deregulating neutrophil chemotaxis toward MCPyV-infected Merkel cells. Additional potential targets for this vmiRNA include PIK3CD and PSME3, which are pivotal for antigen presentation within the host cell. Furthermore, it is noteworthy that RUNX1 can either stimulate or suppress tumorigenesis (Lee et al. 2011).

12.3.1.4 Epstein-Barr Virus

The Epstein-Barr virus is linked to several malignancies, encompassing Burkitt's lymphoma (BL), Hodgkin's lymphoma (HL), various B- and T-cell lymphomas, gastric carcinoma (GC), nasopharyngeal carcinomas (NPC), and colorectal cancer (CRC) (Francies and Dlamini 2021; Dauros-Singorenko et al. 2020; Martin-Perez et al. 2012).

Some studies reported that miR-BARTs 19-3p and 13-3 were overexpressed in HL cells, the latter being the most abundant (Dauros-Singorenko et al. 2020). Additionally, it has been documented that miR-BART2-5p is released via exosomes in EBV-positive HL cells. This miRNA effectively inhibits B-cell receptor (BCR)-mediated NF-kB activation, which is active during the lytic cycle. This inhibition contributes to the reduction of the protein level of BALF5 and the quantity of viral particles, facilitating the maintenance of a latent viral state. Furthermore, it has been observed that miR-BART2-5p also targets MICB, thereby assisting in the evasion of recognition and attack by natural killer cells (Dauros-Singorenko et al. 2020).

In Burkitt's lymphoma, it has also been reported that the expression of BART-6-3p vmiRNA is inversely related to the expression of protein phosphatase and tensin homolog receptor (PTEN) and IL-6, genes involved in tumor suppression and regulation of cell proliferation, respectively (Ambrosio et al. 2014). In addition, vmiRNAs from the BHRF1 gene have been found to enhance the ability of EBV to infect the organism by promoting acute infection and enhancing B-cell transformation (Wahl et al. 2013).

Approximately 10% of all gastric cancer cases are EBV-related. Again vmiRNAs, originating from miR-BARTs are overexpressed in these EBV-associated tumors. EBV-miR-BART20-5p has been associated to an unfavorable prognosis in individuals diagnosed with EBV-related gastric cancer (Nachmani et al. 2009). Furthermore, miR-BART17-5p influences the activity of the transcription factor KLF2. miR-BART3-3p hinders tumor infiltration by natural killer cells and macrophages by modulating the senescence-associated secretory phenotype (SASP), a phenomenon involving temporary or permanent cell cycle arrest. This modulation is achieved by direct interactions with TP53 and p21. Conversely, miR-BART5-5p also targets TP53, thereby promoting cell survival (Dauros-Singorenko et al. 2020). Moreover, miR-BART7-3p has been reported to contribute to tumor growth and invasion by inhibiting PTEN (Shinozaki-Ushiku et al. 2015).

miR-BARTs have been the subject of extensive investigation in the context of NPC. miR-BART13-3p has been found to enhance tumor cell migration and drive metastasis by inducing epithelial-mesenchymal transition through the downregulation of the tumor suppressor gene AB12 (De Re et al. 2020). miR-BART3, on the other hand, reduces the expression of DICE1, another tumor suppressor gene, thereby promoting increased cell proliferation and transformation (Lei et al. 2013). Furthermore, it has been noted that EBV-miR-BART1-5p instigates the activation of the AMPK/mTOR/HIF1 pathway, facilitating glycolysis and angiogenesis in NPC (Lyu et al. 2018).

The effect of vmiRNAs in CRC has also been reported. EBV-miR-BART18-3p regulates LDHA- and FASN-mediated metabolism by affecting lipogenesis (Meng et al. 2022).

Finally, it is remarkable that there are miRNAs from BART participating in tumor metastasis previously discussed as they are able to regulate E-cadherin and promoting cancer cell migration and invasion (Gao et al. 2012; Cai et al. 2015; Meng et al. 2022).

12.3.1.5 Kaposi's Sarcoma Herpesvirus/Human Herpesvirus-8 (KSHV/HHV8)

Kaposi's sarcoma-associated herpesvirus (KSHV) or *Human herpesvirus type 8* (HHV-8) exhibits the ability to initiate chronic infections primarily in lymphocytes, its principal reservoir, along with macrophages, keratinocytes, and endothelial cells, while also eliciting cellular transformation (Chang et al. 1994). KSHV is identified as the causative agent in the aetiology of primary effusion lymphoma (PEL) and multicentric Castleman's disease, with a heightened severity in immunosuppressed patients (Nador et al. 1996). Specifically, kshv-miR-K12-1, -3, and -4-3p suppress the apoptosis-inducing factor, Casp3, resulting in the inhibition of apoptotic processes. Conversely, kshv-K12-1 acts as an oncogene by activating the NF-κB/IL-6/STAT3 signalling pathway (Chen et al. 2016). Additionally, kshv-mir-K12-3 has demonstrated its role in promoting cell migration and invasion by targeting GRK2/CXCR2/AKT signalling (Hu et al. 2015), while kshv-miR-k12-1-5p enhances proliferation, migration, and invasion of Kaposi's sarcoma cells by suppressing SOCS6 signalling (Zhang et al. 2019). Furthermore, kshv-miR-K1-5p and kshv-miR-K4-5p have been identified as direct regulators of CASTOR1, leading to its inhibition and activation of the mTORC1 pathway, ultimately contributing to tumorigenesis (Li et al. 2019). kshv-mir-K10a exerts its influence preventing TWEAK-induced apoptosis and the expression of the inflammatory cytokine IL8 (Abend et al. 2010). Notably, HHV-8 encodes miRNAs with perfectly homologous seed sequences to cellular oncomiRs, as exemplified by kshv-mir-K12-10, an orthologue of hsa-mir-142-3p (Qin et al. 2014). Both of these miRNAs have demonstrated their capacity to inhibit the TGF-β pathway by targeting the TGF-β type II receptor (Lei et al. 2012). Furthermore, kshv-mir-K12-11, a homologue of hsa-mir-155, targets IKKε, BACH-1, and SMAD5, resulting in the downregulation of the transcription factor C/EBPβ, which serves as a regulator of IL6 (Gottwein et al. 2007). As for kshv-mir-K3, a homologue of hsa-mir-23, it shares anti-apoptotic effects through its targeting of caspase 3 and 77 (Manzano et al. 2013). In addition, two KSHV/HHV8-derived miRNAs, kshv-mir-K12-5 and kshv-mir-K12-9, impact inflammatory cytokine secretion by targeting MYD88 and IRAK1, key components of TLR/IL-1R-mediated signalling (Abend et al. 2012). On the other hand, Kshv-mir-K12-11 exerts an influence on the primary response to antiviral immunity through its targeting of IKKε, resulting in a reduction of I-IFN-like signalling. Meanwhile, Kshv-mir-K12-10 diminishes the production of IL-6 and IL-10 by targeting TWEAKR (Abend et al. 2012). Furthermore, Kshv-mir-K12-3 and Kshv-mir-K12-7, by their specific targeting of C/EBPβ, modulate cytokine secretion in monocytes and NK lymphocytes (Liang et al. 2011).

12.3.1.6 Human Cytomegalovirus (HCMV)

Human cytomegalovirus (HCMV) has also been associated with glioblastoma multiforme (GBM), a particularly aggressive cancer affecting the central nervous system. The v-miRNA miR-UL112-3p is significantly upregulated in this tumor type and targets tumor suppressor factor candidate 3 (TUSC3) (Liang et al. 2017). It has also been reported that hcmv-miR-UL70-3p increasing the migratory and invasive capacity of cancer cells could potentially promote tumor growth and metastasis (Ulasov et al. 2017).

12.3.1.7 Role of Bacterial sRNAs in Cancer

To date, little is known about how bacterial sRNAs regulate genes in human cancer cells and their impact on disease development and treatment. Nonetheless, there is a growing body of evidence to support the notion that persistent tissue inflammation, induced by bacterial infections, may promote the initiation and progression of tumors. Several recent studies have indicated that OMVs also exert a pivotal influence in the context of gastric adenocarcinoma, oral squamous cell carcinomas, and colorectal cancer (Liu et al. 2021; Zhang et al. 2020; Lamprinaki et al. 2021). A recent study has shown that *P. gingivalis* OMVs containing sRNA-23392 play a role in facilitating the invasion and movement of cells associated with oral squamous cell carcinoma through the regulation of desmocollin-2 (Liu et al. 2021). *H. pylori* OMVs containing sRNA-2509025 and sRNA-989262 have been also found to reduce the production of IL-8 in human gastric adenocarcinoma cell lines (Zhang et al. 2020).

12.3.2 Role of Fungi milRNAs in Cancer

Fungi appear to play a relevant role in cancer, as they have been detected in 35 different types, and are often found inside cells (Narunsky-Haziza et al. 2022). However, little attention has been paid to fungi that produce miRNAs to interact with hosts. Among the few papers that probe the subject, the 14 miRNAs identified with significant dysregulation in patients with invasive aspergillosis (IA) arising as a severe complication in patients with hematological malignancies. Target prediction of these miRNAs revealed their involvement in biological pathways related to tumorigenesis, apoptosis, cell differentiation, cell cycle, and immune response (Fidler et al. 2022). In other tumor types, no evidence of direct interaction has been found to date, although evidence has been found for miRNAs produced by intestinal fungi and the human gene expression machinery, corroborate the idea that RNAi mechanisms occur in several domains and have the potential to affect the expression of host genes (Pös et al. 2023).

12.4 Conclusion and Future Perspectives

The study of miRNAs, milRNAs, vmiRNAs and sRNAs in microorganisms is of crucial importance at the forefront of scientific research. These molecules serve as a key player in the control of gene expression, which has a significant influence on pathogenesis, therapy development, and prevention of infectious diseases. A thorough understanding of these sncRNAs in microorganisms is critical for identifying novel therapeutic targets and developing effective intervention strategies against emerging pathogens and tropical diseases. In addition, the increasing exploration of the human microbiome in diverse tissues and cavities highlights the complexity of pathogen-host interactions, offering a more realistic perspective on health and disease. This knowledge will contribute to the prevention of pandemics and the development of more targeted and effective cancer therapies, marking a milestone in the search for advanced medical solutions in the twenty-first century.

However, achieving this knowledge is not without considerable challenges. First, much of the mechanisms underlying gene regulation by pathogens still remain largely unknown, making it difficult to identify the specific strategies used by microorganisms to manipulate cellular responses. Furthermore, the precise identification of the targets of these sncRNAs in numerous pathologies represents a significant challenge, as genetic variability and multifactorial interaction complicate the assignment of precise functions. Finally, specificity of analytical methodologies is essential, as sncRNAs are small and diverse molecules, requiring advanced technological approaches for their detection and quantification. Successfully transitioning this knowledge from the laboratory to the clinic is also a crucial hurdle, involving validation and standardization of techniques, as well as understanding the specific clinical contexts in which these therapeutics could be applied. Overcoming these challenges is essential to harness the full therapeutic potential that the study of miRNAs promises, opening new perspectives in disease prevention and treatment, as well as in the management of future global health threats.

Acknowledgments E.A.-L. was the recipient of a postdoctoral fellowship from the regional Andalusian Government (POSTDOC_20_00541). L.C.T.-C. was also the recipient of a postdoctoral fellowship from the regional Andalusian Government (POSTDOC_21_00394).

Compliance with Ethical Standards

Conflict of Interest: The authors declare that they have no conflict of interest.

Ethical Approval: This chapter contains no studies with human/animal participants performed by authors.

References

Abdullah Z, Schlee M, Roth S, Mraheil MA, Barchet W, Böttcher J, Hain T, Geiger S, Hayakawa Y, Fritz JH (2012) RIG-I detects infection with live Listeria by sensing secreted bacterial nucleic acids. EMBO J 31(21):4153–4164

Abend JR, Uldrick T, Ziegelbauer JM (2010) Regulation of tumor necrosis factor-like weak inducer of apoptosis receptor protein (TWEAKR) expression by Kaposi's sarcoma-associated herpesvirus microRNA prevents TWEAK-induced apoptosis and inflammatory cytokine expression. J Virol 84(23):12139–12151

Abend JR, Ramalingam D, Kieffer-Kwon P, Uldrick TS, Yarchoan R, Ziegelbauer JM (2012) Kaposi's sarcoma-associated herpesvirus microRNAs target IRAK1 and MYD88, two components of the toll-like receptor/interleukin-1R signaling cascade, to reduce inflammatory-cytokine expression. J Virol 86(21):11663–11674

Acuña SM, Floeter-Winter LM, Muxel SM (2020) MicroRNAs: Biological regulators in pathogen–host interactions. Cells 9(1):113

Aguilar C, Mano M, Eulalio A (2019) Multifaceted roles of microRNAs in host-bacterial pathogen interaction. Microbiol Spectr 7(3). https://doi.org/10.1128/microbiolspec.bai-0002-2019

Ahmadi Badi S, Moshiri A, Fateh A, Rahimi Jamnani F, Sarshar M, Vaziri F, Siadat SD (2017) Microbiota-derived extracellular vesicles as new systemic regulators. Front Microbiol 8:1610

Ahmadi Badi S, Bruno SP, Moshiri A, Tarashi S, Siadat SD, Masotti A (2020) Small RNAs in outer membrane vesicles and their function in host-microbe interactions. Front Microbiol 11:1209

Ambrosio MR, Navari M, Di Lisio L, Leon EA, Onnis A, Gazaneo S, Mundo L, Ulivieri C, Gomez G, Lazzi S (2014) The Epstein-Barr-encoded BART-6-3p microRNA affects regulation of cell growth and immuno response in Burkitt lymphoma. Infect Agent Cancer 9(1):1–12

Argaman L, Hershberg R, Vogel J, Bejerano G, Wagner EGH, Margalit H, Altuvia S (2001) Novel small RNA-encoding genes in the intergenic regions of Escherichia coli. Curr Biol 11(12):941–950

Auvinen E (2017) Diagnostic and prognostic value of microRNA in viral diseases. Mol Diagn Ther 21(1):45–57

Bartel DP (2004) MicroRNAs: genomics, biogenesis, mechanism, and function. Cell 116(2):281–297

Barrow P, Dujardin JC, Fasel N, Greenwood AD, Osterrieder K, Lomonossoff G, Fiori PL, Atterbury R, Rossi M, Lalle M (2020) Viruses of protozoan parasites and viral therapy: is the time now right?. Virol J 17(1):142. https://doi.org/10.1186/s12985-020-01410-1.

Bartel DP (2018) Metazoan micrornas. Cell 173(1):20–51

Bhaskaran M, Mohan M (2014) MicroRNAs: history, biogenesis, and their evolving role in animal development and disease. Vet Pathol 51(4):759–774

Bitto NJ, Cheng L, Johnston EL, Pathirana R, Phan TK, Poon IK, O'Brien-Simpson NM, Hill AF, Stinear TP, Kaparakis-Liaskos M (2021) Staphylococcus aureus membrane vesicles contain immunostimulatory DNA, RNA and peptidoglycan that activate innate immune receptors and induce autophagy. J Extracell Vesicles 10(6):e12080

Boehm A, Vogel J (2012) The csgD mRNA as a hub for signal integration via multiple small RNAs. Mol Microbiol 84(1):1–5

Boudreault S, Roy P, Lemay G, Bisaillon M (2019) Viral modulation of cellular RNA alternative splicing: A new key player in virus–host interactions? Wiley Interdiscip Rev RNA 10(5):e1543

Bruscella P, Bottini S, Baudesson C, Pawlotsky J-M, Feray C, Trabucchi M (2017) Viruses and miRNAs: more friends than foes. Front Microbiol 8:824

Buck CB, Van Doorslaer K, Peretti A, Geoghegan EM, Tisza MJ, An P, Katz JP, Pipas JM, McBride AA, Camus AC (2016) The ancient evolutionary history of polyomaviruses. PLoS Pathog 12(4):e1005574

Cai L, Lyu X, Luo W, Cui X, Ye Y, Yuan C, Peng Q, Wu D, Liu T, Wang E (2015) EBV-miR-BART7-3p promotes the EMT and metastasis of nasopharyngeal carcinoma cells by suppressing the tumor suppressor PTEN. Oncogene 34(17):2156–2166

Cai Q, He B, Weiberg A, Buck AH, Jin H (2019) Small RNAs and extracellular vesicles: New mechanisms of cross-species communication and innovative tools for disease control. PLoS Pathog 15(12):e1008090

Chandra Sahoo G, Yousuf Ansari M, Ranjan Dikhit M, Gupta N, Rana S, Das P (2013) Computational identification of microRNA-like elements in Leishmania major. Microrna 2(3):225–230

Chang Y, Cesarman E, Pessin MS, Lee F, Culpepper J, Knowles DM, Moore PS (1994) Identification of herpesvirus-like DNA sequences in AIDS-associated Kaposi's sarcoma. Science 266(5192):1865–1869

Chao Y, Papenfort K, Reinhardt R, Sharma CM, Vogel J (2012) An atlas of Hfq-bound transcripts reveals 3′ UTRs as a genomic reservoir of regulatory small RNAs. EMBO J 31(20):4005–4019

Chen M, Sun F, Han L, Qu Z (2016) Kaposi's sarcoma herpesvirus (KSHV) microRNA K12-1 functions as an oncogene by activating NF-κB/IL-6/STAT3 signaling. Oncotarget 7(22):33363

Chen S, Lei Q, Zou X, Ma D (2023) The role and mechanisms of gram-negative bacterial outer membrane vesicles in inflammatory diseases. Front Immunol 14:1157813

Chirayil R, Kincaid RP, Dahlke C, Kuny CV, Dälken N, Spohn M, Lawson B, Grundhoff A, Sullivan CS (2018) Identification of virus-encoded microRNAs in divergent Papillomaviruses. PLoS Pathog 14(7):e1007156

Choi J-W, Kim S-C, Hong S-H, Lee H-J (2017) Secretable small RNAs via outer membrane vesicles in periodontal pathogens. J Dent Res 96(4):458–466

Christie PJ (2019) The rich tapestry of bacterial protein translocation systems. Protein J 38:389–408

Clarke CA, Robbins HA, Tatalovich Z, Lynch CF, Pawlish KS, Finch JL, Hernandez BY, Fraumeni JF Jr, Madeleine MM, Engels EA (2015) Risk of Merkel cell carcinoma after solid organ transplantation. J Natl Cancer Inst 107(2):dju382

Cui C, Wang Y, Liu J, Zhao J, Sun P, Wang S (2019) A fungal pathogen deploys a small silencing RNA that attenuates mosquito immunity and facilitates infection. Nat Commun 10(1):4298

Dang Y, Yang Q, Xue Z, Liu Y (2011) RNA interference in fungi: pathways, functions, and applications. Eukaryot Cell 10(9):1148–1155

Dass D, Dhotre K, Chakraborty M, Nath A, Banerjee A, Bagchi P, Mukherjee A (2023) miRNAs in herpesvirus infection: powerful regulators in small packages. Viruses 15(2):429

Dauros-Singorenko P, Hong J, Swift S, Phillips A, Blenkiron C (2020) Effect of the extracellular vesicle RNA cargo from uropathogenic Escherichia coli on bladder cells. Front Mol Biosci 7:580913

De Re V, Caggiari L, De Zorzi M, Fanotto V, Miolo G, Puglisi F, Cannizzaro R, Canzonieri V, Steffan A, Farruggia P (2020) Epstein-Barr virus BART microRNAs in EBV-associated Hodgkin lymphoma and gastric cancer. Infect Agent Cancer 15(1):1–9

Diallo I, Ho J, Lalaouna D, Massé E, Provost P (2022a) RNA sequencing unveils very small RNAs with potential regulatory functions in bacteria. Front Mol Biosci 9:914991

Diallo I, Husseini Z, Guellal S, Vion E, Ho J, Kozak RA, Kobinger GP, Provost P (2022b) Ebola virus encodes two microRNAs in Huh7-infected cells. Int J Mol Sci 23(9):5228

Dong Z, Zheng N, Hu C, Deng B, Fang W, Wu Q, Chen P, Huang X, Gao N, Lu C (2021) Nosema bombycis microRNA-like RNA 8 (Nb-milR8) increases fungal pathogenicity by modulating BmPEX16 gene expression in its host, Bombyx mori. Microbiol Spectr 9(2):e01048-01021

Duan F, Liao J, Huang Q, Nie Y, Wu K (2012) HSV-1 miR-H6 inhibits HSV-1 replication and IL-6 expression in human corneal epithelial cells in vitro. Clin Dev Immunol. https://doi.org/10.1155/2012/192791

Duan Y, Zeng J, Fan S, Liao Y, Feng M, Wang L, Zhang Y, Li Q (2019) Herpes simplex virus type 1-encoded miR-H2-3p manipulates cytosolic DNA-stimulated antiviral innate immune response by targeting DDX41. Viruses 11(8):756

Duy J, Honko AN, Altamura LA, Bixler SL, Wollen-Roberts S, Wauquier N, O'Hearn A, Mucker EM, Johnson JC, Shamblin JD (2018) Virus-encoded miRNAs in Ebola virus disease. Sci Rep 8(1):6480

Enk J, Levi A, Weisblum Y, Yamin R, Charpak-Amikam Y, Wolf DG, Mandelboim O (2016) HSV1 MicroRNA modulation of GPI anchoring and downstream immune evasion. Cell Rep 17(4):949–956
Esteso G, Luzón E, Sarmiento E, Gómez-Caro R, Steinle A, Murphy G, Carbone J, Valés-Gómez M, Reyburn HT (2014) Altered microRNA expression after infection with human cytomegalovirus leads to TIMP3 downregulation and increased shedding of metalloprotease substrates, including MICA. J Immunol 193(3):1344–1352
Eulalio A, Rehwinkel J, Stricker M, Huntzinger E, Yang S-F, Doerks T, Dorner S, Bork P, Boutros M, Izaurralde E (2007) Target-specific requirements for enhancers of decapping in miRNA-mediated gene silencing. Genes Dev 21(20):2558–2570
Fidler G, Szilágyi-Rácz AA, Dávid P, Tolnai E, Rejtő L, Szász R, Póliska S, Biró S, Paholcsek M (2022) Circulating microRNA sequencing revealed miRNome patterns in hematology and oncology patients aiding the prognosis of invasive aspergillosis. Sci Rep 12(1):7144
Francies FZ, Dlamini Z (2021) Aberrant splicing events and epigenetics in viral oncogenomics: current therapeutic strategies. Cells 10(2):239
Frantz R, Teubner L, Schultze T, La Pietra L, Müller C, Gwozdzinski K, Pillich H, Hain T, Weber-Gerlach M, Panagiotidis G-D (2019) The secRNome of Listeria monocytogenes harbors small noncoding RNAs that are potent inducers of beta interferon. MBio 10(5). https://doi.org/10.1128/mbio.01223-01219
Fruci D, Rota R, Gallo A (2017) The role of HCMV and HIV-1 microRNAs: processing, and mechanisms of action during viral infection. Front Microbiol 8:689
Fu G, Brkić J, Hayder H, Peng C (2013) MicroRNAs in human placental development and pregnancy complications. Int J Mol Sci 14(3):5519–5544
Furuse Y, Finethy R, Saka HA, Xet-Mull AM, Sisk DM, Smith KLJ, Lee S, Coers J, Valdivia RH, Tobin DM (2014) Search for microRNAs expressed by intracellular bacterial pathogens in infected mammalian cells. PLoS One 9(9):e106434
Gallo A, Miceli V, Bulati M, Iannolo G, Contino F, Conaldi PG (2020) Viral miRNAs as active players and participants in tumorigenesis. Cancers (Basel) 12(2):358
Gao W, Li JZ, Ho WK, Chan JY, Wong TS (2012) Biomarkers for use in monitoring responses of nasopharyngeal carcinoma cells to ionizing radiation. Sensors 12(7):8832–8846
Ghosal A (2018) Secreted bacterial RNA: an unexplored avenue. FEMS Microbiol Lett 365(7):fny036
Gottwein E, Mukherjee N, Sachse C, Frenzel C, Majoros WH, Chi J-TA, Braich R, Manoharan M, Soutschek J, Ohler U (2007) A viral microRNA functions as an orthologue of cellular miR-155. Nature 450(7172):1096–1099
Goymer P (2007) Introducing the mirtron. Nat Rev Genet 8(8):569–569
Grimson A, Farh KK-H, Johnston WK, Garrett-Engele P, Lim LP, Bartel DP (2007) MicroRNA targeting specificity in mammals: determinants beyond seed pairing. Mol Cell 27(1):91–105
Grulich AE, Van Leeuwen MT, Falster MO, Vajdic CM (2007) Incidence of cancers in people with HIV/AIDS compared with immunosuppressed transplant recipients: a meta-analysis. Lancet 370(9581):59–67
Grundhoff A, Sullivan CS (2011) Virus-encoded microRNAs. Virology 411(2):325–343
Gupta MD, Fliesser M, Springer J, Breitschopf T, Schlossnagel H, Schmitt A-L, Kurzai O, Hünniger K, Einsele H, Löffler J (2014) Aspergillus fumigatus induces microRNA-132 in human monocytes and dendritic cells. Int J Med Microbiol 304(5–6):592–596
Ha M, Kim VN (2014) Regulation of microRNA biogenesis. Nat Rev Mol Cell Biol 15(8):509–524
Ha JY, Choi S-Y, Lee JH, Hong S-H, Lee H-J (2020) Delivery of periodontopathogenic extracellular vesicles to brain monocytes and microglial IL-6 promotion by RNA cargo. Front Mol Biosci 7:596366
Halder LD, Babych S, Palme DI, Mansouri-Ghahnavieh E, Ivanov L, Ashonibare V, Langenhorst D, Prusty B, Rambach G, Wich M (2022) Candida albicans induces cross-kingdom miRNA trafficking in human monocytes to promote fungal growth. MBio 13(1):e03563–e03521

Han E-C, Choi S-Y, Lee Y, Park J-W, Hong S-H, Lee H-J (2019) Extracellular RNAs in periodontopathogenic outer membrane vesicles promote TNF-α production in human macrophages and cross the blood–brain barrier in mice. FASEB J 33(12):13412

Hansen A, Henderson S, Lagos D, Nikitenko L, Coulter E, Roberts S, Gratrix F, Plaisance K, Renne R, Bower M (2010) KSHV-encoded miRNAs target MAF to induce endothelial cell reprogramming. Genes Dev 24(2):195–205

Harwig A, Das AT, Berkhout B (2014) Retroviral micrornas. Curr Opin Virol 7:47–54

He X, Pan W (2022) Host–parasite interactions mediated by cross-species microRNAs. Trends Parasitol 38(6):478–488

Hombach S, Kretz M (2016) Non-coding RNAs: classification, biology and functioning. Noncoding RNAs Colorectal Cancer 3–17

Hook L, Hancock M, Landais I, Grabski R, Britt W, Nelson JA (2014) Cytomegalovirus microRNAs. Curr Opin Virol 7:40–46

Hu M, Wang C, Li W, Lu W, Bai Z, Qin D, Yan Q, Zhu J, Krueger BJ, Renne R (2015) A KSHV microRNA directly targets G protein-coupled receptor kinase 2 to promote the migration and invasion of endothelial cells by inducing CXCR2 and activating AKT signaling. PLoS Pathog 11(9):e1005171

Hull R, Marima R, Alaouna M, Demetriou D, Reis RM, Molefi T, Dlamini Z (2022) Viral encoded miRNAs in tumorigenesis: theranostic opportunities in precision oncology. Microorganisms 10(7):1448

Ishiji T, Lace M, Parkkinen S, Anderson RD, Haugen T, Cripe T, Xiao J, Davidson I, Chambon P, Turek L (1992) Transcriptional enhancer factor (TEF)-1 and its cell-specific co-activator activate human papillomavirus-16 E6 and E7 oncogene transcription in keratinocytes and cervical carcinoma cells. EMBO J 11(6):2271–2281

Jiang Y, Han Q, Zhao H, Zhang J (2021) The mechanisms of HBV-induced hepatocellular carcinoma. J Hepatocell Carcinoma 8:435–450

Jin Y, Zhao J-H, Zhao P, Zhang T, Wang S, Guo H-S (2019) A fungal milRNA mediates epigenetic repression of a virulence gene in Verticillium dahliae. Philos Trans Royal Soc B 374(1767):20180309

Jørgensen MG, Pettersen JS, Kallipolitis BH (2020) sRNA-mediated control in bacteria: An increasing diversity of regulatory mechanisms. Biochim Biophys Acta (BBA) Gene Regul Mech 1863(5):194504

Kamel W, Segerman B, Punga T, Akusjärvi G (2014) Small RNA sequence analysis of adenovirus VA RNA-derived miRNAs reveals an unexpected serotype-specific difference in structure and abundance. PLoS One 9(8):e105746

Kawano Y, Iwata S, Kawada J-i, Gotoh K, Suzuki M, Torii Y, Kojima S, Kimura H, Ito Y (2013) Plasma viral microRNA profiles reveal potential biomarkers for chronic active Epstein-Barr virus infection. J Infect Dis 208(5):771–779

Keenan J, Day T, Neal S, Cook B, Perez-Perez G, Allardyce R, Bagshaw P (2000) A role for the bacterial outer membrane in the pathogenesis of Helicobacter pylori infection. FEMS Microbiol Lett 182(2):259–264

Khan MA-A-K, Sany MRU, Islam MS, Islam ABMMK (2020) Epigenetic regulator miRNA pattern differences among SARS-CoV, SARS-CoV-2, and SARS-CoV-2 world-wide isolates delineated the mystery behind the epic pathogenicity and distinct clinical characteristics of pandemic COVID-19. Front Genet 11:765

Kim Y, Lee S, Kim S, Kim D, Ahn J-H, Ahn K (2012) Human cytomegalovirus clinical strain-specific microRNA miR-UL148D targets the human chemokine RANTES during infection. PLoS Pathog 8(3):e1002577

King KA, Caudill MT, Caswell CC (2022) A comprehensive review of small regulatory RNAs in Brucella spp. Front Vet Sci 9:1026220

Koeppen K, Hampton TH, Jarek M, Scharfe M, Gerber SA, Mielcarz DW, Demers EG, Dolben EL, Hammond JH, Hogan DA (2016) A novel mechanism of host-pathogen interaction through sRNA in bacterial outer membrane vesicles. PLoS Pathog 12(6):e1005672

Krump NA, You J (2018) Molecular mechanisms of viral oncogenesis in humans. Nat Rev Microbiol 16(11):684–698
Kumar P, Anaya J, Mudunuri SB, Dutta A (2014) Meta-analysis of tRNA derived RNA fragments reveals that they are evolutionarily conserved and associate with AGO proteins to recognize specific RNA targets. BMC Biol 12(1):1–14
Lagatie O, Tritsmans L, Stuyver LJ (2013) The miRNA world of polyomaviruses. Virol J 10:1–9
Lamprinaki D, Garcia-Vello P, Marchetti R, Hellmich C, McCord KA, Bowles KM, Macauley MS, Silipo A, De Castro C, Crocker PR (2021) Siglec-7 mediates immunomodulation by colorectal cancer-associated Fusobacterium nucleatum ssp. animalis. Front Immunol 12:744184
Lee RC, Feinbaum RL, Ambros V (1993) The C. elegans heterochronic gene lin-4 encodes small RNAs with antisense complementarity to lin-14. Cell 75(5):843–854
Lee H-C, Li L, Gu W, Xue Z, Crosthwaite SK, Pertsemlidis A, Lewis ZA, Freitag M, Selker EU, Mello CC (2010) Diverse pathways generate microRNA-like RNAs and Dicer-independent small interfering RNAs in fungi. Mol Cell 38(6):803–814
Lee S, Paulson KG, Murchison EP, Afanasiev OK, Alkan C, Leonard JH, Byrd DR, Hannon GJ, Nghiem P (2011) Identification and validation of a novel mature microRNA encoded by the Merkel cell polyomavirus in human Merkel cell carcinomas. J Clin Virol 52(3):272–275
Lei X, Zhu Y, Jones T, Bai Z, Huang Y, Gao S-J (2012) A Kaposi's sarcoma-associated herpesvirus microRNA and its variants target the transforming growth factor β pathway to promote cell survival. J Virol 86(21):11698–11711
Lei T, Yuen K-S, Tsao SW, Chen H, Kok K-H, Jin D-Y (2013) Perturbation of biogenesis and targeting of Epstein-Barr virus-encoded miR-BART3 microRNA by adenosine-to-inosine editing. J Gen Virol 94(12):2739–2744
Leitão AL, Enguita FJ (2022) A structural view of miRNA biogenesis and function. Noncoding RNA 8(1):10
Li T, Ju E, Gao S-J (2019) Kaposi sarcoma-associated herpesvirus miRNAs suppress CASTOR1-mediated mTORC1 inhibition to promote tumorigenesis. J Clin Invest 129(8):3310–3323
Liang D, Gao Y, Lin X, He Z, Zhao Q, Deng Q, Lan K (2011) A human herpesvirus miRNA attenuates interferon signaling and contributes to maintenance of viral latency by targeting IKKε. Cell Res 21(5):793–806
Liang Q, Wang K, Wang B, Cai Q (2017) HCMV-encoded miR-UL112-3p promotes glioblastoma progression via tumour suppressor candidate 3. Sci Rep 7(1):44705
Liu D, Liu S, Liu J, Miao L, Zhang S, Pan Y (2021) sRNA23392 packaged by Porphyromonas gingivalis outer membrane vesicles promotes oral squamous cell carcinomas migration and invasion by targeting desmocollin-2. Mol Oral Microbiol 36(3):182–191
Lodde V, Floris M, Muroni MR, Cucca F, Idda ML (2022) Non-coding RNAs in malaria infection. Wiley Interdiscip Rev RNA 13(3):e1697
Lopez-Jimenez E, Andres-Leon E (2021) The implications of ncRNAs in the development of human diseases. Noncoding RNA 7(1). https://doi.org/10.3390/ncrna7010017
Luna JM, Scheel TK, Danino T, Shaw KS, Mele A, Fak JJ, Nishiuchi E, Takacs CN, Catanese MT, de Jong YP (2015) Hepatitis C virus RNA functionally sequesters miR-122. Cell 160(6):1099–1110
Lung RW-M, Tong JH-M, Sung Y-M, Leung P-S, Ng DC-H, Chau S-L, Chan AW-H, Ng EK-O, Lo K-W, To K-F (2009) Modulation of LMP2A expression by a newly identified Epstein-Barr virus-encoded microRNA miR-BART22. Neoplasia 11(11):1174–IN1117
Lyu X, Wang J, Guo X, Wu G, Jiao Y, Faleti OD, Liu P, Liu T, Long Y, Chong T (2018) EBV-miR-BART1-5P activates AMPK/mTOR/HIF1 pathway via a PTEN independent manner to promote glycolysis and angiogenesis in nasopharyngeal carcinoma. PLoS Pathog 14(12):e1007484
Makarova JA, Shkurnikov MU, Wicklein D, Lange T, Samatov TR, Turchinovich AA, Tonevitsky AG (2016) Intracellular and extracellular microRNA: An update on localization and biological role. Prog Histochem Cytochem 51(3-4):33–49
Mallick B, Ghosh Z, Chakrabarti J (2008) MicroRNA switches in Trypanosoma brucei. Biochem Biophys Res Commun 372(3):459–463

Manzano M, Shamulailatpam P, Raja AN, Gottwein E (2013) Kaposi's sarcoma-associated herpesvirus encodes a mimic of cellular miR-23. J Virol 87(21):11821–11830
Marin FR, Dávalos A, Kiltschewskij D, Crespo MC, Cairns M, Andres-Leon E, Soler-Rivas C (2022) RNA-seq, bioinformatic identification of potential MicroRNA-Like Small RNAs in the edible mushroom Agaricus bisporus and experimental approach for their validation. Int J Mol Sci 23(9):4923
Martinez-Zapien D, Ruiz FX, Poirson J, Mitschler A, Ramirez J, Forster A, Cousido-Siah A, Masson M, Pol SV, Podjarny A (2016) Structure of the E6/E6AP/p53 complex required for HPV-mediated degradation of p53. Nature 529(7587):541–545
Martin-Perez D, Vargiu P, Montes-Moreno S, Leon E, Rodriguez-Pinilla S, Lisio L, Martinez N, Rodriguez R, Mollejo M, Castellvi J (2012) Epstein-Barr virus microRNAs repress BCL6 expression in diffuse large B-cell lymphoma. Leukemia 26(1):180–183
Mattick JS (2003) Challenging the dogma: the hidden layer of non-protein-coding RNAs in complex organisms. Bioessays 25(10):930–939. https://doi.org/10.1002/bies.10332
Menard KL, Haskins BE, Denkers EY (2019) Impact of Toxoplasma gondii infection on host non-coding RNA responses. Front Cell Infect Microbiol 9:132
Meng F, Siu GK-H, Mok BW-Y, Sun J, Fung KS, Lam JY-W, Wong NK, Gedefaw L, Luo S, Lee TM (2021) Viral microRNAs encoded by nucleocapsid gene of SARS-CoV-2 are detected during infection, and targeting metabolic pathways in host cells. Cells 10(7):1762
Meng Q, Sun H, Wu S, Familiari G, Relucenti M, Aschner M, Li X, Chen R (2022) Epstein-Barr virus-encoded microRNA-BART18-3p promotes colorectal cancer progression by targeting de novo lipogenesis. Adv Sci 9(35):2202116
Morozova N, Zinovyev A, Nonne N, Pritchard L-L, Gorban AN, Harel-Bellan A (2012) Kinetic signatures of microRNA modes of action. RNA 18(9):1635–1655
Nachmani D, Stern-Ginossar N, Sarid R, Mandelboim O (2009) Diverse herpesvirus microRNAs target the stress-induced immune ligand MICB to escape recognition by natural killer cells. Cell Host Microbe 5(4):376–385
Nador RG, Cesarman E, Chadburn A, Dawson DB, Ansari MQ, Sald J, Knowles DM (1996) Primary effusion lymphoma: a distinct clinicopathologic entity associated with the Kaposi's sarcoma-associated herpes virus. Blood 8(2):645–656
Nanbo A, Furuyama W, Lin Z (2021) RNA virus-encoded miRNAs: current insights and future challenges. Front Microbiol 12:679210
Narunsky-Haziza L, Sepich-Poore GD, Livyatan I, Asraf O, Martino C, Nejman D, Gavert N, Stajich JE, Amit G, González A (2022) Pan-cancer analyses reveal cancer-type-specific fungal ecologies and bacteriome interactions. Cell 185(20):3789–3806.e3717
O'Brien J, Hayder H, Zayed Y, Peng C (2018) Overview of microRNA biogenesis, mechanisms of actions, and circulation. Front Endocrinol (Lausanne) 9:402
Pan A, Tan Y, Wang Z, Xu G (2019) STAT4 silencing underlies a novel inhibitory role of microRNA-141-3p in inflammation response of mice with experimental autoimmune myocarditis. Am J Physiol Heart Circ Physiol 317(3):H531–H540
Panda M, Kalita E, Singh S, Kumar K, Rao A, Prajapati VK (2022) MiRNA-SARS-CoV-2 dialogue and prospective anti-COVID-19 therapies. Life Sci 305:120761
Panni S, Lovering RC, Porras P, Orchard S (2020) Non-coding RNA regulatory networks. Biochim Biophys Acta (BBA) Gene Regul Mech 1863(6):194417
Piedade D, Azevedo-Pereira JM (2016) The role of microRNAs in the pathogenesis of herpesvirus infection. Viruses 8(6):156
Pillai RS, Artus CG, Filipowicz W (2004) Tethering of human Ago proteins to mRNA mimics the miRNA-mediated repression of protein synthesis. RNA 10(10):1518–1525
Ponath F, Hör J, Vogel J (2022) An overview of gene regulation in bacteria by small RNAs derived from mRNA 3′ ends. FEMS Microbiol Rev 46(5):fuac017
Pös O, Styk J, Buglyó G, Zeman M, Lukyova L, Bernatova K, Hrckova Turnova E, Rendek T, Csók Á, Repiska V (2023) Cross-kingdom interaction of miRNAs and gut microbiota with

non-invasive diagnostic and therapeutic implications in colorectal cancer. Int J Mol Sci 24(13):10520
Punga T, Darweesh M, Akusjärvi G (2020) Synthesis, structure, and function of human adenovirus small non-coding RNAs. Viruses 12(10):1182
Qi M, Qi Y, Ma Y, He R, Ji Y, Sun Z, Ruan Q (2013) Over-expression of human cytomegalovirus miR-US25-2-3p downregulates eIF4A1 and inhibits HCMV replication. FEBS Lett 587(14):2266–2271
Qian K, Pietilä T, Rönty M, Michon F, Frilander MJ, Ritari J, Tarkkanen J, Paulín L, Auvinen P, Auvinen E (2013) Identification and validation of human papillomavirus encoded microRNAs. PLoS One 8(7):e70202
Qin Z, Freitas E, Sullivan R, Mohan S, Bacelieri R, Branch D, Romano M, Kearney P, Oates J, Plaisance K (2010) Upregulation of xCT by KSHV-encoded microRNAs facilitates KSHV dissemination and persistence in an environment of oxidative stress. PLoS Pathog 6(1):e1000742
Qin Z, Jakymiw A, Findlay V, Parsons C (2012) KSHV-encoded microRNAs: lessons for viral cancer pathogenesis and emerging concepts. Int J Cell Biol. https://doi.org/10.1155/2012/603961
Qin Z, Peruzzi F, Reiss K, Dai L (2014) Role of host microRNAs in Kaposi's sarcoma-associated herpesvirus pathogenesis. Viruses 6(11):4571–4580. https://doi.org/10.3390/v6114571
Qin J, Li W, Gao S-J, Lu C (2017) KSHV microRNAs: tricks of the devil. Trends Microbiol 25(8):648–661
Quendera AP, Seixas AF, Dos Santos RF, Santos I, Silva JP, Arraiano CM, Andrade JM (2020) RNA-binding proteins driving the regulatory activity of small non-coding RNAs in bacteria. Front Mol Biosci 7:78
Reza Aghasadeghi M, Sharifat Salmani A, Mehdi Sadat S, Javadi F, Memarnejadian A, Vahabpour R, Zabihollahi R, Moshiri A, Davar Siadat S (2011) Application of outer membrane vesicle of Neisseria meningitidis serogroup B as a new adjuvant to induce strongly Th1-oriented responses against HIV-1. Curr HIV Res 9(8):630–635
Riahi Rad Z, Riahi Rad Z, Goudarzi H, Goudarzi M, Mahmoudi M, Yasbolaghi Sharahi J, Hashemi A (2021) MicroRNAs in the interaction between host–bacterial pathogens: A new perspective. J Cell Physiol 236(9):6249–6270
Rojas-Pirela M, Andrade-Alviárez D, Medina L, Castillo C, Liempi A, Guerrero-Muñoz J, Ortega Y, Maya JD, Rojas V, Quiñones W (2022) MicroRNAs: master regulators in host–parasitic protist interactions. Open Biol 12(6):210395
Rolhion N, Barnich N, Claret L, Darfeuille-Michaud A (2005) Strong decrease in invasive ability and outer membrane vesicle release in Crohn's disease-associated adherent-invasive Escherichia coli strain LF82 with the yfgL gene deleted. J Bacteriol 187(7):2286–2296
Rorbach G, Unold O, Konopka BM (2018) Distinguishing mirtrons from canonical miRNAs with data exploration and machine learning methods. Sci Rep 8(1):7560
Saçar MD, Bağcı C, Allmer J (2014) Computational prediction of microRNAs from Toxoplasma gondii potentially regulating the hosts' gene expression. Genomics Proteomics Bioinformatics 12(5):228–238
Sahr T, Escoll P, Rusniok C, Bui S, Pehau-Arnaudet G, Lavieu G, Buchrieser C (2022) Translocated Legionella pneumophila small RNAs mimic eukaryotic microRNAs targeting the host immune response. Nat Commun 13(1):762
Saraiya AA, Wang CC (2008) snoRNA, a novel precursor of microRNA in Giardia lamblia. PLoS Pathog 4(11):e1000224
Sarshar M, Scribano D, Palamara AT, Ambrosi C, Masotti A (2022) The Acinetobacter baumannii model can explain the role of small non-coding RNAs as potential mediators of host-pathogen interactions. Front Mol Biosci 9:1088783
Sedaghat M, Siadat SD, Mirabzadeh E, Keramati M, Vaziri F, Shafiei M, Shahcheraghi F (2019) Evaluation of antibody responses to outer membrane vesicles (OMVs) and killed whole cell of Vibrio cholerae O1 El Tor in immunized mice. Iran J Microbiol 11(3):212

Shinozaki-Ushiku A, Kunita A, Isogai M, Hibiya T, Ushiku T, Takada K, Fukayama M (2015) Profiling of virus-encoded microRNAs in Epstein-Barr virus-associated gastric carcinoma and their roles in gastric carcinogenesis. J Virol 89(10):5581–5591

Skalsky RL, Cullen BR (2010) Viruses, microRNAs, and host interactions. Annu Rev Microbiol 64:123–141

Song T, Mika F, Lindmark B, Liu Z, Schild S, Bishop A, Zhu J, Camilli A, Johansson J, Vogel J (2008) A new Vibrio cholerae sRNA modulates colonization and affects release of outer membrane vesicles. Mol Microbiol 70(1):100–111

Song L, Zhang W, Chang Z, Pan Y, Zong H, Fan Q, Wang L (2017) miR-4417 targets tripartite motif-containing 35 (TRIM35) and regulates pyruvate kinase muscle 2 (PKM2) phosphorylation to promote proliferation and suppress apoptosis in hepatocellular carcinoma cells. Med Sci Monit 23:1741

Stanton BA (2021) Extracellular vesicles and host–pathogen interactions: a review of inter-kingdom signaling by small noncoding RNA. Genes 12(7):1010

Suffert G, Malterer G, Hausser J, Viiliäinen J, Fender A, Contrant M, Ivacevic T, Benes V, Gros F, Voinnet O (2011) Kaposi's sarcoma herpesvirus microRNAs target caspase 3 and regulate apoptosis. PLoS Pathog 7(12):e1002405

Theiss JM, Günther T, Alawi M, Neumann F, Tessmer U, Fischer N, Grundhoff A (2015) A comprehensive analysis of replicating Merkel cell polyomavirus genomes delineates the viral transcription program and suggests a role for mcv-miR-M1 in episomal persistence. PLoS Pathog 11(7):e1004974

Ulasov IV, Kaverina NV, Ghosh D, Baryshnikova MA, Kadagidze ZG, Karseladze AI, Baryshnikov AY, Cobbs CS (2017) CMV70-3P miRNA contributes to the CMV mediated glioma stemness and represents a target for glioma experimental therapy. Oncotarget 8(16):25989

Venook AP, Papandreou C, Furuse J, Ladrón de Guevara L (2010) The incidence and epidemiology of hepatocellular carcinoma: a global and regional perspective. Oncologist 15(S4):5–13

Virtanen E, Pietilä T, Nieminen P, Qian K, Auvinen E (2016) Low expression levels of putative HPV encoded microRNAs in cervical samples. SpringerPlus 5(1):1–8

Wagner EGH, Romby P (2015) Small RNAs in bacteria and archaea: who they are, what they do, and how they do it. Adv Genet 90:133–208

Wahl A, Linnstaedt SD, Esoda C, Krisko JF, Martinez-Torres F, Delecluse H-J, Cullen BR, Garcia JV (2013) A cluster of virus-encoded microRNAs accelerates acute systemic Epstein-Barr virus infection but does not significantly enhance virus-induced oncogenesis in vivo. J Virol 87(10):5437–5446

Walsh D, Mathews MB, Mohr I (2013) Tinkering with translation: protein synthesis in virus-infected cells. Cold Spring Harb Perspect Biol 5(1):a012351

Wang YF, Fu J (2019) Secretory and circulating bacterial small RNAs: a mini-review of the literature. ExRNA 1:1–5

Wang J, Liu X, Jia B, Lu H, Peng S, Piao X, Hou N, Cai P, Yin J, Jiang N (2012) A comparative study of small RNAs in Toxoplasma gondii of distinct genotypes. Parasit Vectors 5:1–9

Wang G, Dong F, Xu Z, Sharma S, Hu X, Chen D, Zhang L, Zhang J, Dong Q (2017) MicroRNA profile in HBV-induced infection and hepatocellular carcinoma. BMC Cancer 17(1):1–11

Weiberg A, Wang M, Lin F-M, Zhao H, Zhang Z, Kaloshian I, Huang H-D, Jin H (2013) Fungal small RNAs suppress plant immunity by hijacking host RNA interference pathways. Science 342(6154):118–123

Weng S-L, Huang K-Y, Weng JT-Y, Hung F-Y, Chang T-H, Lee T-Y (2018) Genome-wide discovery of viral microRNAs based on phylogenetic analysis and structural evolution of various human papillomavirus subtypes. Brief Bioinform 19(6):1102–1114

Wightman B, Ha I, Ruvkun G (1993) Posttranscriptional regulation of the heterochronic gene lin-14 by lin-4 mediates temporal pattern formation in C. elegans. Cell 75(5):855–862

Wu W, Guo Z, Zhang X, Guo L, Liu L, Liao Y, Wang J, Wang L, Li Q (2013) A microRNA encoded by HSV-1 inhibits a cellular transcriptional repressor of viral immediate early and early genes. Sci China Life Sci 56:373–383

Yang X, Li H, Sun H, Fan H, Hu Y, Liu M, Li X, Tang H (2017) Hepatitis B virus-encoded microRNA controls viral replication. J Virol 91(10). https://doi.org/10.1128/jvi.01919-01916

Yao L, Zhou Y, Sui Z, Zhang Y, Liu Y, Xie H, Gao H, Fan H, Zhang Y, Liu M (2019) RETRACTED: HBV-encoded miR-2 functions as an oncogene by downregulating TRIM35 but upregulating RAN in liver cancer cells. EBioMedicine 48:117–129

Zhang H, Kolb FA, Jaskiewicz L, Westhof E, Filipowicz W (2004) Single processing center models for human Dicer and bacterial RNase III. Cell 118(1):57–68

Zhang J, Pu X-M, Xiong Y (2019) kshv-mir-k12-1-5p promotes cell growth and metastasis by targeting SOCS6 in Kaposi's sarcoma cells. Cancer Manag Res 11:4985–4995

Zhang H, Zhang Y, Song Z, Li R, Ruan H, Liu Q, Huang X (2020) sncRNAs packaged by Helicobacter pylori outer membrane vesicles attenuate IL-8 secretion in human cells. Int J Med Microbiol 310(1):151356

Zheng Y, Cai X, Bradley JE (2013) microRNAs in parasites and parasite infection. RNA Biol 10(3):371–379

Zou W, Imperiale MJ (2021) Biology of polyomavirus miRNA. Front Microbiol 12:662892